# ENGINEERING GRAPHICS with AutoCAD

REVISED EDITION



**Dhananjay M. Kulkarni**
Professor, Mechanical Engineering
Birla Institute of Technology and Science (BITS) Pilani (Goa Campus)

**A.P. Rastogi**
Former Lecturer, Department of Civil Engineering
Birla Institute of Technology and Science (BITS) Pilani (Pilani Campus)

**Ashoke K. Sarkar**
Professor, Department of Civil Engineering
and
Director, Birla Institute of Technology and Science (BITS) Pilani (Pilani Campus)

**PHI Learning Private Limited**

Delhi-110092

2025

*In fond memory of **Shri Asoke K. Ghosh** (October 1942 – February 2024), Founder Chairman and Managing Director of PHI Learning, whose vision endlessly inspires.*

*The Legacy Continues....*

Published by Pushpita Ghosh, PHI Learning Private Limited, Rimjhim House, 111, Patparganj Industrial Estate, Delhi-110092 and Printed by Star Print-O-Bind, F-31, Okhla Industrial Area Phase 1, New Delhi-110020.

**₹895.00**

**ENGINEERING GRAPHICS WITH AUTOCAD, Revised Edition**
Dhananjay M. Kulkarni, A.P. Rastogi, and Ashoke K. Sarkar



**ISBN-978-81-203-3783-1 (Print Book)**
**ISBN-978-93-5443-493-8 (e-Book)**

To

*Our parents*

for their motivation

# Contents

# Preface

*Engineering Graphics* is the language of the engineers. An engineer requires skill in three areas of engineering or technical drawing: firstly, he must be proficient to use the tools to produce the drawings. Secondly, he must be able to draw freehand technical sketches clearly and rapidly. Thirdly, he must be able to read a drawing.

The need for writing internal course notes on *Engineering Graphics* was felt in 1998, when the course was made fully online at BITS Pilani. It was given a new orientation and did not follow the usual sequence of topics followed in the traditional method of teaching the course. It had been found to be very successful over the years. The internal course notes were produced in the way it is being taught at BITS. This revised edition of the book is based on those course notes. The first edition was published in 2009 with AutoCAD 10. This revised edition is presented with AutoCAD 2010. However, AutoCAD version will not matter much because we have preferred AutoCAD commands instead of graphical user interface. Though the book is written on AutoCAD as a drafting tool, it can be successfully used by the readers following traditional drafting instruments.

The contents in this volume have been divided into three parts. Part I deals with the basic commands of AutoCAD. It is a widely used drafting package for engineers and architects. It is, therefore, believed that some hands-on experience in AutoCAD will not only help the students in learning *Engineering Graphics*, but also will help them to maintain their cutting edge in the professional job market. Part II contains the various projection techniques used in engineering for technical drawings. These techniques have been explained with a number of graphical representations so that the students can understand the techniques easily. Part III contains descriptive geometry which mainly deals with three dimensional objects that require imagination.

In this book, the model solutions and exercises have been treated in the third-angle as well as the first-angle methods of projection according to the revised code of Indian Standard Code of Practice for General Drawing. All drawings and the textual matter are in conformity with the specifications of Bureau of Indian Standards (B.I.S.), adopted from I.S.O.

*Engineering graphics* is not only engineering and science, but also an art of imagination. The engineering and science can be taught in the class, but the art of imagination can be inculcated by practising as many exercises as possible. For best results, it is suggested that the exercises should be solved in practical classes under the supervision of the instructors.

The complex concepts and imaginations are made simpler in this revised edition to help the students for better understanding.

The authors would welcome any constructive suggestions from the readers of this book at *kulkarnidm@gmail.com*.

**Dhananjay M. Kulkarni**
**A.P. Rastogi**
**Ashoke K. Sarkar**

# Acknowledgements

This book, *Engineering Graphics with AutoCAD*, is the final outcome of an idea that was originated in year 2002, when we undertook the task of writing the subject course notes. During the last seven years we have put in our efforts to produce the revised versions of the course notes, with the help and support of many well-wishers. Thanks are due for all who have directly or indirectly contributed to this work.

We thank Prof. L.K. Maheshwari, Vice Chancellor, Prof. K.E. Raman, Director, BITS Pilani, Goa campus, Prof. V.S. Rao, Director, BITS Pilani, Hyderabad campus and Prof. M. Ramachandran, Director, BITS Pilani, Dubai campus for providing us the opportunity to bring out this book. Sincere appreciation is extended to Prof. R.K. Patnaik, former Dean, Instruction Division, BITS Pilani, who suggested us to write a book.

We acknowledge the inspiration and support of the students and colleagues over the years that have led to the writing of this book. We are grateful to the colleagues who have been teaching the course in BITS Pilani, Dubai, Goa and Hyderabad campuses. Their solicitous comments and feedback were of great assistance in planning and developing the contents of this text.

We are also grateful to Prof. M.S. Dasgupta for permitting us to use the handouts and problem sheets that he developed along with Prof. A.P. Rastogi after the course was made fully online at BITS Pilani.

We are also indebted to the Editor, Ms. Shivani Garg; Copyeditor, Ms. Koyel Chakraborty and Assistant Production Manager, Mr. Pankaj Manohar of PHI Learning for their guidance and assistance in developing and producing this work.

Special thanks are extended to Mr. Shreenivasa Paramahamsa and Mr. Avinash Sonee, students of BITS Pilani, Goa campus, who had initially conceived the concept of the cover page.

Finally, we extend profound and heartfelt thanks to our family members. Without their encouragement, patience and understanding this endeavour would not have been possible. The first author would like to express his deep affection and thanks to his parents-in-law, elder brother, wife Reshma and daughters Vedashree and Sanika for their constant persuasion and moral support.

**Dhananjay M. Kulkarni**
**A.P. Rastogi**
**Ashoke K. Sarkar**

# 1

# Engineering Graphics: An Overview

## Objectives

- Knowing Engineering Graphics as a subject.
- Learning the graphic language.
- Need of Engineering Graphics.

## Introduction

Engineering Graphics is the field of solving graphical problem and includes two areas of specialisation: working drawings and descriptive geometry. Other areas that can be used for a wide variety of applications are nomography, graphical mathematics, empirical equations, technical illustration, vector analysis, data analysis and computer graphics. Graphics is one of the designer's primary methods of thinking, solving problems and communicating ideas.

Gaspard Monge (1746–1818)

Gaspard Monge (1746–1818) is the "father of descriptive geometry". While a military student in France, young Monge used the graphical methods to solve design problems related to fortifications and battlements. He was scolded by his headmaster for solving problems graphically instead of the usual (long and tedious) mathematical process. Only after long explanations and comparisons of both methods' solutions, Mongue was able to convince that his graphical methods solved problems in considerably less time. Descriptive geometry was such an improvement upon the mathematical method that it was kept a military secret for fifteen years before it was allowed to be taught as part of the technical curriculum. Monge became a scientific and mathematical aide to Napolean during his reign as the Emperor of France.

## 1.1 The Graphic Language

People learned to draw pictures of the objects around them long before they learned to write. Drawings carved by primitive people on rocks, the walls of caves and so forth have survived to our day. Many drawings of human beings, animals, fish and so on, made by our forefathers thousands of years ago, have been found on the eastern shore of lake Onega and on the shores of the White Sea. These drawings were carved on granite rocks with stone (flint) tools. The ability to make simple drawings helped man to develop his first written language. There were no words or characters in ancient writing. Ideas of things were conveyed by pictures of the things themselves.

Engineering drawing is a graphic language, a universal language, a language of use and of ever-increasing value. Like every other systematised language it has its rules of grammar. To express ourselves in a literate fashion, we must master, as it were, laws of syntax and of concord. Just as in our English books we are warned against the improper use of words in sentences, likewise, in our study of engineering drawing we shall be warned against the improper use of lines in drawings. Each word in a well-constructed sentence has its function; each line on a well-executed drawing has its function too. Drawings, then, to those initiated into the language, become articulate vehicles of expression between the drawing-office and workshop, between the textbook and the student.

With a very little application we shall master the rules of the universal graphic language, and thus be able to not only express ourselves in it on drawing paper, but also to understand the expression of it by another. In a word, we set out to learn to draw so that our ideas may readily be interpreted by others and to *read drawings* so as to gather a clear mental picture of what another wishes to convey to us.

## 1.2 Need of Engineering Graphics

Without a good knowledge of engineering graphics, engineers and scientists are nowhere. An engineer or a scientist needs skill in three areas: firstly, he must be able to imagine; secondly, he must be able to draw clearly and rapidly; thirdly, he must be able to read the drawings drawn by others. This course develops imagination skill, drawing skill and reading skill in the early part of engineers' education, which is repeatedly used in successive courses.

Engineering design uses engineering graphics as the way to communicate and document ideas. By studying engineering graphics, student becomes aware of how the industry communicates technical information. Engineering graphics teaches the principles of accuracy and clarity in presenting the information necessary to produce products. It offers students an insight into the methods of attacking engineering and scientific problems. Its lessons teach the principles of accuracy, exactness and positiveness with regard to the information necessary for the production of non-existing object or structure. Finally, it develops the imagination skill that is essential to the creation of successful design.

## 1.3 Objectives of Course

Engineers and scientists are constantly engaged in creative and productive work. Rapid advances in technological world demand engineers and scientists to invent new devices, to improve existing ones, to optimise the use of resources and to reduce the time and cost of production. Proficiency in Engineering Graphics will be of considerable aid in achieving these aims.

The objective of this course is to develop a clear understanding of the basic principles involved in engineering drawings. The basic principles of Engineering Graphics help an engineer and scientist in communicating their ideas through engineering drawings. Apart from basic principles, this course teaches the use of standardised practices and conventions in the preparation of engineering drawings.

# PART I

# AutoCAD

# 2

# Computer Aided Drafting

## Objectives

- Study of AutoCAD software for the Engineering Graphics course.
- Know the required AutoCAD commands for the course.
- Practice AutoCAD commands to acquire a drawing skill.

## Introduction

An engineering drawing is prepared by using conventional tools. Traditionally, drawing board, drafter, T-square, drawing sheet and other drafting instruments have been used. The popular alternative now is to prepare engineering drawing with the aid of computer. This method is known as Computer Aided Drafting and is abbreviated as CAD. The CAD requires computer hardware and drawing software. The user needs a good training on the drawing software, which can reduce the time of drawing. The advantages, applications of computer aided drafting system and the required commands for Engineering Graphics course are covered in this Chapter.

## 2.1 Computer Aided Drafting System

A computer aided drafting system consists of computer hardware (CPU, monitor, key board an mouse) and drawing software. This system totally eliminates the traditional equipments like drawin board, drafter, scales, various grades of pencils etc. More important is this system complete eliminates a human skill of drawing. The well known software used for engineering drawing i AutoCAD developed by a company AutoDesk.

An engineer designing a new engineering element (machine/building/electronic circuit) can vie a drawing on the screen. He can make the drawing to turn slowly, thus providing different views. H can quickly make changes, which will be instantaneously shown on the screen. The changes wi also get recorded in the storage. He can reduce or enlarge any part of the drawing and can sho the various components in different colours. He can also test the performance of the design throug a simulation exercise. Overall, the design cycle is reduced drastically and the new car can be o the road in a remarkably short time.

### 2.1.1 Advantages of Computer Aided Drafting

There are many advantages of computer aided drafting system over the traditional drafting system A few advantages are listed over here.

1. The computer has tremendous speed and it has almost unlimited storage and rapid reca capabilities. This results in reduction of drafting labour and drafting time.
2. The drawing can be stored in 'database' for future use by different programmers for variet of applications.
3. With the ability to interact with the computer, we can quickly correct a drawing error and se a revised picture on the graphic screen.
4. Visual modeling of any object or engineering component is possible.
5. When we use computer aided drawing for a drawing containing the same component a several places, we can draw the component once and then insert it wherever needed.
6. Colour graphics help to display more distinct information on the screen, highlight certai important features, etc.
7. The computer aided drawing can be exported to commercial analysis software for analysi purpose.
8. Computer aided design and drafting is used in the creation, modification, analysis an optimization of designs for improved engineering productivity. An engineer can try ou different design ideas by just sitting at the terminal without making any prototype.

### 2.1.2 Applications of Computer Aided Drafting

Computer aided drawing applications in various fields of engineering are given here.

(i) *Mechanical:* Design of machine elements, CNC machine tools, Robotics.
(ii) *Automotive:* Kinematics, Hydraulics, Steering.
(iii) *Electrical:* Circuit layout, Panel design, control system.
(iv) *Electronics:* Schematic diagrams of PCs, Ics, etc.
(v) *Communication:* Communication network, satellite transmitting pictures, T.V. telecasting.
(vi) *Civil:* Mapping, contour plotting, building drawing, structural design.
(vii) *Architectural:* Town planning, interior decorations, multi storyed complex.
(viii) *Aerospace:* Design of spacecraft, flight simulator, lofting, etc.

## 2.2 AutoCAD: Drafting Software

AutoDesk, the developer of 'AutoCAD software' is the world-leading supplier of computer aided design and drawing software packages. The AutoCAD automates repetitive design and drawing tasks. We can create and erase drawing with ease. Since its inception in 1982, the company has introduced a family of software packages for use in a wide range of industries. It has hit the technical world like a shock wave, revolutionizing the manner of drawing operation. Any technical person can use this software with ease and without bothering to learn computer programming.

In 1982, AutoDesk launched DOS based AutoCAD release 1.0. Thereafter, 2.0, 3.0, ..., 10.0 were launched out of which R10 became very popular in DOS based system. Thereafter, Autodesk launched windows based R11, R13, R14, R2000, R2002, R2004, R2005, R2006, R2007 and R2008 versions. The present version is 'AutoCAD 2010'.

### 2.2.1 AutoCAD 2010

Recently launched (March-2009) R2010 has flexible features and customized to designer's needs and suitable for the basic course of engineering graphics. This graphic screen helps to accommodate number of drawings for practical assignments as well as for online examinations.

Students are advised to follow command mode though the required icons (features) are available on the screen. The commands which are required only for course on Engineering Graphics are explained here as it appears in the AutoCAD software. The commands are shown with the help of capital letters throughout the textbook wherever these are required to mention. The features of response of required command during command operation are explained here. Figure 2.1 shows AutoCAD 2010 graphic screen.

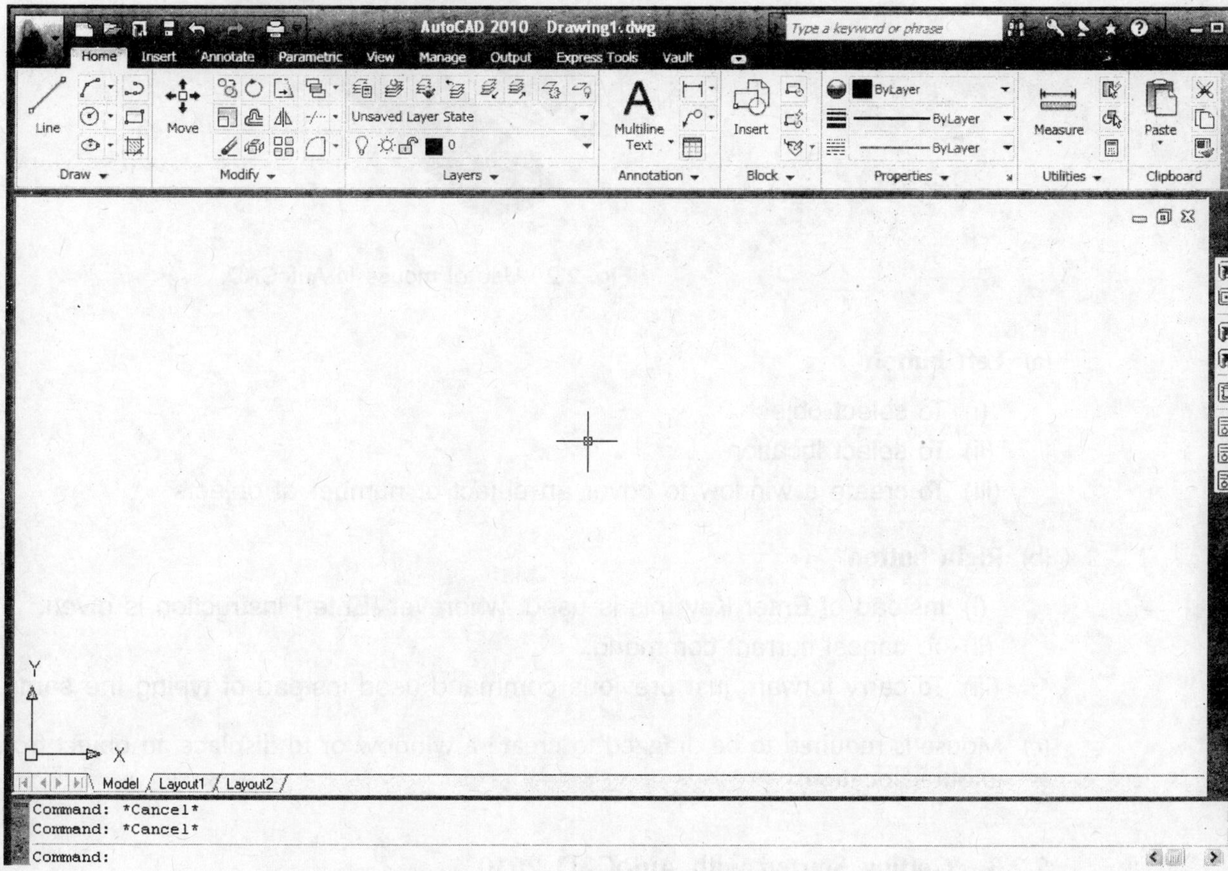


**Fig. 2.1** AutoCAD 2010 graphic screen.

### 2.2.2 Key Points and Important Guidelines During Command Operation

1. AutoCAD does not have any unit like meter or millimeter. It has got its own unit called a AutoCAD unit.
2. Prompt means an instruction or symbol asking user to implement the next operation.
3. The input at the command prompt is case insensitive (capital letters or small letters can b input).

   There are only two important instructions for user. These instructions are enclosed i square bracket as shown in (4) and (5).
4. [Enter] to instruct 'Press Enter Key'
5. [Select ..........] to instruct 'Select Object'. The cursor is converted into a target box ☐ t select entity (object) in AutoCAD.
6. The parenthesis is used for user (information) during command steps.
7. Text in angle bracket e.g. <Diameter> gives a default option.
8. Other options are shown separated by an oblique sign: Xyz/Pqr/Abc <lmn>:

   In this response shown above, the lmn option is a default option. To activate other option first capital letter is required to be typed. e.g. to activate Xyz option, X needs to be typed an to activate Pqr option, P needs to be typed.
9. Use of Mouse for AutoCAD software (Fig. 2.2).

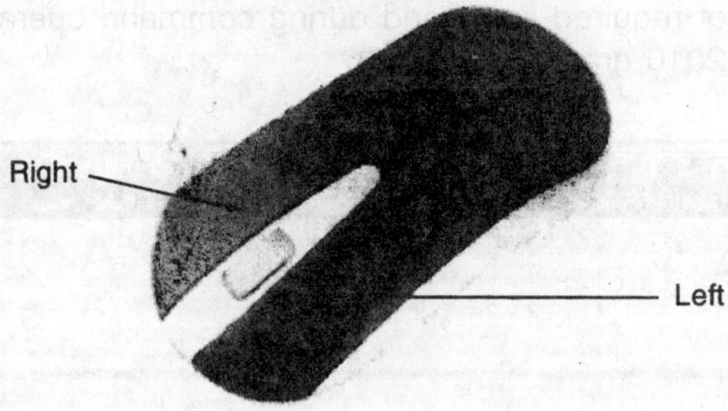


**Fig. 2.2** Use of mouse in AutoCAD.

(a) **Left button**

(i) To select object
(ii) To select location
(iii) To create a window to cover an object or number of objects

(b) **Right button**

(i) Instead of Enter Key this is used, wherever [Enter] instruction is given.
(ii) To cancel current command.
(iii) To carry forward just previous command used instead of typing the same.

(c) Mouse is required to be dragged to create a window or to displace an object from one location t another location.

### 2.2.3 Getting Started with AutoCAD 2010

Figure 2.3 shows division of graphic screen. There are three major divisions: drawing area fc drawing objects, flexible features (icons) and command prompt for command input. There are tw

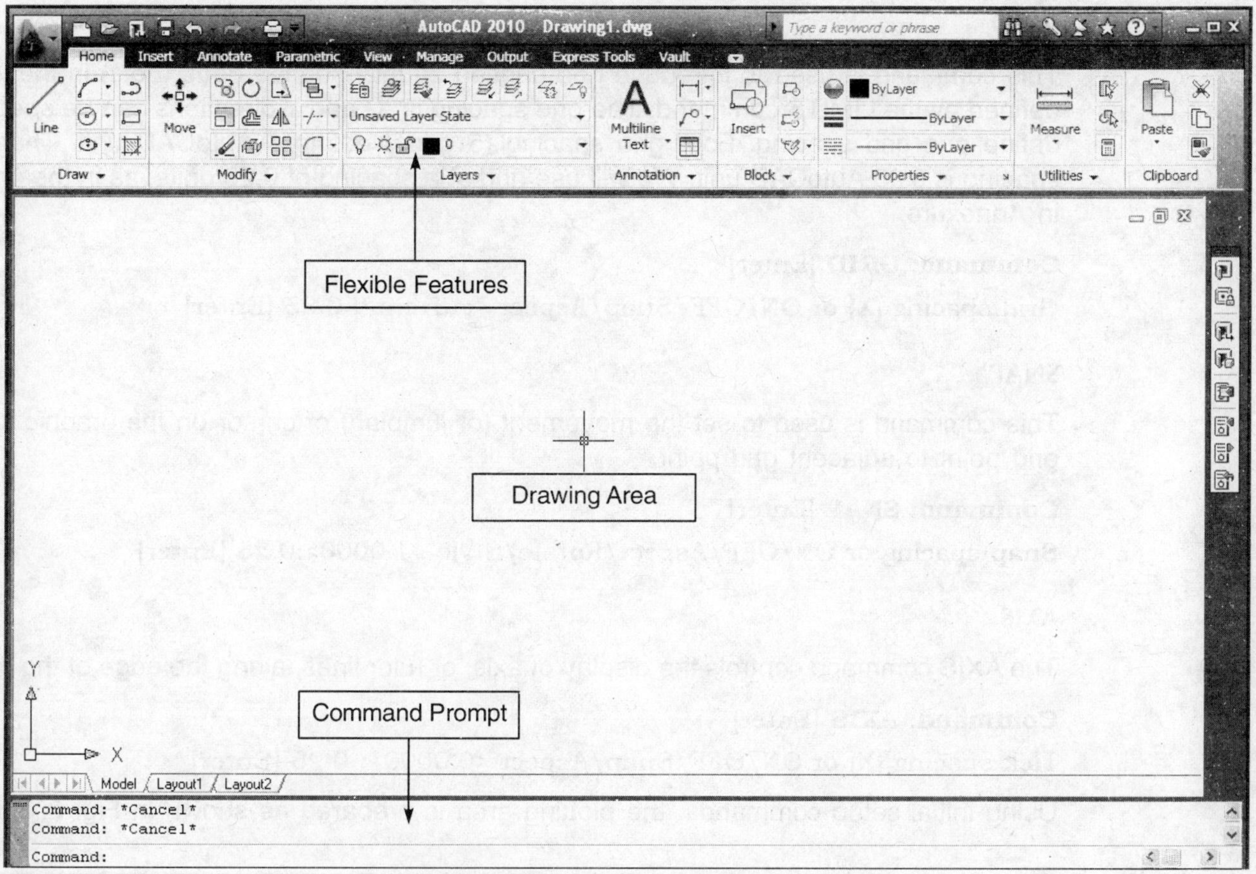


**Fig. 2.3** Division of graphic screen.

options for drawing. We can use menu to reach to a required command or its options. Otherwise, we can use a required AutoCAD command directly at the command prompt. AutoCAD users are encouraged to use command at the command prompt than the menu option for Engineering Graphics. In menu option, one has to enter into depth to search for a particular option. The command option reduces the search time to a great extent.

## 2.3 Initial Setup Commands

Initial setup commands are used to set technical aspect of drawing sheet. Various technical aspects of drawing sheet are discussed in Chapter-4 in detail.

### LIMITS

This command is used to set the limits of the rectangular graphic screen area i.e. the X and Y coordinate limits. For all drawing assignments we will be using full screen limits to create maximum drawing area.

**Command: LIMITS** [Enter]

ON/OFF/<Lower left corner> <0.0000,0.0000>: [Enter]
Upper right corner <11.5, 8.00>: [select upper-right corner]

The limits shown in angle bracket are defined in AutoCAD unit. The limits of rectangular area are set as per the drawing sheet size. This is discussed in Chapter-4 in detail.

### GRID

This command is used to introduce non-printing equally spaced grids (dots) in the workspace area defined by the LIMITS command. The grid spacing in X- and Y-directions can be specified. Two grids define one grid spacing. Four grid spacing (5 grids) define 1 AutoCAD unit. Therefore, one grid spacing is 0.25 AutoCAD unit. We will use grid at a spacing of 0.25 units, as in the grid paper shown in Annexure.

**Command: GRID** [Enter]

Grid spacing (X) or ON/OFF/Snap/Aspect <current>: 0.25 [Enter]

### SNAP

This command is used to set the movement (or jumping) of cursor on the graphic screen from one grid point to adjacent grid point.

**Command: SNAP** [Enter]

Snap spacing or ON/OFF/Aspect/Rotate/Style <1.0000>:0.25 [Enter]

### AXIS

The AXIS command controls the display of axis, or ruler lines, along the edge of the graphics display.

**Command: AXIS** [Enter]

Tick spacing (X) or ON/OFF/Snap/Aspect <0.0000>: 0.25 [Enter]

Using initial setup commands, the plotting area is prepared as shown in Fig. 2.4.

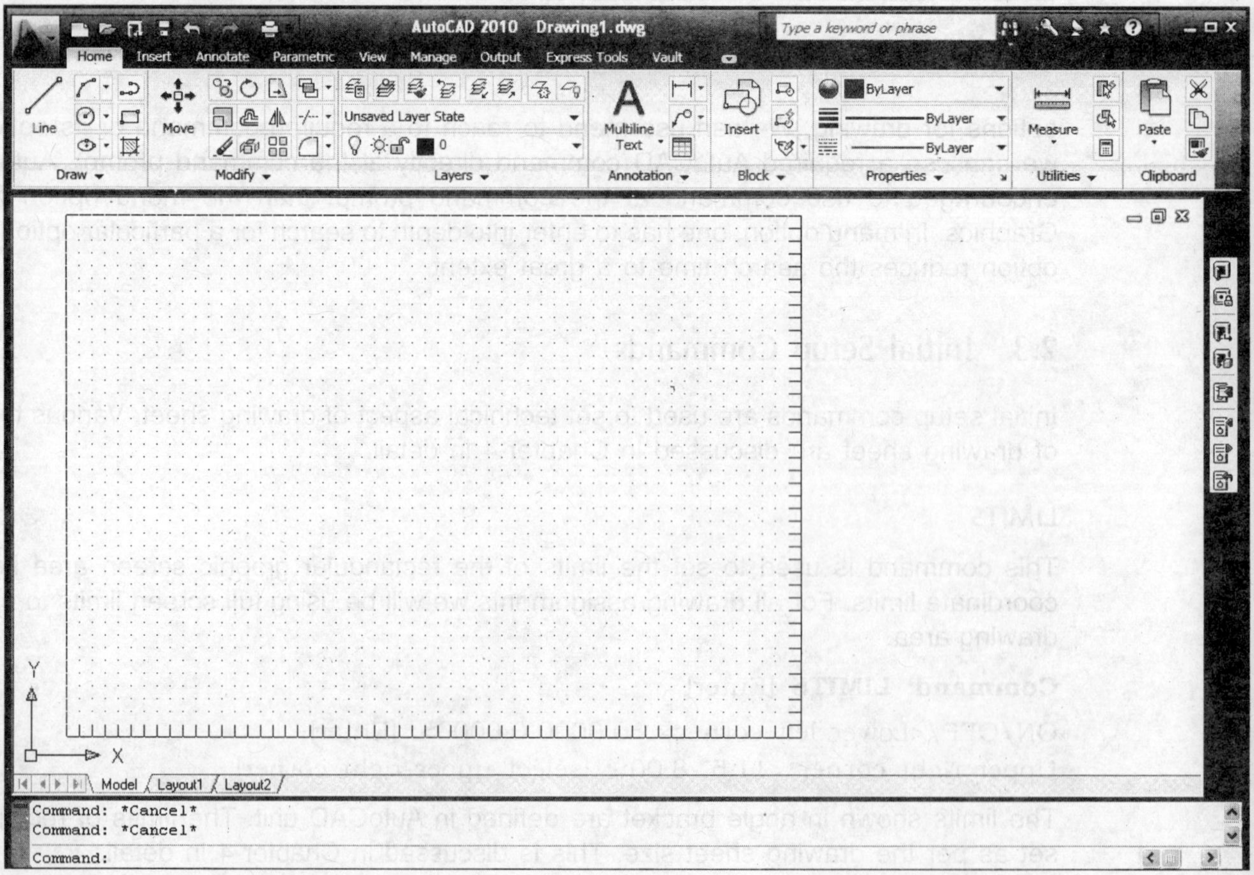


**Fig. 2.4** AutoCAD 2010 graphic screen.

## 2.4 Utility Commands

These are general commands which are used to operate the drawing file.

### HELP

This command is used to know more about a specific command. Type HELP followed by command name for the on-line help about that command. The help will be displayed on to a text screen, which is at the back of graphic screen. In order to flip back to the graphics screen, function key F1 is pressed. Following command explanation shows help for a DRAW command.

**Command: HELP DRAW** [Enter]

(Now you will reach to back of graphics screen where the explanation of DRAW command will be available.

### QUIT Command

The QUIT command exits from the Drawing Editor, discarding all updates to the current drawing, and returns you to the main menu. Two responses are expected. If QUIT command is used following SAVE command then no response. If not following is the response.

**Command: QUIT** [Enter]

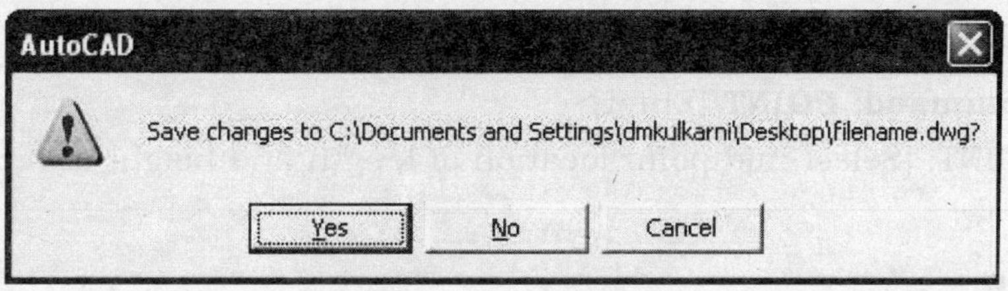


Yes: to save changes. No: to discard changes. Cancel: to cancel the QUIT command and continue drawing process. The SAVE command is always preferred before the QUIT command is executed.

### SAVE

This command is used to save the drawing file. This command should be used frequently in order to save changes and to avoid power failure problem if any.

**Command: SAVE** [Enter]

SAVE<current file name>: filename [Enter]

Make use of either right button of mouse OR Ctrl + C keys together to interrupt the drawing process and return to the command prompt.

## 2.5 Drawing Aids

There are various control modes, which are frequently required while drawing. These are operated by function keys, which are also called as toggle keys.

| *Function key* | *Use of function or toggle keys* |
|---|---|
| F1 | To flip back to text screen from graphic screen and vice a versa. |
| F6 | To turn the coordinates ON or OFF |
| F7 | To turn the grid ON or OFF |

| *Function key* | *Use of function or toggle keys* |
|---|---|
| F8 | To turn Ortho ON or OFF. In Ortho ON mode, only orthogonal (horizontal or vertical) cursor movement is possible. The default setting is OFF. If OFF, angular movement of cursor is responded. |
| F9 | To turn the snap ON or OFF. If ON, the cursor jump from one grid to another grid. If OFF, the cursor can move with a fine movement. |

## 2.6 Entity Draw Commands

### POINT

This command is used to specify a point on an object. The appearance of point in drawing is governed by the PDMODE & PDSIZE system variable. PDMODE command sets the mode of point as per requirement. Two important modes of POINT are cross (x) mode and plus (+) mode. PDSIZE command sets the size (height and width) of point entity.

**Command: PDMODE** [Enter]
PDMODE: 3 [Enter]
**Command: PDSIZE** [Enter]
PDSIZE: 0.25 [Enter]
**Command: POINT** [Enter]
POINT: [Select mid point location of length and height of rectangle]

### LINE

The LINE command is used to draw a straight line. Line can be drawn by providing with the coordinate of a point (we will be using 2 D points only) or by selecting x-y location with the left button of mouse.

**Command: LINE** [Enter]
From point: [select P1]
To point: [select P2]
To point: [select P3]
To point: [select P4]

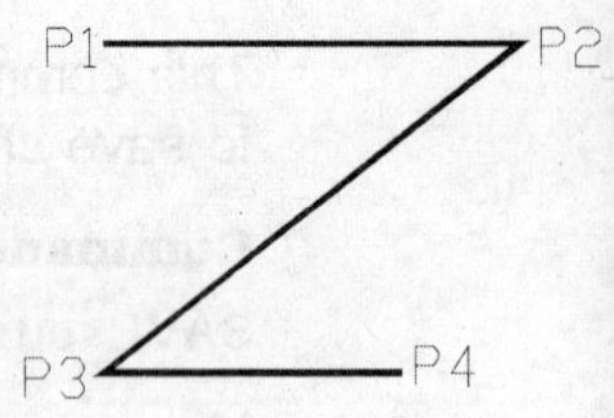


*Other options to draw line:*

1. ***C:*** This option draws a line from the last point to the first point entered when the LINE command was invoked and makes a polygon. Previous To point prompt is continued here to explain this option.
   To point: [select point say P4]
   To point: C [Enter]

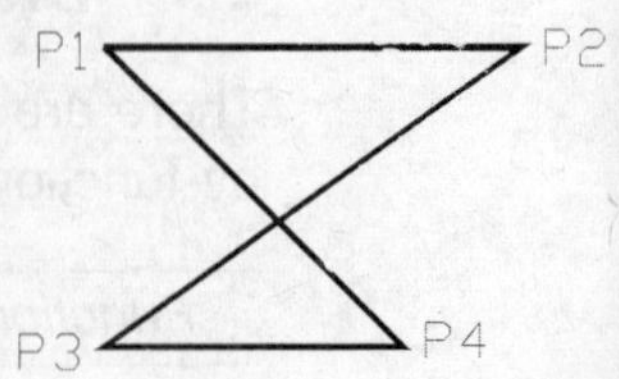


2. ***U:*** While LINE command is in operation (i.e. at To Point prompt), use of UNDO command undoes the last line drawn.
   To point: U [Enter]

3. ***LINE from or to specific point***

**Command: LINE** [Enter]
From point: [select A]
To point: [select B]
To point: [select C]
To point: **Mid** (for mid point location of AB) [Enter]
Mid point of: [select line 'AB' with target box]
To point: **Per** (for perpendicular location) [Enter]
Per to: [select line 'BC' with target box]
To point: **End** (for end point location) [Enter]
End point of: [select line AB with target box, near to A]
To point: **Tan** (for tangent location) [Enter]
Tan to: [select arc with target box]
To point: **Cen** (for center location) [Enter]
Center of: [select arc with target box]
To point: **Int** (for intersection location) [Enter]
Intersection of: [select intersection point of CM & AP]
To point: [Ctrl + C or click right button of mouse to come to command prompt]

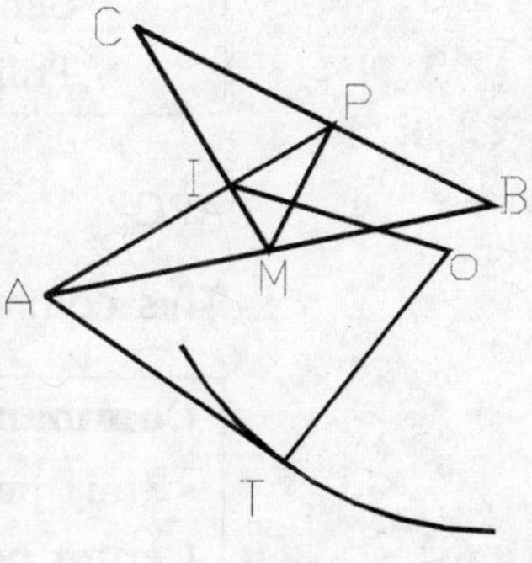


4. ***Using length and angle***

This option is used to draw a line of specific length at specific angle. E.g. it is required to draw a line of 2.25 AutoCAD unit at an angle of 30°.

The angle is measured from +ve X-axis in anticlockwise direction.

**Command: LINE** [Enter]
From point: [select P]
To point: @2.25<30 [Enter]

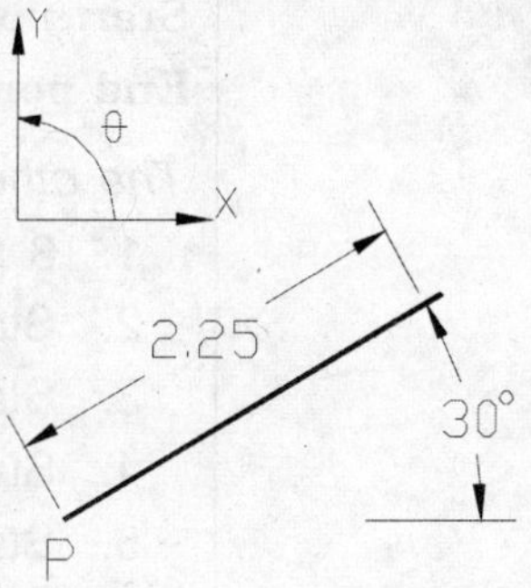


## CIRCLE

This command is used to draw a circle based on input of a center point and diameter or radius. Three more options to draw a circle are also explained here.

**Command: CIRCLE** [Enter]
3P/2P/TTR/<Center Point>: [select center point location]
Diameter/<Radius>: [drag the mouse for required radius and click left button]

In this example, radius of circle is 1 AutoCAD unit (4 grid spacings).

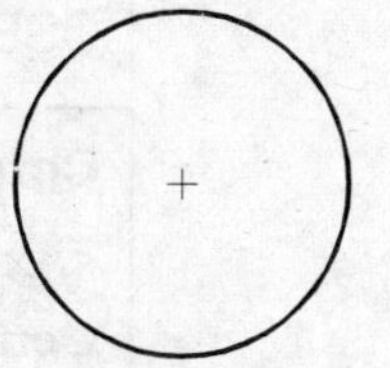

1. ***TTR option***

Circle C1 & C2 are drawn. It is required to draw a circle C3 with 0.75 AutoCAD unit radius and tangent to C1 & C2.

**Command: CIRCLE** [Enter]
3P/2P/TTR/<Center Point>: TTR [Enter]
Enter tangent specification: [select circle C1]
Enter second tangent specification: [select circle C2]
Radius: 0.75 [Enter]

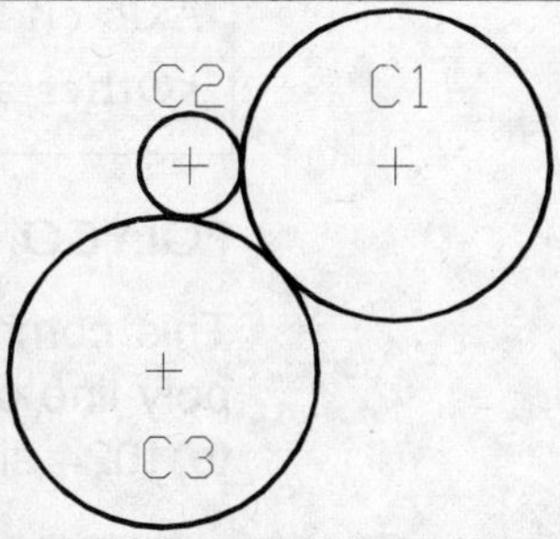

2. ***Use of 2P (2 points) option to draw a circle***

**Command: CIRCLE** [Enter]
3P/2P/TTR/<Center Point>: 2P [Enter]
First point on diameter: [select P]
Second point on diameter: [select Q]

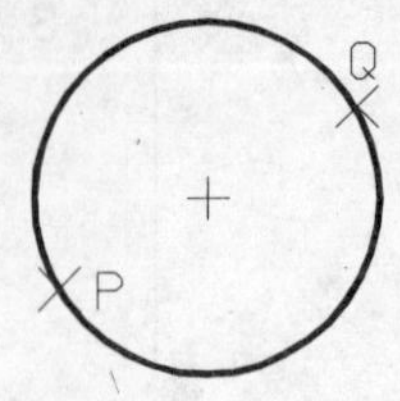


3. ***Use of 3P (3 points) option to draw a circle***

**Command: CIRCLE** [Enter]
3P/2P/TTR/<Center Point>: 3P [Enter]
First point: [select P]
Second point: [select Q]
Third point: [select R]

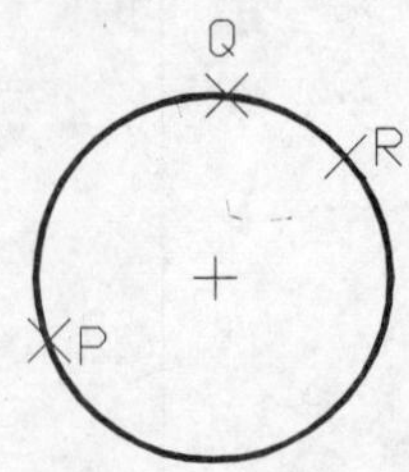


**ARC**

This command is used to draw an arc based on input of center point and diameter or radius.

**Command: ARC** [Enter]
<Start point>/center: C (for center option) [Enter]
Center point: [select O]
Start point: [select P1]
End point: [select P2]

*The other available options are:*

1. 3 points
2. Start, Center, End (SCE)
3. Start, Center, Angle (SCA)
4. Start, Center, Length (SCL)
5. Start, End, Angle (SEA)
6. Start, End, Direction (SED)
7. Start, End, Radius (SER)
8. Center, Start, End (CSE) explained here
9. Center, Start, Angle (CSA)
10. Center, Start, Length (CSL)

CSE is a frequently used option in Engineering Graphics.

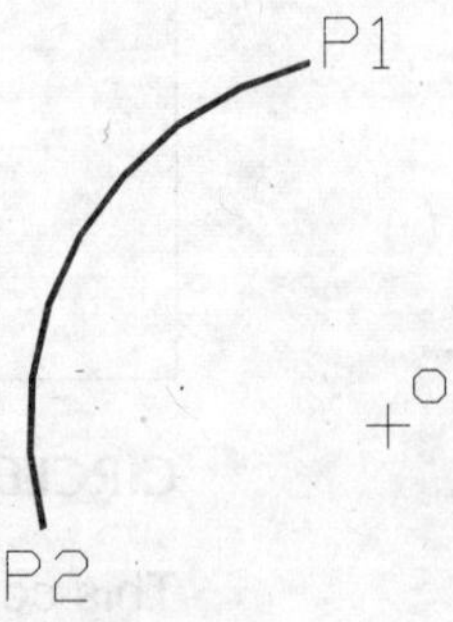


**Command: ELLIPSE** [Enter]
<Axis endpoint 1>/center: C (for center option) [Enter]
Center of ellipse: [select + location]
Axis endpoint: [select P1]
<Other axis distance>/Rotation: [select P2]

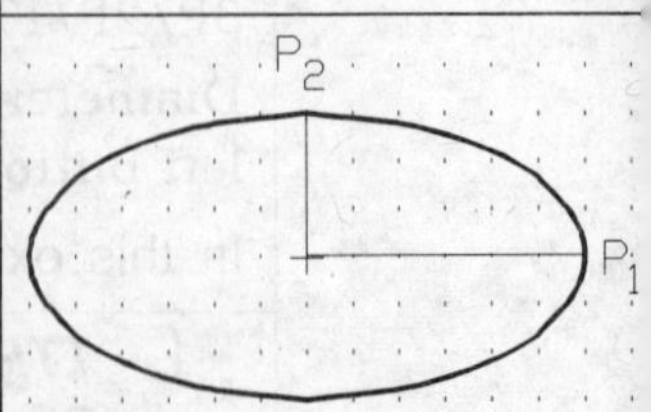


**POLYGON**

This command is used to draw a regular triangle, square, pentagon and hexagon. A polygon is a poly-line object. AutoCAD draws poly-lines with zero width. It can draw regular 2-D polygons with 3 to 1024 sides.

**Command: POLYGON** [Enter]
Number of sides: 3 [Enter]
Edge/<Center of Polygon>: E [Enter]
First endpoint of edge: [select point P1]
Second endpoint of edge: [select point P2]
Depending upon the selection of location of P1 and P2, polygon with various orientations can be drawn.

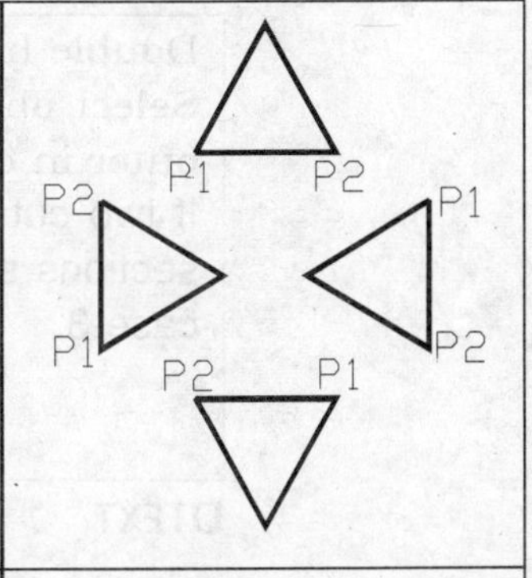


To draw a POLYGON of specific edge length and of specific orientation.

**Command: POLYGON** [Enter]
Number of sides: 4 [Enter]
Edge/<Center of Polygon>: E [Enter]
First endpoint of edge: [select P1]
Second endpoint of edge: @1.5<30 [Enter]

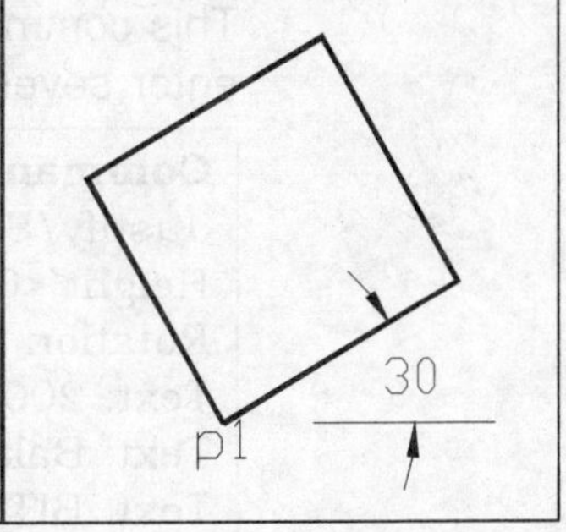


## PLINE

This command is used to draw a poly-line i.e. continuous line segments. The number of segments drawn in one operation becomes a single entity. User can control the width of PLINE.

**Command: PLINE** [Enter]
From point: [select P1]
Current line width is 0.0000 [Enter]
Arc/Close/Half width/Length/Undo/Width/<Endpoint of line>: W [Enter]
Starting width <0.0000>: 0.025 [Enter]
Ending width <0.000>: 0.025 [Enter]
Arc/Close/Half width/Length/Undo/width/<Endpoint of line>: [select P2]

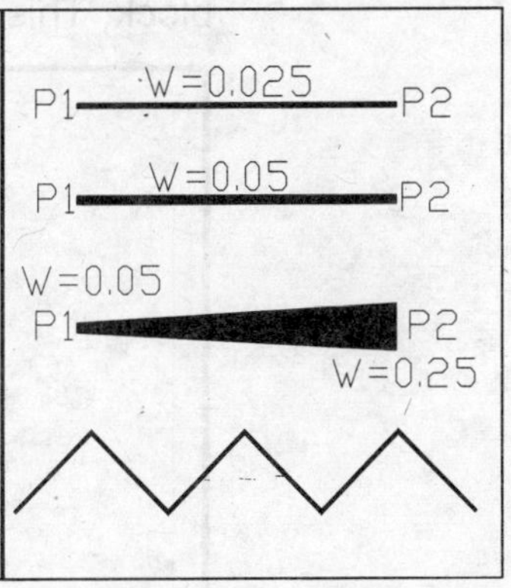


Similarly, a PLINE with width 0.05 AutoCAD unit is drawn. The PLINE command can be used to draw a taper line as shown. The PLINE command is used to draw a series of lines as shown. The PEDIT command is used to fit a smooth curve passing through all key points of series of lines.

## HATCH

This command is used to show a cut section of an engineering solid. This command is also used to HATCH two adjacent areas which are made from different materials. You can fill a boundary that consists of an object or objects that completely enclose an area. The area to be hatched should have closed boundary. There are various patterns to hatch the area. In Engineering Graphics, we use a Universal (U) pattern to show the sections of solids. Other patterns are shown in Plate 2.9.

**Command: HATCH** [Enter]
Enter pattern name [?/Solid/User defined] <U>: U (for universal pattern) [Enter]
Angle for crosshatch lines <0>: 45 [Enter]
Spacing between lines <1.00>: 0.125 [Enter]

Double hatch area? <N>: [Enter] (for No option)
Select objects: [select inner square in case 1 and both, inner and outer in case 2]
If two cut sections are adjacent to each other then hatch lines in two sections should be mutually perpendicular to each other as shown in case-3.

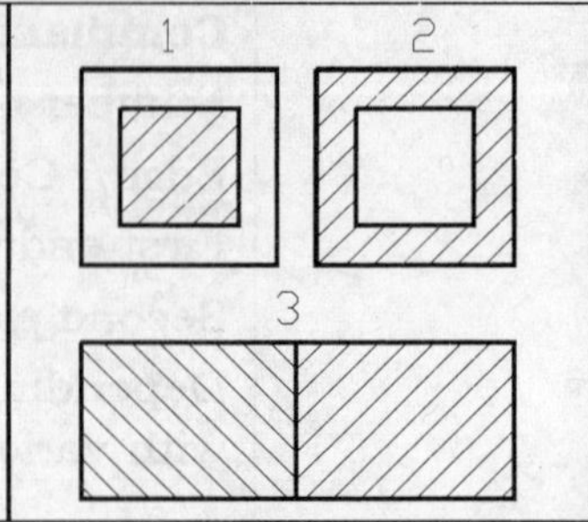


### DTEXT

This command is used to describe an object or specify notes on electronic drawing sheet. You can enter several lines of text using DTEXT. Each line of text is a separate object.

**Command: DTEXT** [Enter]
Justify/Style/ <Start point>: [select location]
Height <0.200>: [specify required value]
Rotation angle <0>: [specify required]
Text: 2001C6PS337 [Enter]
Text: Balaji Rao K [Enter]
Text: BITS-Pilani
Text: Rajasthan [press enter to end the text command]

With the DTEXT command, we can prepare the title block. Now we are ready to prepare our standard file. In Fig. 2.5, we can draw a border line using a LINE command and ORTHO ON option. We can draw a title block with the help of a LINE command. We can type required text in the title block. This is shown in Fig. 2.5.

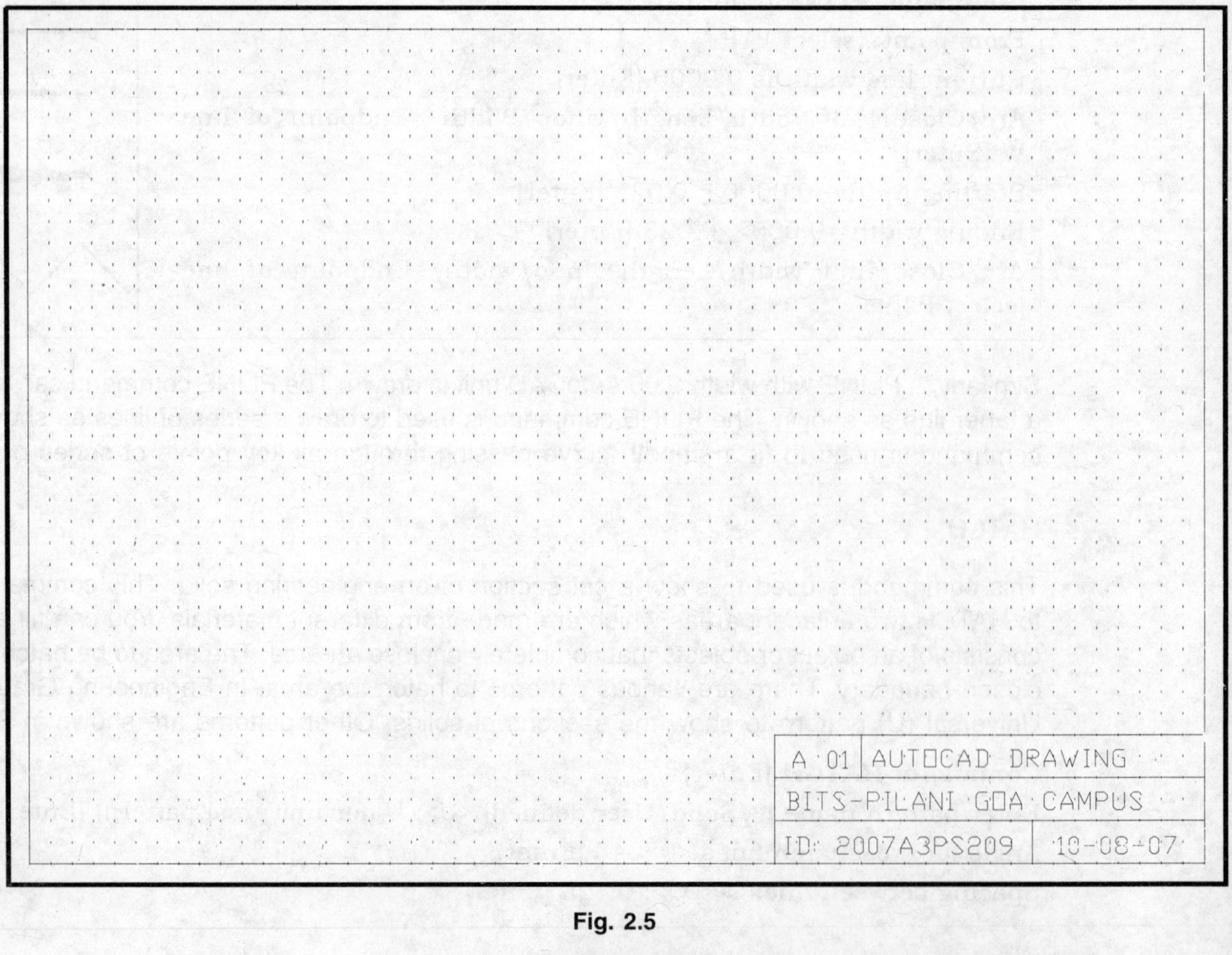


Fig. 2.5

## 2.7 Display Commands

### REDRAW

This command is used to redraw the screen and to remove any non-printing marks left by previous commands. This command should be used frequently to keep the drawing file clear.

**Command: REDRAW**

Instead of REDRAW, R can be used as a short form to activate the command.

### ZOOM

This command is used to increase the apparent size of object in the current view port.

**Command: ZOOM** [Enter]
All/Center/Dynamic/Extents/Left/Previous/Window/
<Scale (x)>: W [Enter]
First corner: [select upper left location outside the Fig.]
Other corner: [select lower right location outside the Fig.]

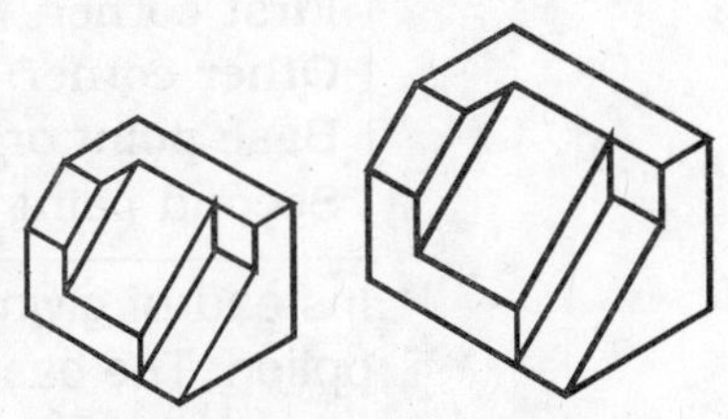

Normal view    Zoomed view

Three options are useful for Engineering Graphics. A is for All, P is for Previous and W is for Windows. The ***P***revious option is used to reach to the previous view port. The ***A***ll option is used to reach a normal screen view. The user is always advised to edit a particular object in zoomed view in order to be able to draw comfortably with AutoCAD and thereby reduce the time of drawing.

## 2.8 Edit Commands

### ERASE

AutoCAD removes objects from the drawing. If it is a single entity then a target box (□) can be used to select a particular object to be removed. Otherwise a window is created to cover number of objects.

**Command: ERASE** [Enter]
Select objects: [select desired objects to erase with target box]
Select objects: [select another object to erase]
Select objects: [press enter to end the object selection process if it is over]

**Command: ERASE** (with windows option for more than one object)
Select objects: W [Enter]
First corner: [select point nearest to top-left of the object]
Other corner: [select point nearest to bottom-right of the object]
Select objects: [press enter here to end the selection process or click a right button of mouse]

### OOPS

OOPS restores objects erased by the last ERASE command.

**Command: OOPS** [Enter]

### UNDO (U for UNDO)

With UNDO command, whatever is done in previous command is undone. e.g. if an object is drawn in the previous command, UNDO command removes the same. On the other hand, if an object is erased in the last command, the UNDO command restores the same. However OOPS command restores only the last erased object.

### MOVE

This command displaces object from current location to the desired location with a specified distance in a specified direction without changing the orientation and size.

**Command: MOVE** [Enter]
Select objects: W [Enter]
First corner: [select]
Other corner: [select]
Base point or displacement: [select O1]
Second point of displacement: [select O2]

Instead of giving displacement in terms of numerical value in AutoCAD unit, we use a base point option. The base point is to be carefully selected, as this is the point with respect to which the object is moving. Generally the base point should be either corner or center of the object.

### COPY

This command is used to duplicate object(s). The response steps of this COPY command are similar to the MOVE command. In this case, an original object remains in its position and only a replica of an object is moved to the second point of displacement. There are two options: single copy and multiple copies.

1. ***Single copy***

**Command: COPY** [Enter]
Select objects: W [Enter]
First corner: [select top-left corner outside the object]
Other corner: [select bottom-right corner outside the object]
Select object : [Enter here if the selection is over]
<Base point or displacement>/Multiple: [select O1]
Second point of displacement: [select O2]

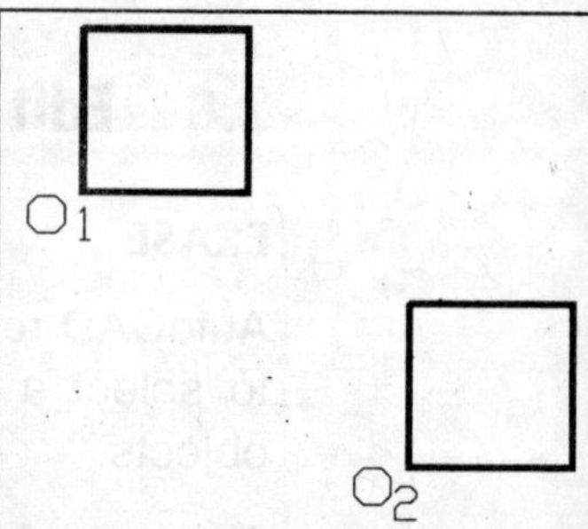


2. ***Multiple copy***

**Command: COPY** [Enter]
Select objects: W [Enter]
First corner: [select top-left corner outside the object]
Other corner: [select bottom-right corner outside the object]
Select objects: [Enter]
<Base point or displacement>/Multiple: M [Enter]
Base point: [select O1]
Second point of displacement: [select O2]
Second point of displacement: [select O3]
Second point of displacement: [select O4]
Second point of displacement: [Enter]

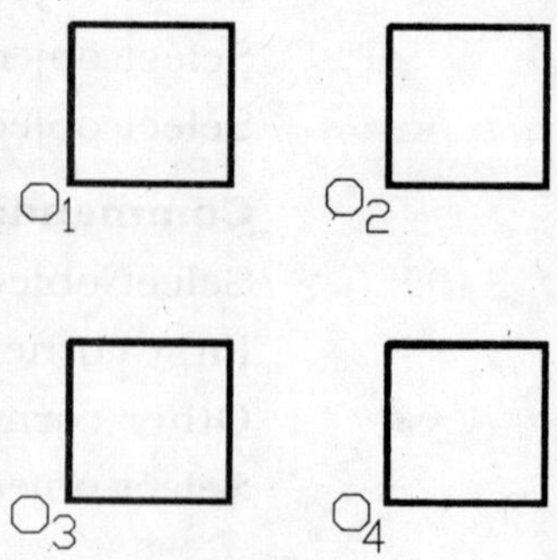


### ROTATE

This command is used to move an object to a new angular position (orientation) about a base point. The rotation angle can be entered relative to a line or as an absolute angle measured counter-clockwise from the original poison. Following example shows, the rotation of a square through 45° in ccw direction about base point 'B'.

**Command: ROTATE** [Enter]
Select objects: [select object by target box or W option)
Select objects: [Enter]
Base point: [select B]
<Rotation angle>/reference: 45 [Enter]

The angular position can also be adjusted with the help of mouse.

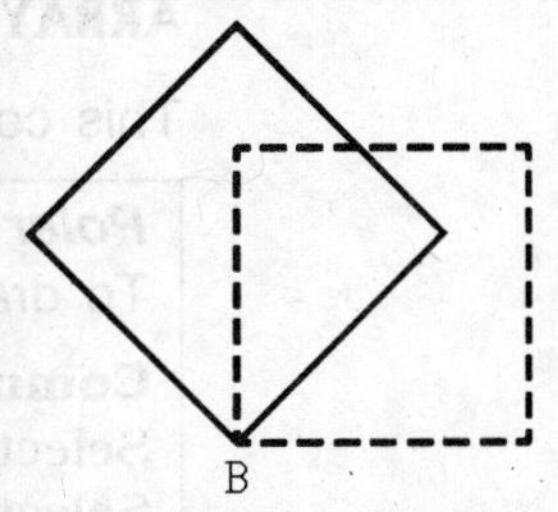


**MIRROR**

This command is used to form a mirror image of an object. The result of this command is an object, symmetric to the original object and place equally back to the mirror. The mirror command involves first selecting the object to mirror and then defining a mirror location by entering the starting and ending point of a mirror line.

**Command: MIRROR** [Enter]
Select object: W [Enter]
First corner: [select]
Other corner: [select]
First point of mirror line: [select P]
Second point: [select Q]
Delete old object? N: [Enter] (Y, if yes)

The starting and end points of MIRROR line are shown as cross mode point

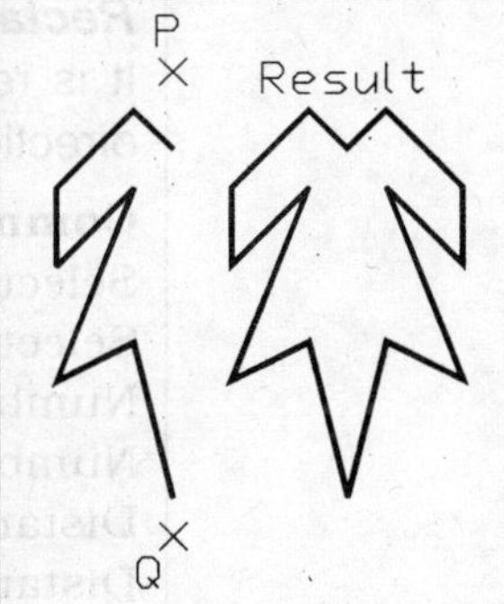


**CHANGE**

This command is used to change the properties of existing objects. We will be often required to change the type of line and colour of objects.

***Change of Line Type***

To change a solid continuous line (default type bylayer) to various types of lines, following are the command steps.

**Command: CHANGE**
Select objects: [select line AB]
Select objects: [Enter]
Properties/<Change point>: P [Enter]
Change what property (Color/Elev/Ltype/LTscale/Thickness)?
Lt [Enter]
New linetype <BYLAYER>: hidden [Enter]
Change what property (Color/Elev/Ltype/LTscale/Thickness)?: [Enter]

Following similar steps, the bylayer line can be changed into various options shown.

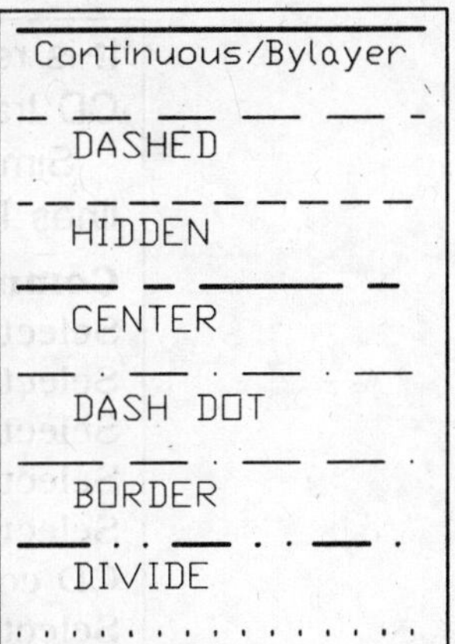


Following similar steps, the default colour can be changed into various colours.

**Command: CHANGE** [Enter]
Select objects: [select line AB]
Select objects: [Enter]
Properties/<Change point>: P
Change what property Color/Elev/Ltype/LTscale/Thickness)?
C [Enter]
New colour <BYLAYER>: red [Enter]
Change what property (Color/Elev/Ltype/LTscale/Thickness)?: [Enter]

## ARRAY

This command is used to create multiple copies of object(s) in a rectangular/polar pattern.

***Polar ARRAY***

To draw 12 radial lines in a circle following are the command steps.

**Command: ARRAY** [Enter]
Select object: [select line OP]
Select object: [Enter]
Rectangular or Polar (R/<P>): P [Enter]
Base/<Specify center point of array>: cen [Enter]
Centre of: [click circle with target box]
Number of items: 12 [Enter]
Angle to fill (+=ccw, -=cw) <360>: 360 [Enter]
Rotate objects as they are copied? <Y>: [Enter]

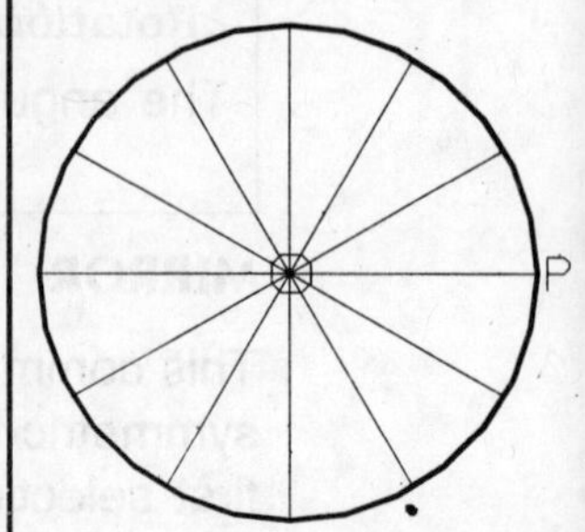


***Rectangular ARRAY***

It is required to create multiple copies of a square object in X & Y directions. Following are the command steps.

**Command: ARRAY** [Enter]
Select objects: (R/P): R [Enter]
Select objects: [Enter]
Number of rows (---)<1>: 6 [Enter]
Numbers of Columns ( | | | ) <1>: 5 [Enter]
Distance between rows (---): 0.75 [Enter]
Distance between columns (iii): 0.75 [Enter]

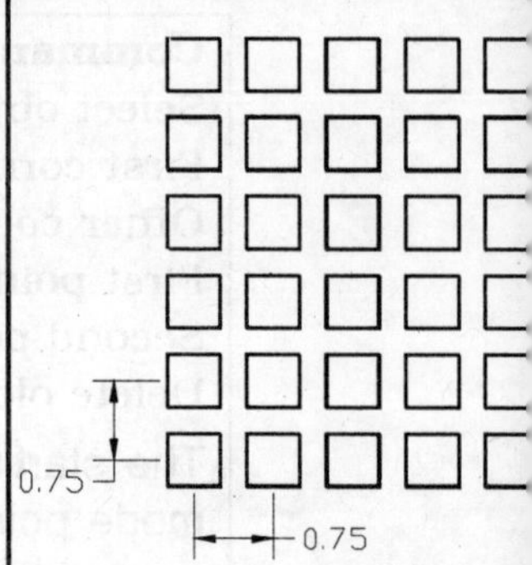


## TRIM

This command is used to trim an object so that it ends precisely at a cutting edge (intersectio points).

It is required to remove segments of lines AB & CD trapped by line PQ & RS.

Similarly, it is required to remove segments of lines PQ & RS trapped by line AB & CD

**Command: TRIM** [Enter]
Select cutting edge(s)
Select objects: [select PQ]
Select objects: [select RS]
Select objects: [Enter]
Select object to trim: [select the portion of AB & CD covered by PQ & RS]
Select object to trim: [Enter]
Following similar steps, the portion of PQ & RS covered by AB & CD is trimmed.

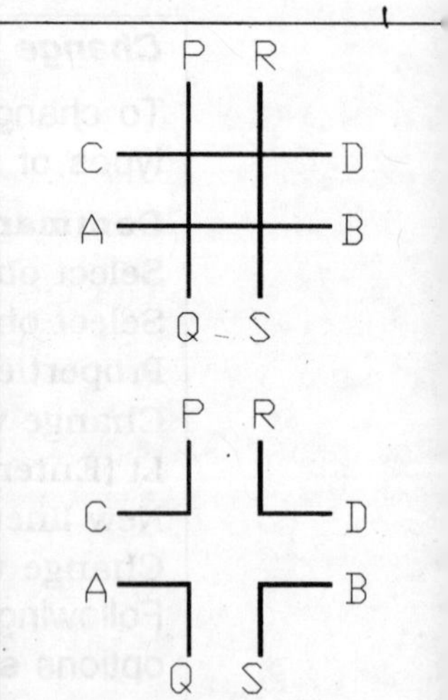


## BREAK

This command is used to erase part of object or splits an object in to two at required location. W will often require changing part of a line to hidden. In order to do that the line needs to be broke at specified location and redraw in the same portion.

**Command: BREAK** [Enter]
Select object: [specify object to be broken] (line AB)]
Enter second point (or F for first point): F [Enter]
Enter first point (point to one end of deletion): [Enter]
Enter second point (point to other end of deletion): [select S]

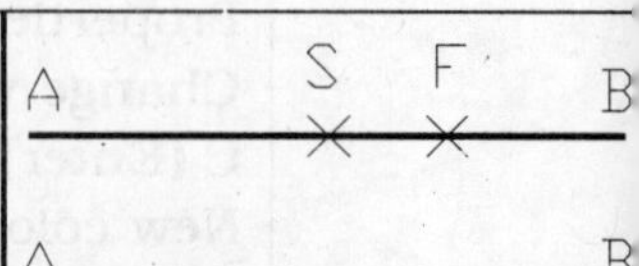

## EXTEND

This command is used to lengthen a line/arc so that it precisely touches a boundary defined by another line/arc. The steps are self-explanatory.

It is required to extend a line PQ so that it touches line AB.

**Command: EXTEND** [Enter]
Select boundary edge(s)
Select objects: [select line AB as a boundary edge]
Select objects: [Enter]
Select object to extend: [select line PQ near to end point Q]
Following similar steps, arc PQ also can be extended to touch line AB as shown in adjacent Fig.

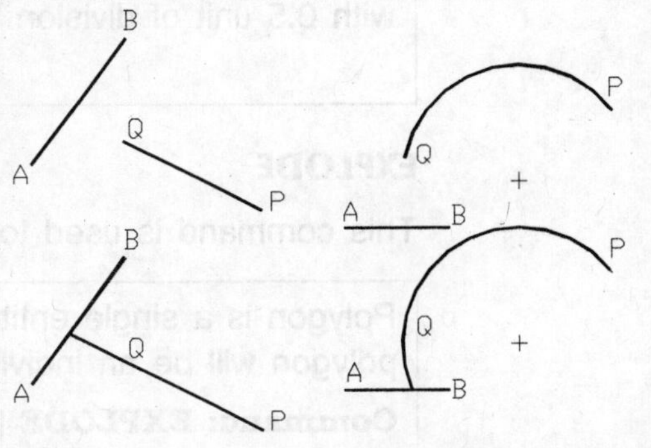


## FILLET

This command is used to fit a smooth curve between two lines /arcs. It involves two stages, first to define a fillet radius, second filleting.

**Command: FILLET** [Enter]
Polyline/Radius/trim/<select first object>: R [Enter]
Enter fillet radius <0.25>: 0.5 [Enter]
**Command: FILLET** [Enter]
Polyline/Radius/trim/<select first object>: [select L1]
Select second object: [select L2]

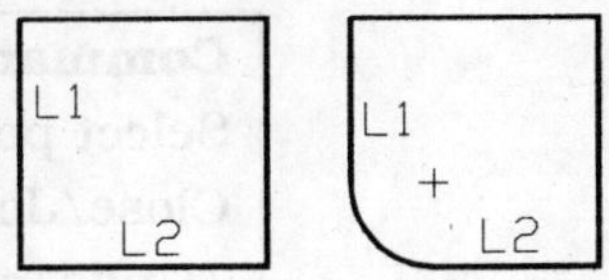


## OFFSET

This command creates concentric circles, parallel lines, and parallel curves. The concentric circles and concentric squares are drawn using offset command. The entity to be offset should be a single entity.

**Command: OFFSET** [Enter]
Offset distance or through <Through>: T [Enter]
Select object to offset: [Select outer circle]
Through point: [select point inside the outer circle]

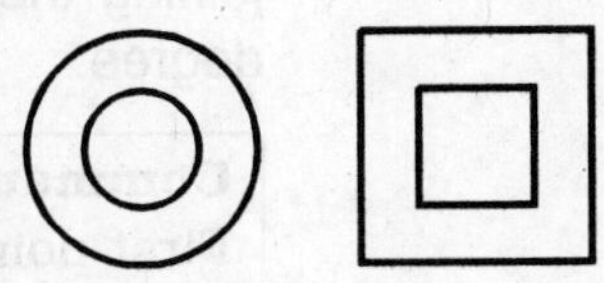

## DIVIDE

This command lets you to divide an entity into specified number of equal parts. Before use of DIVIDE command, the point should be set with PDMODE: 3 and PDSIZE: 0.125

**Command: DIVIDE** [Enter]
Select object to divide: [select line AB]
<Number of segments>/Block: 6 [Enter]
Following similar steps, of command, circle is divided into 6 segments.

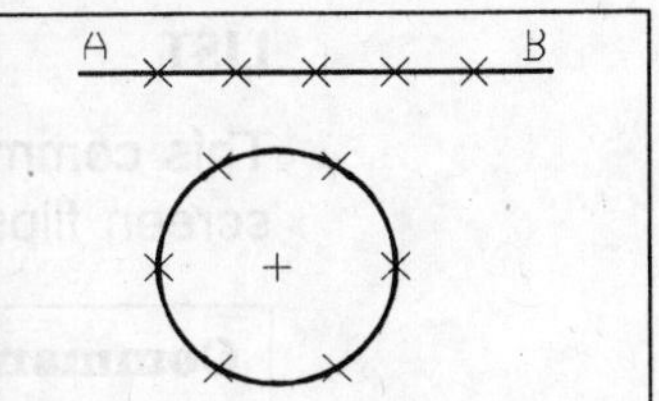


## MEASURE

This command lets you to divide an entity with a specified distance, starting from the end, which is nearer to the target box placed for selection. Before use of MEASURE command, the point should be set with PDMODE: 3 and PDSIZE: 0.125.

**Command: MEASURE** [Enter]
**Select object to measure: [select line AB]**
**<Segment length>/Block: 0.5 [Enter]**
Following similar steps, of command, arc is divided into 6 segments with 0.5 unit of division.

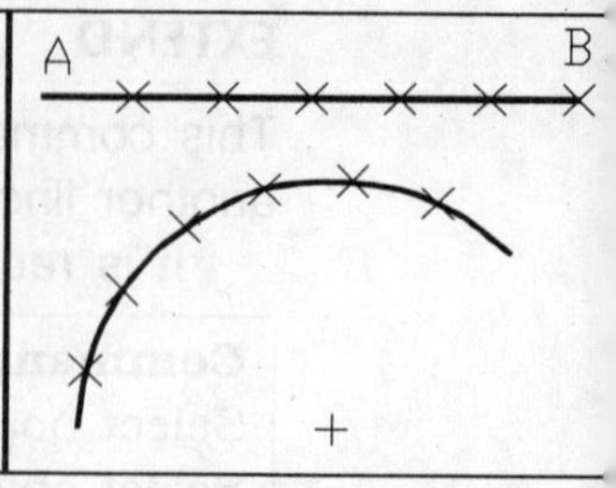


## EXPLODE

This command is used to break entities say a polyline or polygons into individual lines.

Polygon is a single entity. By EXPLODE command the sides of the polygon will be an individual entity.
**Command: EXPLODE** [Enter]
Select objects: [select polygon]

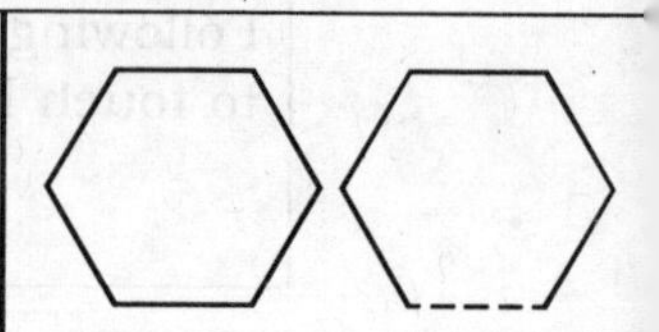

## PEDIT

This command is used to edit polyline. In engineering graphics, this is used to fit a smooth curve passing through the key points of entity.

**Command: PEDIT** [Enter]
Select polyline: [select PLINE]
Close/Join/Width/Edit vertex/Fit/Spline/Decurve/Ltype gen /Undo/eXit <X>: F (for fit)
Following similar steps, of command, PLINE is converted into a spline.

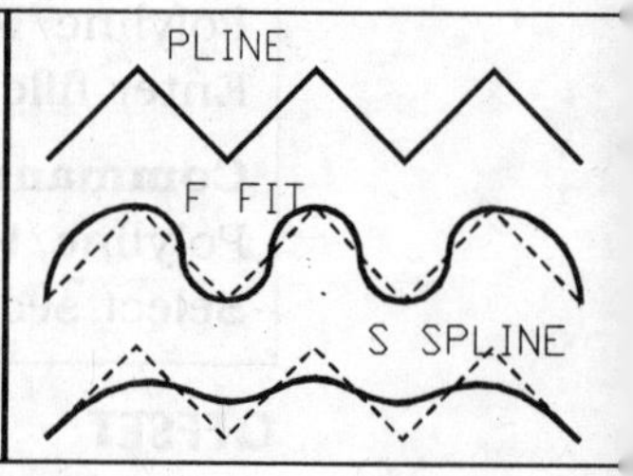


## DIST

This command is used to display the distance between two selected points and the angle of the line joining the first point to the second. Note that the scale, displaying angle is rounded to nearest degree.

**Command: DIST** [Enter]
First point: [select A]
Second point: [select B]
Length = 2.5739, Angle in XY Plane = 29, Angle from XY Plane = 0
Delta X = 2.2500, Delta Y = 1.2500, Delta Z = 0.0000

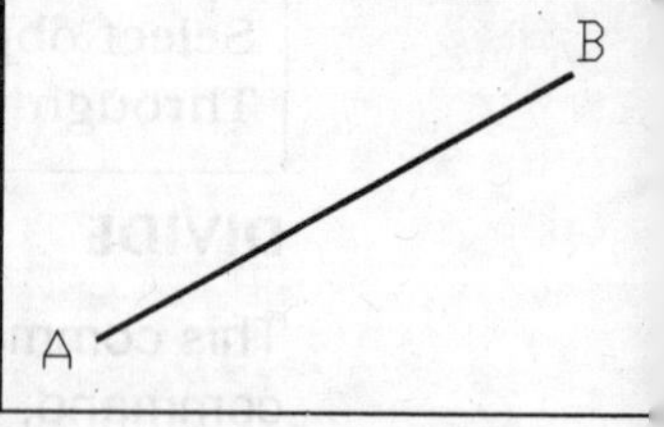


## LIST

This command is used to display exhaustive list of data stored about the selected object(s). The screen flips to the text mode.

**Command: LIST** [Enter]
Select objects: [select circle]
Center point, X = 7.2500 Y = 2.0000 Z = 0.0000
Radius = 1.0000; Circumference = 6.2832
Area = 3.1416

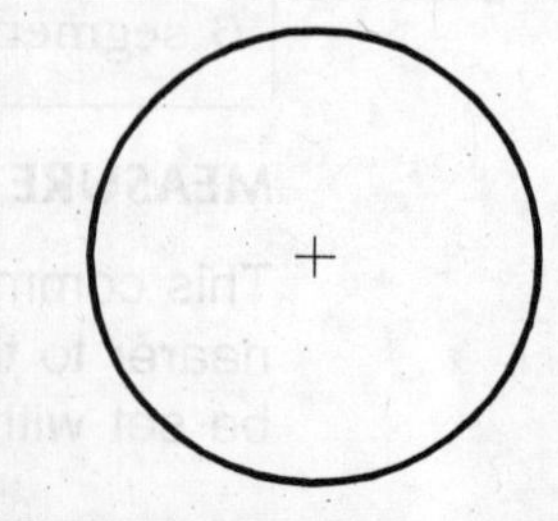

## Exercise (Practice of AutoCAD Commands)

**1 AutoCAD Unit = 4 grid spacing = 100 mm**

**2.1** Draw Figs. shown in Plate 2.1 using LINE command.

**2.2** Draw Figs. shown in Plate 2.2 (P.1 to P.9) using LINE and CIRCLE command.

**2.3** Draw Figs. shown in Plate 2.2 (P.10 to P.27) using LINE, CIRCLE and ARC command.

**2.4** Draw Figs. shown in Plate 2.2 (P.28 to P.45) using LINE, CIRCLE, ARC and ELLIPSE command.

**2.5** Draw Figs. shown in Plate 2.2 (P.46 to P.54). Every Fig. has two views aligned vertically.

**2.6** Two lines AB, BC along with an arc exist as shown in P.1/Plate 2.3. Draw lines CM, MP, PE and ET using Mid, Perpendicular, End and Tangent, options of LINE command.

**2.7** Two lines AB, BC along with a circle exist as shown in P.2/Plate 2.3. Draw lines CM, MP, PE and ET, TO and OI using Mid, Perpendicular, End, Tangent, Center and Intersection options of LINE command.

**2.8** Draw a line AB, 200 mm long and at an angle of 30° as shown in P.3/Plate 2.3.

**2.9** Draw a PLINE with width 0.0, 0.01, 0.05, 0.1, 0.2, 0.5 and 1.0. Draw a taper line with width 0.2 at one end and width 0.75 at another end. This is shown in P.4/Plate 2.3.

**2.10** Two smaller circles exist as shown in P.5/Plate 2.3. Draw a third circle touching the smaller two with TTR option of CIRCLE command and radius R = 1.50 AutoCAD unit as shown.

**2.11** Two smaller circles exist as shown in P.6/Plate 2.3. Draw a third circle touching the smaller two with TTR option of CIRCLE command and radius R = 1.25 AutoCAD unit as shown.

**2.12** Draw a bylayer (continuous) line of 400 mm length. DIVIDE it into 4 numbers of segments with cross mode of point as shown in P.1/Plate 2.4. (POINT with cross mode and size 0.25 should be set in your file before use of this command).

**2.13** Draw a bylayer (continuous) line of 400 mm length. Divide it into equal number of segments using MEASURE command and cross mode of point. Each segment has length of 60 mm.

**2.14** Draw an arc and DIVIDE it into 6 numbers of segments with cross mode of point as shown in P.3/Plate 2.4.

**2.15** Draw an arc and divide it into equal number of segments using MEASURE command and with cross mode of point. Each segment has length of 75 mm as shown in P.4/Plate 2.4.

**2.16** Lines AB, BC and PQ are existing as shown in P.5/Plate 2.4. EXTEND the line PQ to meet line AB. Also, EXTEND the line BC to meet the line PQ.

**2.17** Line PQ and an arc AB are existing as shown in P.6/Plate 2.4. EXTEND the arc AB to meet the line PQ.

**2.18** Draw a circle with 175 mm diameter. DIVIDE it into 6 equal numbers of parts as shown in P.7/Plate 2.4.

**2.19** Draw an ellipse with 125 mm major radius and 75 mm minor radius. DIVIDE it into 8 equal numbers of parts as shown in P.8/Plate 2.4.

**2.20** A circle and an arc are existing as shown in P.9/Plate 2.4. EXTEND the arc to intersect the circle.

**2.21** Draw a square with 250 mm diagonal length. DIVIDE it into 12 equal numbers of parts as shown in P.10/Plate 2.4.

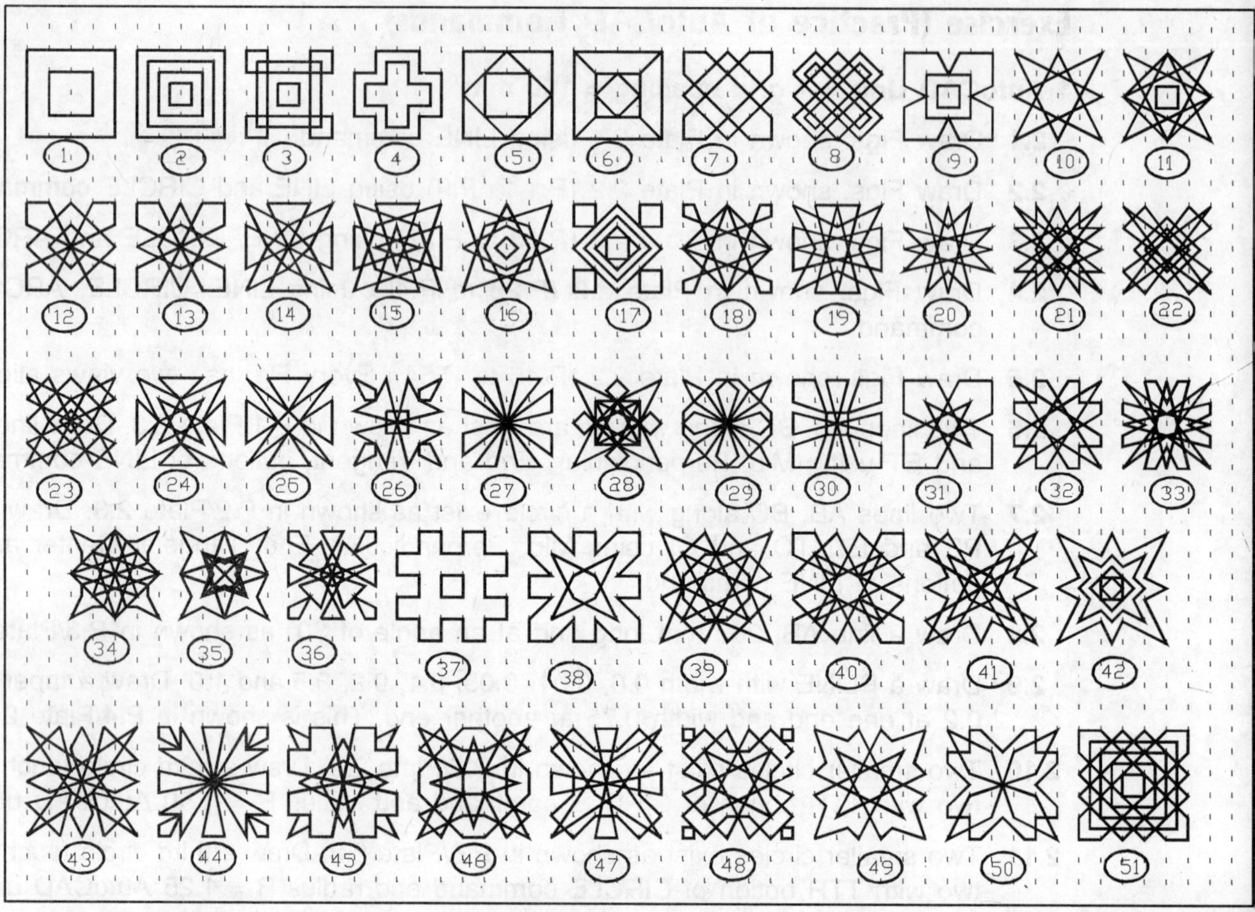


**PLATE 2.1**

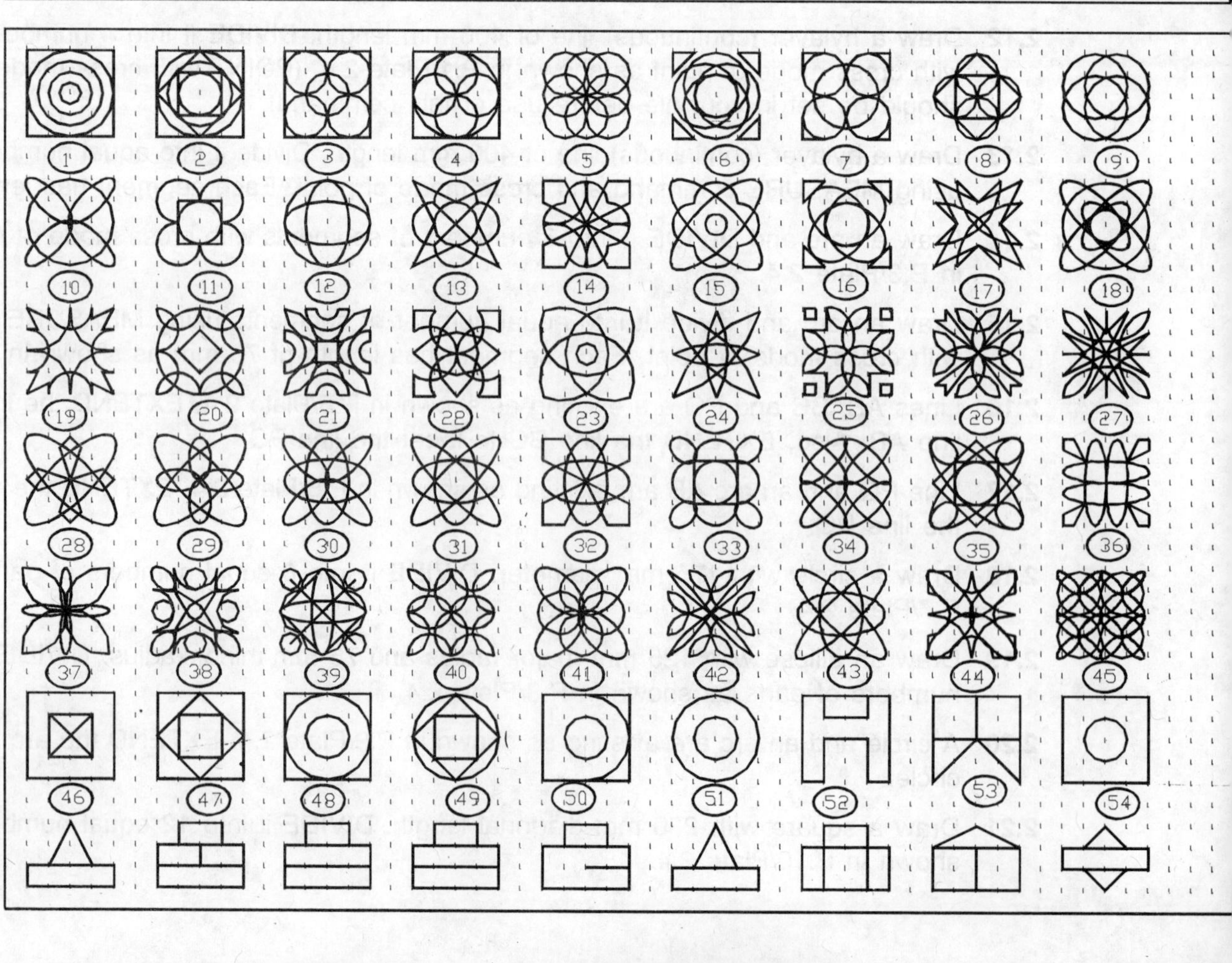


**PLATE 2.2**

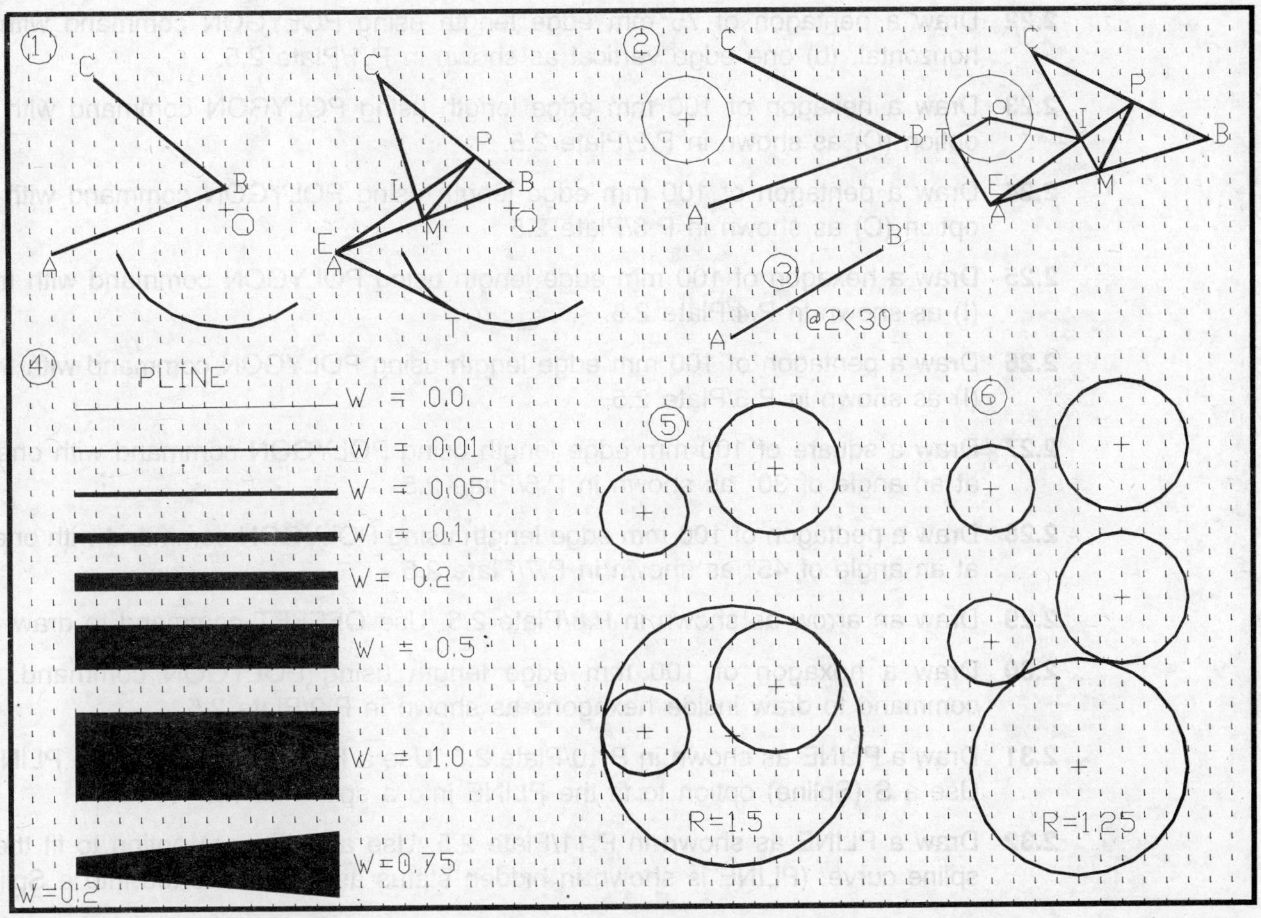


**PLATE 2.3**

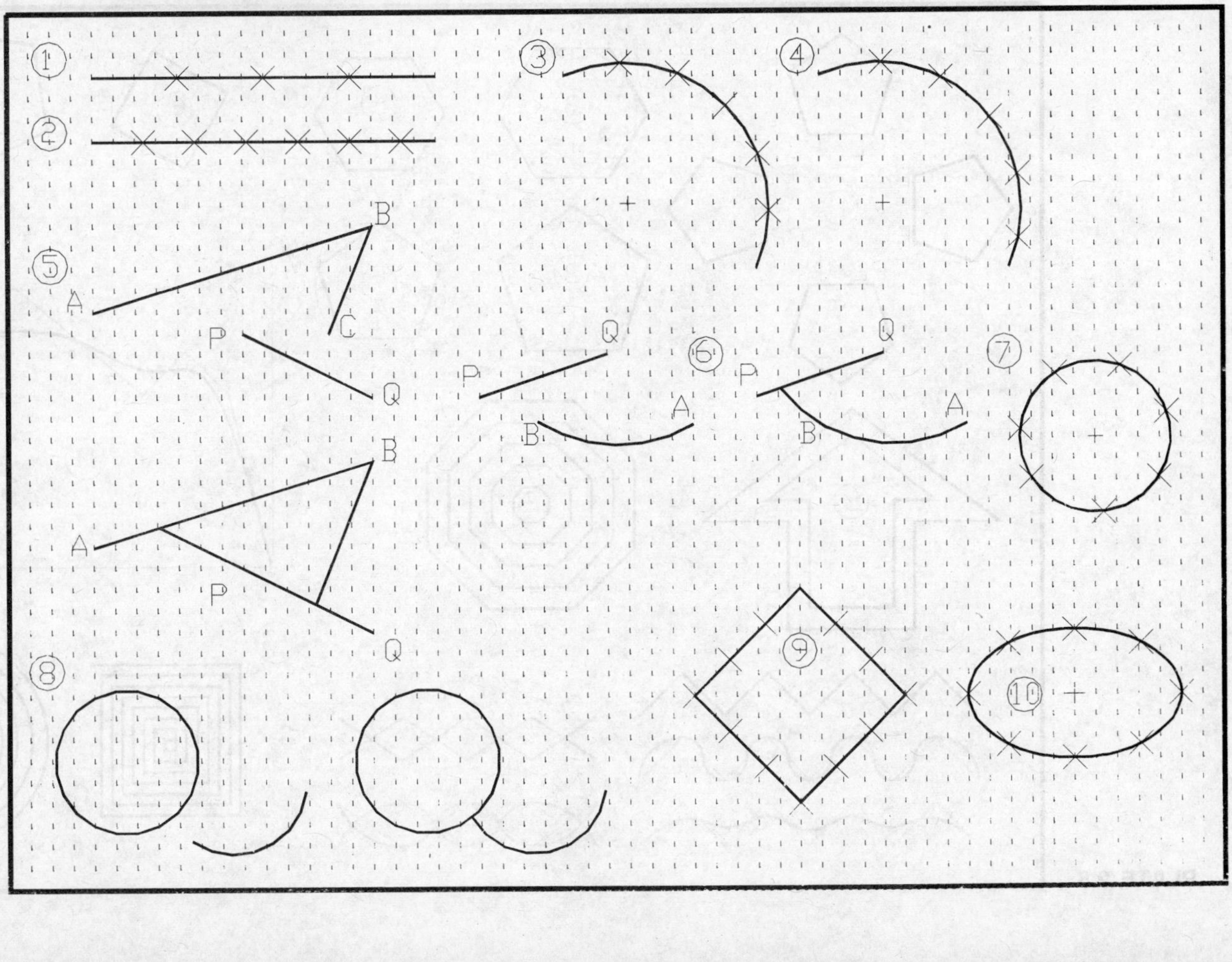

**PLATE 2.4**

**2.22** Draw a pentagon of 75 mm edge length using POLYGON command with (a) one edge horizontal, (b) one edge vertical as shown in P.1/Plate 2.5.

**2.23** Draw a hexagon of 100 mm edge length using POLYGON command with Circumscribed option (C) as shown in P.2/Plate 2.5.

**2.24** Draw a pentagon of 100 mm edge length using POLYGON command with Circumscribed option (C) as shown in P.3/Plate 2.5.

**2.25** Draw a hexagon of 100 mm edge length using POLYGON command with Inscribed option (I) as shown in P.4/Plate 2.5.

**2.26** Draw a pentagon of 100 mm edge length using POLYGON command with Inscribed option (I) as shown in P.5/Plate 2.5

**2.27** Draw a square of 100 mm edge length using POLYGON command with one edge (P1-P2) at an angle of 30° as shown in P.6/Plate 2.5.

**2.28** Draw a pentagon of 100 mm edge length using POLYGON command with one edge (P1-P2) at an angle of 45° as shown in P.7/Plate 2.5.

**2.29** Draw an arrow as shown in P.8/Plate 2.5. Use OFFSET command to draw outside arrow.

**2.30** Draw a hexagon of 100 mm edge length using POLYGON command. Use OFFSET command to draw inside hexagons as shown in P.9/Plate 2.5.

**2.31** Draw a PLINE as shown in P.10/Plate 2.5. Use a F (Fit) option to fit the PLINE into a curve. Use a S (Spline) option to fit the PLINE into a spline curve.

**2.32** Draw a PLINE as shown in P.11/Plate 2.5. Use a S (Spline) option to fit the PLINE into a spline curve. (PLINE is shown in hidden status as it is converted into a Spline curve).

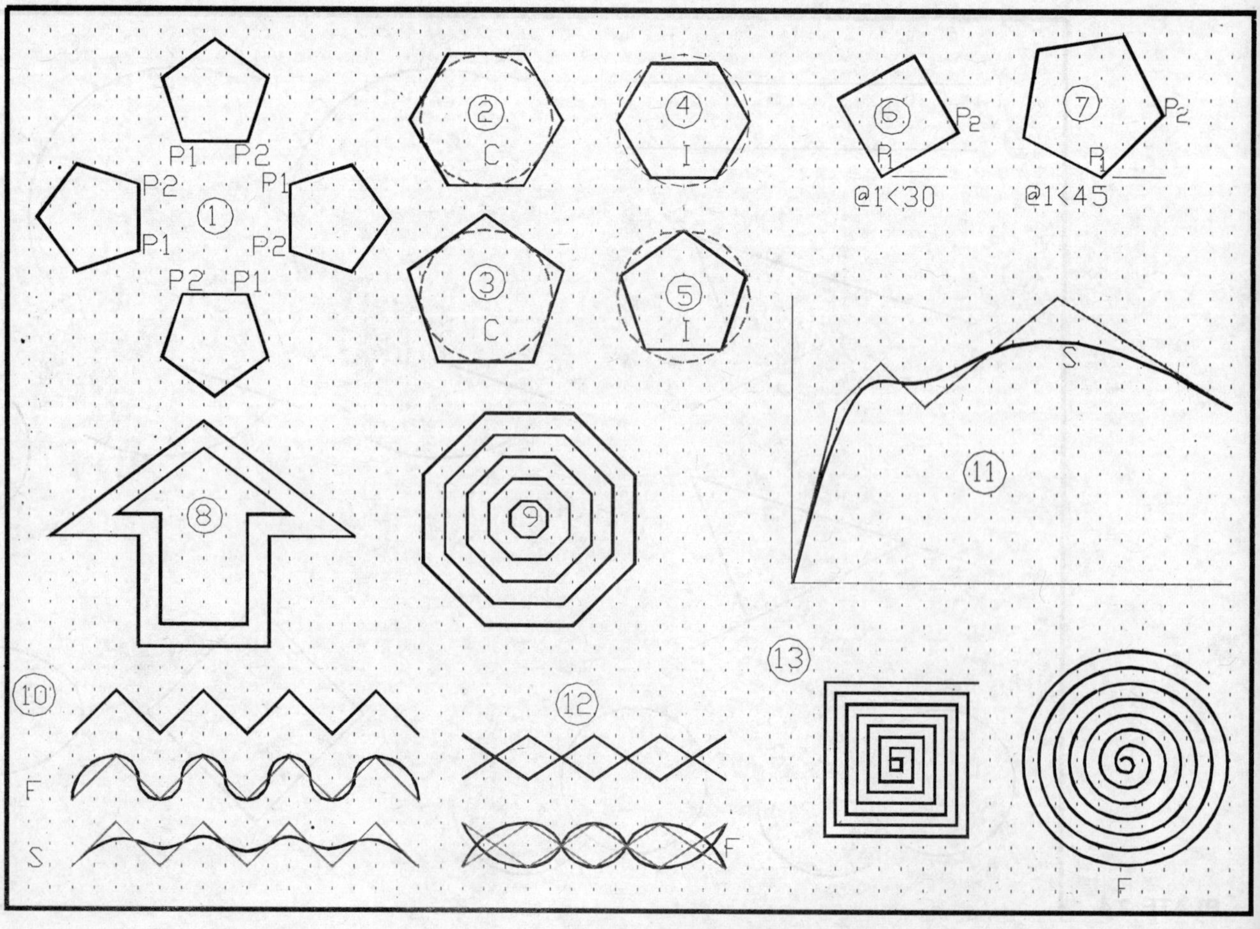


**PLATE 2.5**

**2.33** Draw two PLINEs crossing each other as shown in P.12/Plate 2.5. Use a F (Fit) option to fit the PLINEs into curves.

**2.34** Draw a PLINE as shown in P.13/Plate 2.5. Use a F (Fit) option to fit the PLINE into a spiral.

**2.35** Draw a design as shown in P.1/Plate 2.6.

**2.36** Draw a design as shown in P.2/Plate 2.6.

**2.37** Draw a design as shown in P.3/Plate 2.6.

**2.38** Draw a design as shown in P.4/Plate 2.6.

**2.39** Draw a design as shown in P.5/Plate 2.6.

**2.40** Draw a design as shown in P.6/Plate 2.6.

**2.41** Draw a design as shown in P.7/Plate 2.6.

**2.42** DIVIDE a circular sector into 12 equal numbers of parts as shown in P.8/Plate 2.6.

**2.43** Draw a design as shown in P.9/Plate 2.6.

**2.44** Draw a wall with brick pattern as shown in P.10/Plate 2.6.

**2.45** Draw a table as shown in P.11/Plate 2.6.

**PLATE 2.6**

**2.46** Draw two sets of two concentric squares as shown in P.1/Plate 2.7. HATCH them as shown in the same Figure.

**2.47** Draw three sets of two concentric ellipses as shown in P.2/Plate 2.7 using polar ARRAY and TRIM command.

**2.48** Draw three sets of three concentric hexagons as shown in P.3/Plate 2.7 using COPY and TRIM command.

**2.49** COPY the Fig. drawn in P.3/Plate 2.7 as shown in P.4/Plate 2.7.

**2.50** Draw a design as shown in P.5/Plate 2.7 using MIRROR command.

**2.51** Draw a design as shown in P.6/Plate 2.7 using MIRROR command.

**2.52** Draw a design as shown in P.7/Plate 2.7 using MIRROR command.

**2.53** Draw a design as shown in P.8/Plate 2.7 using MIRROR command.

**2.54** Use multiple COPY option to COPY the Fig. drawn in P.2/Plate 2.7 as shown in P.9/Plate 2.7 at bottom left and bottom right locations.

**2.55** Draw a design as shown in P.10/Plate 2.7.

**2.56** Draw a design as shown in P.11/Plate 2.7.

**PLATE 2.7**

**2.57** Draw four bylayer (continuous) lines of 500 mm length. CHANGE it into dashed, hidden, center, dash-dot, border, divide dot, phantom type line as shown in P.1/Plate 2.8.

**2.58** Draw a square of edge 150 mm. HATCH the area with universal pattern, 45°, 0.125 spacing, as shown in P.2/Plate 2.8.

**2.59** Draw a square of edge 150 mm. HATCH the area with universal pattern, 45°, 0.125 spacing and double hatch option, as shown in P.3/Plate 2.8.

**2.60** Draw a circle of diameter 150 mm. HATCH the area with universal pattern, 45°, 0.125 spacing as shown in P.4/2.8.

**2.61** Draw two squares of edge length 150 mm and adjacent to each other. HATCH one square with Universal pattern, 45°, 0.125 spacing and other square Universal pattern, 135°, 0.125 spacing as shown in P.5/Plate 2.8.

**2.62** Draw a rectangle of 300 × 150 mm. Use OFFSET command to draw inside rectangle 200 mm × 50 mm. Hatch the portion in between two rectangular boundaries with universal pattern, 45°, 0.125 spacing, as shown in P.7/Plate 2.8.

**2.63** COPY solution of P.7/Plate 2.8. HATCH the inside square with universal pattern, 135°, 0.125 spacing, as shown in P.8/Plate 2.8.

**2.64** Draw an ellipse and two concentric circles as shown in P.9/Plate 2.8. HATCH the portion in between boundaries of ellipse and outer circle with Universal pattern, 45°, 0.125 spacing, and inner circle with Universal pattern, 135°, 0.125 spacing.

**2.65** Draw two circles as shown in P.10/Plate 2.8. HATCH the portion in between intersecting boundaries with Universal pattern, 45°, 0.125 spacing.

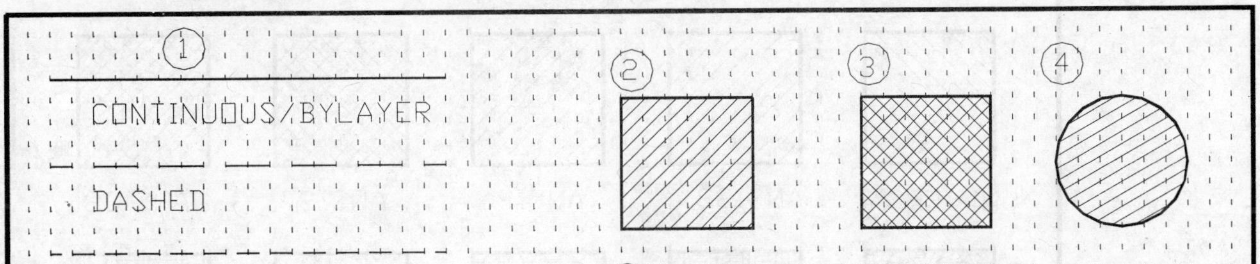


**PLATE 2.8**

**2.66** Draw a circle with 150 mm diameter. Draw another concentric circle with 100 mm diameter.
(a) HATCH the inner circle with universal pattern, 45° and 0.125 spacing.
(b) HACTH the portion confined by two circular boundaries with universal pattern, 135° and 0.125 spacing.

**2.67** Draw a circle with 150 mm diameter. Draw another concentric circle with 100 mm diameter. HATCH the inner circle with universal pattern, 45° and 0.125 spacing. HATCH the annular portion with universal pattern, 135° and 0.125 spacing.

**2.68** Draw a circle with 150 mm diameter. Draw a line in horizontal direction passing through center of circle as a diameter. HATCH upper semicircle universal pattern, 45° and 0.125 spacing.

**2.69** Draw a circle with 150 mm diameter. Draw two lines one in horizontal direction and one in vertical directions passing through center of circle as diameters. Hatch any quarter-circle using universal pattern, 45° and 0.125 spacing.

**2.70** Draw a circle with 150 mm diameter. Draw a line in horizontal direction passing through center of circle as a diameter. Draw a line in vertical direction passing through center of circle as a diameter. HATCH any two opposite quarter portion with universal pattern, 45° and 0.125 spacing and remaining two quarter portion with universal pattern, 135° and 0.125 spacing.

**2.71** Draw a hexagon with 75 mm edge. Draw another concentric hexagon with 125 mm edge. HATCH the portion confined by two hexagonal boundaries with universal pattern, 45° and 0.125 spacing.

**2.72** Draw a square of side 150 mm. Use variety of hatch pattern shown in Plate 2.9 to HATCH the closed area. Mention the name of pattern below each type of pattern.

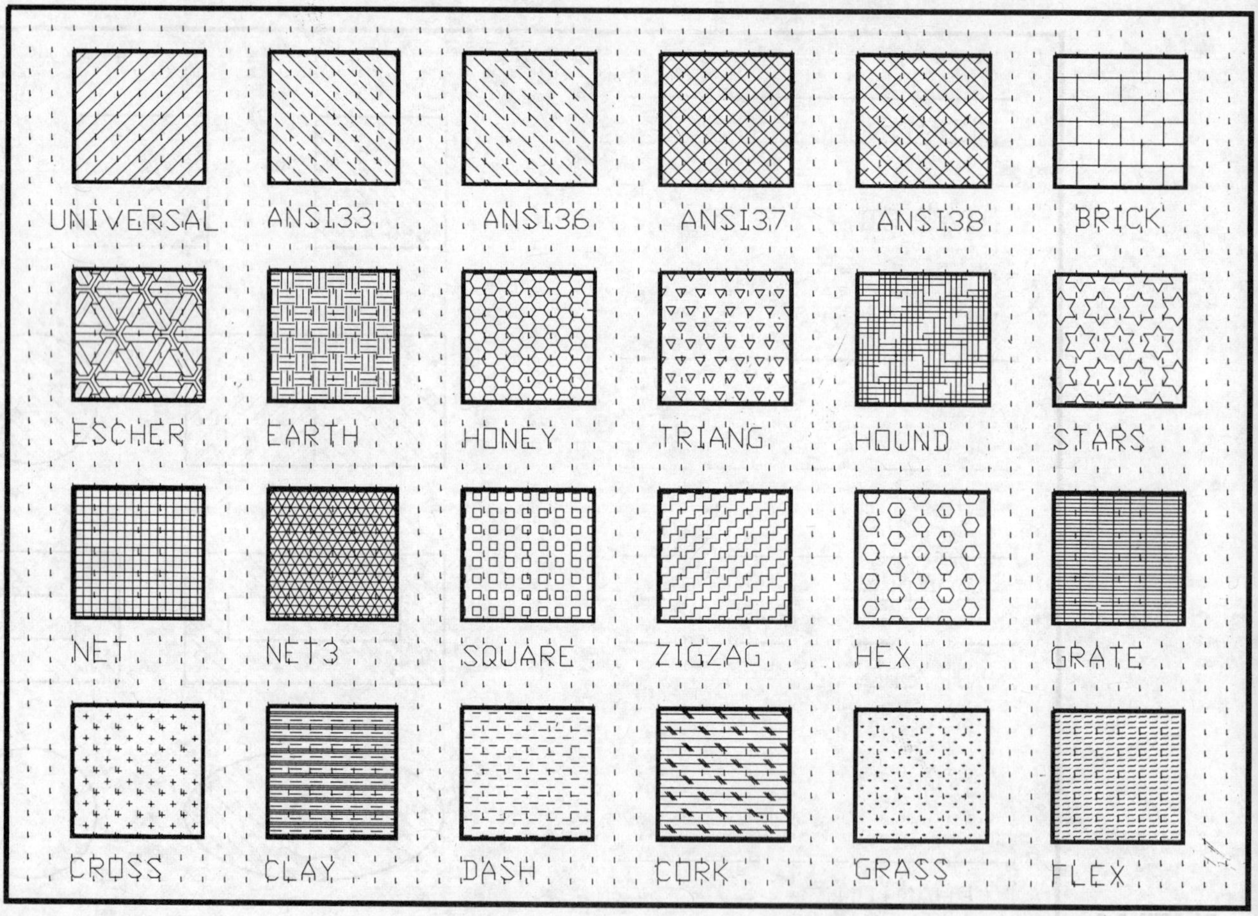


**PLATE 2.9**

**2.73** Draw a square with edge 100 mm long and in horizontal direction. ROTATE the square by 45°.

**2.74** Draw a hexagon with edge 100 mm long, in horizontal direction. ROTATE the hexagon by 45°.

**2.75** Draw a circle with 150 mm diameter. Divide the circular sector into ten equal numbers of parts by using straight lines.

**2.76** Draw a circle with 0.5 AutoCAD unit diameter. Make an ARRAY of the circle in horizontal and vertical direction with unit cell distance in between two rows is 0.5 AutoCAD unit and in between two columns is 0.5 AutoCAD unit.

**2.77** Draw four bylayer (continuous) lines of 500 mm length. CHANGE the status of these lines into hidden, center, dot and phantom.

**2.78** Draw a circle with 100 mm diameter. CHANGE the status into a hidden circle.

**2.79** Draw four bylayer (continuous) lines of 500 mm length. CHANGE the colour to red, yellow, blue and green.

**2.80** Draw a Table of 6 columns and 8 rows. The height of row is 0.5 AutoCAD unit and width of column is 0.75 AutoCAD unit.

**2.81** Draw two circles of 150 mm radii with centers 100 mm apart. TRIM overlapping part of both circles and create a common area.

**2.82** Draw a line of 200 mm length and at 40° to the horizontal. Find out horizontal distance and vertical distance between end points.

**2.83** Draw a hexagon with edge 100 mm long and in horizontal direction. EXPLODE the hexagon to make the edges as separate entities.

**2.84** Draw Fig.s as shown in Plate 2.10.

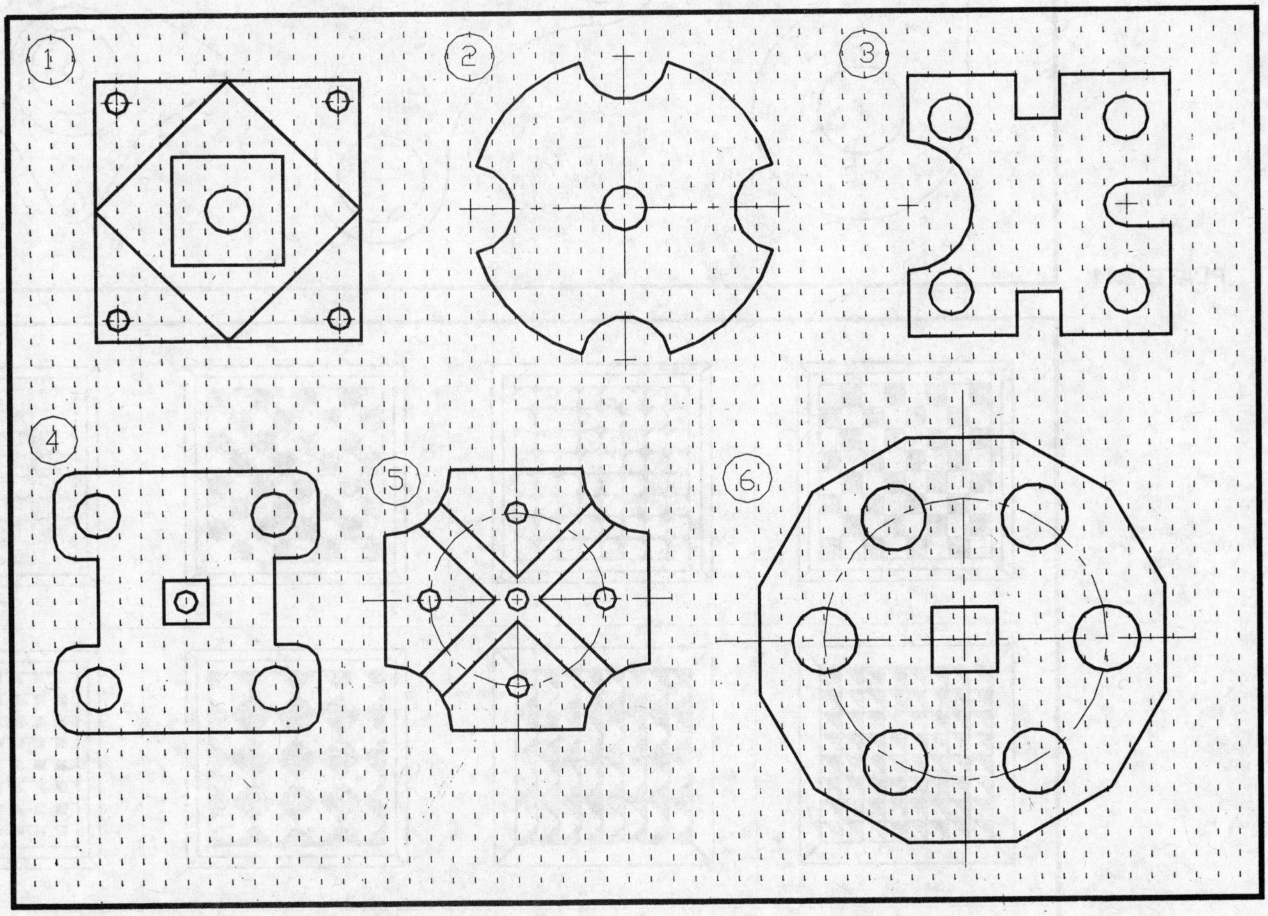


**PLATE 2.10**

**2.85** Draw Figs. as shown in Plate 2.11.

**2.86** Draw Figs. as shown in Plate 2.12.

**2.87** Draw Figs. as shown in Plate 2.13.

**2.88** Draw Figs. as shown in Plate 2.14.

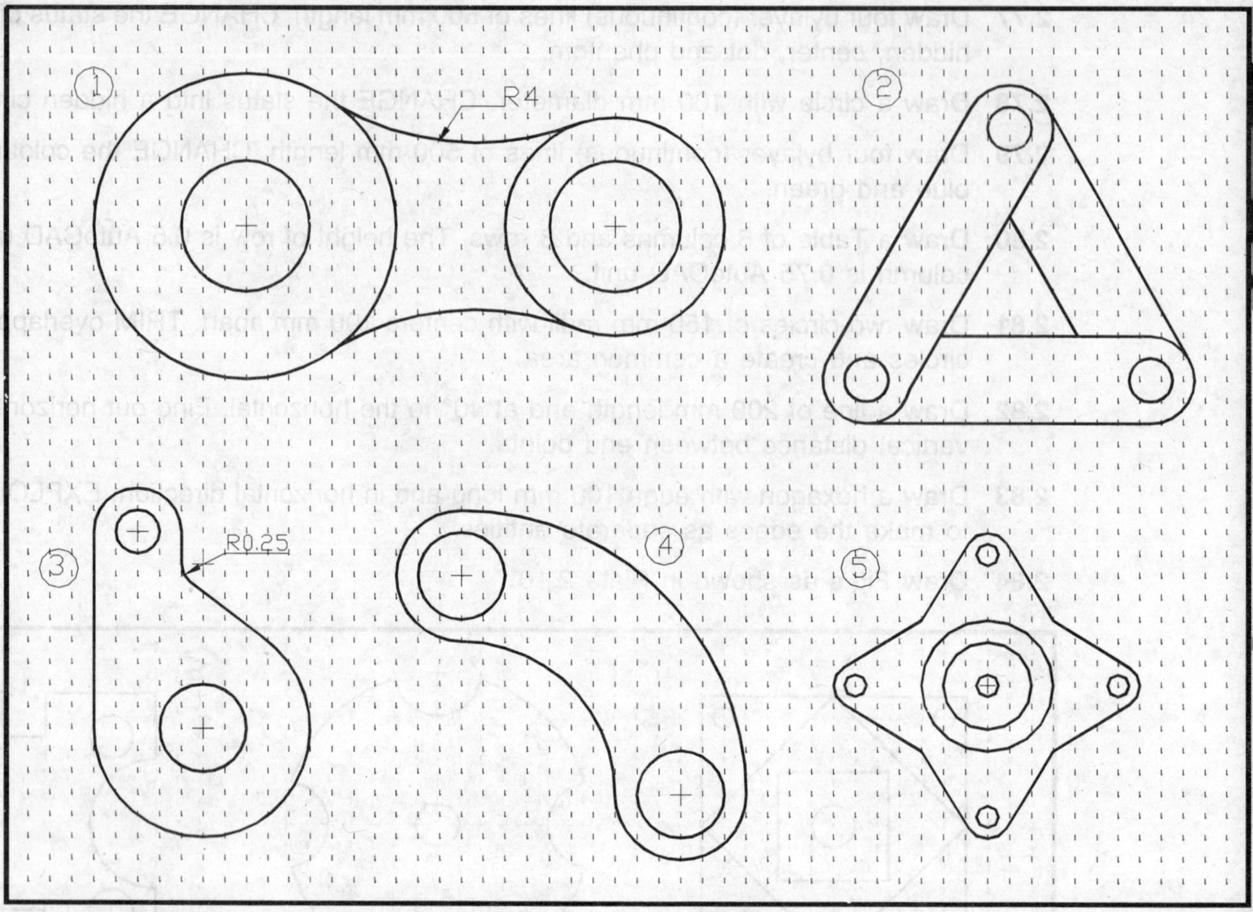


PLATE 2.11

PLATE 2.12

**PLATE 2.13**

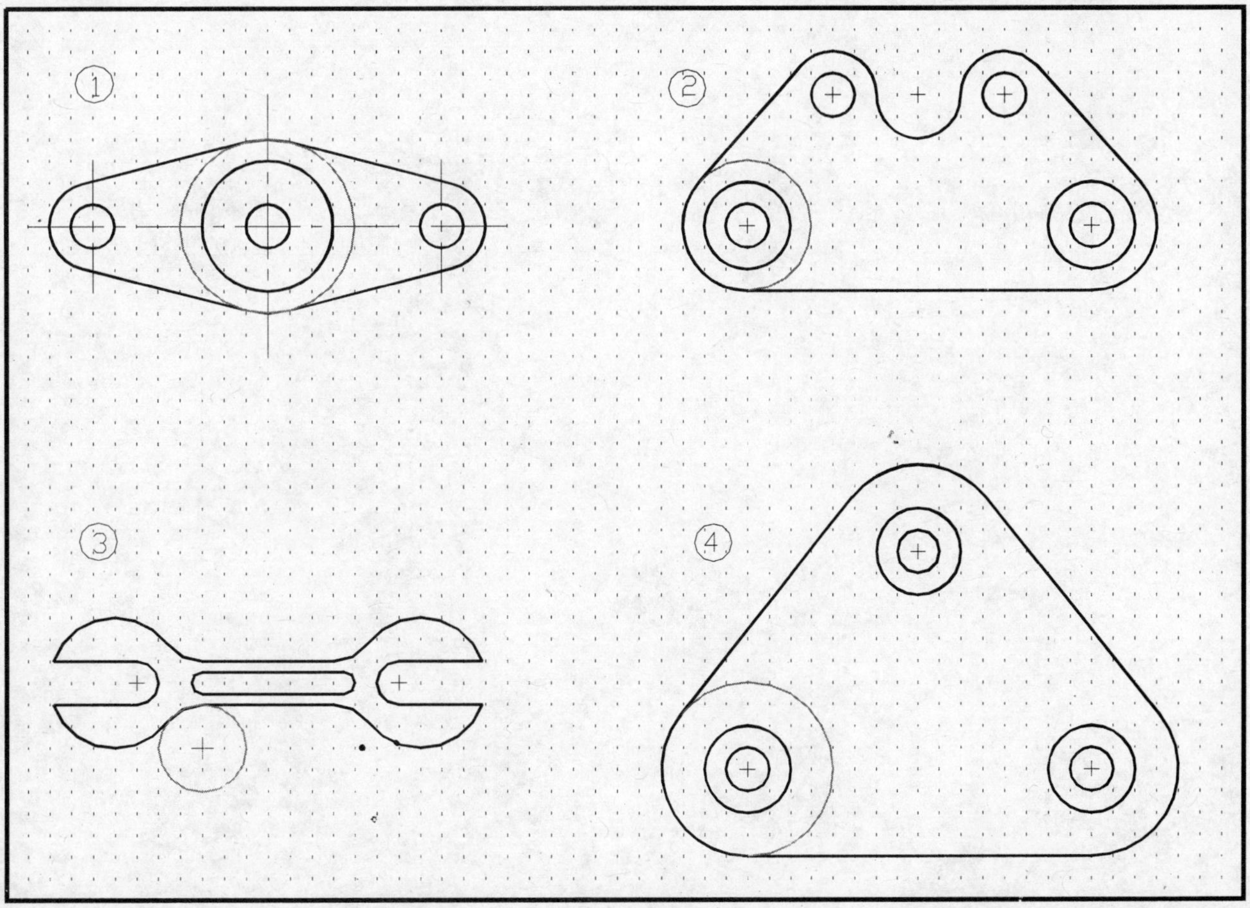


**PLATE 2.14**

# PART II

## Projection Techniques

# 3

# Theory of Projection

## Objectives

- Know the elements of projection.
- Study the systems of projection.
- Learn the methods of projection.
- Focus on the standard practices in projection.

## Introduction

We record the shapes and sizes of three-dimensional (3-D) objects on an electronic sheet or on a drawing paper. The electronic sheet or a drawing paper is the picture plane on which the geometry is projected or drawn. When we represent a 3-D object on a computer screen or on a paper, we should be aware of what type of projection method we are using in representing the object. Our goal should always be to represent an object with:

- least amount of distortion;
- critical features most visible; and
- highest degree of realism.

We will find that these requirements are, in most cases, mutually exclusive. Instead, we will have to decide which projection method will best achieve these goals. This chapter deals with the elements, system and methods of projection.

## 3.1 Elements of Projection

A projection is the representation of an object on a plane. The projections convey all the three dimensions (length, width and height) of the object besides other details. The elements to be considered while drawing a projection are the point of sight (observer's eye position), visual rays, an object and plane of projection. Figures 3.1(a) and (b) show the elements of parallel projection. The parallel projection system is based on the following assumptions:

1. The observer is at infinite distance.
2. Visual rays are parallel to each other and perpendicular to the plane of projection.

The difference between these two figures is the position of the plane of projection and object. In Fig. 3.1(a) the plane of projection is in between the object and an observer while in Fig. 3.1(b) an object is in between an observer and the plane of projection. The rays or lines connecting all these elements are called projectors.

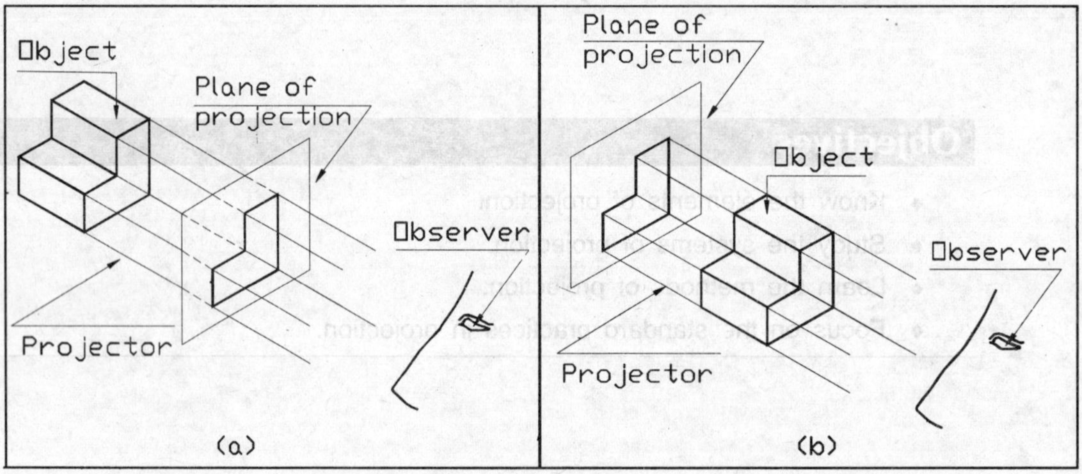


**Fig. 3.1** Elements of parallel projection.

## 3.2 Systems of Projection

The systems of projection used by engineers and technologists to represent shape and size of an object and the theory governing each system should be understood thoroughly before it is used. Basically, projections are classified into two types based on the number of views. The given classification shows the various systems of projection (Fig. 3.2).

- Projection
  - Multiview projection (Orthographic projection)
  - Single view projection (Pictorial projection)
    - Axonometric projection
      - Isometric
      - Dimetric
      - Trimetric
    - Oblique projection
      - Cavalier
      - Cabinet
      - Clinographic
      - General
    - Perspective projection
      - Linear
        - Parallel
        - Angular
        - Oblique
      - Aerial

**Fig. 3.2** Classification of projections.

## 3.3 Multiview Projection

In multiview (multiplanar) projection, exact shape and size of an object is represented by two or more views projected on projection planes, which are at right angles to each other.

The views of an object may be obtained by using either of the two methods:

### Natural method

In this method, each of the necessary view is obtained by looking directly at the particular side of an object. This method is also called the “direct view” method.

Figure 3.3 shows two-dimensional views of an object: front, top and side views. These are obtained by looking directly at front, top and right side, respectively. In application of this method, for heavy/bigger objects, the position of an object is fixed and the position of an observer is shifted (Fig. 3.3) for each view. For smaller objects, observer's position is fixed and the position of an object is changed for each view. The planes of projections are imagined as transparent planes, which are perpendicular to visual rays.

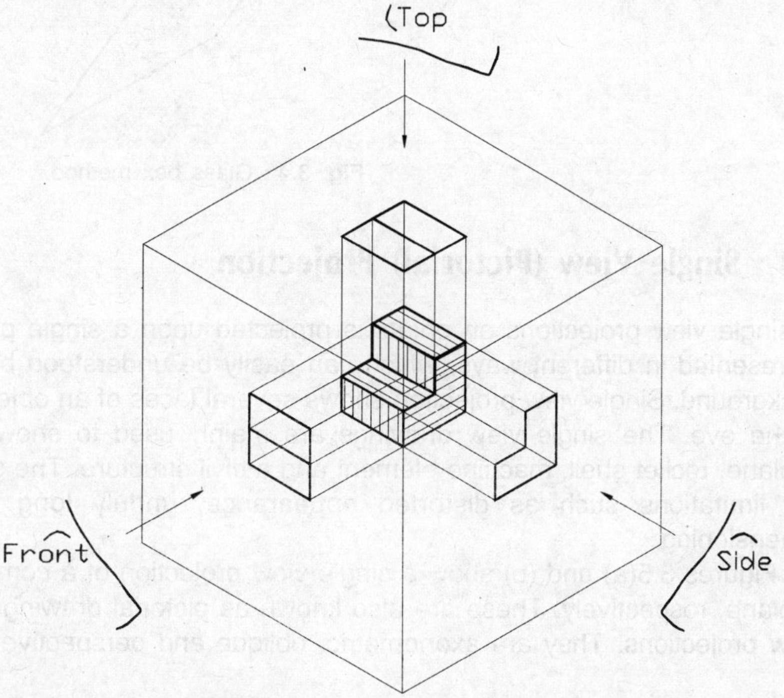


**Fig. 3.3** Two-dimensional views of a 3-D object with natural method.

### Glass box method

An imaginary glass box is used to explain the arrangement of 2-D views. The object is imagined in an open glass box and projections are taken on five faces of glass box. The projection planes (top, bottom, left and right are assumed to be hinged to the frontal plane at AB, CD, AD and BC, respectively, as shown in Fig. 3.4. It is considered that planes of projection, parallel to the six faces of an object, form an enclosing glass box. The observer views the enclosed object from outside. The views are obtained by running projectors from points on the object to the planes.

Multiview projection is also called orthographic projection as the visual rays are perpendicular to the planes of projections. This kind of projection is widely used for the presentation of 2-D views of an object.

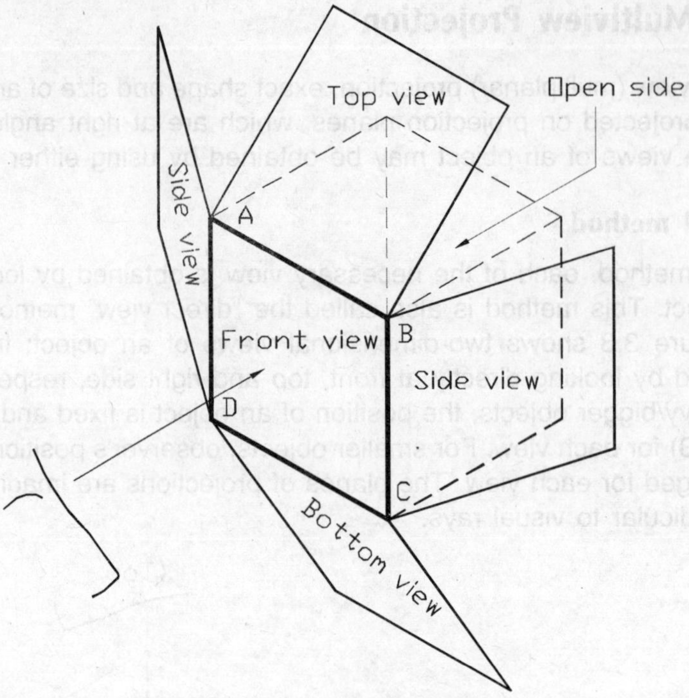


**Fig. 3.4** Glass box method.

## 3.4 Single View (Pictorial) Projection

In single view projections an object is projected upon a single plane. In this system, an object is represented in different ways, which can easily be understood by persons even without technical background. Single view projection shows several faces of an object in a single view as they appear to the eye. The single view drawings are mainly used to show complicated structures such as airplane, rocket shell, machine element and a civil structure. The single view projection suffers from the limitations such as distorted appearance, unduly long time of execution and difficult dimensioning.

Figures 3.5(a) and (b) show a single view projection of a complicated machine element and an airplane, respectively. These are also known as pictorial drawings. There are three types of single view projections. They are axonometric, oblique and perspective projections.

(a) Machine element

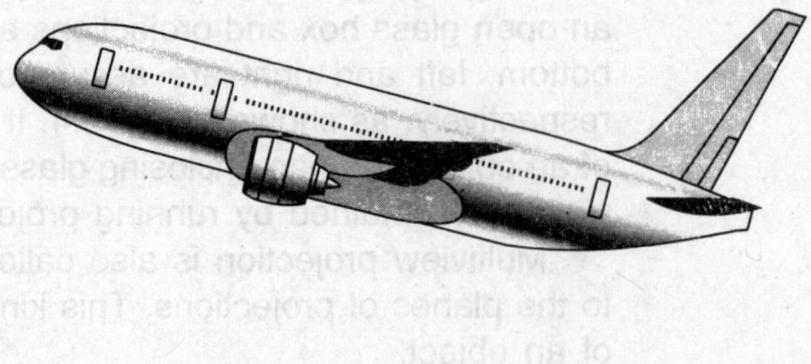

(b) Airplane

**Fig. 3.5** Single view projection.

### 3.4.1 Theory of Axonometric Projection

Figure 3.6 shows a cube placed on its bottom face on a platform with a transparent projection plane. In the first stage, one face of the object is parallel to the plane of projection and the observer can see the true shape and size of that face. In the second stage, the cube is turned on its bottom face such that two faces are equally inclined to the plane of projection and the observer can see the reduced size of these two faces. In the third stage, the cube is tipped forward on the front-most bottom face corner such that three faces are visible to the observer. These three faces are inclined to the plane of projection resulting in an axonometric projection. In this view, the observer can see the reduced size of these three faces.

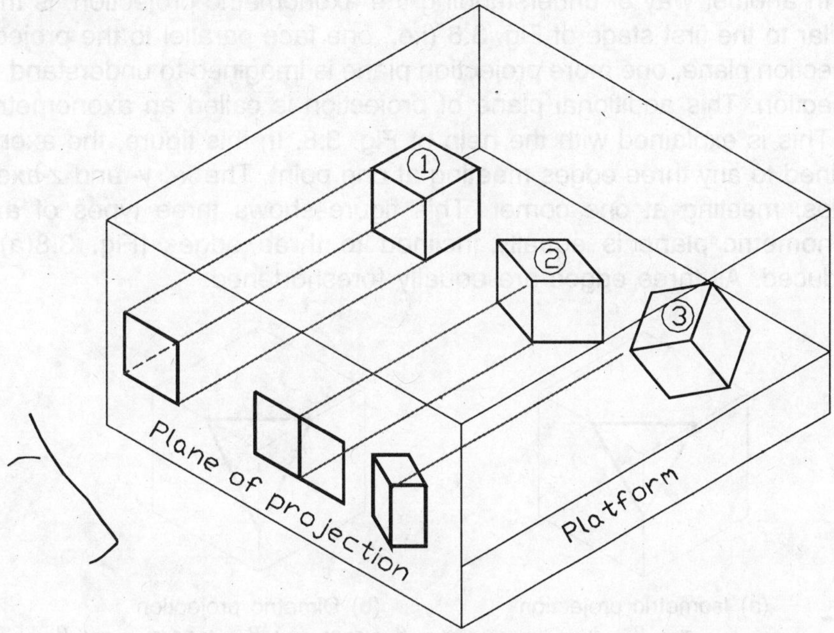


**Fig. 3.6** Theory of axonometric projection.

This is further explained with respect to Cartesian coordinate system. Let us consider a coordinate system *x*, *y* and *z* having origin O as one of the bottom-face corner about which the object was tipped forward. This is shown in Fig. 3.7.

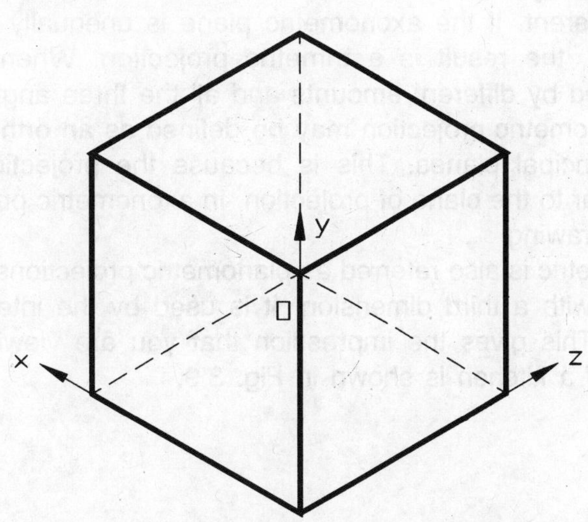


**Fig. 3.7** Object with coordinate system.

In the first stage, in Fig. 3.7, the *x*-axis is parallel to the projection plane, *y*-axis is perpendicular to the platform and *z*-axis is perpendicular to the projection plane. In the second stage, *y*-axis remains perpendicular to the platform; however, *x*- and *z*-axis are equally inclined to the projection plane. In the third stage, *x*-, *y*- and *z*-axes are inclined at an angle of $\alpha$, $\beta$ and *y* to the platform, respectively. Based on the axes inclination to the platform, there are three kinds of projections.

(a) Isometric projection ($\alpha = \beta = \gamma$)
(b) Dimetric projection ($\alpha = \beta \neq \gamma$ or $\alpha \neq \alpha = \gamma$ or $\alpha = \gamma \neq \beta$)
(c) Trimetric (Axonometric) projection ($\alpha \neq \beta \neq \gamma$)

In this process, an object is turned to understand the axonometric projection.

In another way of understanding the axonometric projection, is that the object position is fixed, similar to the first stage of Fig. 3.6 (i.e., one face parallel to the projection plane). In addition to this projection plane, one more projection plane is imagined to understand the three kinds of axonometric projection. This additional plane of projection is called an axonometric plane of projection.

This is explained with the help of Fig. 3.8. In this figure, the axonometric plane of projection is inclined to any three edges meeting at one point. The *x*-, *y*- and *z*-axes can pass through any three edges, meeting at one corner. This figure shows three types of axonometric projections. If the axonometric plane is equally inclined to three edges [Fig. 3.8(a)], an isometric projection is produced. All three edges are equally foreshortened.

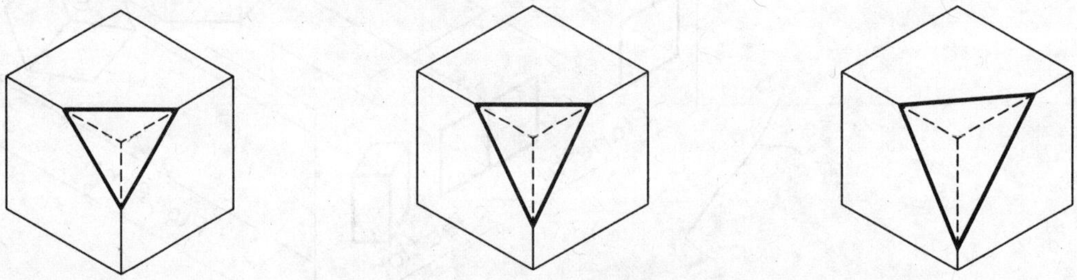

(a) Isometric projection $\alpha = \beta = \gamma$ (b) Dimetric projection $\alpha = \beta \neq \gamma$ or $\alpha \neq \beta = \gamma$ or $\alpha = \gamma \neq \beta$ (c) Trimetric projection $\alpha \neq \beta \neq \gamma$

**Fig. 3.8** Types of axonometric projection.

If the axonometric plane is equally inclined to two of the three edges [Fig. 3.8(b)], a dimetric projection is produced. Two of the edges are equally foreshortened and the third edge is foreshortened by a different amount. Two of the angles between the axes are equal and the third angle is different. If the axonometric plane is unequally inclined to all the three principal planes [Fig. 3.8(c)], the result is a trimetric projection. When this happens all the three edges are foreshortened by different amounts and all the three angles are unequal.

An axonometric projection may be defined as an orthographic projection upon a plane oblique to three principal planes. This is because the projection rays are parallel to each other and perpendicular to the plane of projection. In axonometric projection, parallel lines in an object remain parallel in drawing.

Axonometric is also referred as planometric projections as it is known to be a method of drawing a top view with a third dimension. It is used by the interior designers, architects and landscape gardeners. This gives the impression that you are viewing the objects from above. Planometric projection of a kitchen is shown in Fig. 3.9.

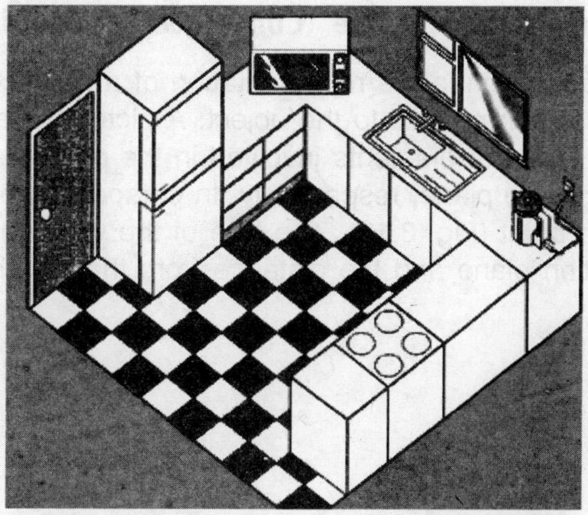

**Fig. 3.9** Planometric projection (a specific type of axonometric projection).

### 3.4.2 Theory of Oblique Projection

Another form of single view projection is an oblique projection. In this type of single view projection, front face of an object is imagined to be parallel to the plane of projection and appears in its true shape and size. The receding lines representing the other two faces are usually drawn at 30°, 45° or 60° to the horizontal. 45° is the most common practice. Obviously, the projectors are not perpendicular to the plane of projection and hence, this kind of projection is not an orthographic projection. Oblique projection provides an easy way of turning an existing orthographic view into a pictorial view as shown in Fig. 3.10(b). Figure 3.10 shows an isometric view and an oblique view together for a comparison. Figure 3.11 shows an oblique view of an office table.

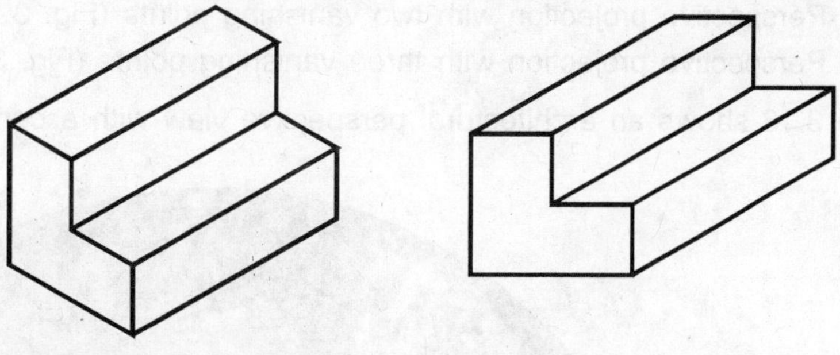

(a) Isometric projection (b) Oblique projection

**Fig. 3.10** Stair case.

**Fig. 3.11** Oblique projection of an office table.

In axonometric and oblique projections all parallel lines in an object appear parallel in the drawing. Vertical lines on the object appear vertical.

### 3.4.3 Theory of Perspective (Convergent) Projection

Perspective projection is the representation of an object as it appears to an observer stationed at a particular position relative to the object. A picture taken by a camera is a real perspective i.e., a natural view. The four elements involved in the photography are, namely: object, visual rays, point of sight and picture plane, respectively. In perspective projection, the projecting lines or visual rays converge at a point (Fig. 3.12). The size of the view depends upon the distance from an observer to the projection plane and the distance from the projection plane to the object.

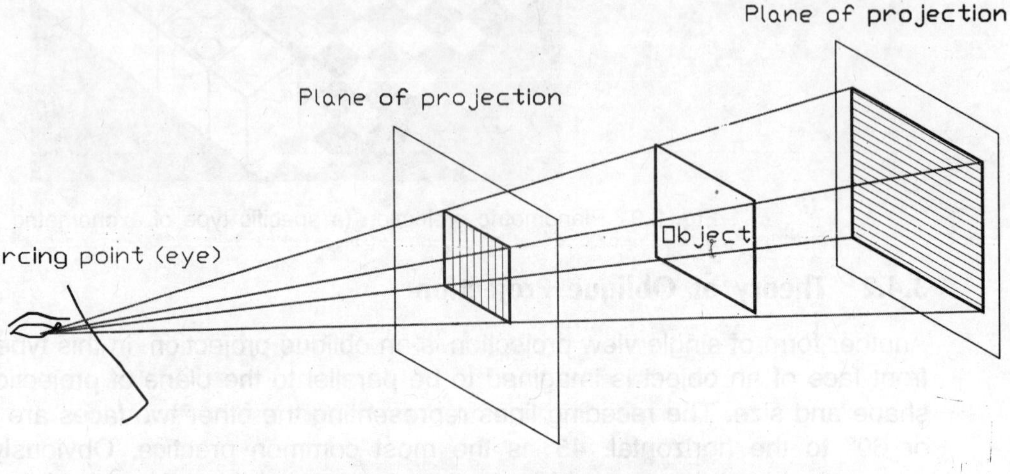


**Fig. 3.12** Elements of perspective (convergent) projection.

Perspective projection has three types of system of projection.

1. Perspective projection with single vanishing point (Fig. 3.13)
2. Perspective projection with two vanishing points (Fig. 3.14)
3. Perspective projection with three vanishing points (Fig. 3.15)

Figure 3.13 shows an architectural perspective view with a central vanishing point.

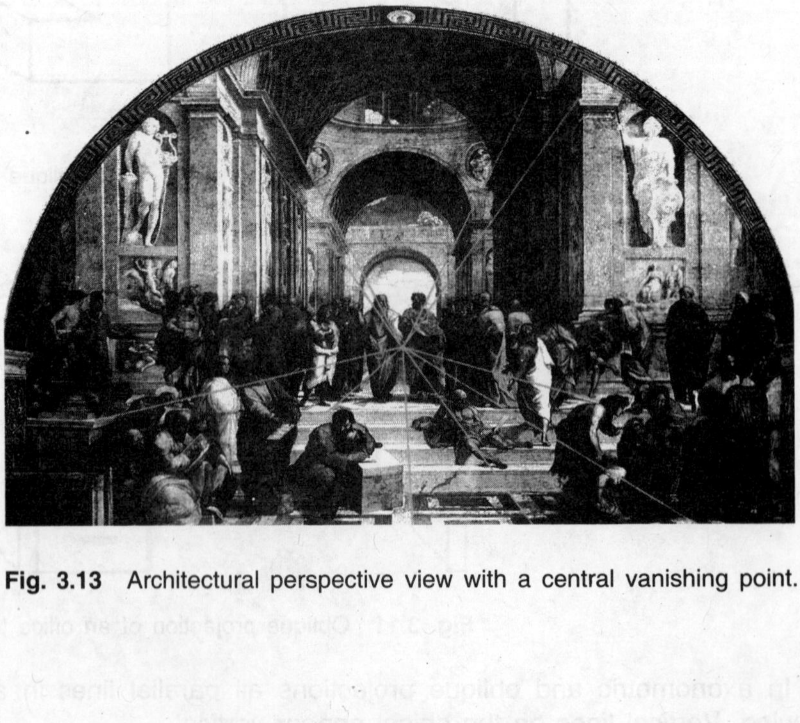

**Fig. 3.13** Architectural perspective view with a central vanishing point.

Figure 3.14 shows an architectural perspective view with two vanishing points. Figure 3.15 shows a perspective view with three vanishing points.

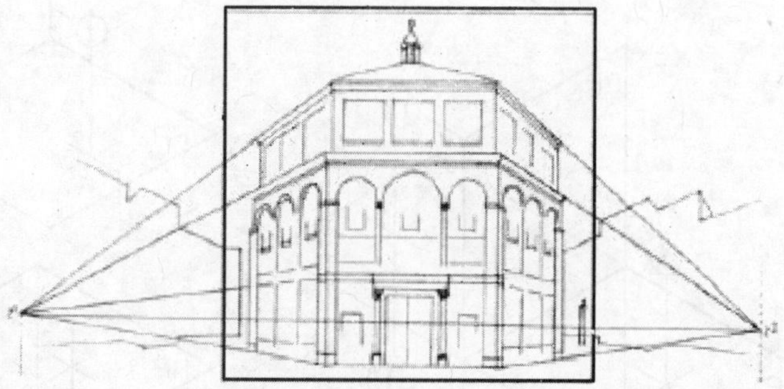

**Fig. 3.14** Architectural perspective view with two vanishing points.

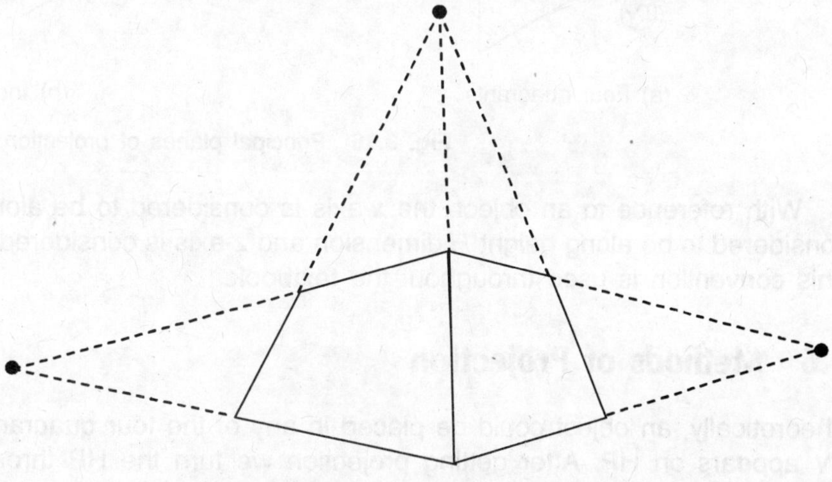

**Fig. 3.15** Perspective projection with three vanishing points.

Perspective view of an object has considerable advantages over the other types of single view projections since it clearly represents the object as seen by the observer. It is often shaded and tastefully coloured, sometimes, to get a more realistic appearance. Perspective projections are not used by engineers for manufacturing and construction because a perspective view does not reveal the exact size and shape. Perspectives may be used in marketing where a natural view of the product may be desirable.

## 3.5 Principal Planes of Projection

If two planes: vertical plane (VP) and horizontal plane (HP) intersect each other at right angle to each other, they form four quadrants as shown in Fig. 3.16(a). With respect to an observer's position, these four quadrants are named as I, II, III and IV. Figure 3.16(b) shows such four individual quadrants.

In the I-quadrant, an object is held above HP and in front of VP. In the II-quadrant, an object is held above HP and behind VP. In the III-quadrant, an object is held below HP and behind VP. In the IV-quadrant, an object is held below HP and in front of VP.

The line of intersection formed by the intersection of HP and VP is taken along the *x*-axis. The line of intersection formed by the intersection of VP and profile plane (PP) is taken along the *y*-axis. The line of meeting of HP and PP is taken along the *z*-axis. Now, we have three principal planes: HP, VP and PP. We have three principal axes: *x*-, *y*- and *z*-axes.

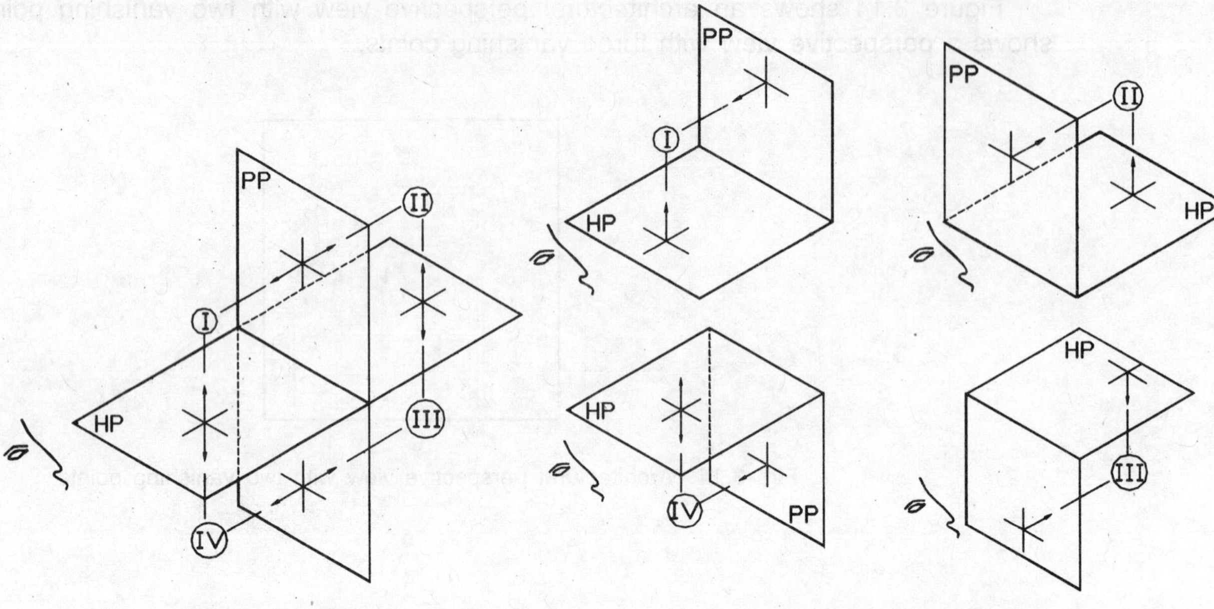


(a) Four quadrants

(b) Individual quadrants

**Fig. 3.16** Principal planes of projection.

With reference to an object, the *x*-axis is considered to be along length *L* dimension, *y*-axis is considered to be along height *H* dimension and *z*-axis is considered to be along depth *D* dimension. This convention is used throughout the textbook.

## 3.6 Methods of Projection

Theoretically, an object could be placed in any of the four quadrants. The FV appears on VP and TV appears on HP. After getting projection we turn the HP through 90° towards the feet of an observer and PP through 90° as shown in Fig. 3.17. With this standard practice, an object cannot be placed in II and IV quadrant as the TV and FV will overlap each other in these quadrants. The options left are first (I) and third (III) quadrant.

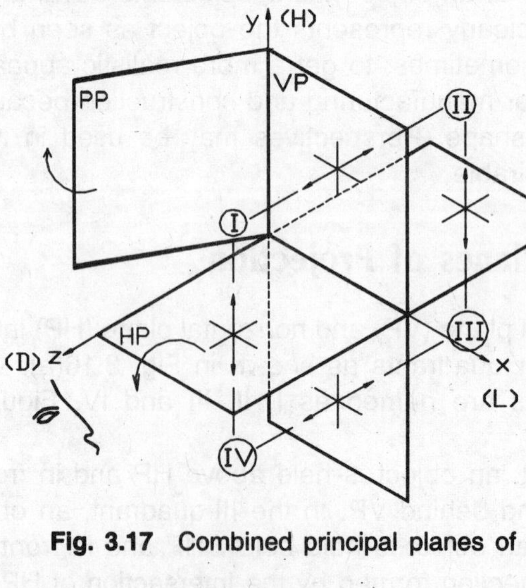


**Fig. 3.17** Combined principal planes of projection.

### 3.6.1 III-Quadrant Method of Projection

The III-quadrant is generally used because when the views (side view and TV) are turned through 90° and brought into the plane of the front view may, they are in their natural positions. That is, the top view (TV) appears above the front view (FV). The side view showing the right side falls to the right side of the front view, and side view showing the left side falls in the left side of the front view. This is shown in Figs. 3.18(a) and (b).

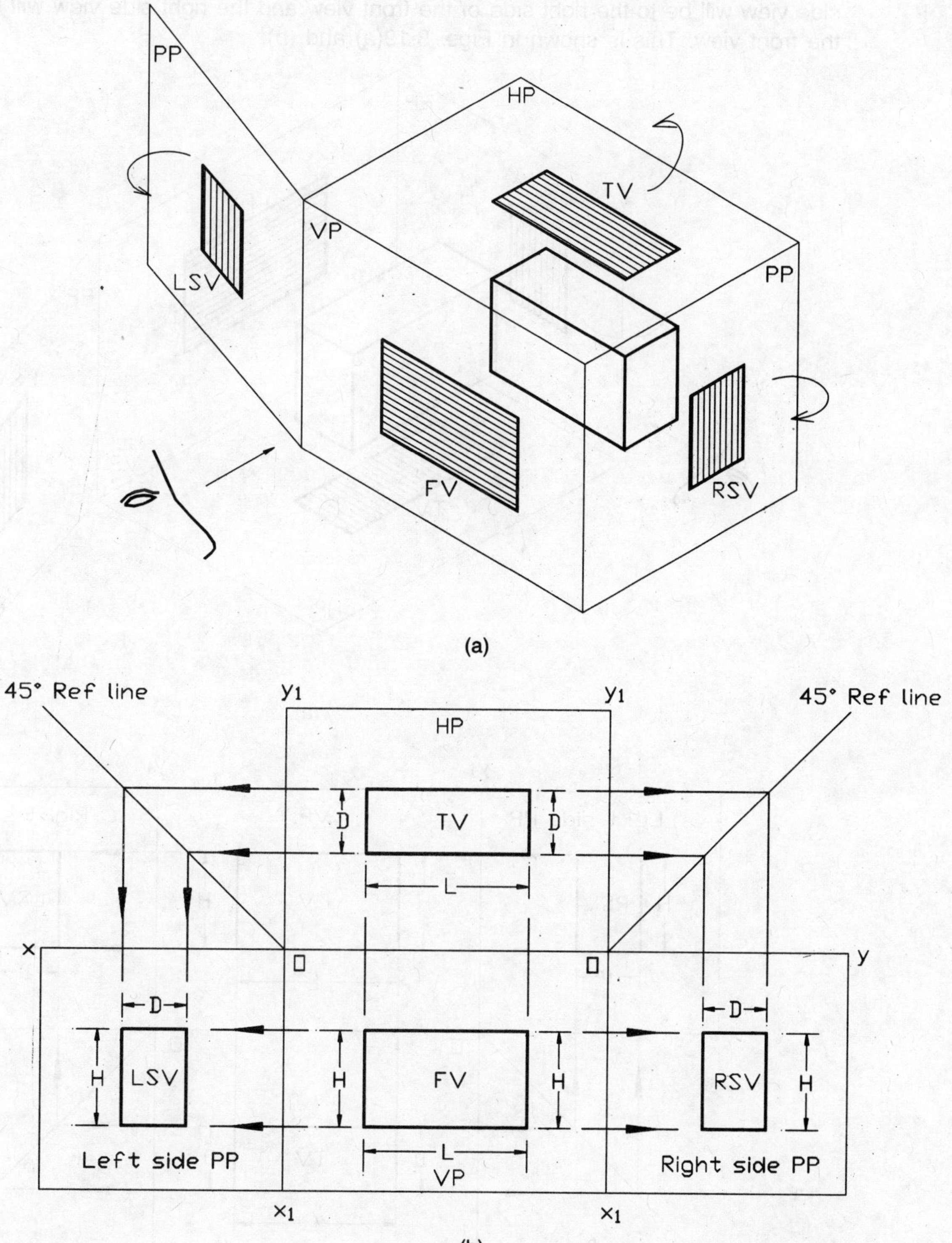


**Fig. 3.18** III-Quadrant method of projection.

### 3.6.2 I-Quadrant Method of Projection

In some countries, the I-quadrant projection is used for engineering drawings. Once we turn the HP and PP through 90° following a standard practice, the top view will be below the front view, the left side view will be to the right side of the front view and the right side view will be to the left side of the front view. This is shown in Figs. 3.19(a) and (b).

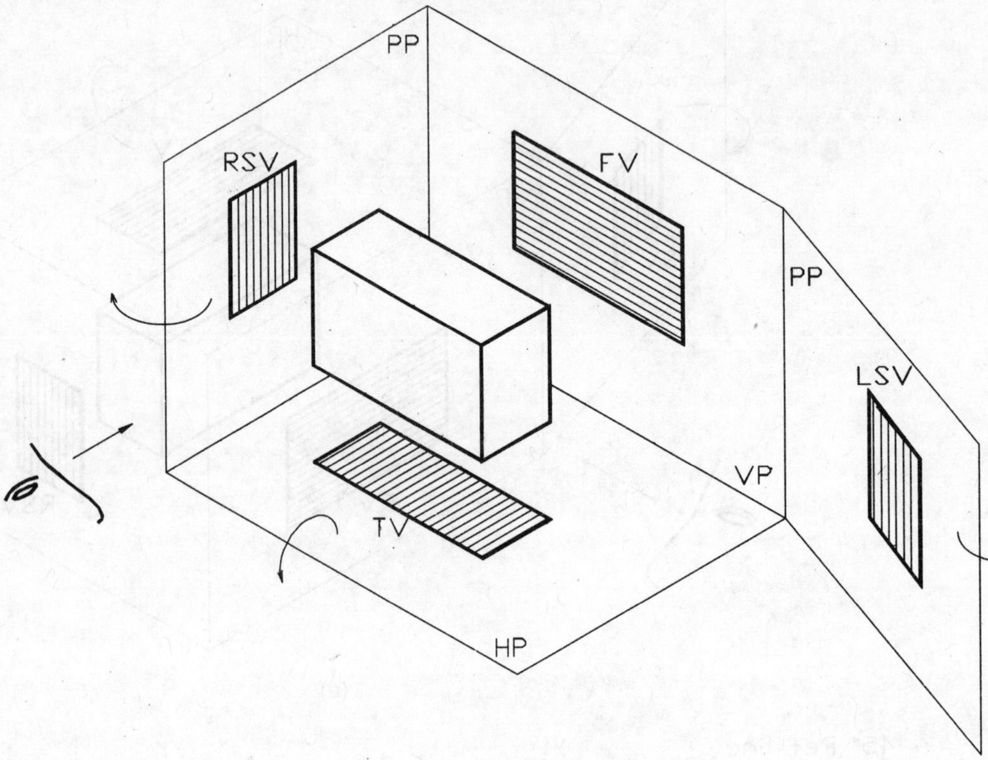


(a)

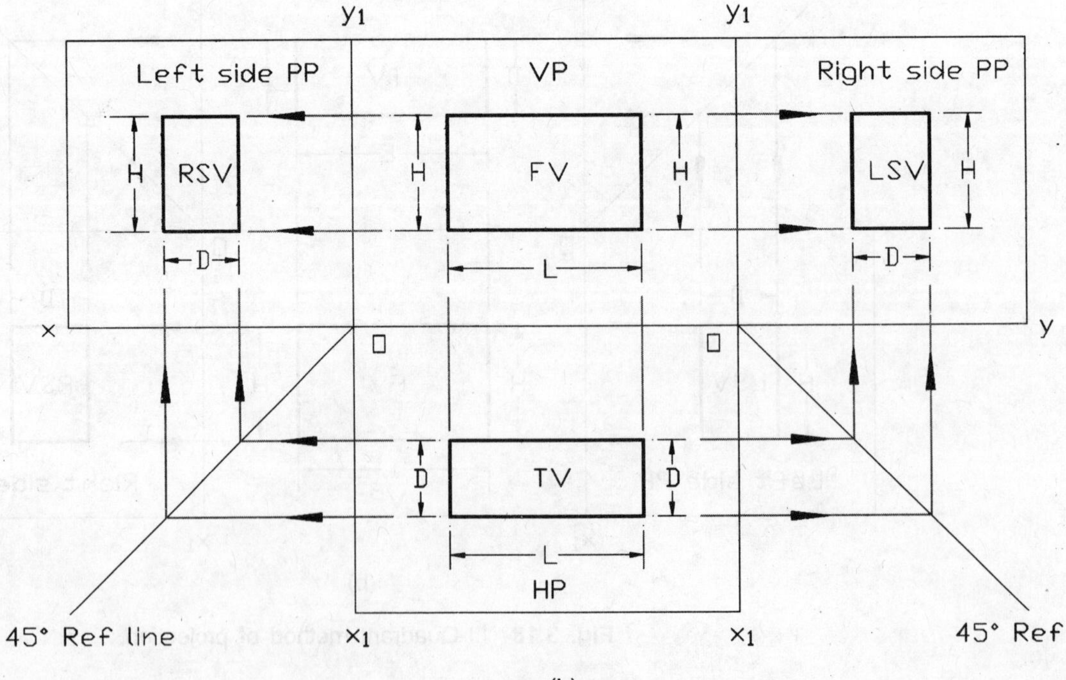


(b)

**Fig. 3.19** I-Quadrant method of projection.

FV shows length $L$ and Height $H$ dimensions. TV shows length $L$ and depth $D$ dimensions. Side view (SV) shows depth $D$ and height $H$ dimensions.

The length is a common dimension between TV and FV. The depth is a common dimension between SV and TV. The height is a common dimension between FV and SV.

### 3.6.3 45° Reference Line Method

This method is used to obtain side views (LSV and RSV) from TV and FV. The 45° reference line method is used in Fig. 3.18 and Fig. 3.19 in order to obtain the side views. The procedure is mentioned here to understand the method.

1. Draw horizontal line $xy$ and vertical lines $x_1y_1$ between TV and FV.
2. Draw a 45° line from the intersection point O of horizontal and vertical lines.
3. Project TV key points on the 45° reference line.
4. Draw vertical lines from the points at which, the projector lines meet the reference line.
5. Project FV key points to intersect the TV projection lines. The intersection points of TV projector lines and FV projector lines, give the key points to draw a side view from TV and FV.

Figure 3.20 shows an example of I- and III-quadrant method of projection together for comparison of the two methods of projection.

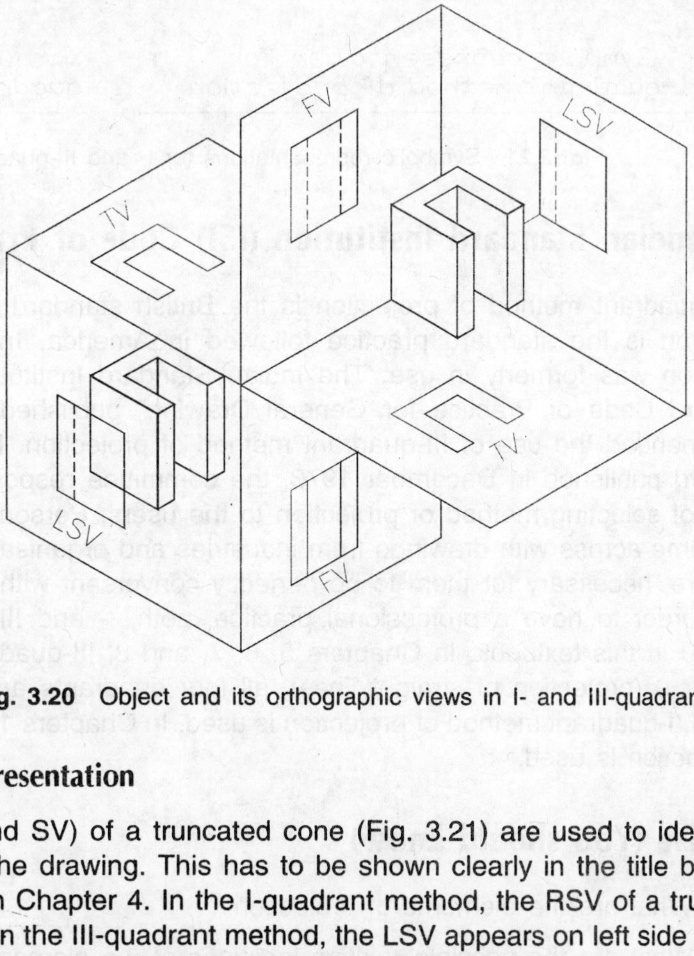


**Fig. 3.20** Object and its orthographic views in I- and III-quadrant.

### 3.6.4 Symbolic Representation

The two views (FV and SV) of a truncated cone (Fig. 3.21) are used to identify which method of projection is used in the drawing. This has to be shown clearly in the title block of drawing sheet as will be discussed in Chapter 4. In the I-quadrant method, the RSV of a truncated cone appears on left side of FV and in the III-quadrant method, the LSV appears on left side of FV. Truncated cone is chosen as an object in order to show that circular feature is drawn first and then projections are taken to draw other view. In the I-quadrant (angle) symbol, the larger end of the truncated cone is towards the donut whereas in III-quadrant (angle) symbol, the smaller end of the truncated cone is towards the donut.

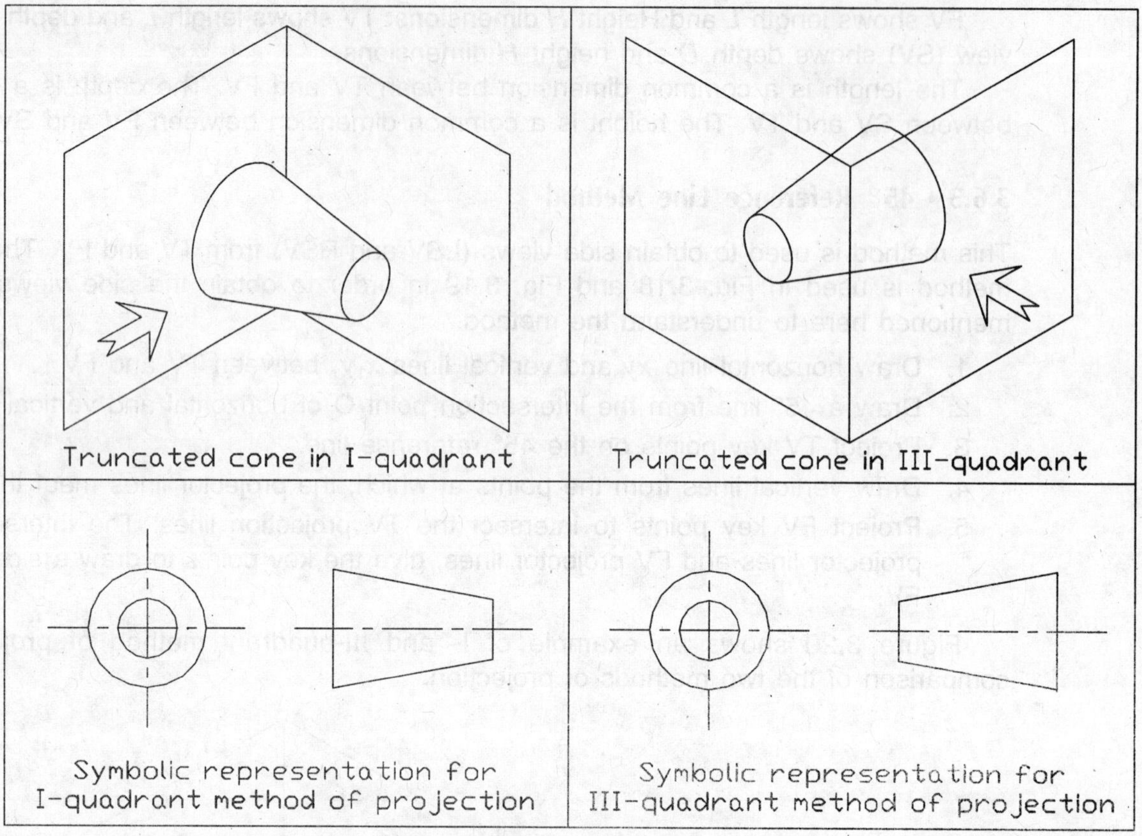


**Fig. 3.21** Symbolic representations for I- and III-quadrant method of projection.

## 3.7 Indian Standard Institution (ISI) Code of Practice

The I-quadrant method of projection is the British standard practice. The III-quadrant method c projection is the standard practice followed in America. In our country, I-quadrant method c projection was formerly in use. The Indian Standard Institution, in its earlier versions of "India Standard Code of Practice for General Drawing" published in 1955 and revised in 1960, ha recommended the use of III-quadrant method of projection. In the second revised version of thi standard published in December 1973, the committee responsible for its preparation has left th option of selecting method of projection to the users. Persons engaged in engineering professio may come across with drawings from industries and organisations following any one method. It is therefore, necessary for them to be perfectly conversant with both the methods.

In order to have a professional practice, both, I- and III-quadrant methods of projection ar followed in this textbook. In Chapters 5, 6, 7, and 8, III-quadrant method of projection is used. I Chapter 9 (projection of straight lines), all four quadrants are used. In Chapter 10 (projection c planes), I-quadrant method of projection is used. In Chapters 12, 13, 14 and 15, III-quadrant metho of projection is used.

## Exercise (You should know)

**3.1** What are the elements of projection?

**3.2** What are the possible relative locations of the elements of projection?

**3.3** Classify the system of projection.

**3.4** What is an axonometric projection? What are the types of axonometric projection?

**3.5** What is a planometric projection? Where is it used?

**3.6** What is an oblique projection?

**3.7** What is a perspective projection? Where is it used?

**3.8** What are the principal planes of projection?

**3.9** What are the principal axes and how they are related to dimensions of an object?

**3.10** What are the methods of projection?

**3.11** What are the relative positions of TV, FV, RSV and LSV in I-quadrant and III-quadrant method of projection?

**3.12** Why cannot we use II- and IV-quadrant methods of projection?

**3.13** What are the relative positions of an object, observer and plane of projection in I- and III-quadrant methods of projection?

**3.14** Why LSV comes on the right side of PP and RSV comes on the left side of PP in case of I-quadrant method of projection?

**3.15** What is the symbol for I- and III-quadrant methods of projection?

# 4

# Introduction to Technical Drawing

## Objectives

- Know the standard practices of technical drawing.
- Observe the standard practices for dimensioning.
- Learn the related AutoCAD commands for setting the electronic drawing sheet and plot th dimensioning.

## Introduction

While producing an engineering drawing, it is important that one should follow some importa rules. Some of these rules, although may not be applicable for Engineering Graphics, will b useful when working in industry. In this chapter, aspects of technical drawing are discussed details.

These include selecting sheet size and setting limits suitable to sheet size. These als include selecting a proper format and preparing a systematic drawing sheet. Various types lines used and dimensioning rules are also discussed in this chapter.

## 4.1 Size of Drawing Sheet

The drawing size typically comply with either ISO standard or US standard. The size of drawing sheet needs to be defined, based on the drawing size and drawing scale. Two popular series ISO 'A' and ISO 'B' are discussed here.

### 4.1.1 ISO 'A' Series

The ISO 'A' series has A0, A1, A2, A3, A4, A5 and A6 family members. The ISO 'A' series of sheet sizes is based on a constant width to length ratio of $1:\sqrt{2}$. The maximum A0 size is defined as having an area of approximately one square metre (1189 × 841 $mm^2$). Each smaller sheet size is exactly half the area of the previous size, i.e., if we cut an A0 sheet in half, we get two A1 sheets; if we cut an A1 sheet in half, we get two A2 sheets; and so on.

This relationship of $1:\sqrt{2}$ is particularly important as all metric equipment including microfilm cameras, microfilm printers, photocopiers and even drawing pen sizes are designed around this ratio. This makes the process of archiving drawings, resizing and modifying drawings easy. Table 4.1 shows dimensions of regular sheets for A series.

**Table 4.1** Dimensions of Regular Sheet

| *Designation* | *Dimensions in* $mm^2$ | *Dimensions in* $inches^2$ |
|---|---|---|
| A0 | 841 × 1189 | 46.81 × 33.11 |
| A1 | 594 × 841 | 33.11 × 23.39 |
| A2 | 420 × 594 | 23.39 × 16.55 |
| A3 | 297 × 420 | 16.55 × 11.69 |
| A4 | 210 × 297 | 11.69 × 8.27 |
| A5 | 149 × 210 | 8.27 × 5.84 |
| A6 | 105 × 149 | 5.84 × 4.13 |

Figures 4.1, 4.2 and 4.3 show comparisons of one sheet size with respect to next sheet size based on the actual size.

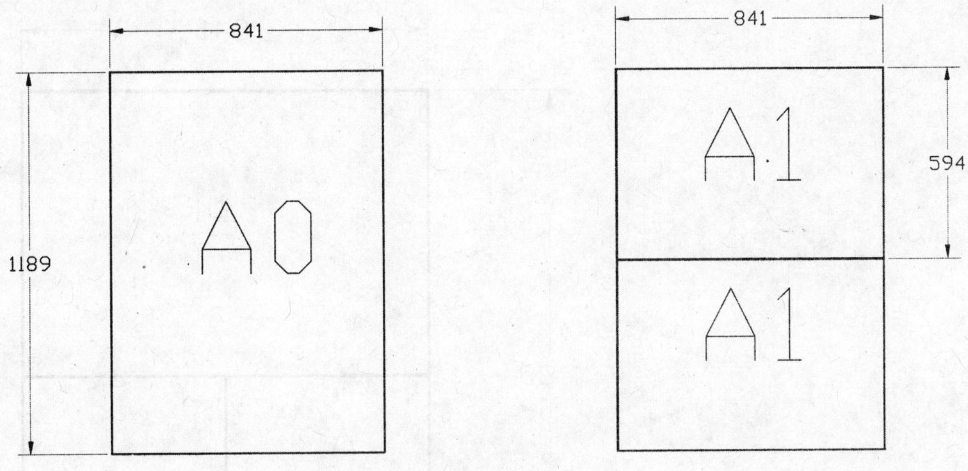


**Fig. 4.1** A0 and A1 drawing sheet.

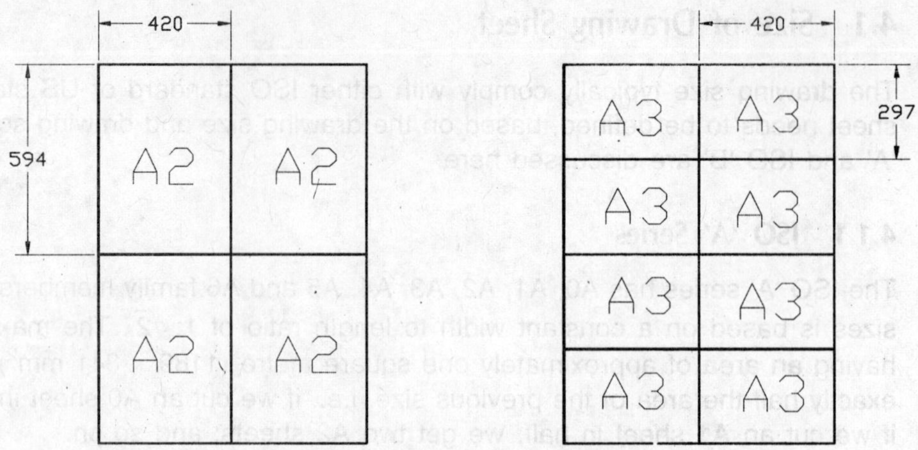


**Fig. 4.2** A2 and A3 drawing sheet.

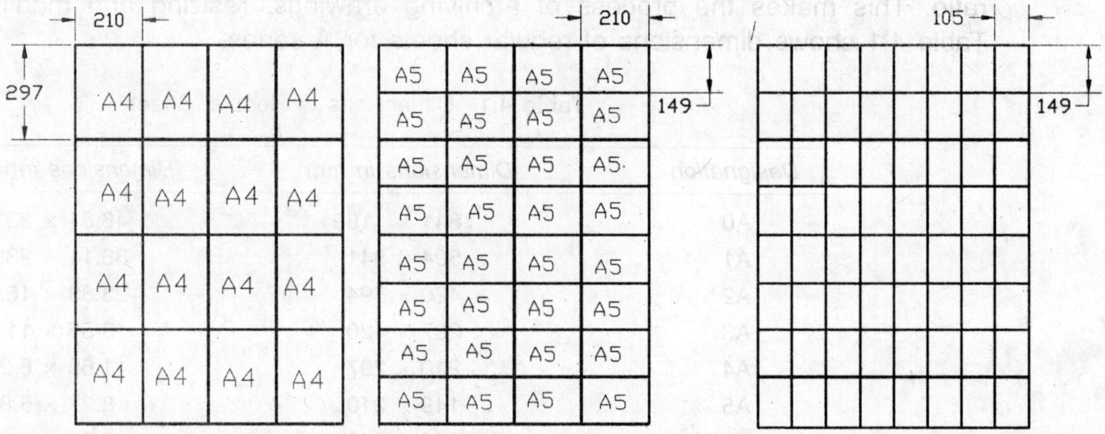


**Fig. 4.3** A4 and A5 drawing sheet.

Figure 4.4 shows a qualitative comparison of all the drawing sheets of 'A' series with respect to eac other. All the types of 'A' sheets are placed in A0 sheet to know the relative size at a glance.

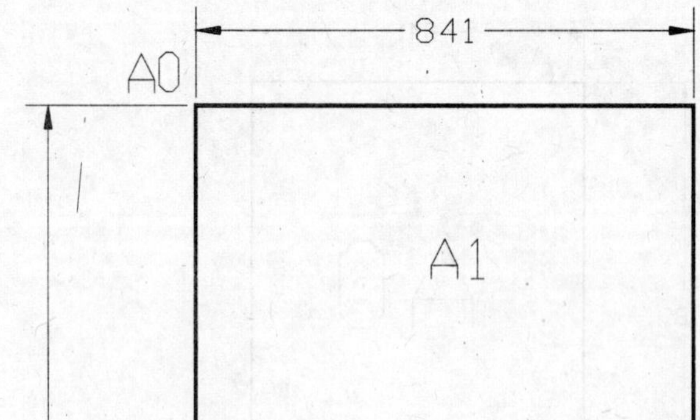


**Fig. 4.4** Relative sheet size of all 'A' sheets.

### 4.1.2 Oversize Sheet

Oversize sheets are used when it is desirable to give extra protection to the drawing sheets to provide a binding or trimming margin. Table 4.2 shows dimensions of oversize sheet.

**Table 4.2** Dimensions of Oversize Sheet

| *Designation* | *Dimensions in* mm$^2$ |
|---|---|
| RA0 | 860 × 1220 |
| RA1 | 610 × 860 |
| RA2 | 430 × 610 |
| RA3 | 305 × 430 |
| RA4 | 215 × 305 |
| RA5 | 153 × 215 |

### 4.1.3 ISO 'B' Series

The ISO 'B' series is defined to satisfy the requirements of drawing offices if the differences between the 'A' series sizes are too high. The area of a B*n* sheet is the area of the A*n* sheet multiplied by the square root of 2. In practice, B series is not commonly used. Table 4.3 shows the dimensions of oversize sheet.

**Table 4.3** Dimensions of Regular Sheets

| *Designation* | *Dimensions in* mm$^2$ |
|---|---|
| B0 | 1000 × 1414 |
| B1 | 707 × 1000 |
| B2 | 500 × 707 |
| B3 | 354 × 500 |
| B4 | 250 × 354 |
| B5 | 177 × 250 |

### 4.1.4 Non-metric Paper Sizes

For reference purposes, the non-metric paper sizes are still in use in some parts of the world. Tables 4.4 and 4.5 show a list of non-metric paper sizes for US Engineering and US Architectural, respectively.

**Table 4.4** Non-metric Paper Sizes (US Engineering)

| *Designation* | *Dimensions in* mm$^2$ | *Dimensions in* inches$^2$ |
|---|---|---|
| **US Engineering:** | | |
| ANSI A | 215.9 × 279.4 | 8.5 × 11 |
| ANSI B | 279.4 × 431.8 | 11 × 17 |
| ANSI C | 431.8 × 558.8 | 17 × 22 |
| ANSI D | 558.8 × 863.6 | 22 × 34 |
| ANSI E | 863.6 × 1117.6 | 34 × 44 |

**Table 4.5** Non-metric Paper Sizes (US Architectural)

| *Designation* | *Dimensions in* mm$^2$ | *Dimensions in* inches$^2$ |
|---|---|---|
| **US Architectural:** | | |
| ARCH A | 228.6 × 304.8 | 9 × 12 |
| ARCH B | 304.8 × 457.2 | 12 × 18 |
| ARCH C | 457.2 × 609.6 | 18 × 24 |
| ARCH D | 609.6 × 914.4 | 24 × 36 |
| ARCH E | 94.4 × 1219.2 | 36 × 48 |

## 4.2 Drawing Formats

Drawing format depends upon the size of views (2-D views or 3-D views). This also depends upo the user or company. Figures 4.5 and 4.6 show two drawing format for same A3 size drawing Figure 4.5 shows drawing format for A3 size sheet in landscape mode and Figure 4.6 shows drawin format for A3 size sheet in the portrait mode.

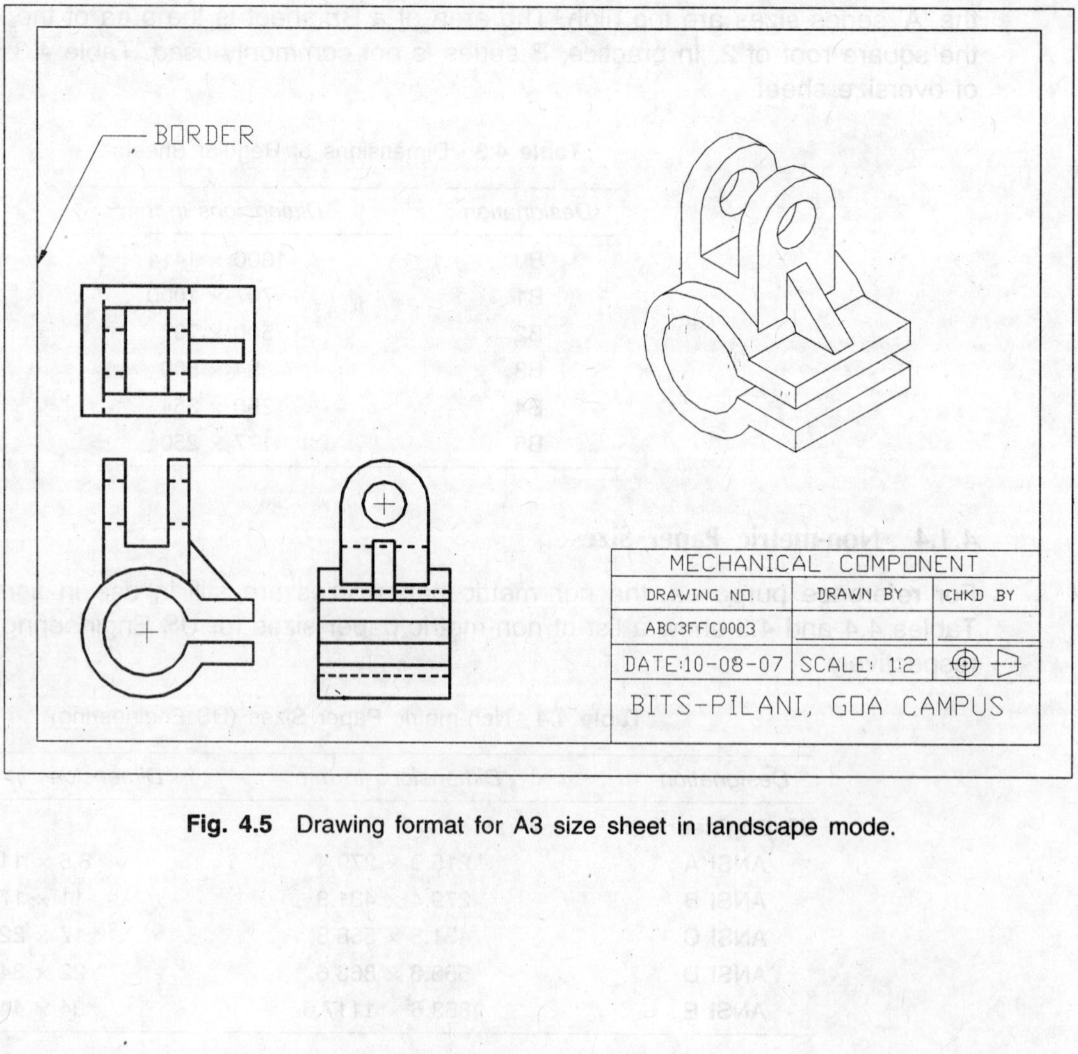


**Fig. 4.5** Drawing format for A3 size sheet in landscape mode.

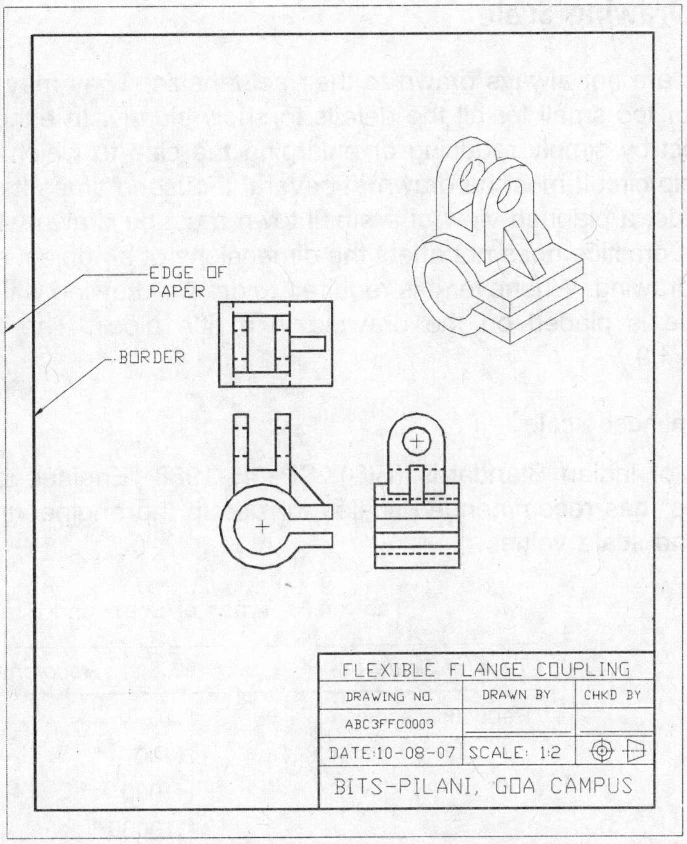


**Fig. 4.6** Drawing format for A3 size sheet in portrait mode.

## 4.3 Setting Limits for Various Sheet Sizes

Before starting your engineering drawing you should plan how you are going to make the best use of the space of drawing sheet. It is important to think about the number of views your drawing will have and how much space of the paper you will use. Important points to remember are: (a) try to make the maximum use of the available space and (b) if you intend to add dimensions to the drawing, remember to leave enough space around the drawing for them to be added later.

Boundaries of a drawing sheet are set with the help of LIMITS and ZOOM command in AutoCAD. LIMITS and ZOOM command set the drawing area to match the particular size of a drawing sheet. Following the five steps of AutoCAD command set the boundary limits for a A2 size of a drawing paper. Table 4.1 shows A2 paper has 594 mm × 420 mm dimensions. However, to provide extra protection to the drawing sheets for binding or trimming margin, RA2 size (610 mm × 430 mm) is selected from Table 4.2 for A2 size drawing.

**Command: LIMITS** [Enter]
ON/OFF/<Lower left corner> <0.00, 0.00>: [Enter]
Upper right corner <12.00, 9.00>: 610, 430 [Enter]

**Command: ZOOM** [Enter]
All/Center/Dynamic/Extents/Previous/Scale(X/XP)/Window/<Realtime>: A [Enter]

The change in the drawing limits brought by these two commands is not visible to the user. However, now the graphic screen is set to align new drawing limits. While drawing, the user is always advised to zoom a particular portion of the drawing to edit or draw the entities in drawing. This is because a normal monitor screen is used to accommodate a larger size drawing. Now, the AutoCAD unit is matched with the drawing unit e.g. 1 AutoCAD unit = 1 mm. Accordingly, grids can be set on the background to produce drawings with a least count of 1 grid spacing = 1 mm.

## 4.4 Drawing Scale

Objects are not always drawn to their actual size. They may be too large to fit a standard drawing sheet, or too small for all the details to show clearly. In either case, it is still possible to represent an object by simply reducing or enlarging the size to which the drawing is made. For example, a microchip circuit must be drawn at several thousand times its actual size that is to be seen. On the other side, a planned view of a small town must be drawn with reduced scale than the actual size.

This practice does not affect the dimensions of an object since the actual dimensions are shown on the drawing. Whenever it is required to draw a drawing with reduced size or enlarged size, a note for scale is placed on the drawing in a title block. The title block features are discussed in Section 4.9.

### Recommended scale

Bureau of Indian Standards (BIS), SP: 46-1988 "Engineering Drawing Practice for Schools and Colleges" has recommended scales for use in the engineering drawing. Table 4.6 shows types of scale and scale values.

**Table 4.6** Types of Scale and Scale Values

| *Type of Scale* | *Recommended Scale* | | |
|---|---|---|---|
| Reduction scale | 1:10 | 1:2 | 1:5 |
| | 1:100 | 1:20 | 1:50 |
| | 1:1000 | 1:200 | 1:500 |
| | 1:10000 | 1:2000 | 1:5000 |
| Enlargement scale | 10:1 | 20:1 | 50:1 |
| | 2:1 | 5:1 | |
| Full scale | | 1:1 | |

In AutoCAD, the scale is set by a SCALE command. The normal drawing file or an individual object can be scaled to the defined scale. Following are the few steps of SCALE command.

**Command: SCALE** [Enter]
**Select Objects:** [select object]
**Select Objects:** [Enter]
**Base point:** [select appropriate]
**<Scale factor>/Reference:** (enter a scale value) [Enter]

## 4.5 Various Types of Lines and Their Usage

Drawings are made up of a combination of lines of different lengths and intensity. These define shape, size and details of an object. It is possible, through correct use of lines, to graphically describe an object so that it can be accurately visualised by any person having a basic understanding of drawing principles. Since lines are the basis for all drawings or sketches, important lines are listed over here which are useful for Engineering Graphics.

**Outline of view**

This line is a continuous thick line used to represent boundaries of view and visible edges of an object.

**Projection line**

This line is a continuous thin line used to construct one figure with reference to another figure. This line is also called as construction line.

**Hidden line**

This line is used to indicate invisible line or hidden surface which is falling into a line view. These consists of short dashes closely and evenly spaced and of medium thickness.

**Centre line**

This line is used to indicate the axis of an object, pitch circles and centreline of circles. These are thin long chain lines with alternate long and short dashes.

**Section line**

This line is used to indicate a section plane. This line is a line view of a section plane. These are thin long chain lines with alternate long dash and two short dashes.

Figure 4.7 shows a model example which includes all types of lines used in Engineering Graphics.

The lines related to the dimensioning are discussed in section 4.7.

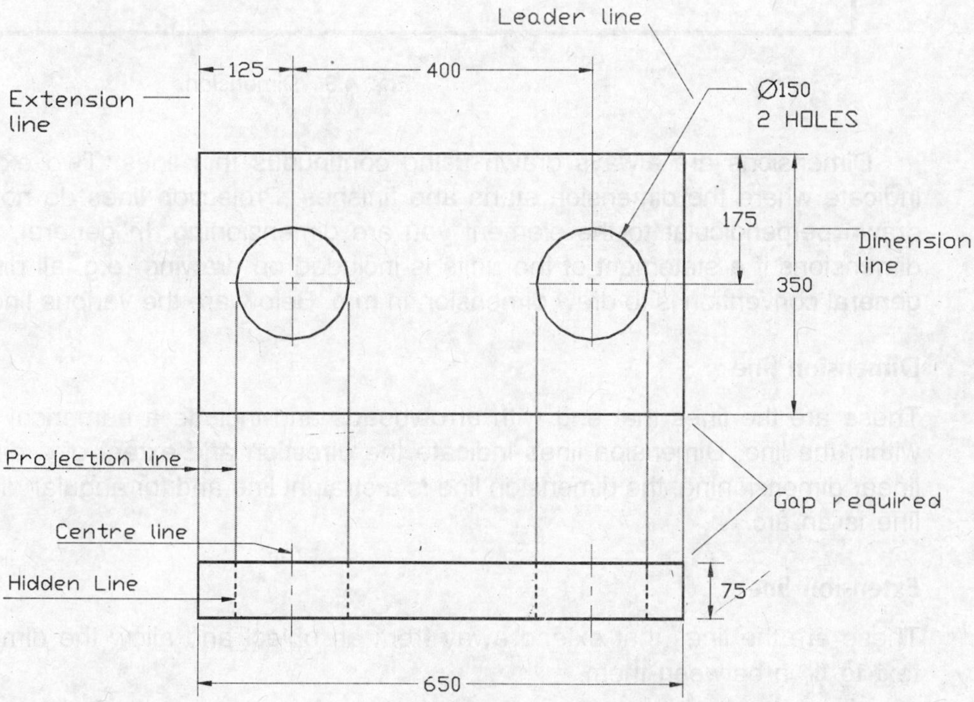


**Fig. 4.7** Lines used in engineering graphics.

## 4.6 Lettering

Lettering is used to convey additional information on engineering drawings. Writing of titles, subtitles dimensions and other details on a drawing is known as lettering.

All notes and dimensions should be clear and easy to read. In general, all notes should be written in capital letters to aid legibility. All lettering should be of the same size and preferably no smaller than 3 mm. An example of typeface is shown below.

## 4.7 Dimensioning

Dimensions show the geometric measurements of objects, the distances or angles between objects or the x- and y-coordinate of a feature. The elements of dimensioning are: projection line or extension line, dimension line, leader line, arrow head and dimension itself. A dimensioned drawing should provide all the information necessary for a product or an object. An example of dimension is shown in Fig. 4.8.

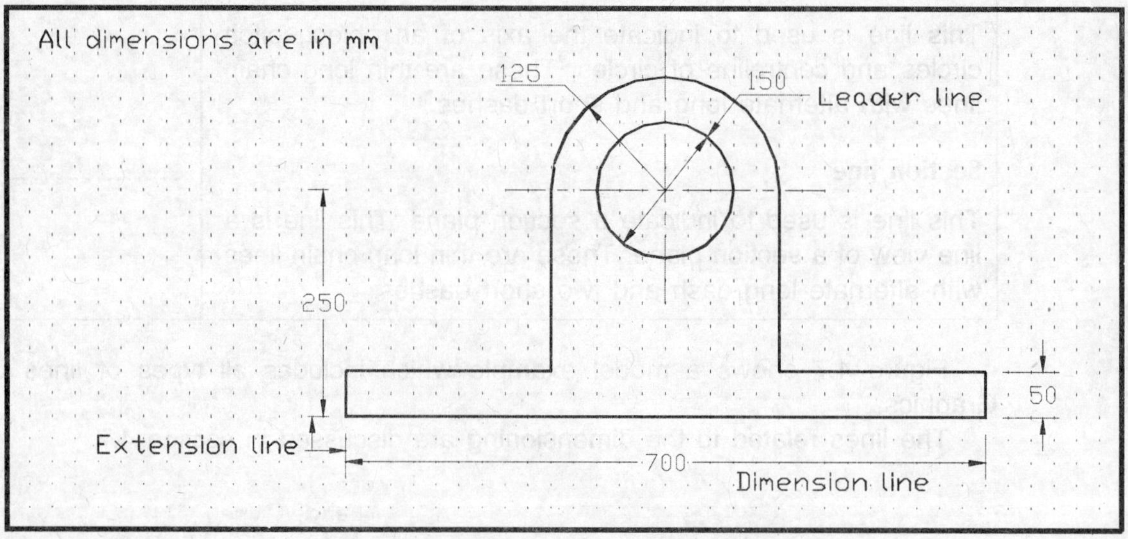


**Fig. 4.8** Dimension.

Dimensions are always drawn using continuous thin lines. Two extension or projection lines indicate where the dimension starts and finishes. Projection lines do not touch the object and are drawn perpendicular to the element you are dimensioning. In general, units can be omitted from dimensions if a statement of the units is included on drawing (e.g. all dimensions are in mm). The general convention is to draw dimension in mm. Below are the various lines related to dimensioning.

### Dimension line

These are the lines that end with arrowheads and include a numerical dimensional value located within the line. Dimension lines indicate the direction and extent of a dimension. For example, for linear dimensioning, the dimension line is a straight line and for angular dimensioning, the dimension line is an arc.

### Extension line

These are the lines that extend away from an object and allow the dimension line and dimension text to fit in between them.

### Leader line

These are the lines drawn at an angle, neither horizontal nor vertical. These are used to dimension specific shapes such as holes. The start point of a leader line includes an arrowhead. Numerical values or text is mentioned at the end, opposite the arrowhead. Refer to the dimension of hole in Fig. 4.8.

The following points are taken care of while dimensioning.

- All dimensions necessary to define an object or component must be clearly marked on the drawing.
- No dimension should be repeated, except where unavoidable.
- The longer dimension should be placed outside all intermediate dimensions.
- Dimensions are placed on the view where corresponding features are shown more clearly.
- Dimension lines are placed in such a way that they do not cross each other. Dimension lines and text are placed outside the drawings except in special cases where marking inside the drawing is readable.

## 4.8 Basic Types of Dimensioning

AutoCAD provides three basic modes of dimensioning:

1. Linear
2. Angular
3. Radial

### 4.8.1 Linear Dimensioning

This mode creates linear dimensions between two points. The DIM command puts the command line in dimensioning mode in which we can use the dimensioning subcommands. The general scale used is 1 AutoCAD unit = 4 grid spacing = 100 mm. The linear dimensioning is used to specify horizontal, vertical and aligned dimensions as shown in Fig. 4.9.

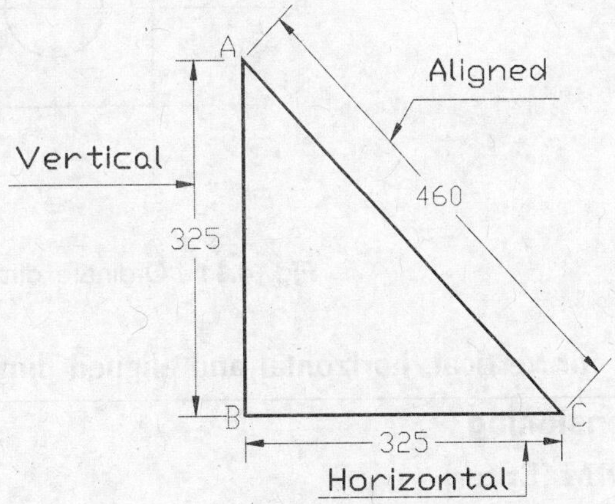


**Fig. 4.9** Linear dimensioning.

***Baseline dimensioning*** The subcommand BASELINE draws a series of related dimensions measured from the same baseline. This subcommand is used to avoid overlaying the previous dimension line. Baseline dimensioning is also known as parallel dimensioning. In Fig. 4.10, all dimensions are with respect to a baseline AB.

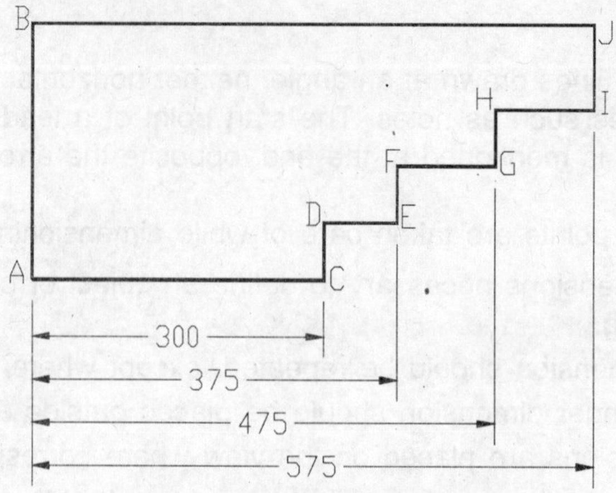


**Fig. 4.10** Baseline dimensioning.

***Ordinate dimensioning*** With ordinate dimensions, one horizontal and one vertical extension line establishes an origin for the entire view. The origin is identified with zeroes placed at the ends of these extension lines. Distances along the *x*- and *y*-axes to other features are specified using other extension lines, with the distances indicated numerically at their ends. Figure 4.11 shows ordinate type of dimensioning.

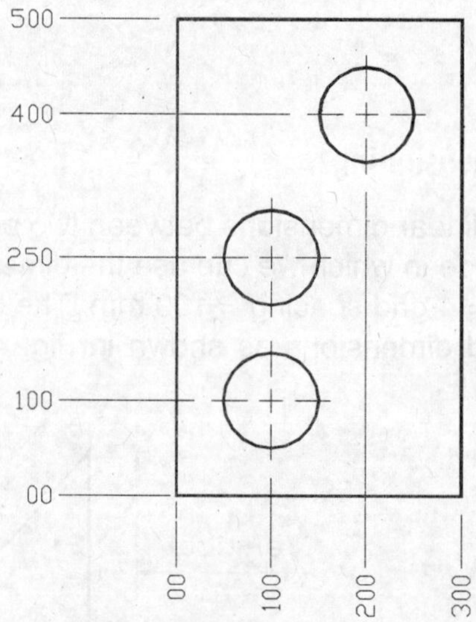


**Fig. 4.11** Ordinate dimensioning.

**AutoCAD steps for vertical, horizontal and aligned dimensioning**

**Vertical Dimensioning**

**Command: DIM** [Enter]
**DIM:** ver (for vertical dimension) [Enter]
First extension line origin or press ENTER to select: [select A]
Second extension line origin: [select B]
Dimension line location (Text/Angle): [select]
Dimension text <1.00>: 100 [Enter]

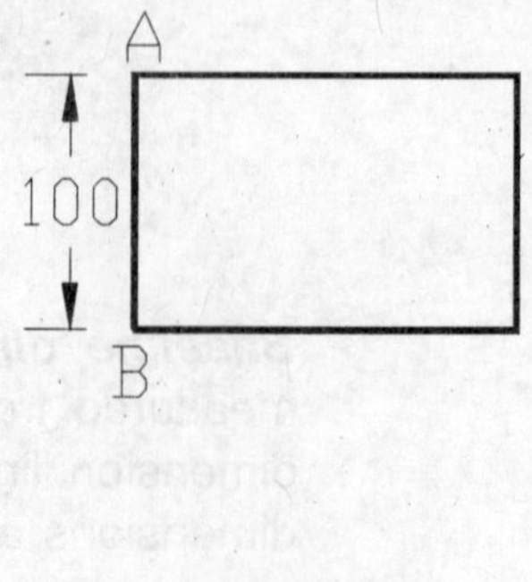


**Horizontal Dimensioning**

**DIM:** hor (for horizontal dimension) [Enter]
First extension line origin or press ENTER to select: [select B]
Second extension line origin: [select C]

| | |
|---|---|
| Dimension line location (Text/Angle): [select]<br>Dimension text <1.50>: 150 [Enter]<br><br>**Aligned Dimensioning**<br>**DIM:** ali (for aligned dimension) [Enter]<br>First extension line origin or press ENTER to select: [select A]<br>Second extension line origin: [select B]<br>Dimension line location (Text/Angle): [select]<br>Dimension text <1.118>: 112 [Enter] | 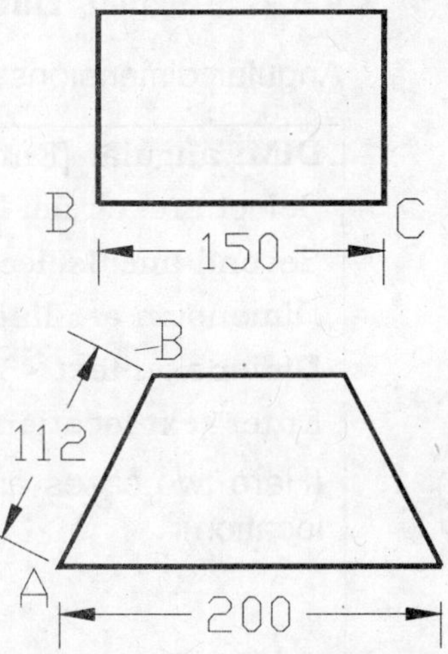 |
| **Baseline Option for Step Dimensioning**<br>**DIM:** ver [Enter]<br>First extension line origin or press ENTER to select: [select A]<br>Second extension line origin: [select B]<br>Dimension line location (Text/Angle): [select]<br>Dimension text <0.50>: 50 [Enter]<br>DIM: baseline [Enter]<br>Specify a second extension line origin or press Enter to select: [select C]<br>Specify a second extension line origin or press Enter to select: [select D] | 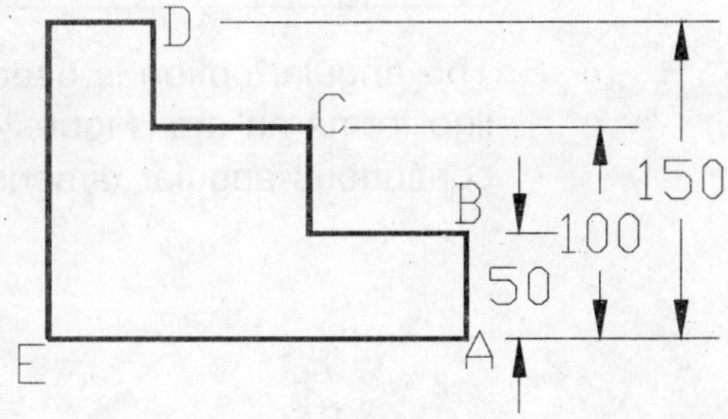 |
| **Ordinate Option for Step Dimensioning**<br>**DIM:** ordinate [Enter]<br>Select feature: [select]<br>Leader endpoint (Xdatum/Ydatum/Text): Y [Enter]<br>Leader endpoint: [select]<br>Dimension text <1.00>: 100 [Enter]<br><br>DIM: ordinate [Enter]<br>Select feature: [select]<br>Leader endpoint (Xdatum/Ydatum/Text):X [Enter]<br>Leader endpoint: [select by mouse]<br>Dimension text <1.00>: 100 [Enter] | 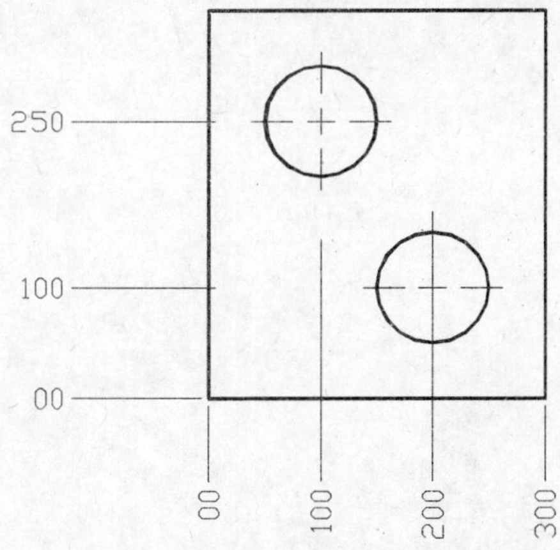 |
| **Leader Option for Specification Details**<br>**DIM:** leader [Enter]<br>Leader to start: [select appropriate point]<br>To point: [select appropriate point]<br>To point: [Enter]<br>Dim Text <1.00>: Hole, 50 [Enter] | 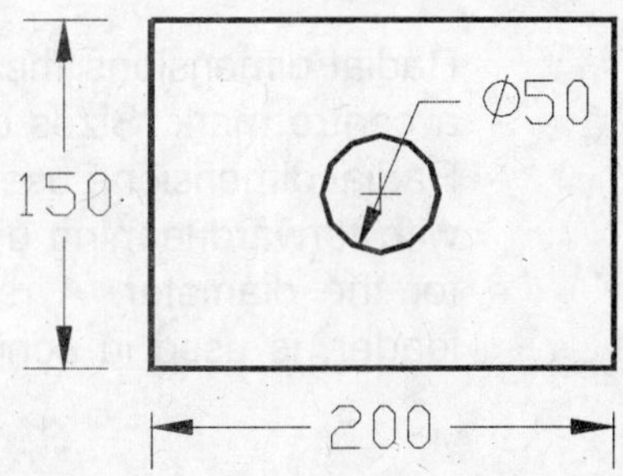 |

### 4.8.2 Angular Dimensioning

Angular dimensions measure the angle between two lines or three points.

**DIM:** angular [Enter]
Select arc, circle, line, or press ENTER: [select line AB]
Second line: [select line AC]
Dimension arc line location (Text/Angle): [select appropriate]
Dimension text <34.00>: [Enter]
Enter text location (or press ENTER): [Enter]
(Here two cases are shown depending upon the text and arc location)

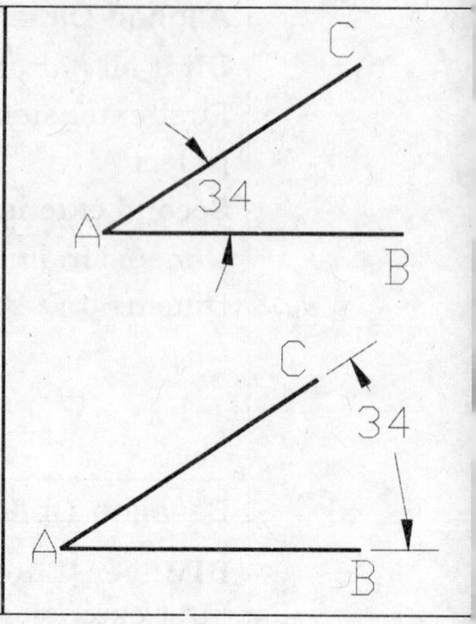


The angular option is used to measure the angle between the two radii of a circle. The dimension line forms an arc. Figure 4.12 shows angular dimension between the radii of a circle. There is a continuous angular dimensioning between the radii of a circle.

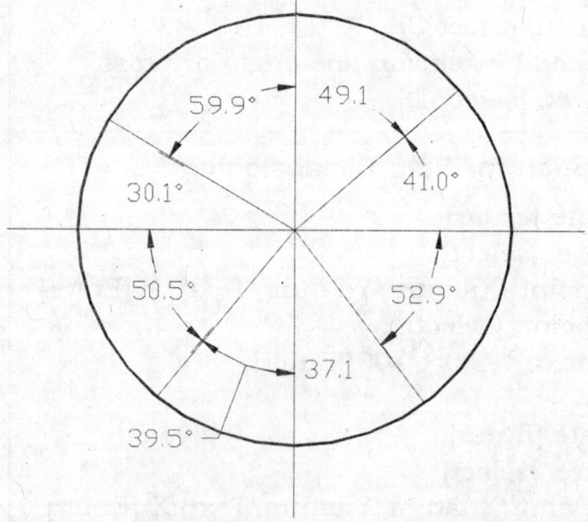


**Fig. 4.12** Angular dimension between radii of a circle.

### 4.8.3 Radial Dimensioning

Radial dimensions measure the radii and diameters of arcs and circles with optional centerlines or a centre mark. Sizes of circular features are indicated using either radial or diametral dimensions. Radial dimensions use an 'R' followed by the value for the radius. Diametral dimensions use a circle with forward-leaning diagonal line through it, called the diameter symbol ($\phi$), followed by the value for the diameter. A radially-aligned line with arrowhead pointing to the circular feature, called a leader, is used in conjunction with both diametral and radial dimensions.

**DIM:** rad (for radial) [Enter]
Select arc or circle: [select]
Dimension text <0.75>: R75 [Enter]
Dimension line location (Text/Angle): [select appropriate]

**DIM:** dia (for diameter) [Enter]
Select arc or circle: [select]
Dimension text <1.500>: 150 [Enter]
Dimension line location (Text/Angle): [select appropriate]

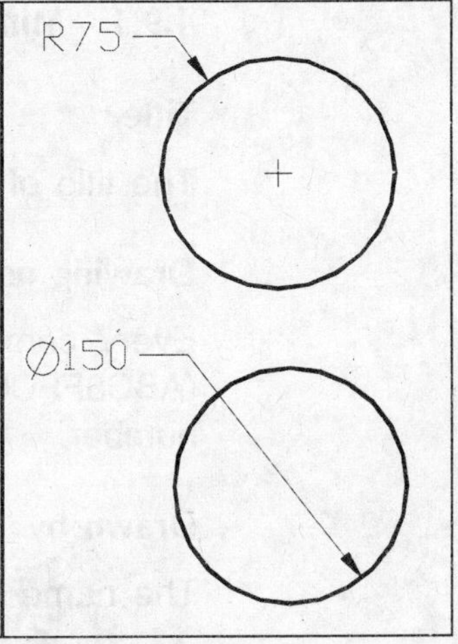


Figure 4.13 shows a model example, which includes all types of dimensions, which are discussed before.

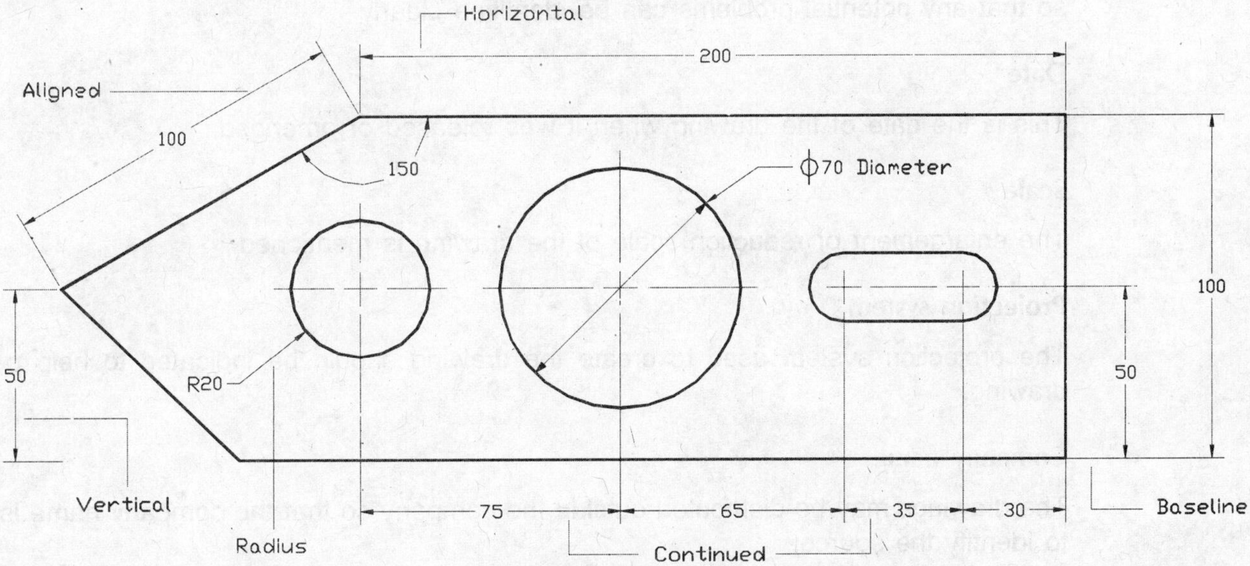


**Fig. 4.13** Types of dimensions.

## 4.9 Title Block

Engineering drawings should feature an information box called title block. One example of title block is shown in Fig. 4.14. The title block can vary from company to company. However, minimum information should be always available for future reference purposes. This title block is not applicable to the first year course on Engineering Graphics. The title block for Engineering Graphics is discussed in Chapter 2.

| FLEXIBLE FLANGE COUPLING | | |
|---|---|---|
| DRAWING NO. | DRAWN BY | CHKD BY |
| ABC3FFC0003 | | |
| DATE:10-08-07 | SCALE: 1:2 | |
| BITS-PILANI, GOA CAMPUS | | |

**Fig. 4.14** Title block.

### 4.9.1 Minimum Common Information Recorded on an Engineering Drawing

**Title**

The title of the drawing should clearly indicate the product.

**Drawing no.**

Every company has its own format to define a drawing number. An example is shown here (ABC3FFC003) with **ABC** company, Division **3**, **F**lexible **F**lange **C**oupling product, **0003** as drawing number.

**Drawn by**

The name of the person who produced the drawing and his signature.

**Chkd by**

The name of the design executive (design engineer), who verifies the drawing and puts his signature so that any potential problems can be identified clearly.

**Date**

This is the date of the drawing when it was released or amended.

**Scale**

The enlargement or reduction scale of the drawing is mentioned.

**Projection system**

The projection system used to create the drawing should be indicated to help people read the drawing.

**Company name**

The drawings may be distributed outside the company so that the company name is usually added to identify the source.

## Exercise

**4.1** What is the series of 'A' sheet size?

**4.2** What is the ratio of length to width dimension for 'A' series?

**4.3** How would you scale an object?

**4.4** How would you set the boundaries of an A0 sheet in AutoCAD?

**4.5** What are the various types of lines used and their usage in engineering drawing?

**4.6** Draw a rectangle and dimension it as shown in P.1/Plate 4.1.

**4.7** Draw two concentric circles and dimension them as shown in P.2/Plate 4.1.

**4.8** Draw two concentric circles and dimension them as shown in P.3/Plate 4.1.

**4.9** Draw a circle and dimension them as shown in P.4/Plate 4.1.

**4.10** Draw two concentric circles and dimension them as shown in P.5/Plate 4.1.

**4.11** Draw a figure and dimension it as shown in P.6/Plate 4.1.

**4.12** Draw a figure and mention the angles as shown in P.7/Plate 4.1.

**4.13** Draw a figure and dimension it as shown in P.8/Plate 4.1.

**4.14** Draw two views and dimension them as shown in P.9/Plate 4.1.

**4.15** Draw two views and dimension them as shown in P.10/Plate 4.1.

**4.16** Draw a figure and dimension it as shown in P.10/Plate 4.1.

**4.17** Draw a figure and dimension it as shown in P.11/Plate 4.1.

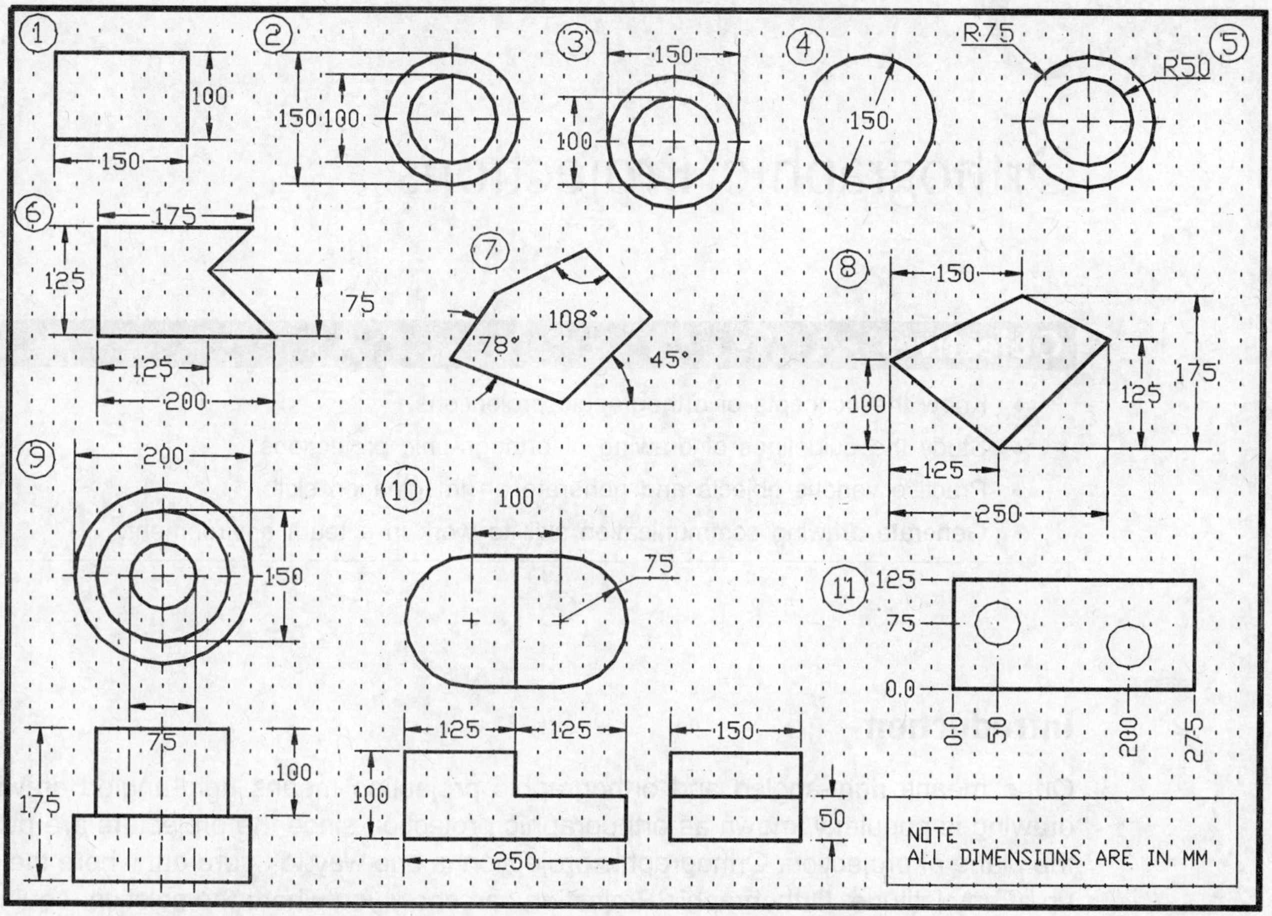


**PLATE 4.1**

# 5

# Orthographic Projections

## Objectives

- Know the concepts of orthographic projections.
- Study the guidelines of drawing of orthographic projections.
- Practise various objects and generate an imagination skill.
- Generate drawing communication skill to work in a team environment.

## Introduction

Ortho means right-angled and orthographic projection means right-angled drawing. Multiview drawing is popularly known as orthographic projection since the projectors are perpendicular to the plane of projection. Orthographic projection is one way to figure out where the sun will come up at any latitude. Orthographic Projection can show you where the solstice, equinox and cross-quarter day sunrises and sunsets will take place anywhere on earth.

## 5.1 Orthographic Projections

In orthographic projections, an observer is assumed to be at a long distance from the object. With this assumption, visual rays of the observer are parallel to each other and perpendicular to the plane of projection. The orthographic projections are also called parallel projections. Figure 5.1 shows object with visual rays and a view in the direction of visual rays.

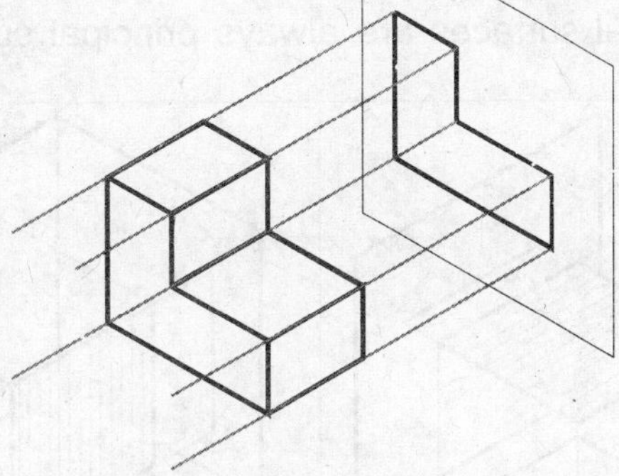

**Fig. 5.1** Parallel projections.

## 5.2 Types of Surfaces

Visualisation is the ability to look at a three-dimensional object and mentally see the appropriate three orthographic views. It is an important skill for a designer and an engineer to develop. Visualisation also includes the ability to look at two-dimensional orthographic views and mentally picture the three-dimensional object the views represent. In order to generate the skill of visualisation let us classify the surfaces (Fig. 5.2) with respect to the principal planes (HP, VP and PP) as:

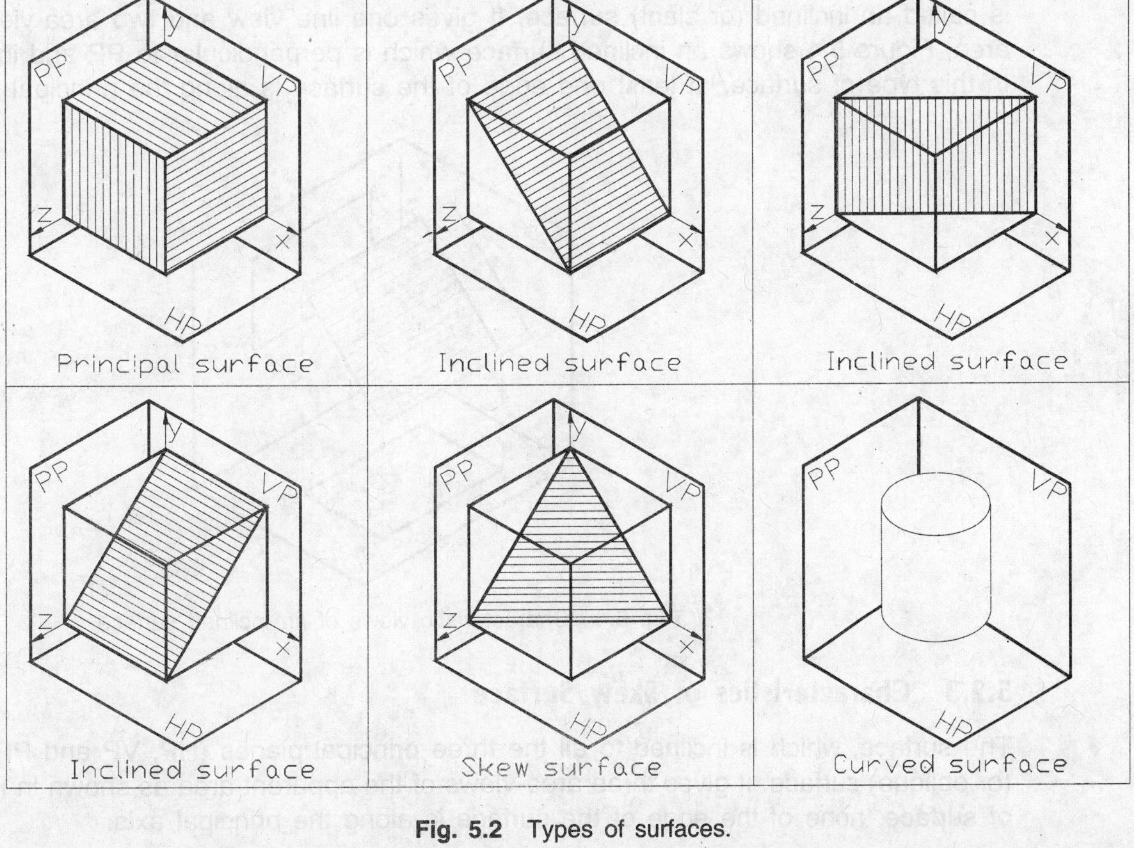


**Fig. 5.2** Types of surfaces.

1. Principal surface
2. Inclined surface
3. Skew surface
4. Curved surface

### 5.2.1 Characteristics of Principal Surface

The surface, which is parallel to one reference plane and perpendicular to the other two, is called a principal surface. It gives two line views and one area view as the true area. This is shown in Fig. 5.3. Horizontal surfaces are always principal surfaces but not the vertical.

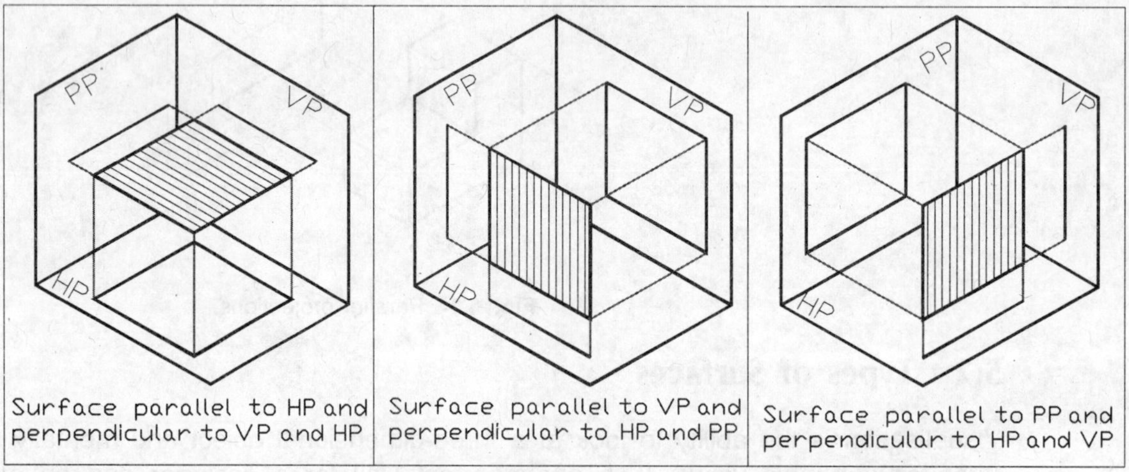


**Fig. 5.3** Characteristic views of a principal surface.

### 5.2.2 Characteristics of Inclined Surface

The surface, which is inclined to two reference planes and perpendicular to a third reference plane, is called an inclined (or slant) surface. It gives one line view and two area-views of the apparent area. Figure 5.4 shows an inclined surface which is perpendicular to PP and inclined to HP & VP. In this type of surface, at least one edge of the surface is along the principal axis.

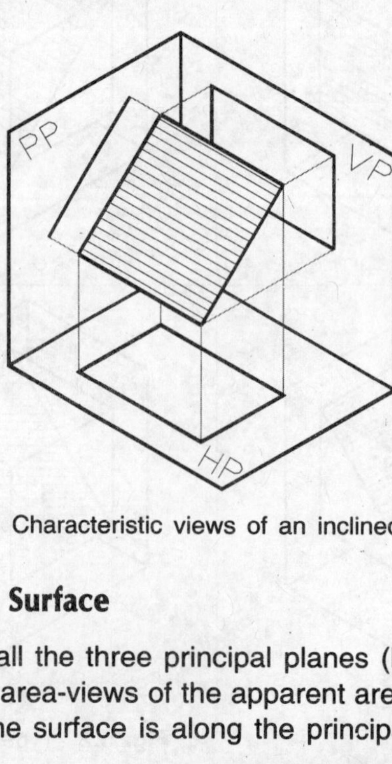


**Fig. 5.4** Characteristic views of an inclined surface.

### 5.2.3 Characteristics of Skew Surface

The surface, which is inclined to all the three principal planes (HP, VP and PP), is called a skew (or oblique) surface. It gives three area-views of the apparent area as shown in Fig. 5.5. In this type of surface, none of the edge of the surface is along the principal axis.

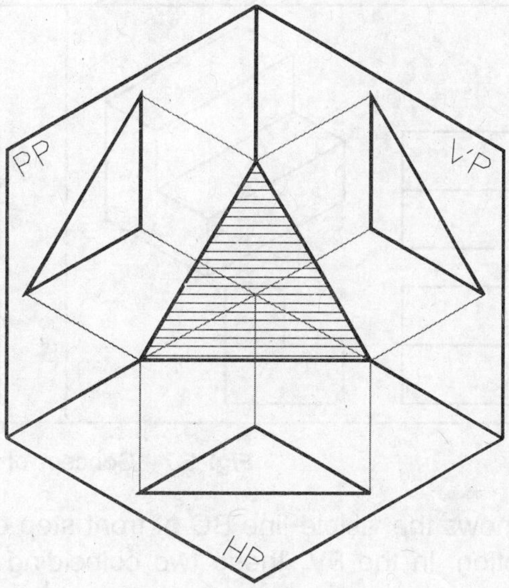


**Fig. 5.5** Characteristic views of a skew surface.

### 5.2.4 Characteristics of Curved Surface

The surface, which has a constant radius such as an arc or a circle, is called a curved (or rounded) surface. The surface that does not have a constant radius is classified as an irregular curved surface, which is out of scope of this course. In this type of surface, the last visible part of the curved surface falls in line view. This line is called as the line of tangency. Figure 5.6 shows the FV and SV as rectangles with the longer edge as the last visible part or the line of tangency.

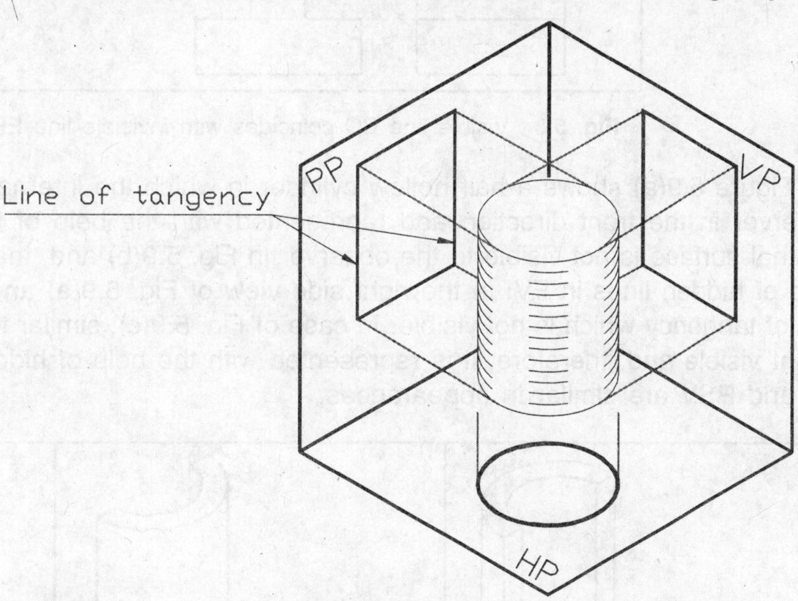


**Fig. 5.6** Characteristic view of a curved surface (solid cylinder).

## 5.3 Concept of Invisible (Hidden) Line

Hidden lines are used to represent surfaces that are not directly visible to the observer and falls into a line view. Figure 5.7 (A) shows a surface ABCD that is visible directly in the FV and falls in a line view, however, the same surface in Fig. 5.7(B) is not directly visible to the observer. In order to satisfy that surface, a hidden line is shown in the FV.

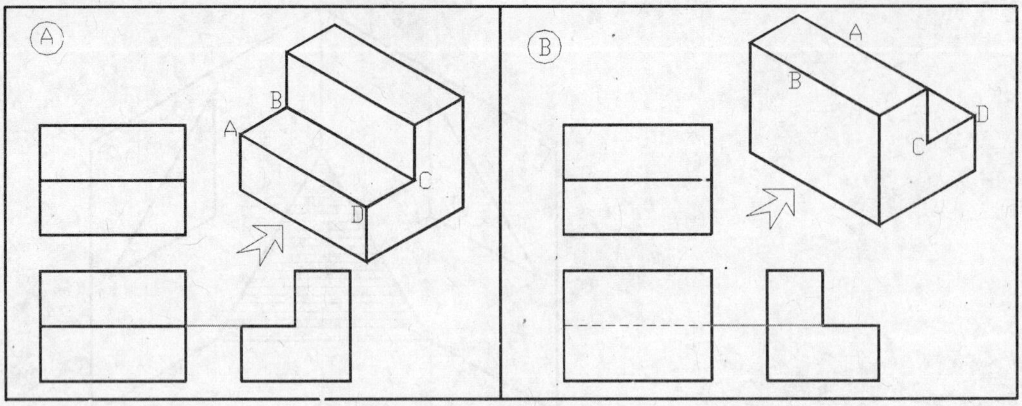


**Fig. 5.7** Concept of hidden line.

Figure 5.8 shows the visible line BC of front step coinciding with an invisible line EF of the back step in FV direction. In the FV, these two coinciding lines are shown as one solid line in FV.

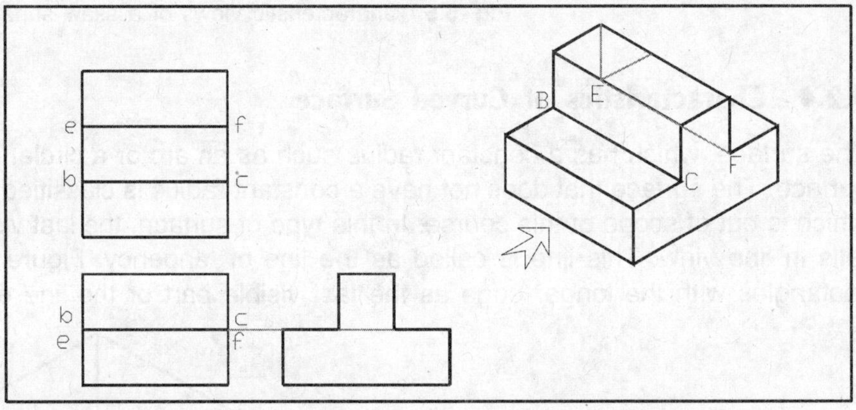


**Fig. 5.8** Visible line BC coincides with invisible line EF in FV direction.

Figure 5.9(a) shows a half hollow cylinder in which the internal curved surface is visible to the observer in the front direction and represented with the help of the solid lines in FV. The same internal surface is not visible to the observer in Fig. 5.9(b) and, therefore, it is represented with the help of hidden lines in FV. In the right side view of Fig. 5.9(a) and (b), a hidden line represents a line of tangency which is not visible. In case of Fig. 5.9(c), similar to Fig. 5.9(b), the internal surface is not visible and, therefore, it is represented with the help of hidden lines in FV. In this case, the FV and RSV are similar in appearances.

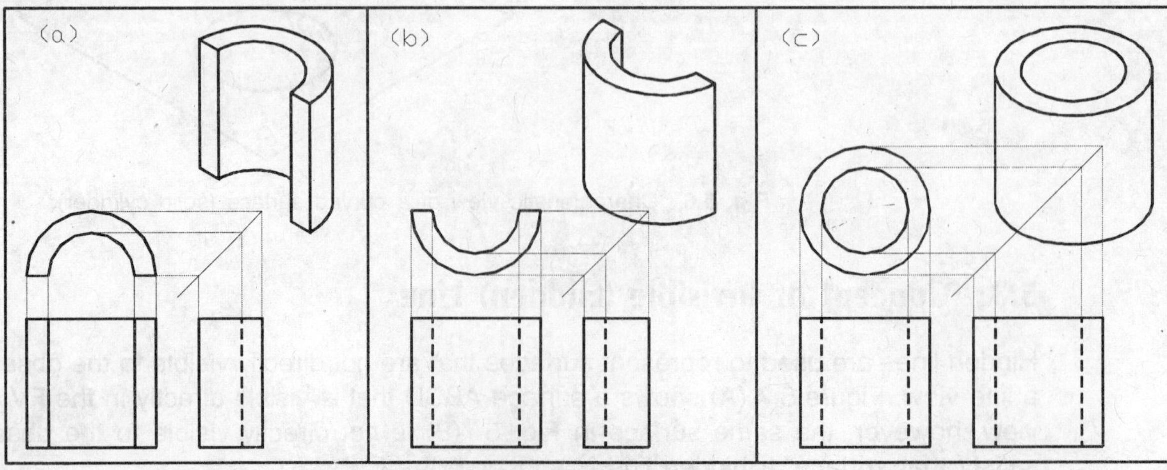


**Fig. 5.9** Concept of hidden line, curved objects.

## 5.4 Representation of Curved Surface

We have seen projection of a complete curved surface in section 5.2.4 with the help of Fig. 5.6. In many objects, we come across with incomplete or partly curved surface in continuation with either a flat surface or another partly curved surface. In order to understand the visualisation and projections four cases are discussed with reference to Fig. 5.10(a), (b), (c) and (d).

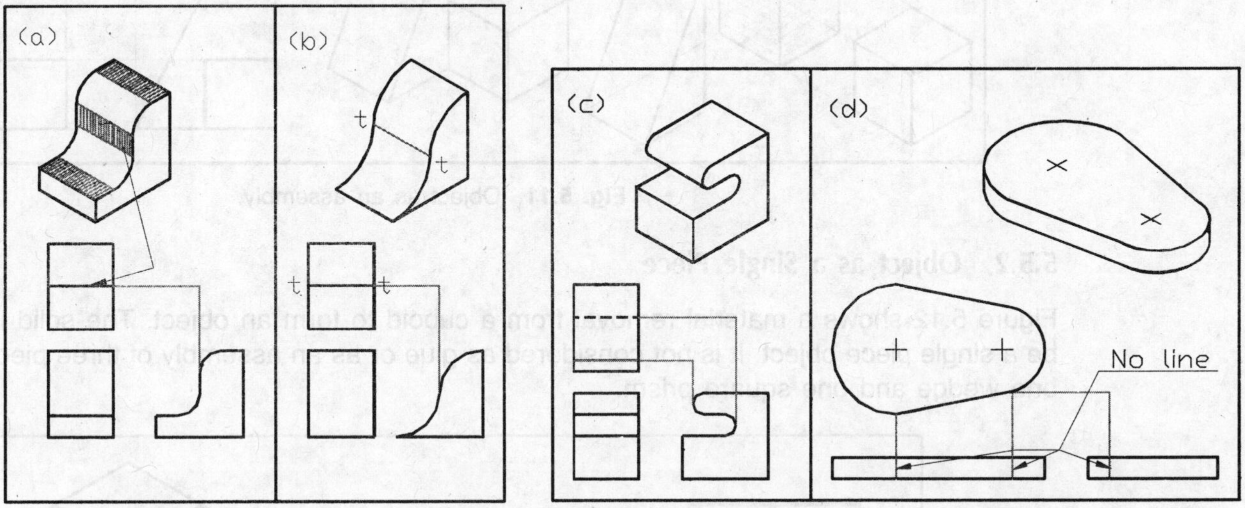


(a) curved-flat-curved surfaces (b) curved-curved (c) visible curved-invisible curved surfaces (d) curved-slant-curved surfaces

**Fig. 5.10** Representation of curved surface.

(a) In case of curved-flat-curved case, a line in TV represents a vertical surface in between two curved surfaces.
(b) In case of curved-curved case, a line in TV represents last visible portion of top curved surface (line of tangency tt in the given object. However, this line of tangency does not exist in FV. In FV, both the curved surfaces appear to be merged at line of tangency in FV.
(c) In case of visible curved-invisible curved case, a solid line in TV represents last visible part of curved surface and a hidden line represents last invisible portion of curved surface.
(d) In case of curved-slant-curved case a curved surface and a flat surface have continuity at their joint. Both the surfaces are blended at their joint. Therefore, there is no line of intersection or line of tangency at their meeting point.

## 5.5 Understanding of Object

In interpretation of solids, one method is to break the object into its basic elementary solids and then the orthographic views of individual components are obtained from the exploded view and superimposed suitably to complete the drawing (orthographic views).

### 5.5.1 Object as an Assembly

Figure 5.11 shows the exploded view of object with its elementary solids and gluing (hatched) area. The same figure shows the generated object and superimposed orthographic views.

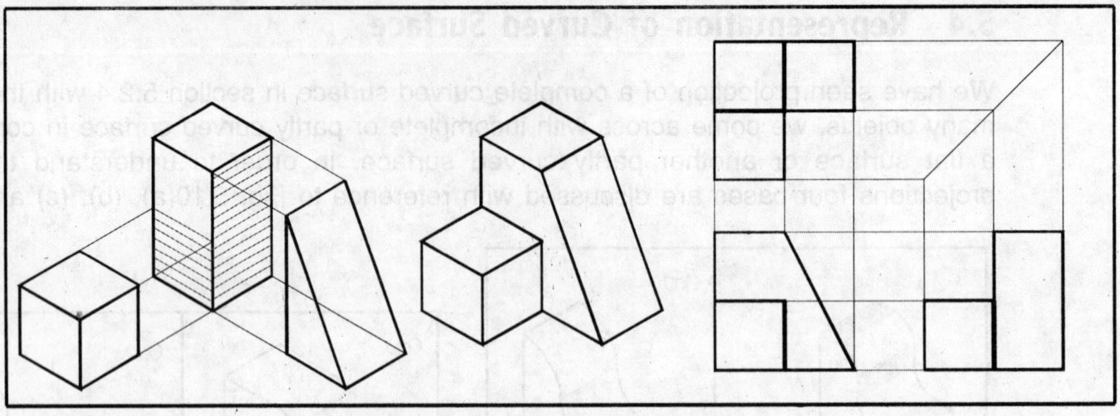

**Fig. 5.11** Object as an assembly.

### 5.5.2 Object as a Single Piece

Figure 5.12 shows a material removal from a cuboid to form an object. The solid is considered to be a single piece object. It is not considered as glue or as an assembly of three pieces: one cuboid, one wedge and one square prism.

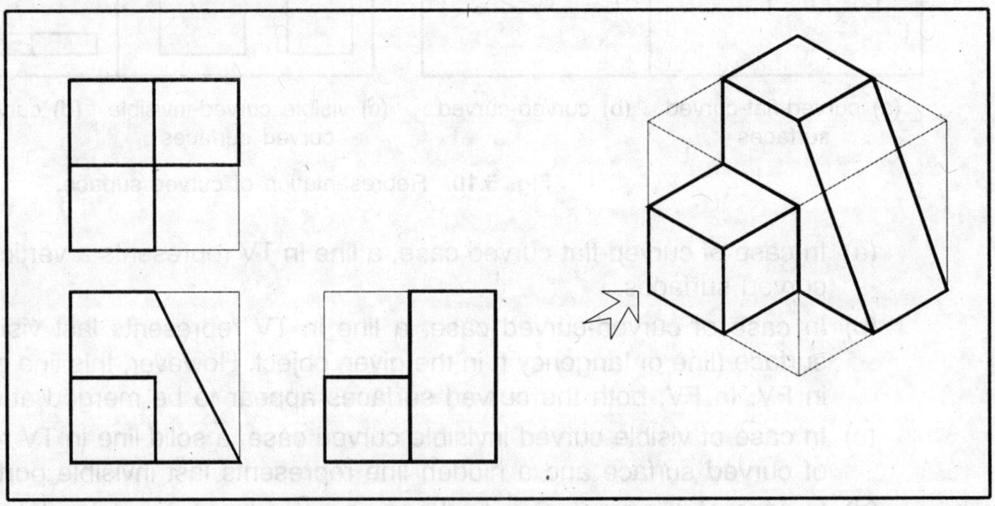

**Fig. 5.12** Object as a single piece.

Figure 5.13 shows a material removal from a solid cylinder to form an object. First a solid cylinder is made as a hollow cylinder and then 50% of the upper half of the hollow cylinder is removed to form a final object. The same figure shows the orthographic view of the formed object.

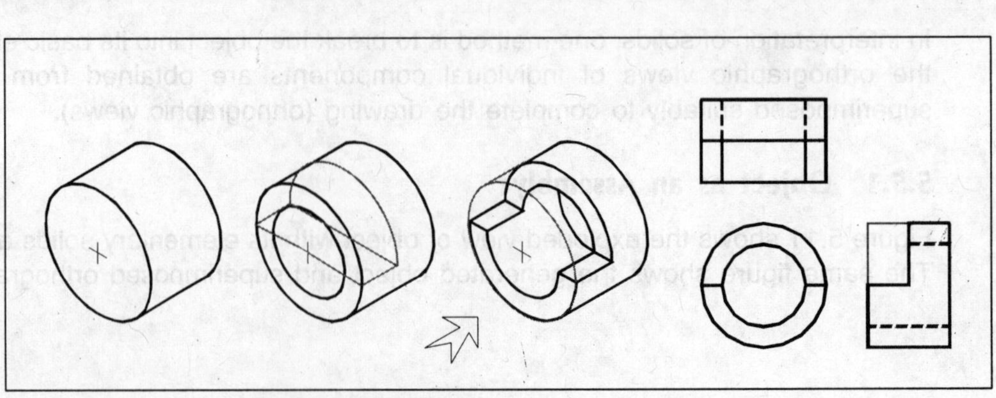

**Fig. 5.13** Concept of a single piece object.

Figure 5.14(A) shows the exploded view of an object along with its assembly and orthographic views. The other way to interpret the solid is as a single piece object. It is cut out from a block to obtain the required shape [Fig. 5.14(B)].

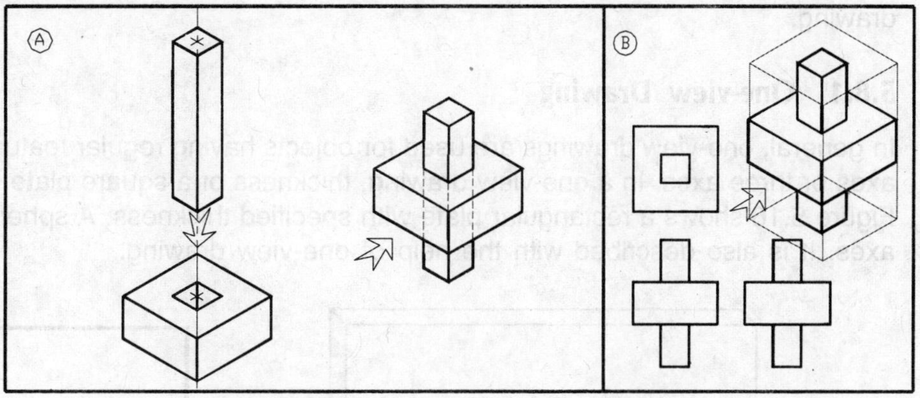


**Fig. 5.14** Assembly versus single piece object.

In an elementary course like Engineering Graphics, an object is considered as a single piece and not an assembly. The reason is at the first year level of all engineering disciplines the objective of the course is to generate an imagination skill. Assembly drawing is a part of the courses like Machine Drawing and Structural Drawing, in specific disciplines.

## 5.6 Precedence of Lines

When orthographic views are prepared, it is required to draw one type of line over another. Drawing convention has established a precedence of lines: a continuous (solid) line takes precedence over a hidden line and a hidden line takes precedence over a centre line. Figure 5.15 shows a solid line coinciding with a hidden line, which coincides with a centre line in the FV of a given object. By using the convention, a solid line has taken precedence over the hidden line; subsequently the hidden line has taken precedence over the centre line.

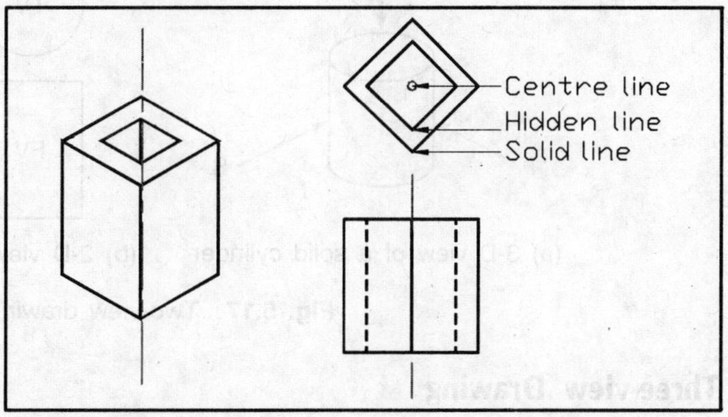


**Fig. 5.15** Example of precedence of lines.

## 5.7 Selection of Views

The selection of views required to describe an object depends upon the complex nature of the object. Every object has three principal dimensions: length L, depth D and height H. As we have seen in Chapter 3, every view shows only two dimensions out of three in true size. The most descriptive view should be selected as the FV. In case of curved objects, a FV direction should be chosen perpendicular to the plane of curved surfaces.

## 5.8 Types of Drawing

In engineering drawing, minimum number of views are drawn to describe an object completely. Based on this the drawings are classified as: one-view drawing, two-view drawing and three-view drawing.

### 5.8.1 One-view Drawing

In general, one-view drawings are used for objects having regular features and symmetry, about two axes or three axes. In a one-view drawing, thickness of a square plate can be mentioned by a note. Figure 5.16 shows a rectangular plate with specified thickness. A sphere is symmetrical about three axes. It is also described with the help of one-view drawing.

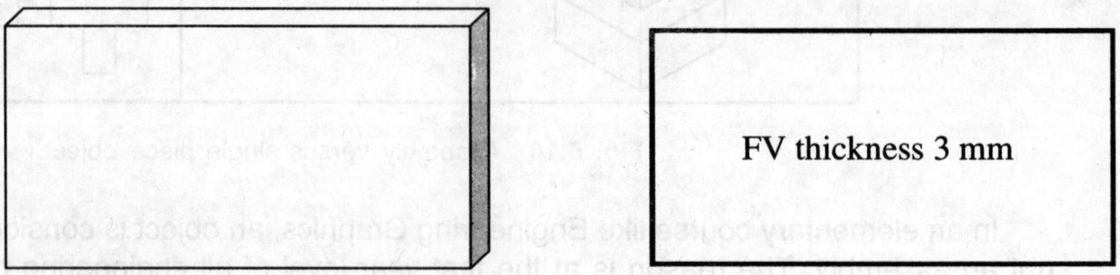


**Fig. 5.16** One-view drawing.

### 5.8.2 Two-view Drawing

In general, two-view drawings are used for objects having symmetry about one axis. The largest face of an object having more visible details is selected as the front view. The second view is either TV or SV. Figure 5.17(a) shows a 3-D view of a solid cylinder with three viewing directions top, front and side. Figure 5.17(b) shows the TV, FV and RSV of the cylinder. The TV and FV shapes are different; however, the shape of FV and any SV (LSV or RSV) is similar. There is a need to draw either a TV or a SV along with the FV in this case.

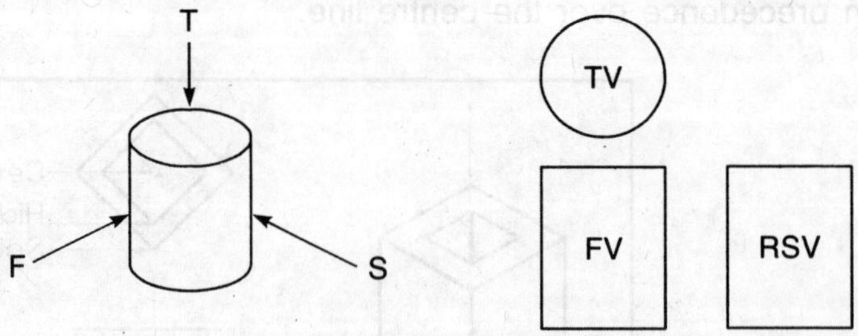


(a) 3-D view of a solid cylinder (b) 2-D views of a solid cylinder

**Fig. 5.17** Two-view drawing.

### 5.8.3 Three-view Drawing

Three-view drawings are used for objects having symmetry about no axes. The largest face of the object having more visible details and least invisible lines is selected as the front view. Three-view drawings are used to describe any object clearly and completely. Figure 5.18 shows (a) 3-D view of a rectangular solid and Fig. 5.18 (b) shows 2-D view of the same. All three views are required to describe the object.

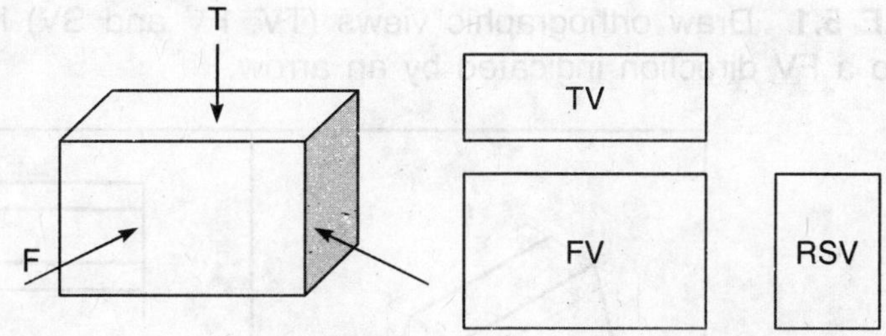


(a) 3-D view of a rectangular solid (b) 2-D views of a rectangular solid

**Fig. 5.18** Three-view drawing.

## 5.9 Principles of Multiview Drawing

The following principles should be studied carefully and understood thoroughly before any attempt is made to prepare an orthographic drawing:

1. The front and top views are always vertically aligned.
2. The front and side views are horizontally aligned.
3. The length of the top view is same as the length of the front view.
4. The height of the side view is the same as the height of the front view.
5. The depth of the side view is same as the depth of the top view.

## 5.10 Steps to Draw Orthographic Views

### 5.10.1 Box Method

1. Find the maximum length L, maximum depth D and maximum height H from a given pictorial view.
2. Accommodate a given object in a box of L × D × H with thin pencil on paper or with any colour on AutoCAD screen. The surfaces of this box serve the purpose of principal planes.
3. Start drawing a front view (as it shows more details). This is optional.
4. While drawing any view, the principal surfaces (which show the true area) should be drawn first. The inclined and skew surfaces cannot be drawn directly. Locate the corners of non-principal surfaces and complete those surfaces by connecting the appropriate points.
5. Satisfy all visible surfaces in all three views.
6. Check for the hidden lines if any.

### 5.10.2 Centre Line Method

1. Locate the centres of the curved parts.
2. Join the centres by a thin line, which is referred here as a centre line.
3. This centre line serves the purpose to find the maximum length L, maximum depth D and maximum height H from the pictorial view. While finding these dimensions move along the centre line and gridlines perpendicular to the centre line.
4. Draw that view first in which circles or curves appear. With circles, it becomes easy to draw other views.
5. Repeat the steps 4 to 6 in the Box Method.

**EXAMPLE 5.1** Draw orthographic views (TV, FV and SV) in Fig. 5.19 of the given object with respect to a FV direction indicated by an arrow.

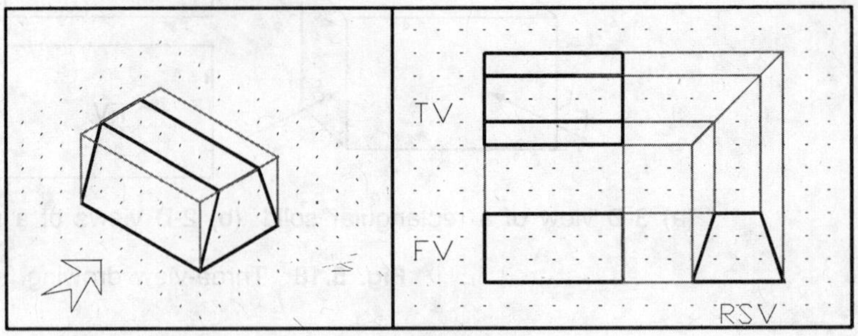


**Fig. 5.19**

***Solution*** The maximum length, depth and height required to accommodate the given object in box are 6 × 4 × 3 grid spacing. There are six surfaces. Out of six surfaces, four are principal and two are inclined surfaces.

**EXAMPLE 5.2** Draw orthographic views (TV, FV and SV) in Fig. 5.20 of the given object with respect to a FV direction indicated by an arrow.

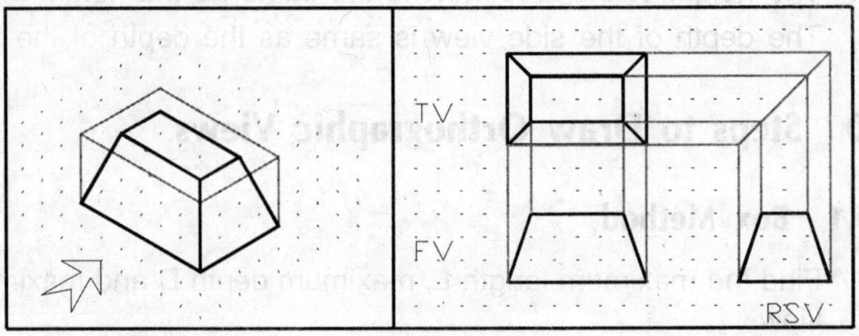


**Fig. 5.20**

***Solution*** The maximum length, depth and height required to accommodate the given object in box are 6 x 4 x 3 grid spacing. There are six surfaces. Out of six surfaces, four are inclined and two are principal surfaces.

**EXAMPLE 5.3** Draw orthographic views (TV, FV and SV) in Fig. 5.21 of the given object with respect to a FV direction indicated by an arrow.

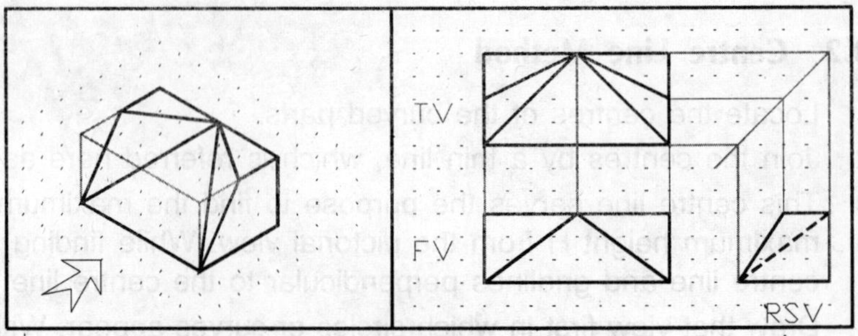


**Fig. 5.21**

***Solution*** The maximum length, depth and height required to accommodate the given object in box are 6 × 4 × 3 grid spacing. There are nine surfaces. Out of nine surfaces, six are principal surfaces, one is inclined and two are skew surfaces.

**EXAMPLE 5.4** Draw orthographic views (TV, FV and SV) in Fig. 5.22 of the given object with respect to a FV direction indicated by an arrow.

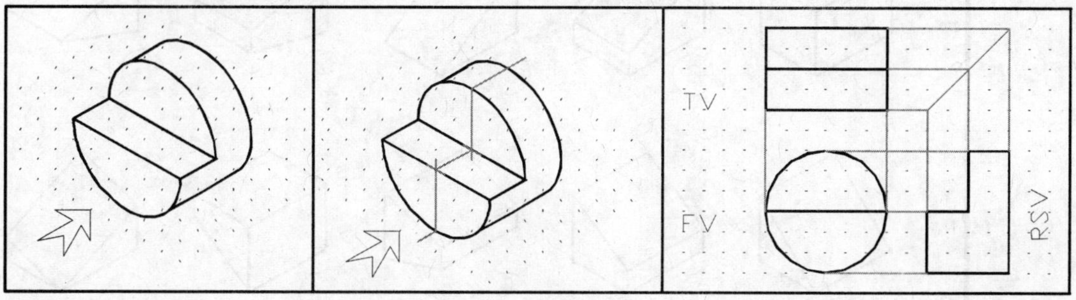


**Fig. 5.22**

***Solution*** Draw a centre line by locating centres of visible semicircle as shown in the solution. The length, height and depth of the object are 6 × 6 × 4 grid spacing. First draw FV in which there is a circular portion. Now draw other two views with reference to projection lines drawn from FV.

**EXAMPLE 5.5** Figure 5.23 shows a tetrahedron ABCO. Name the vertices for a tetrahedron *a-b-c-o*. The FV direction is parallel to an edge *ac*.

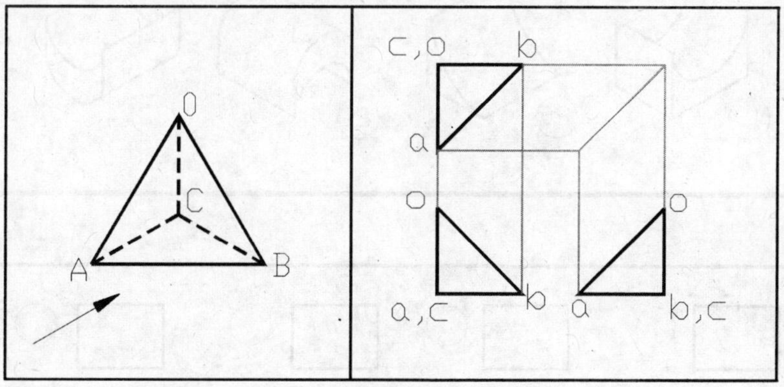


**Fig. 5.23**

***Solution*** In the FV, the triangular surface AOC and the triangular base ABC falls in a line view. In TV, the triangular surface AOC and the triangular surface COB falls in a line view. In RSV, the triangular surface BOC and the triangular base ABC falls in a line view. Triangular skew surface OAB shows the apparent area view in all the three views.

## Exercise

**5.1** One of the prime objectives of the course is to develop imagination skill, which is required for all scientists and engineers. Attempt has been made to generate imagination skill by starting from isometric and orthographic view of a simple block to complex geometric solid model. This exercise will also train the students to think about the type of surfaces which is useful to draw isometric drawing from given orthographic views in Chapter 6.

**5.2** Match the isometric drawing with orthographic views in Plate 5.1–5.22 and draw the orthographic views for the same.

**5.3** Draw orthographic views for the isometric drawings given in Plate 5.23–5.39.

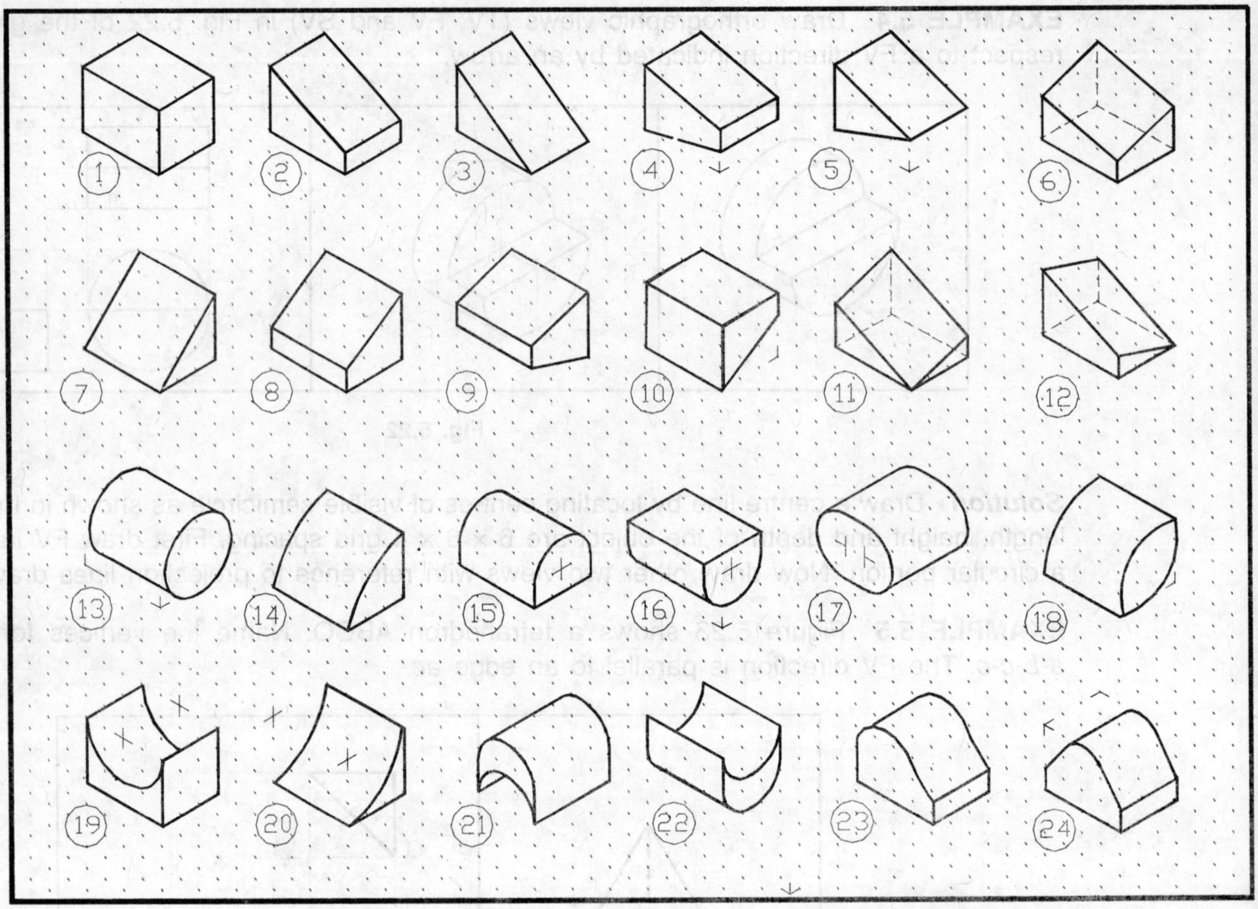


**PLATE 5.1**

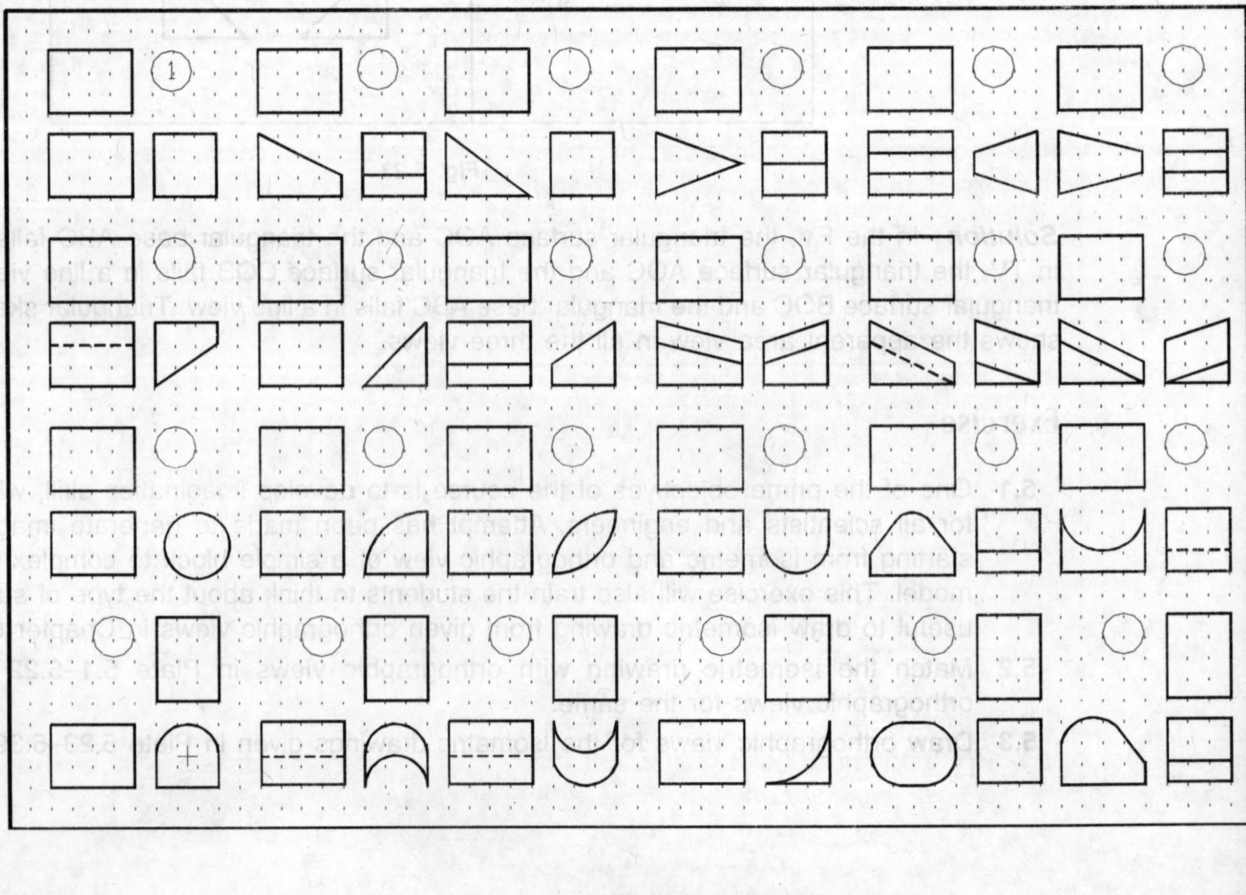


**PLATE 5.2**

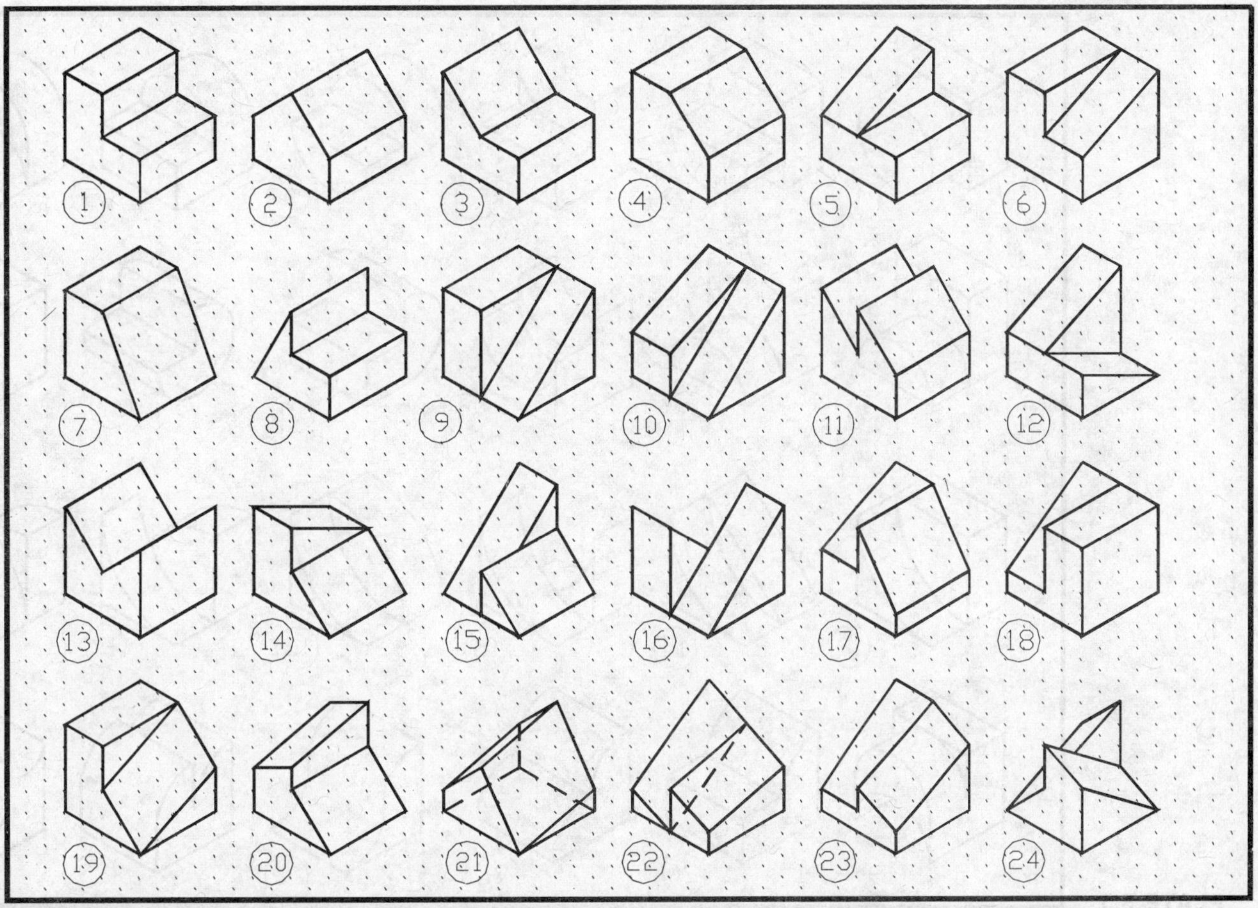


PLATE 5.3

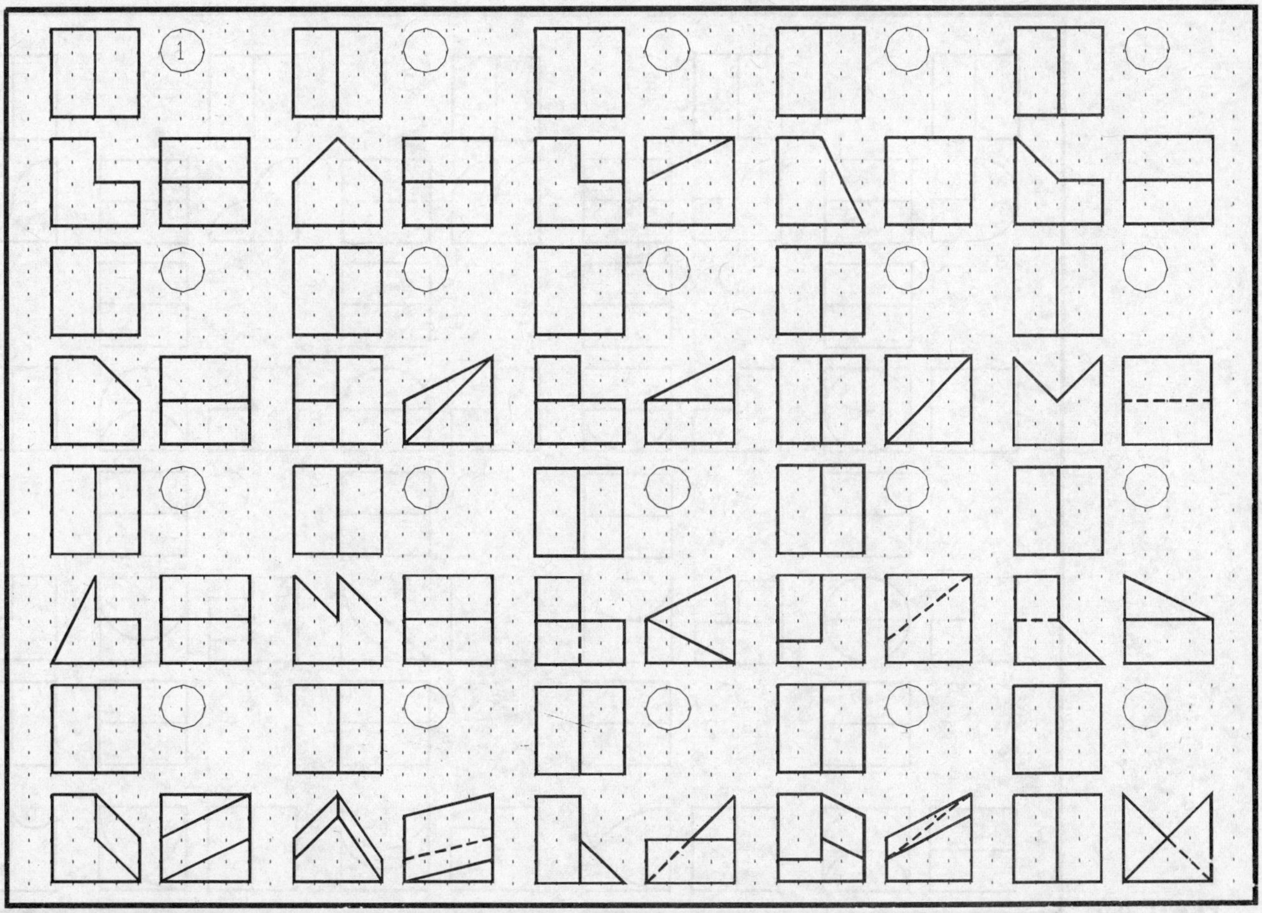


PLATE 5.4

PLATE 5.5

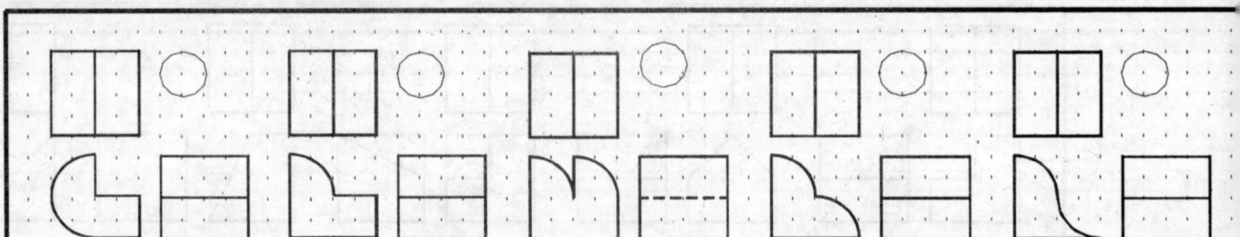

PLATE 5.6

**PLATE 5.7**

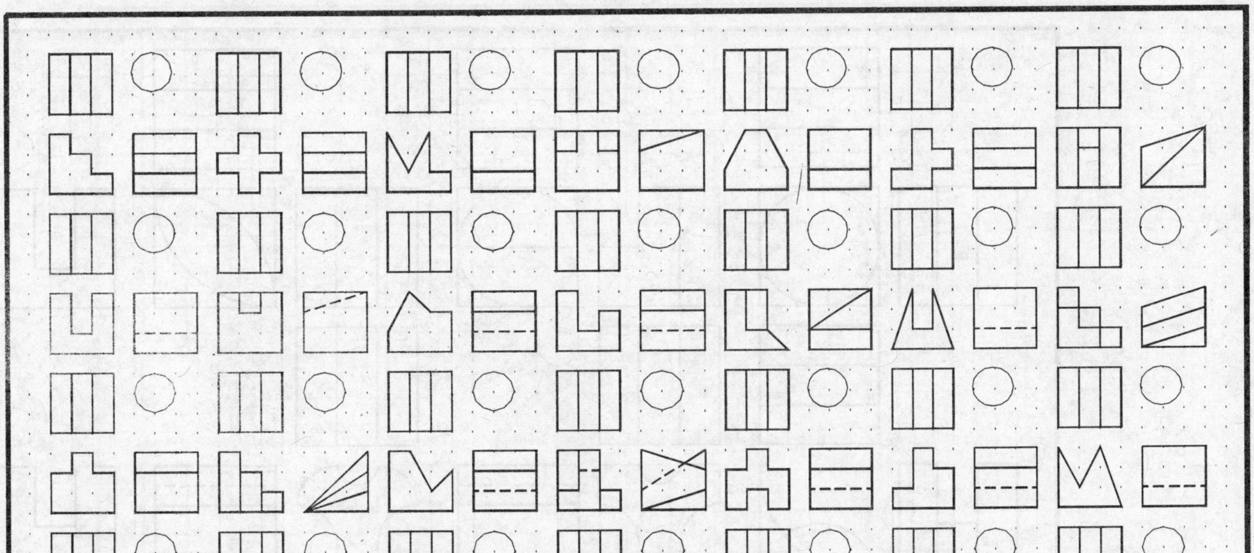

**PLATE 5.8**

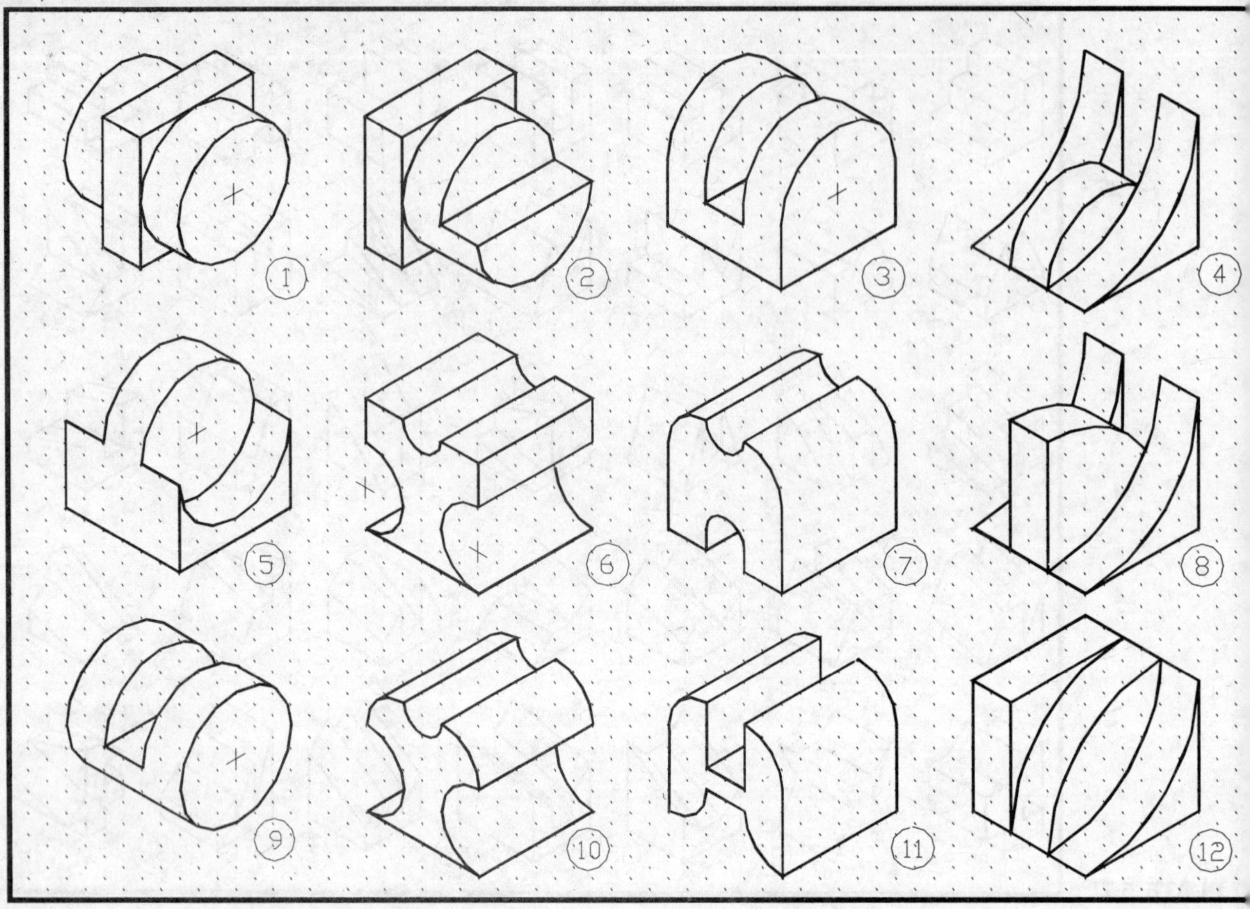


PLATE 5.9

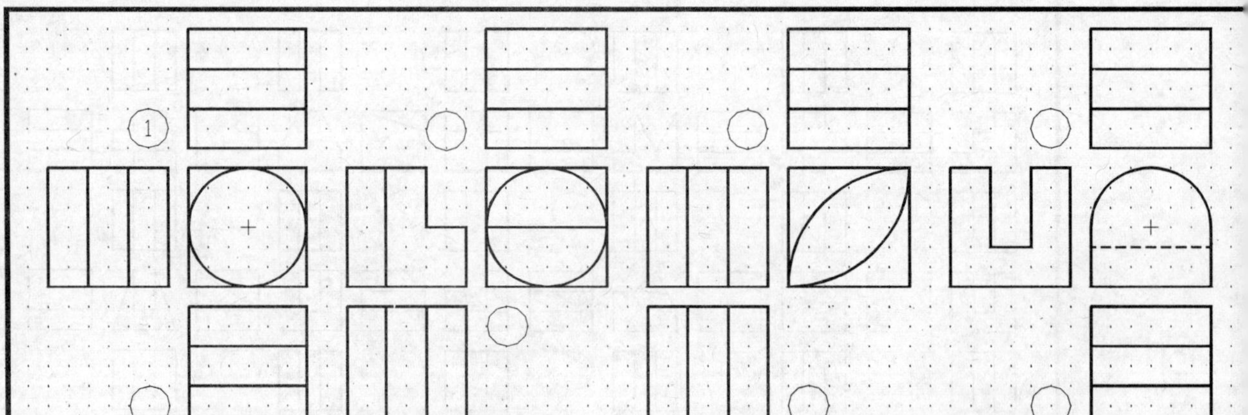


PLATE 5.10

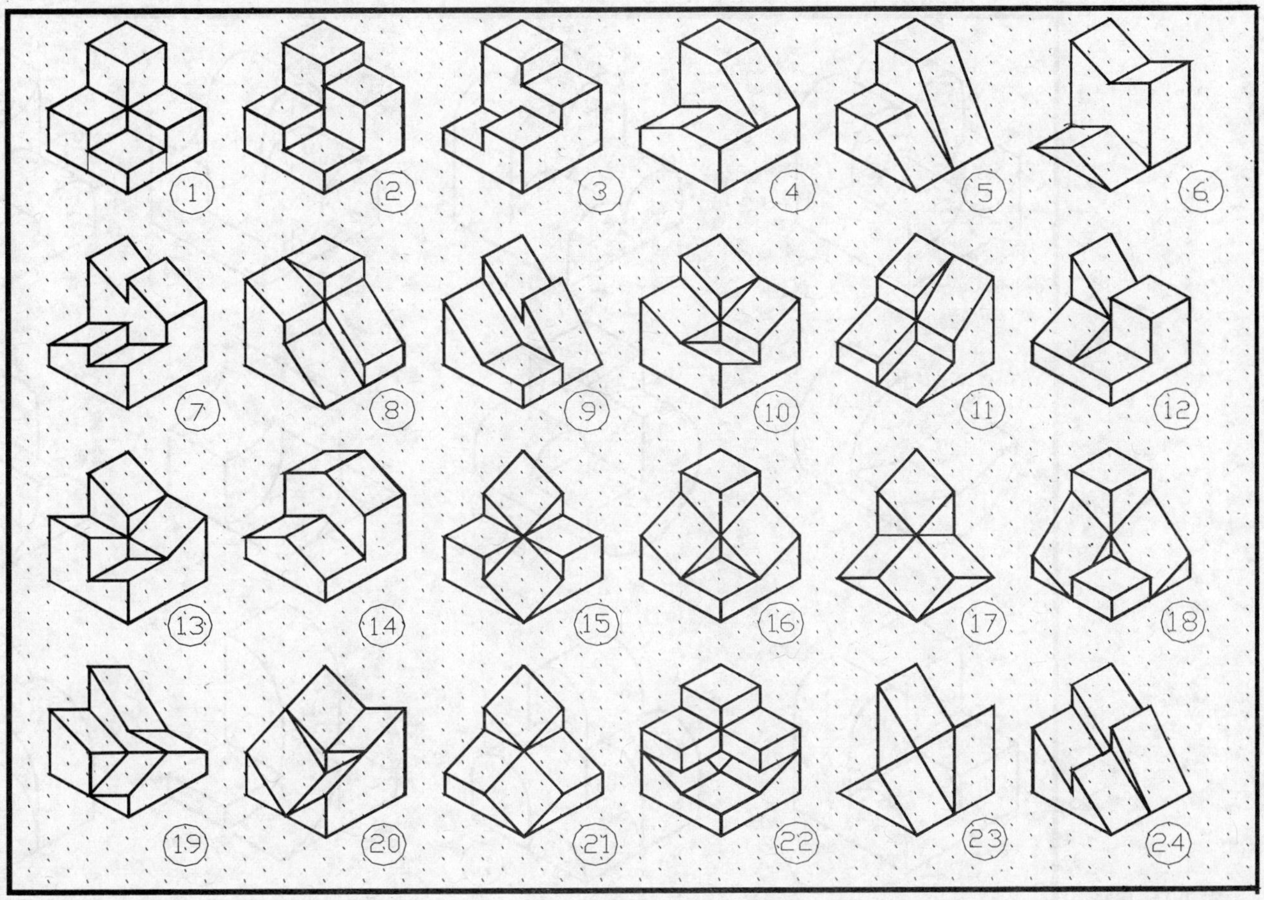


**PLATE 5.11**

**PLATE 5.12**

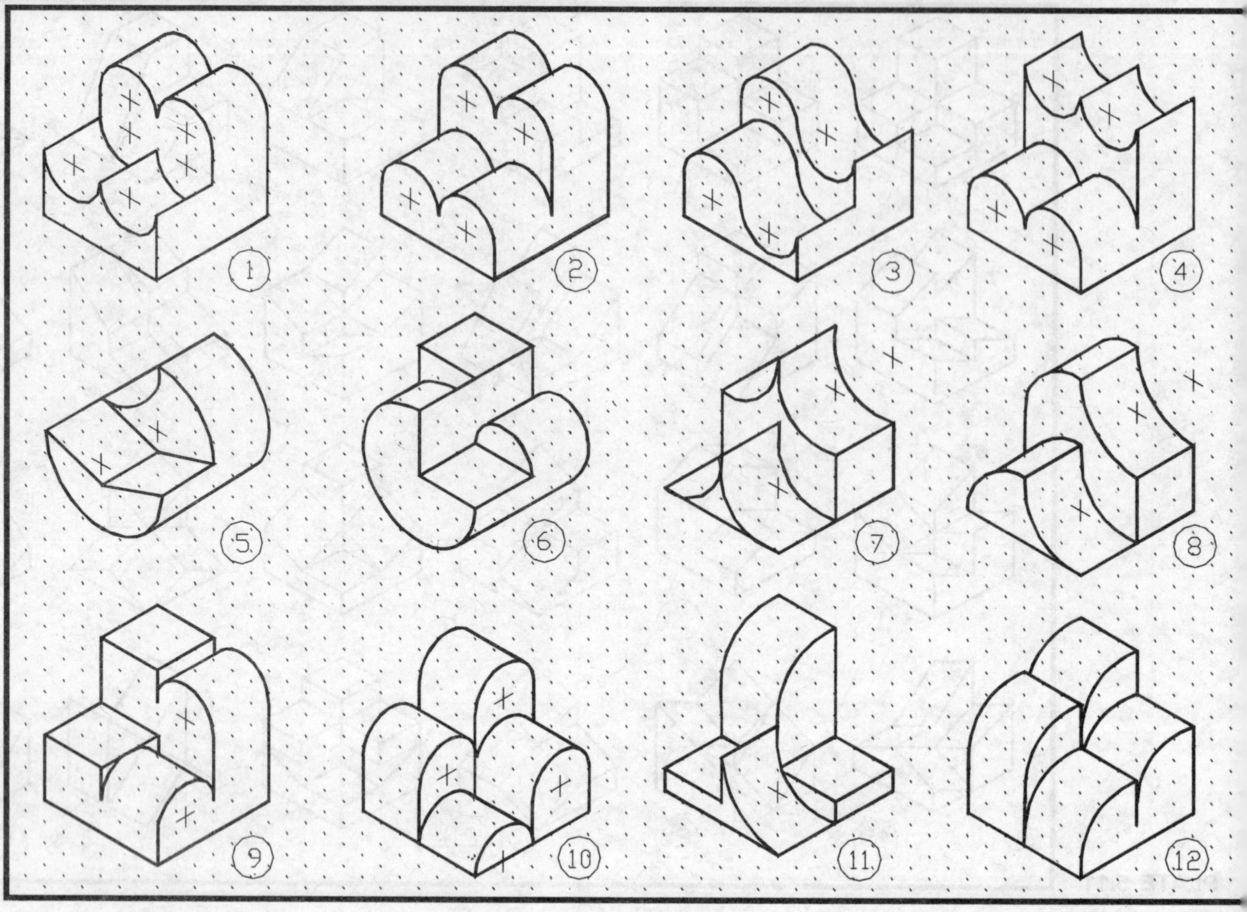


PLATE 5.13

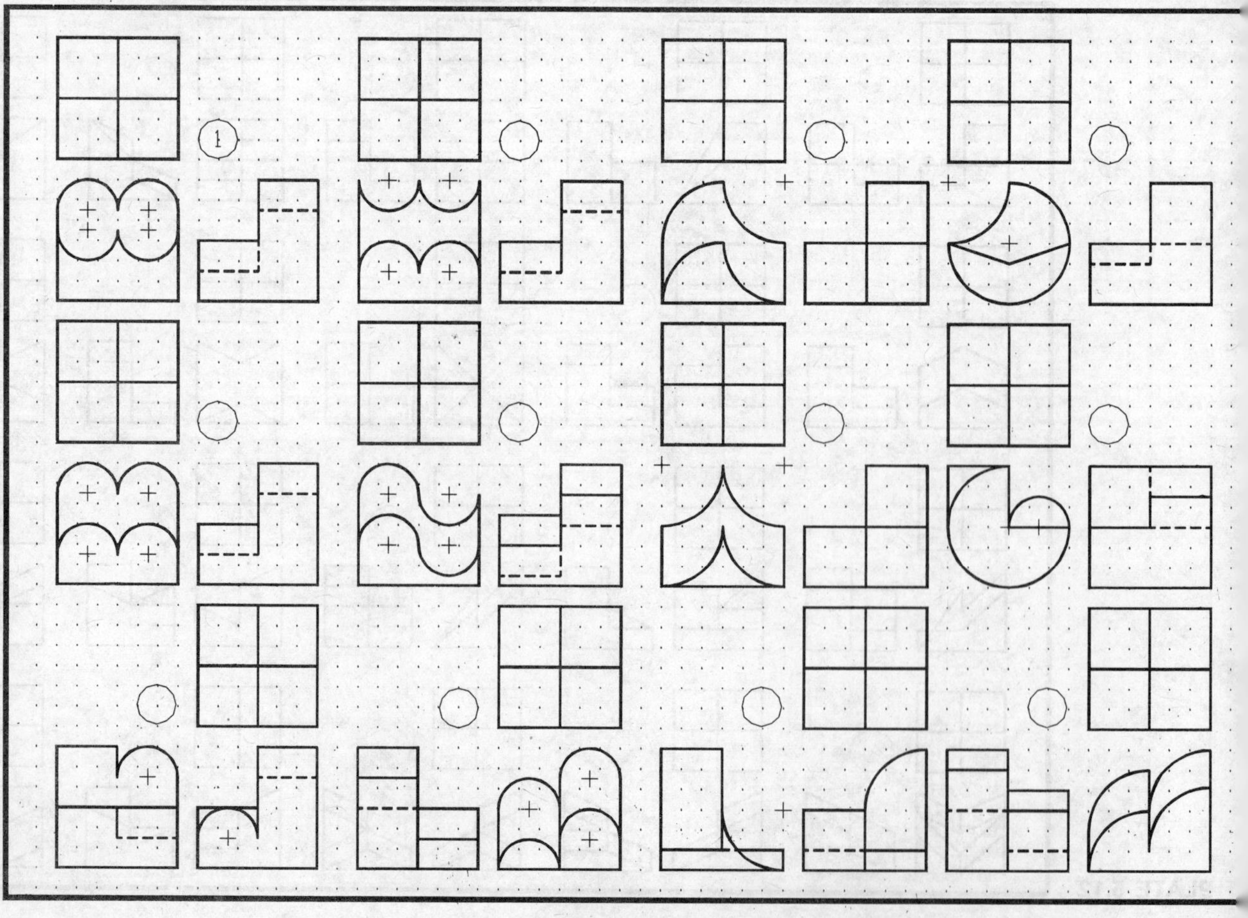


PLATE 5.14

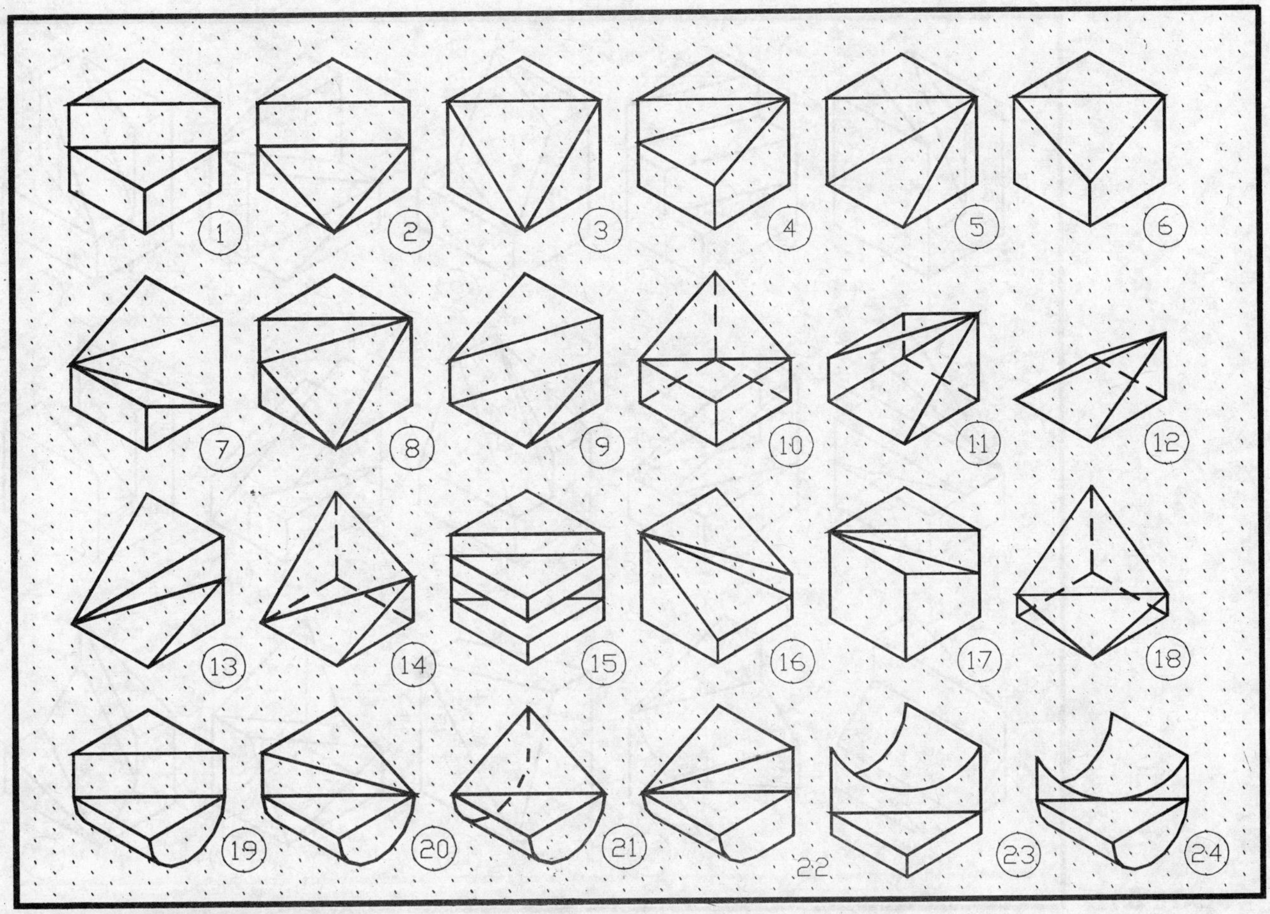


PLATE 5.15

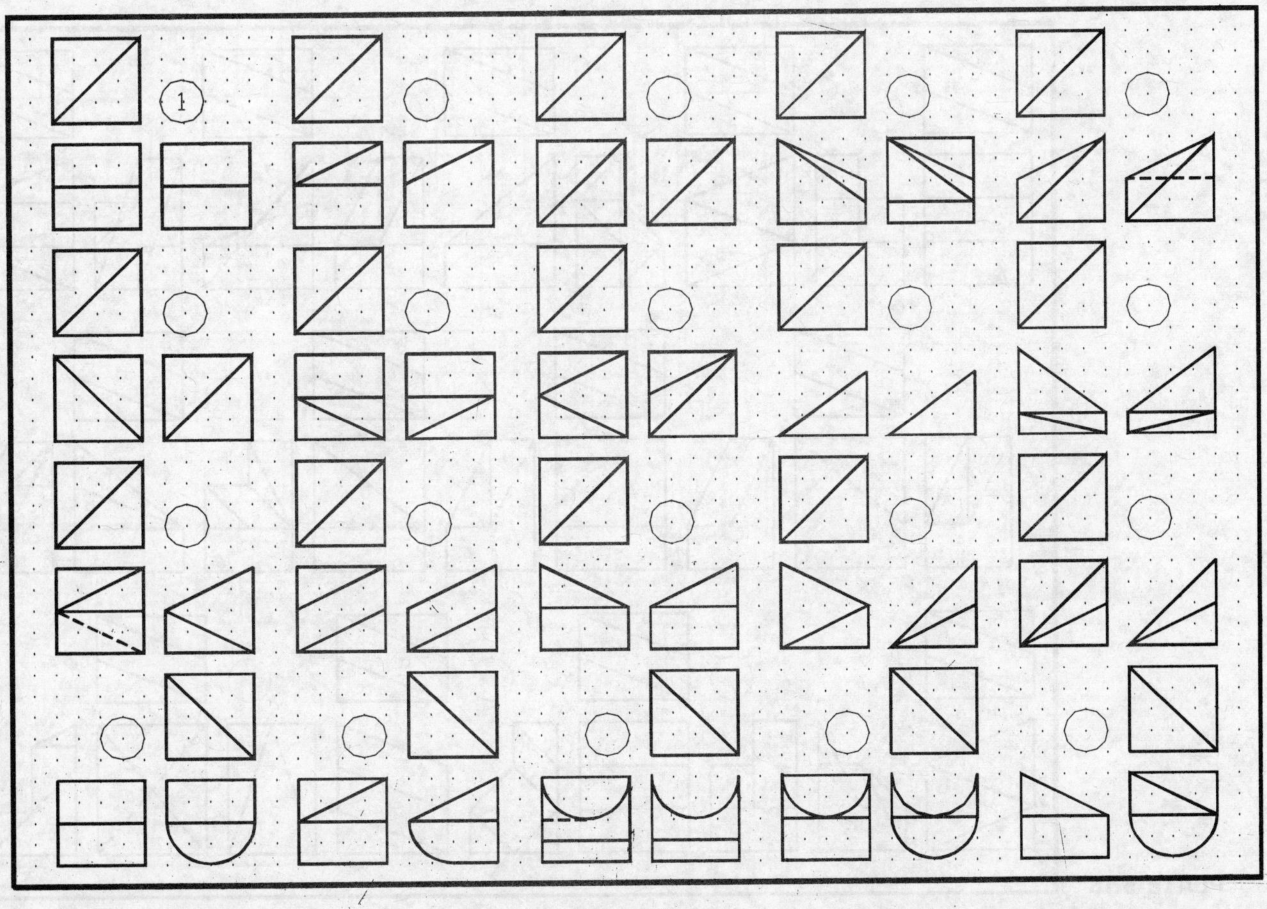


PLATE 5.16

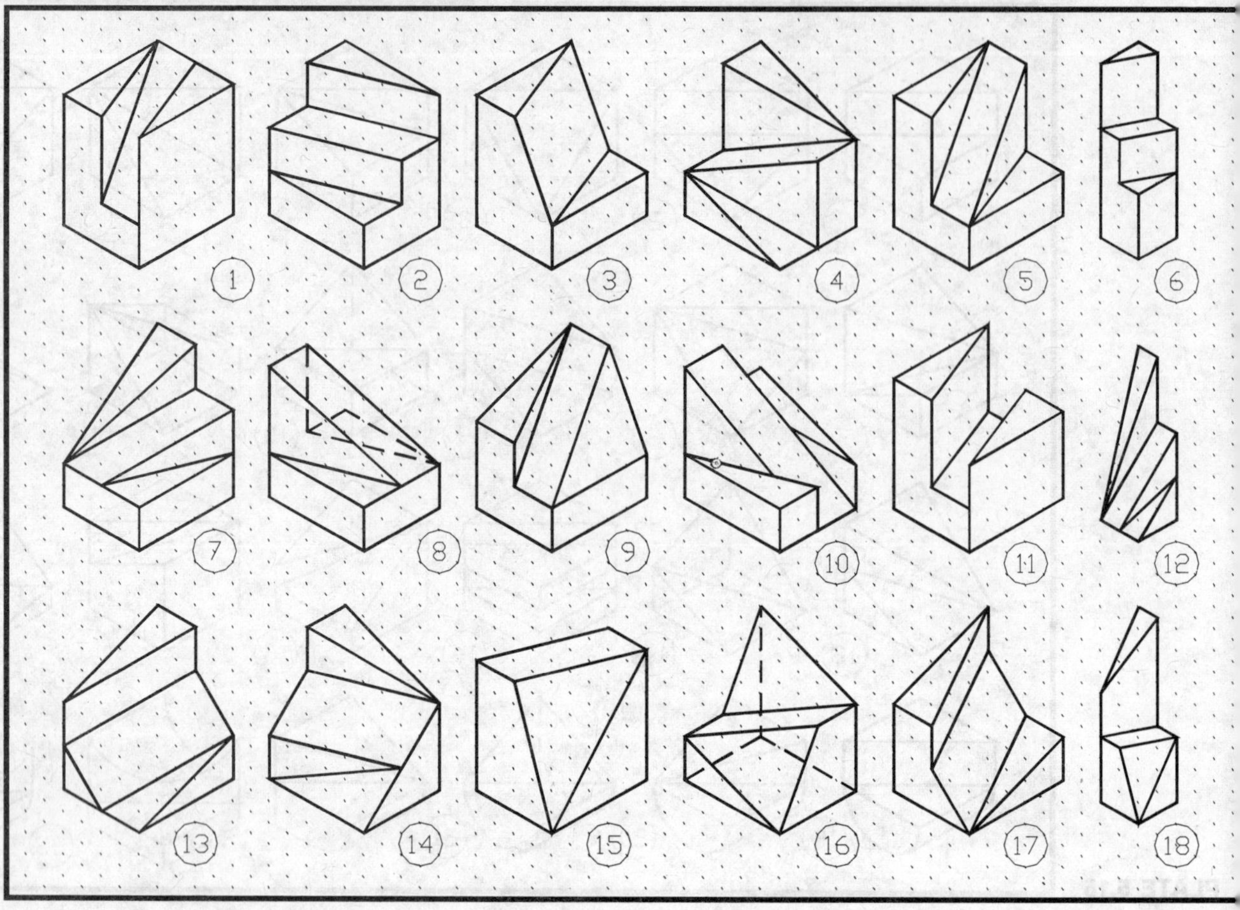


PLATE 5.17

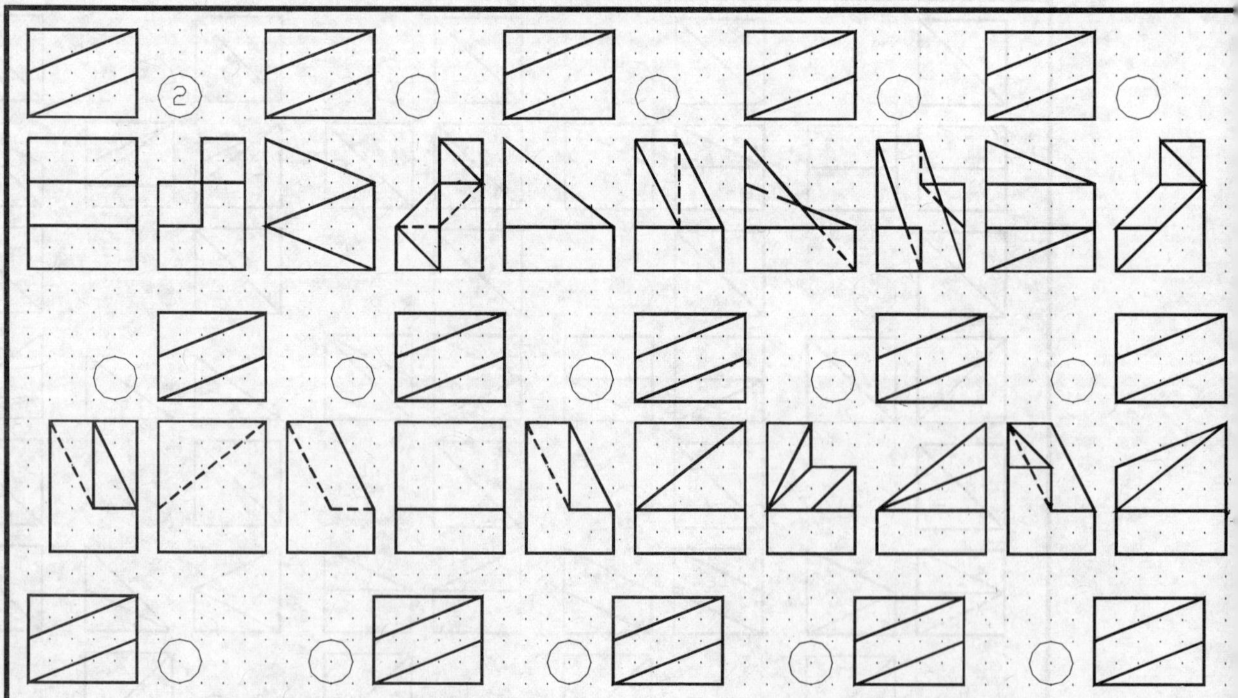


PLATE 5.18

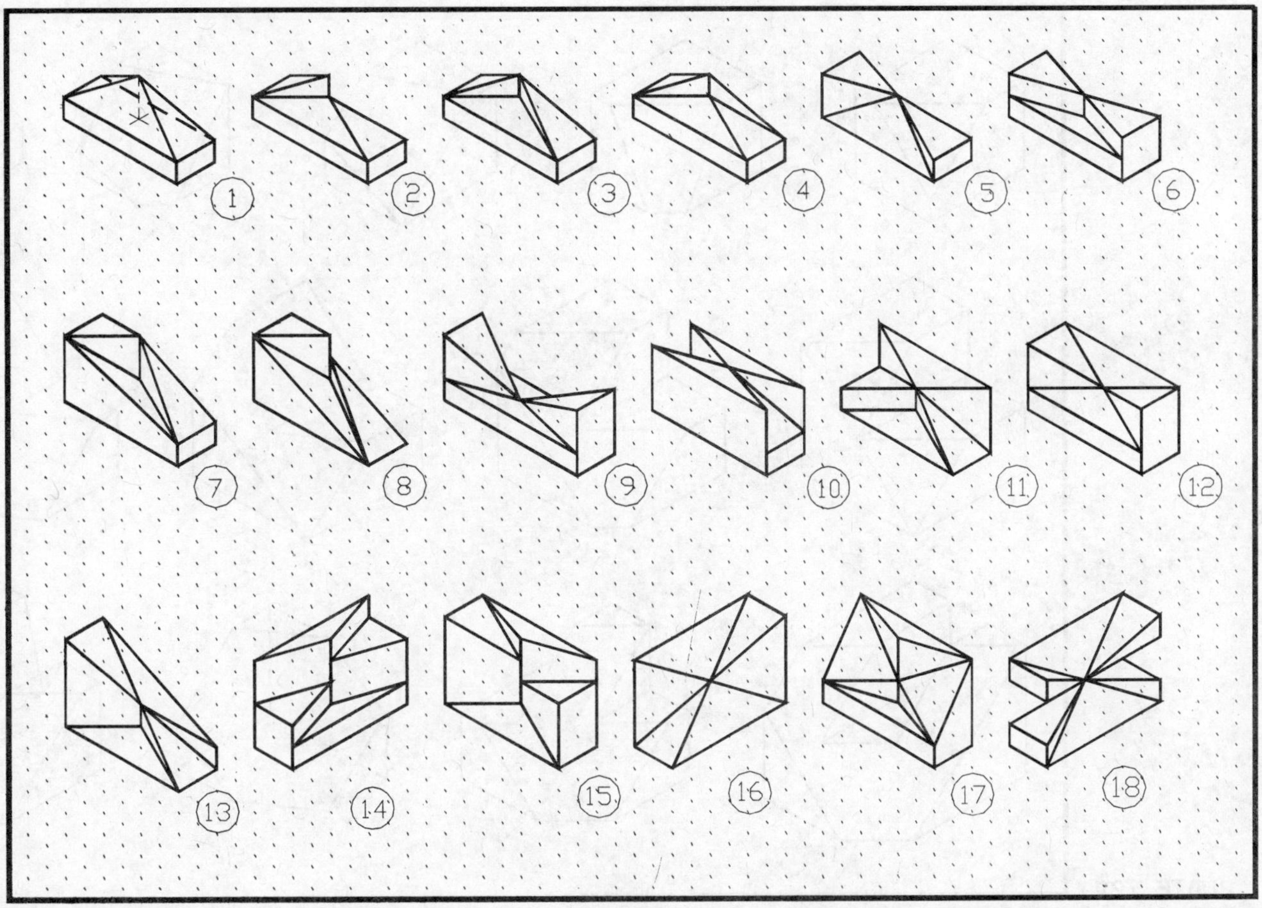


PLATE 5.19

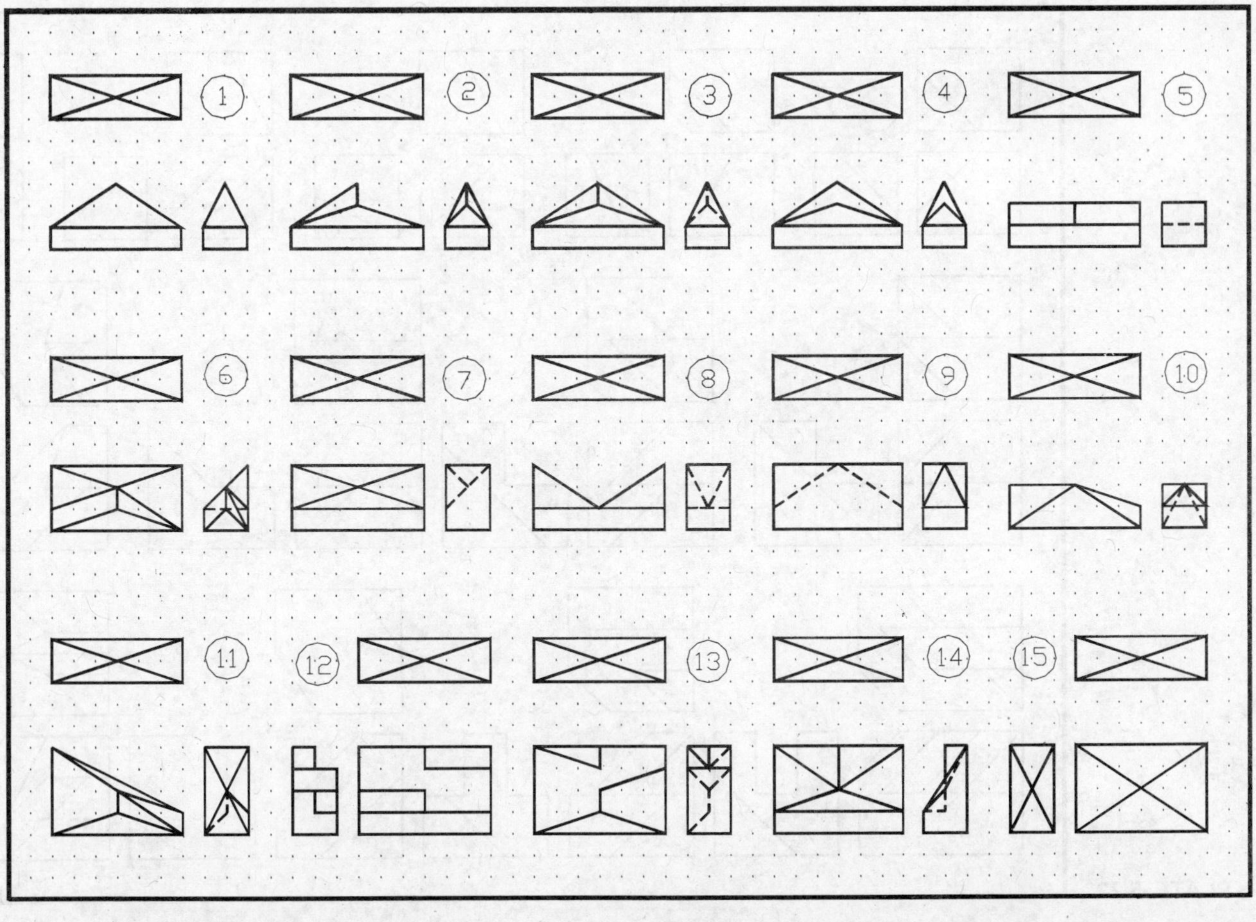


PLATE 5.20

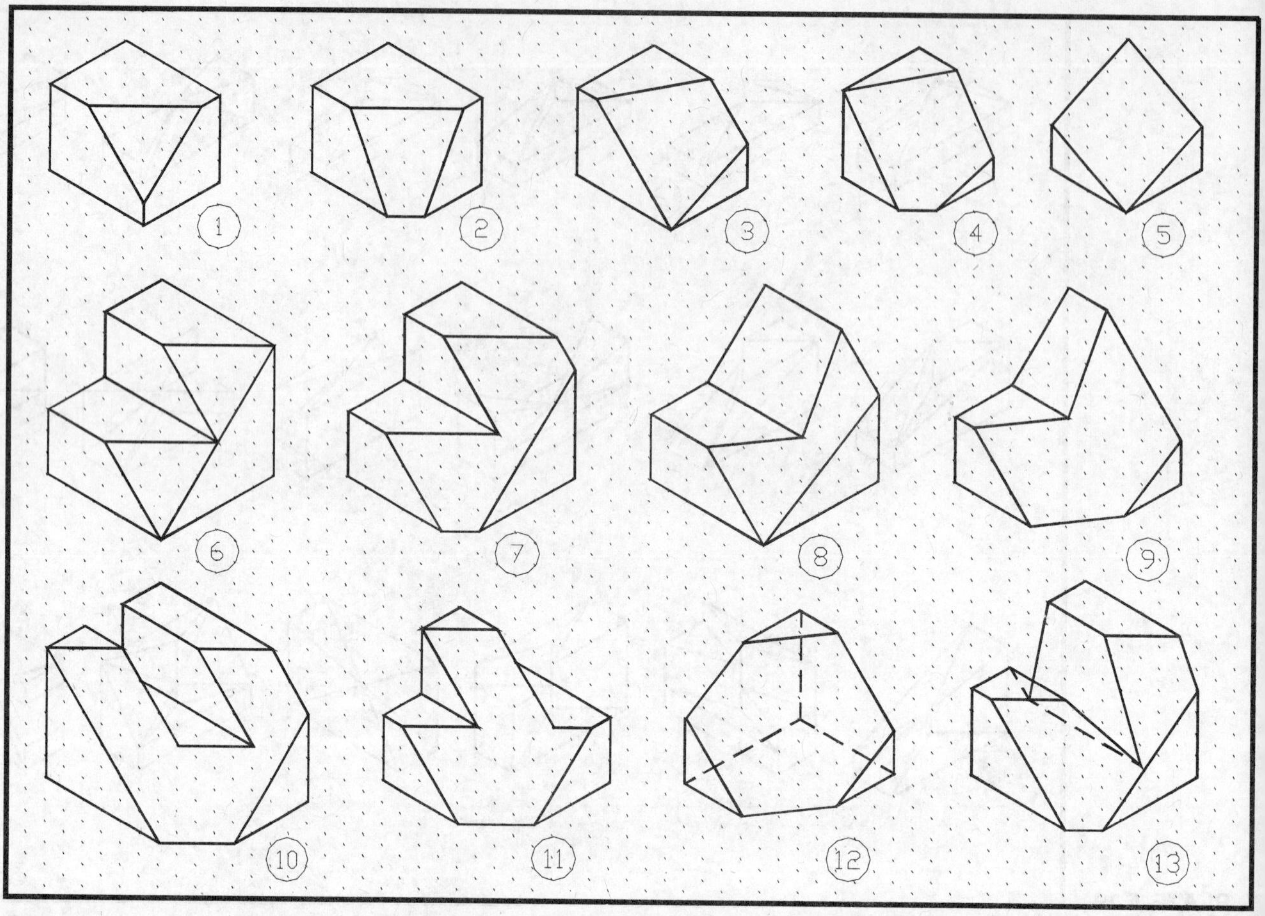


PLATE 5.21

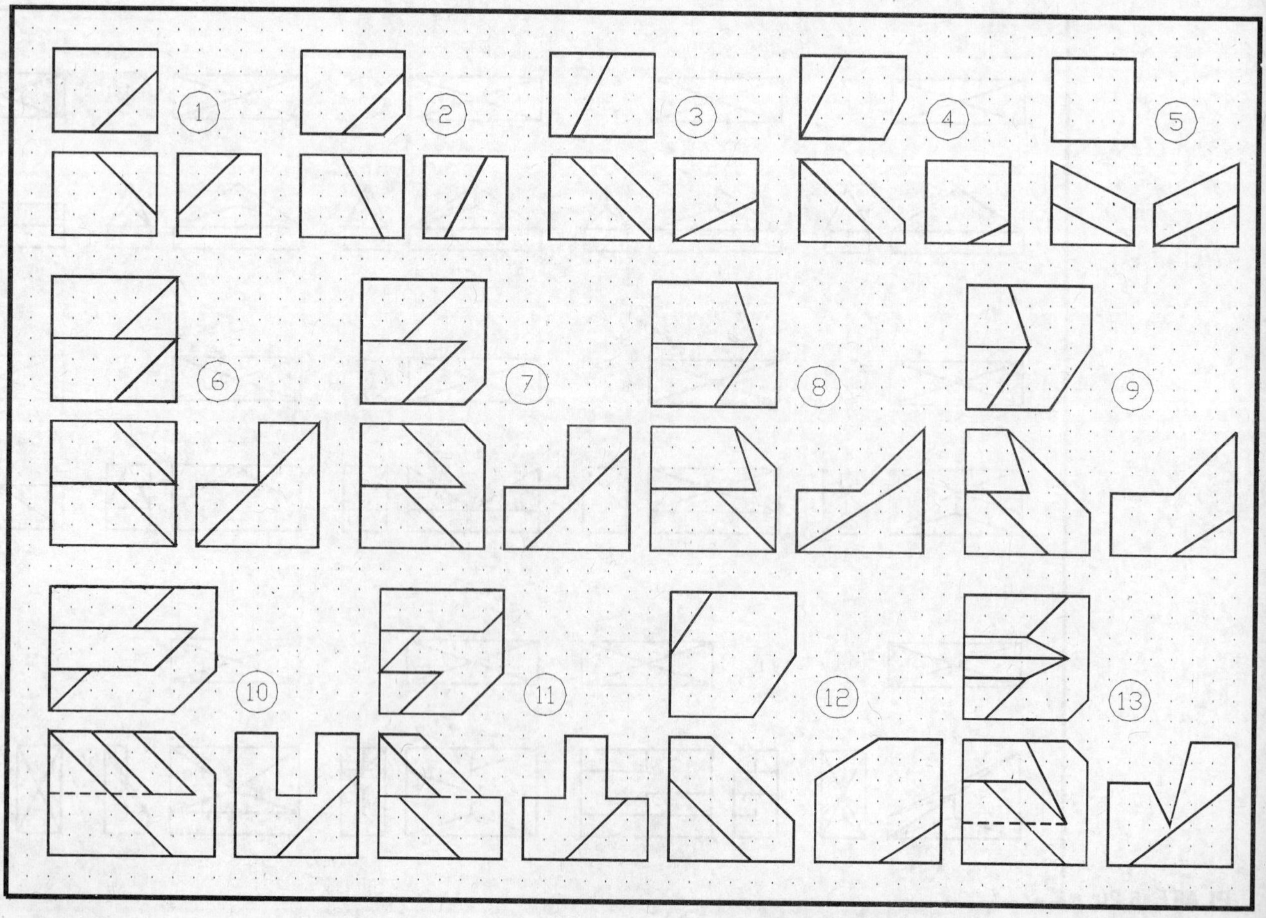


PLATE 5.22

PLATE 5.23

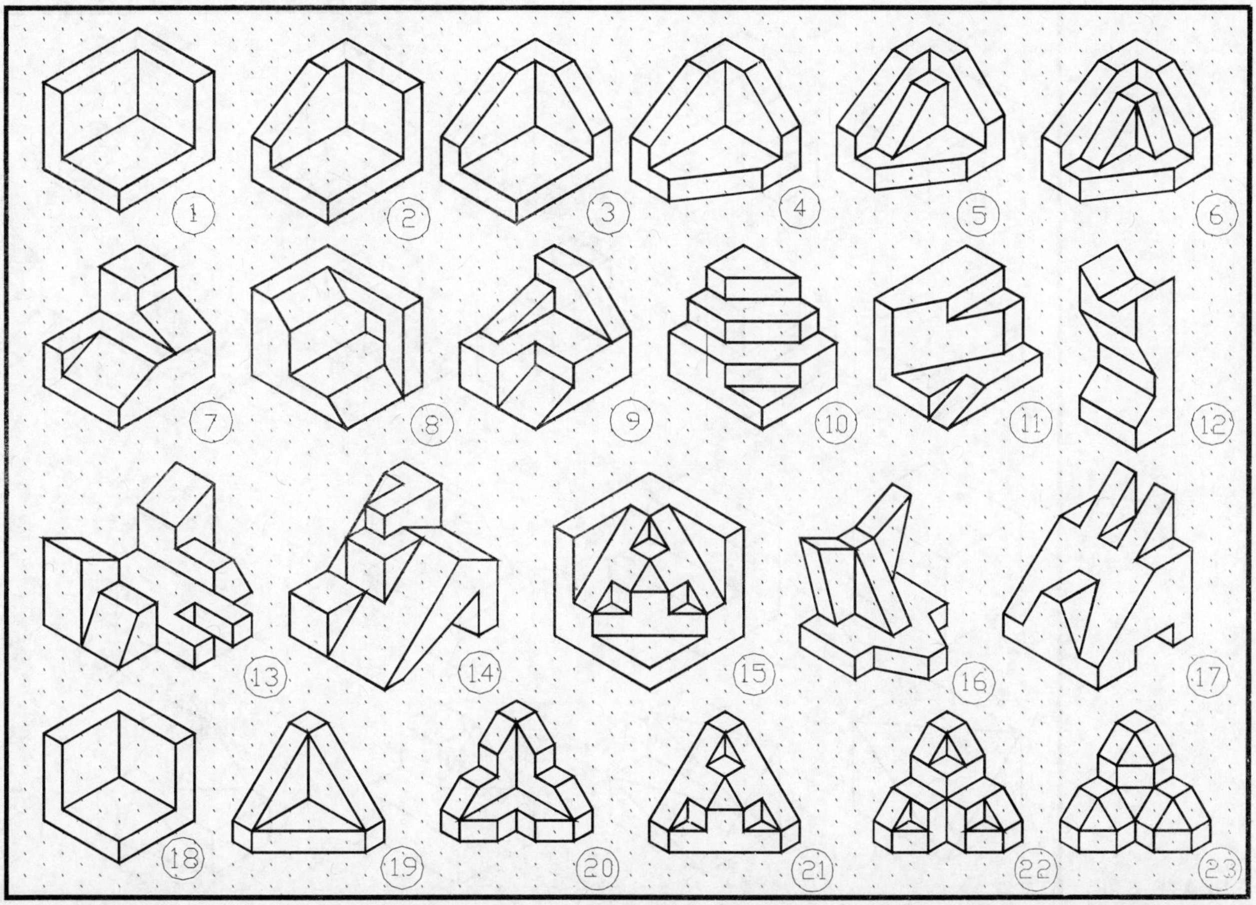


PLATE 5.24

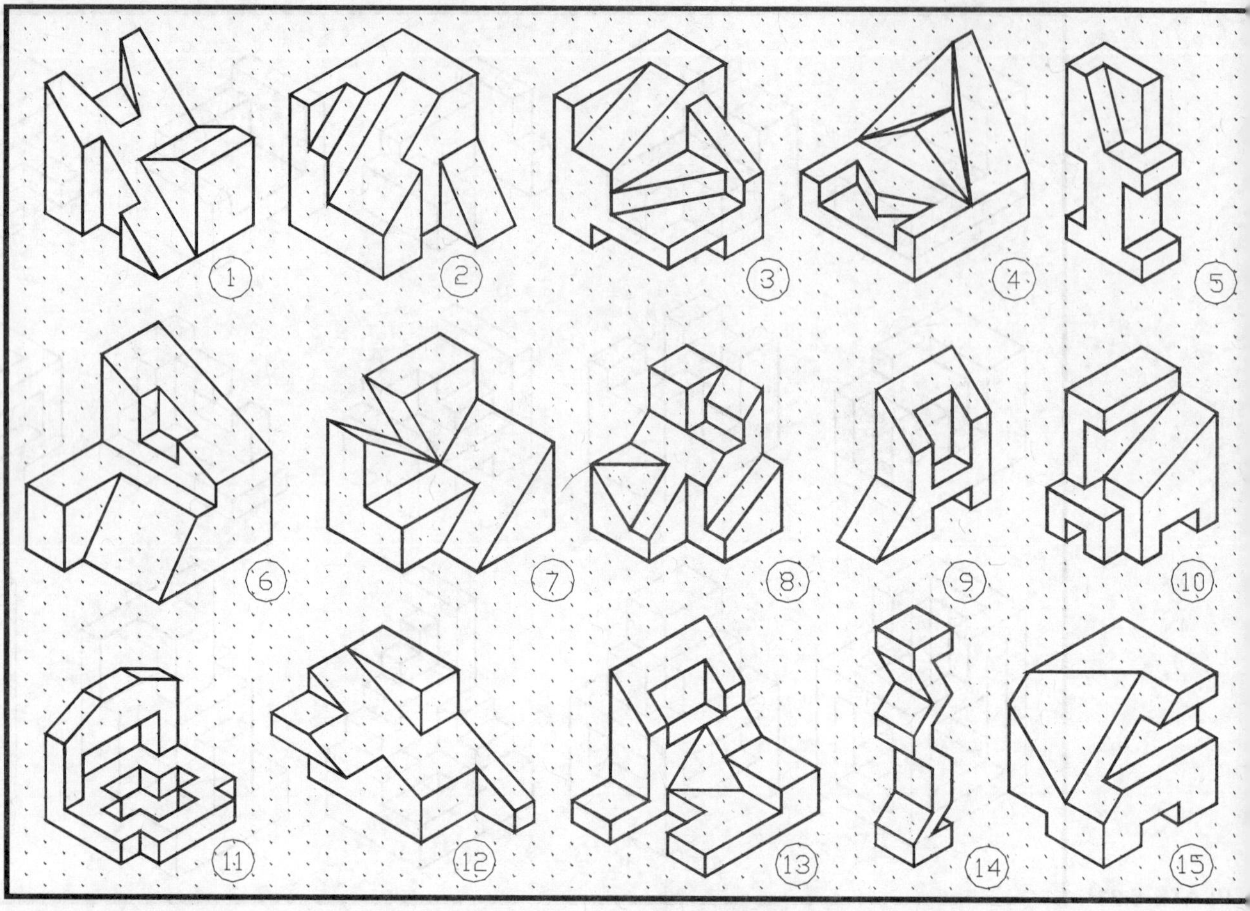


PLATE 5.25

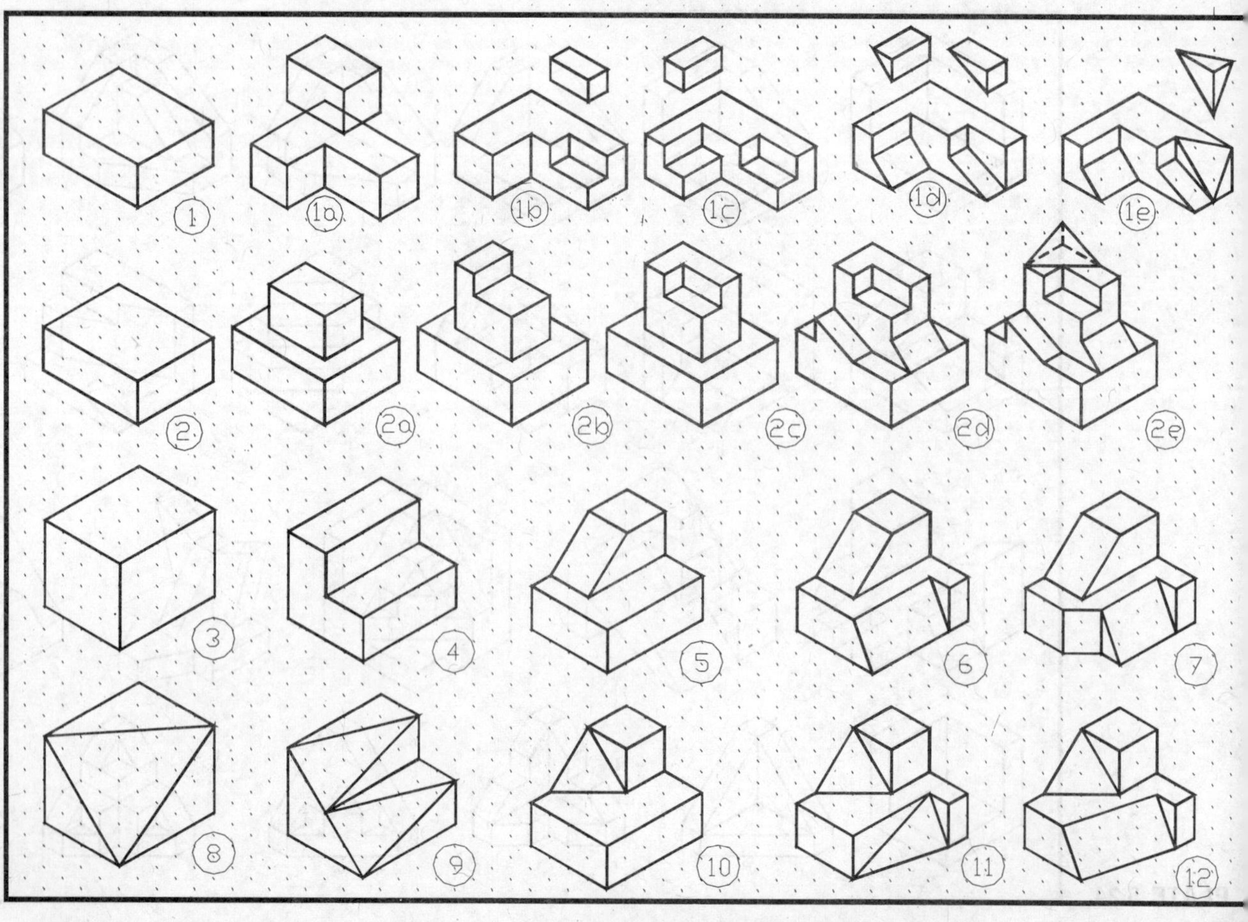


PLATE 5.26

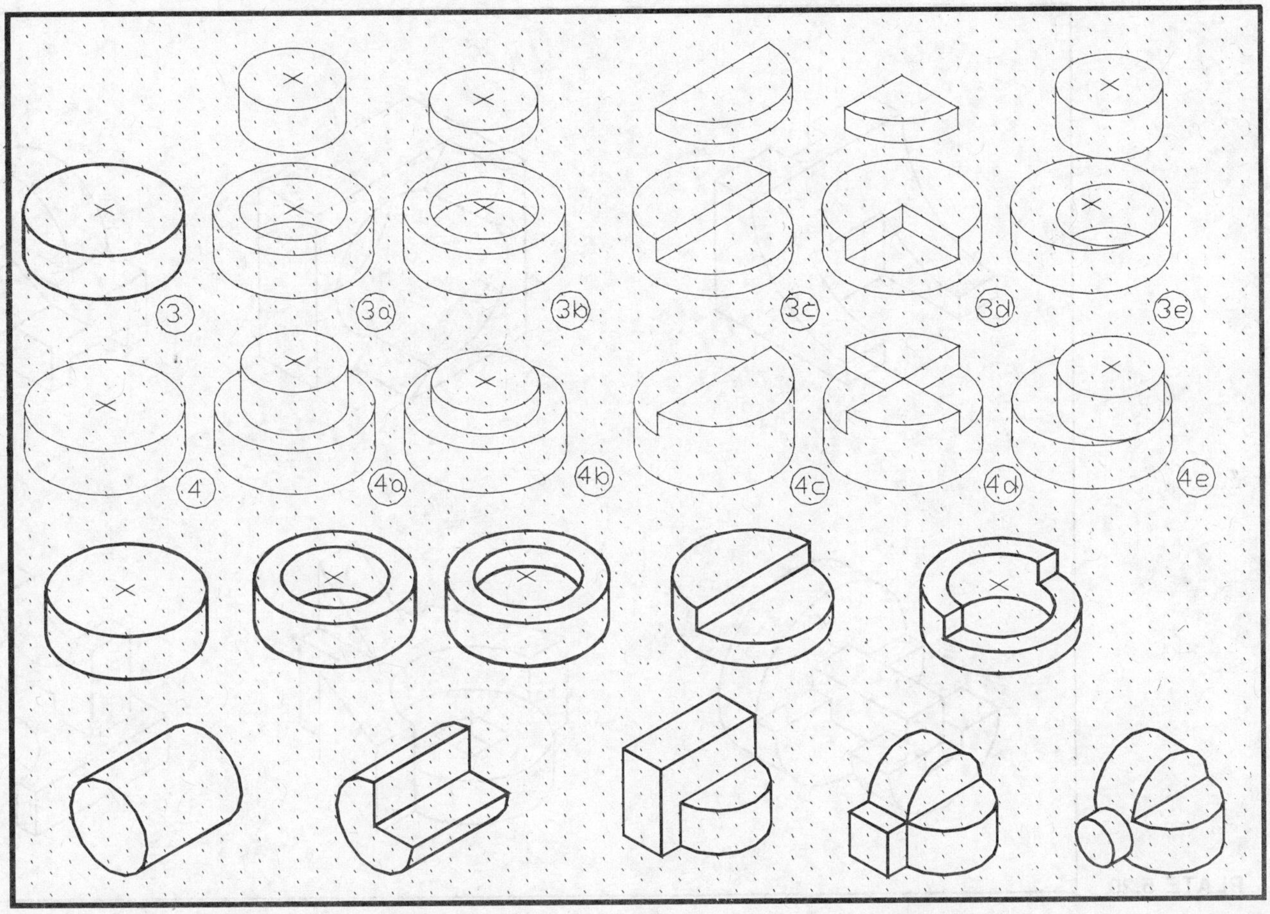


PLATE 5.27

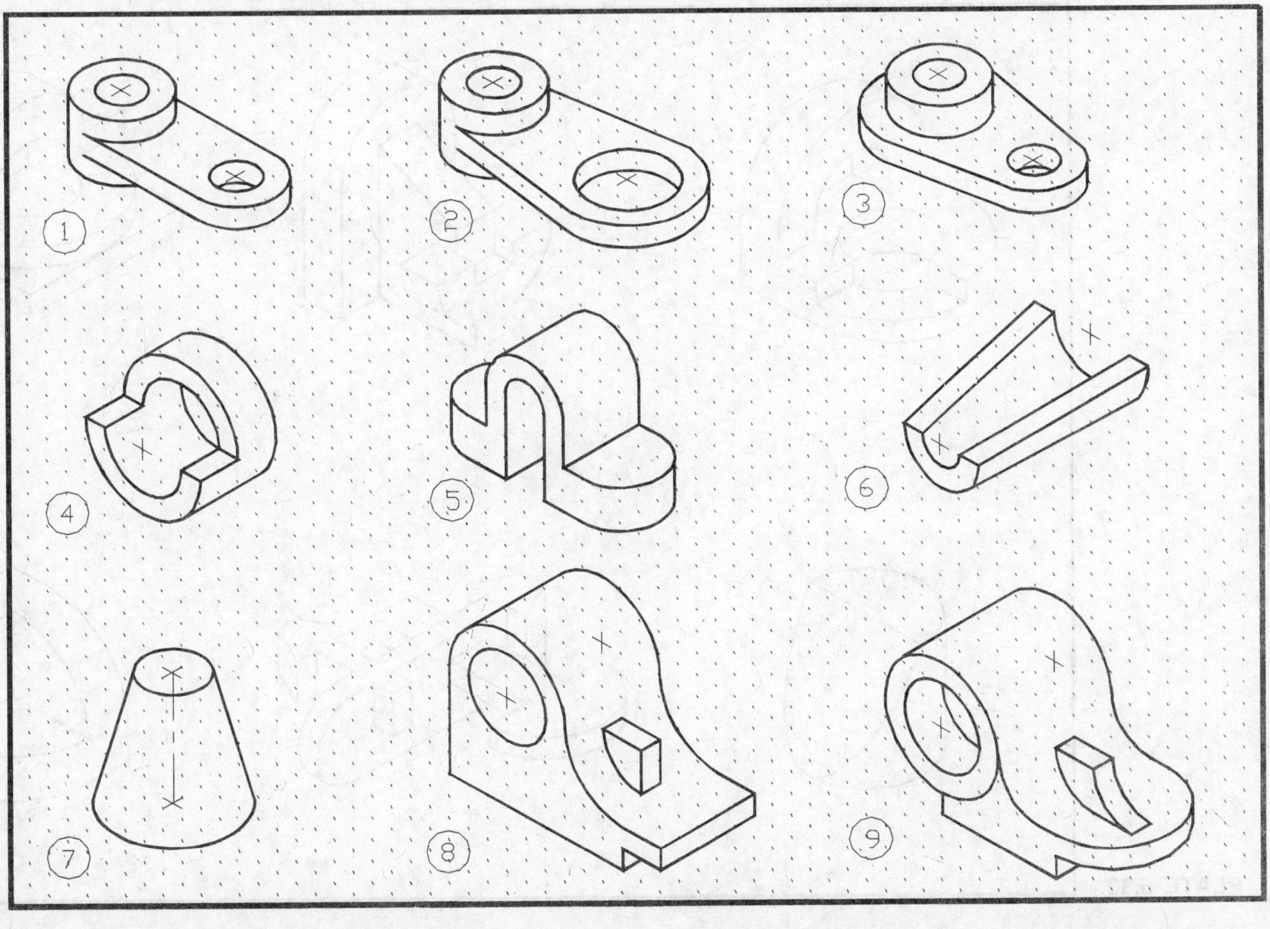


PLATE 5.28

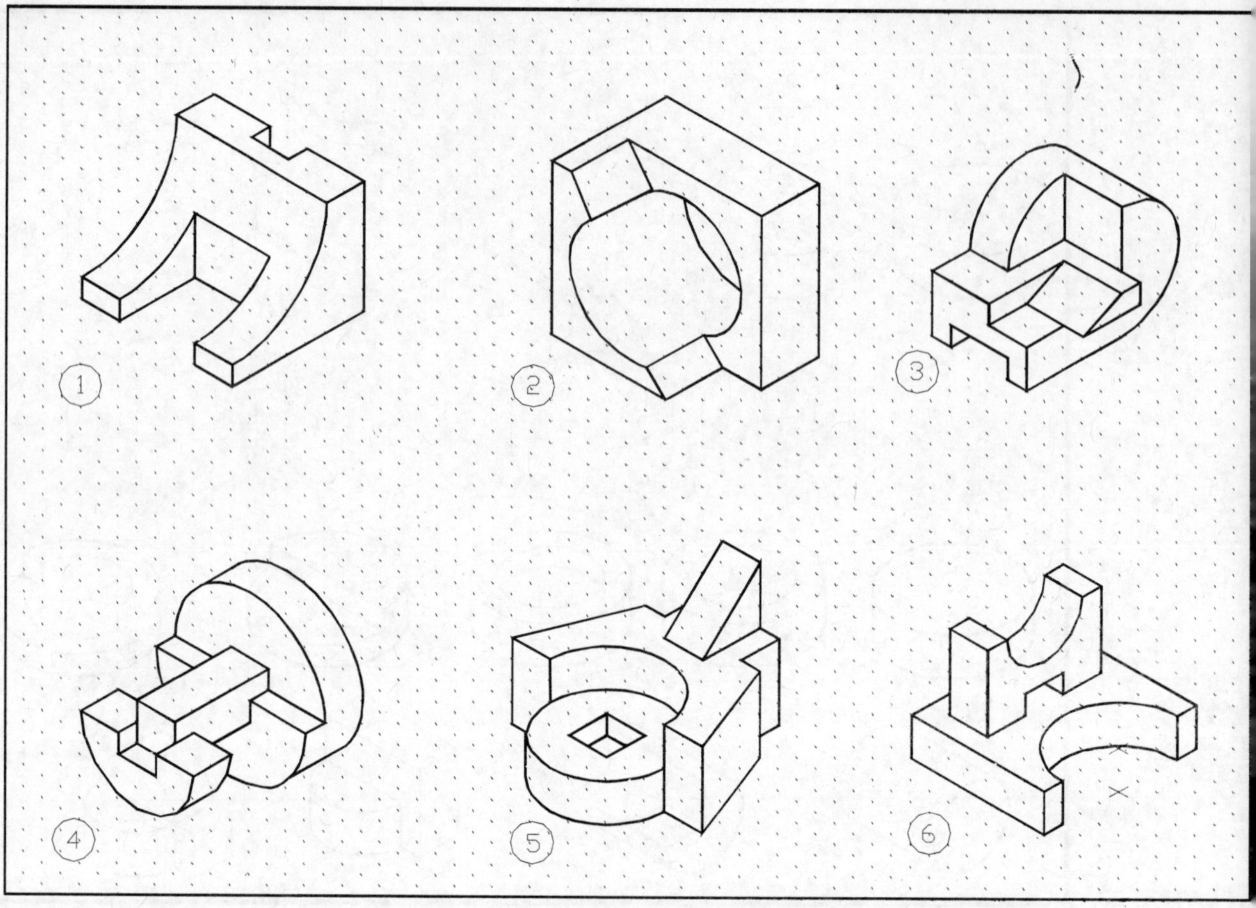


PLATE 5.29

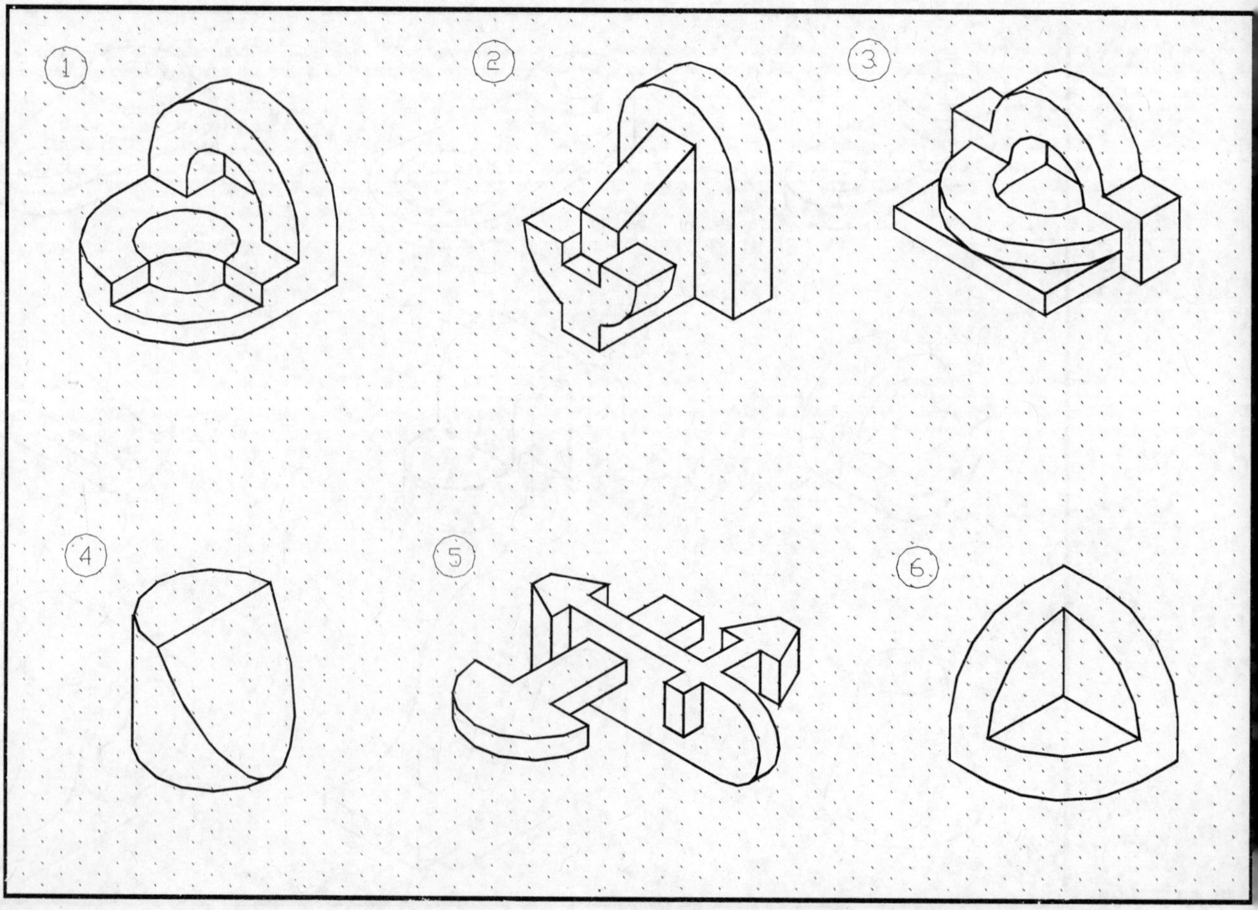


PLATE 5.30

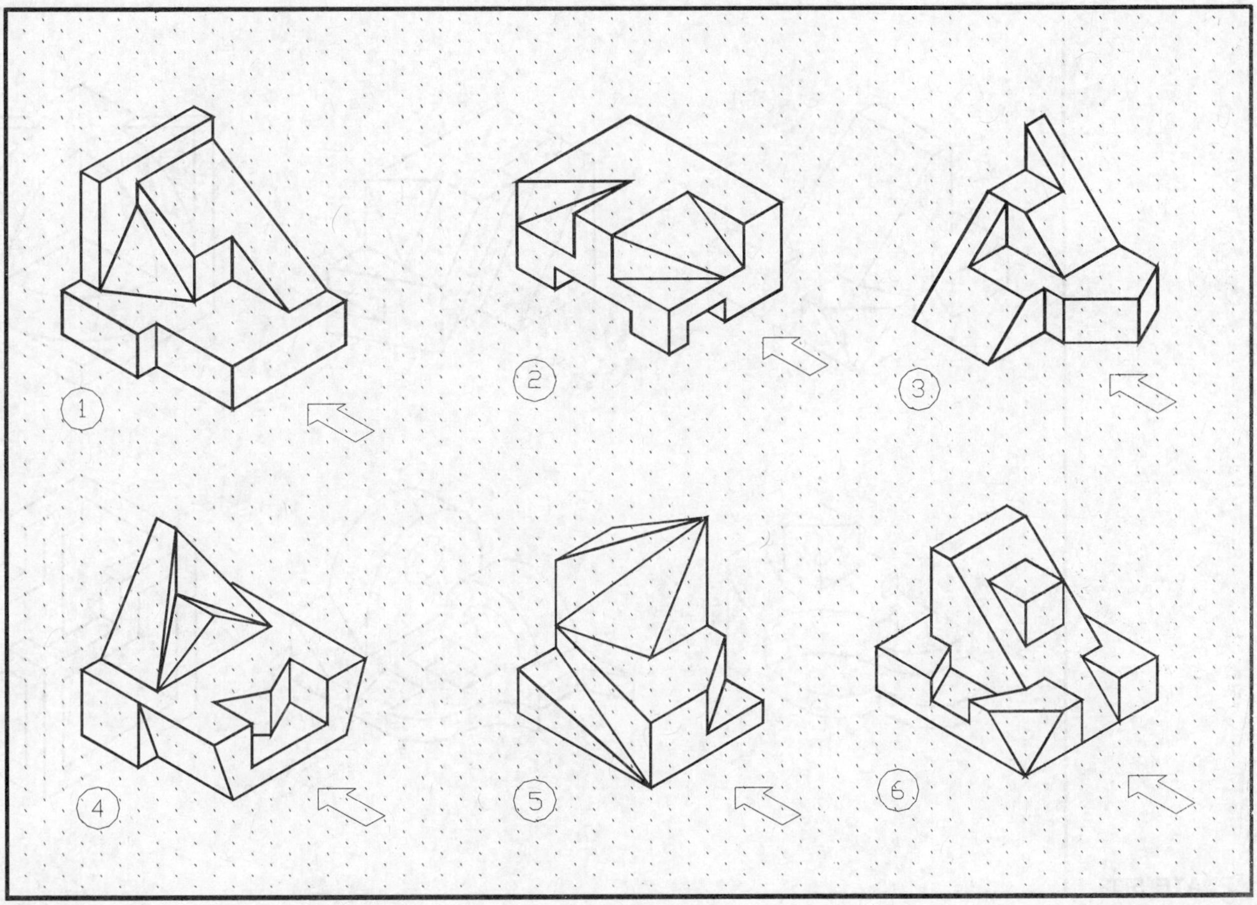


**PLATE 5.31**

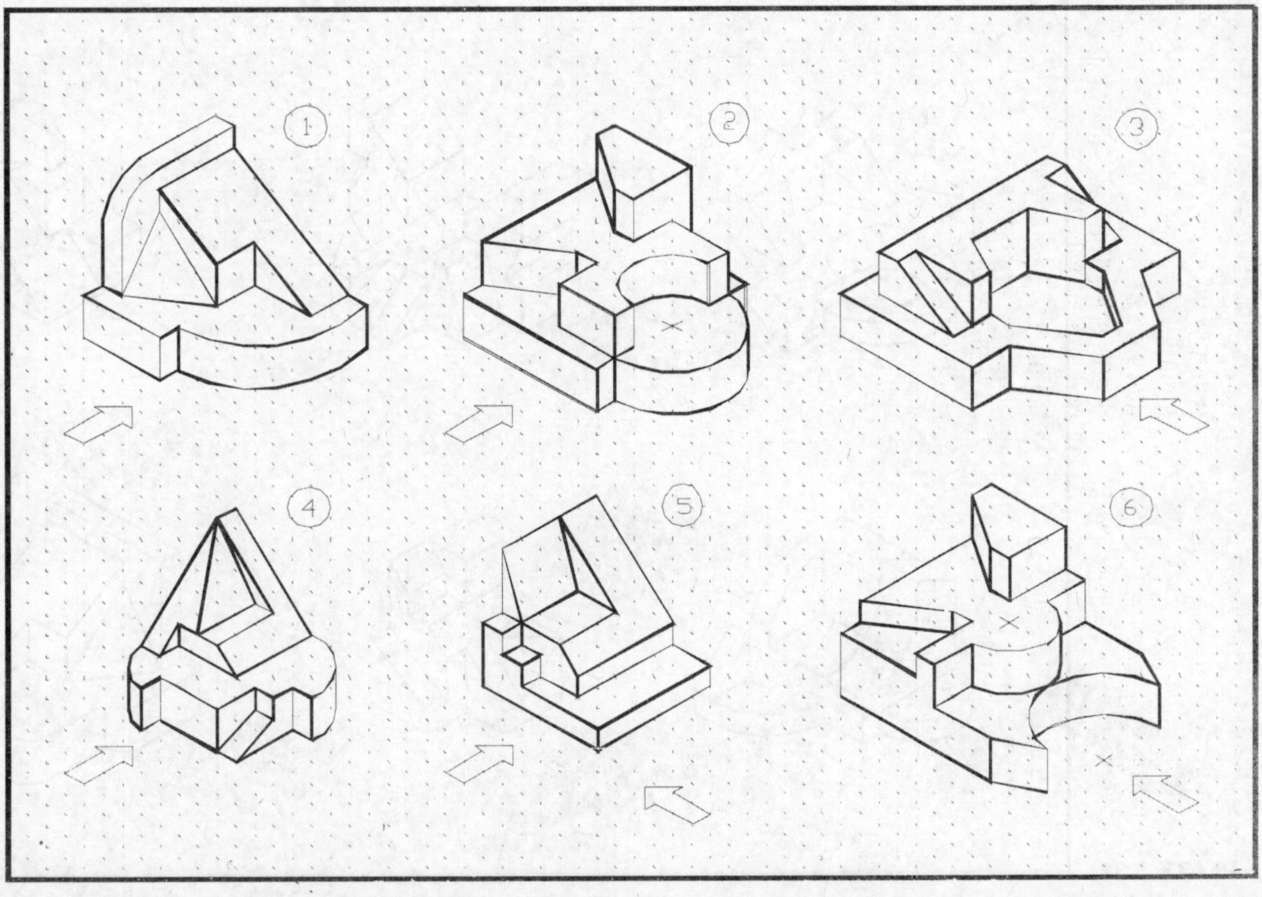


**PLATE 5.32**

PLATE 5.33

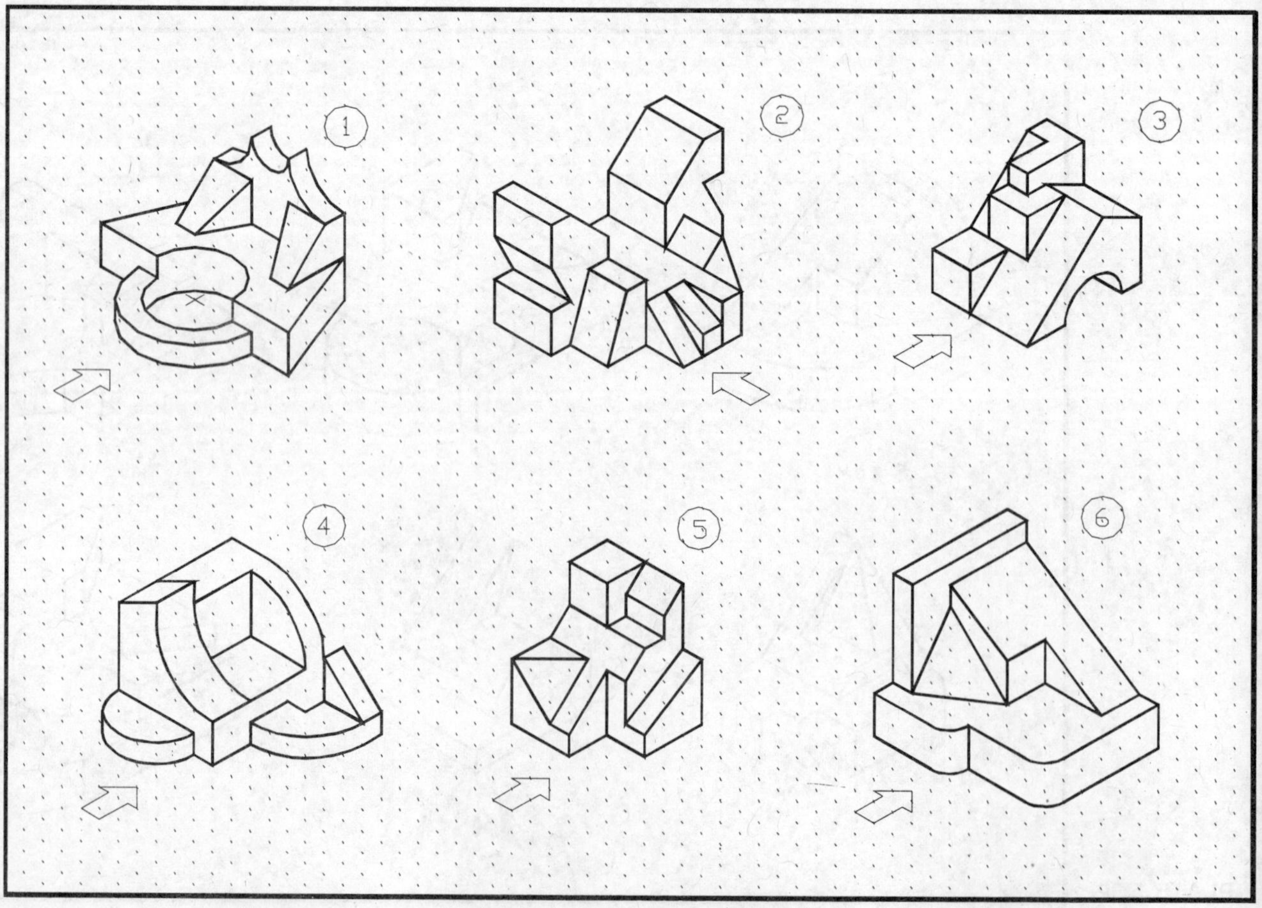


PLATE 5.34

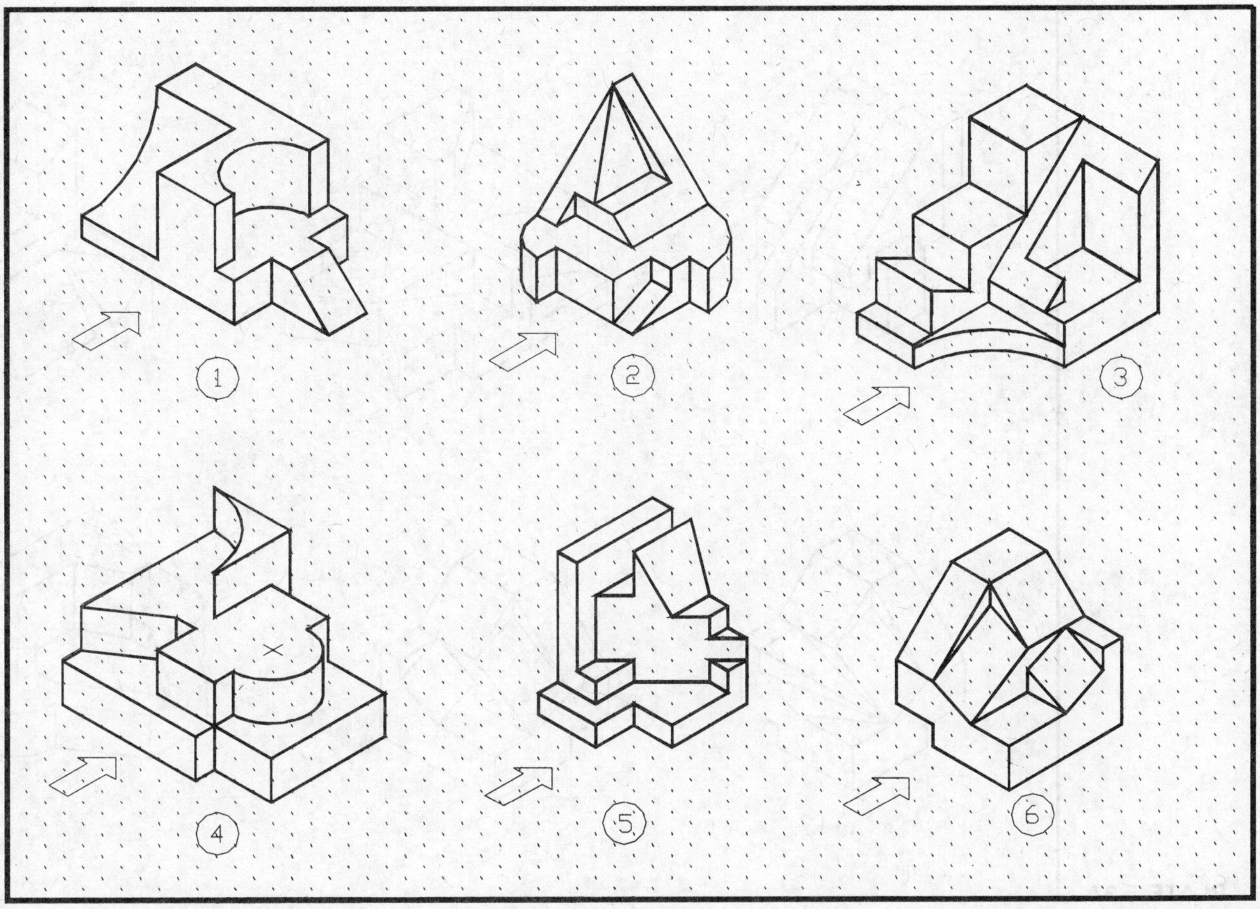


PLATE 5.35

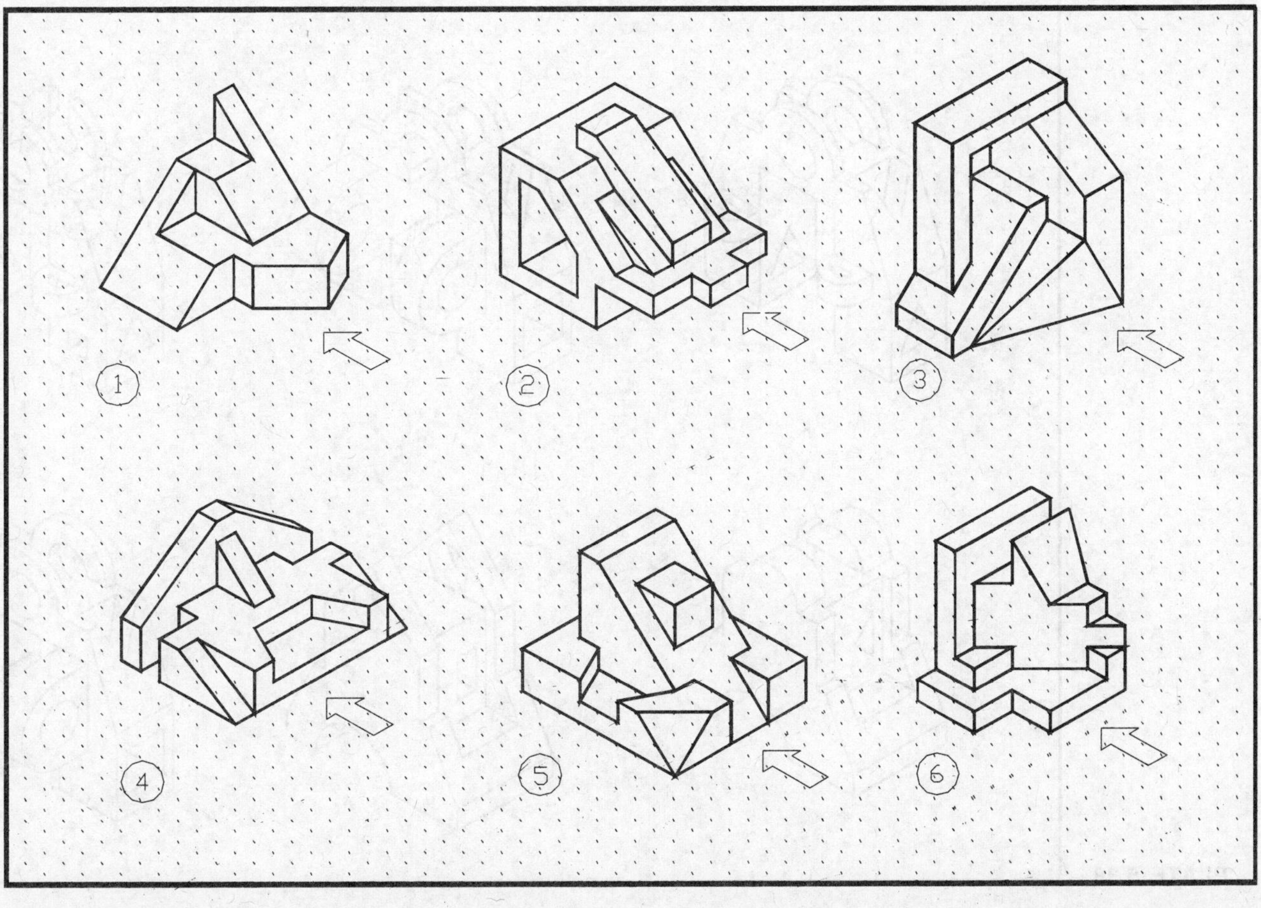


PLATE 5.36

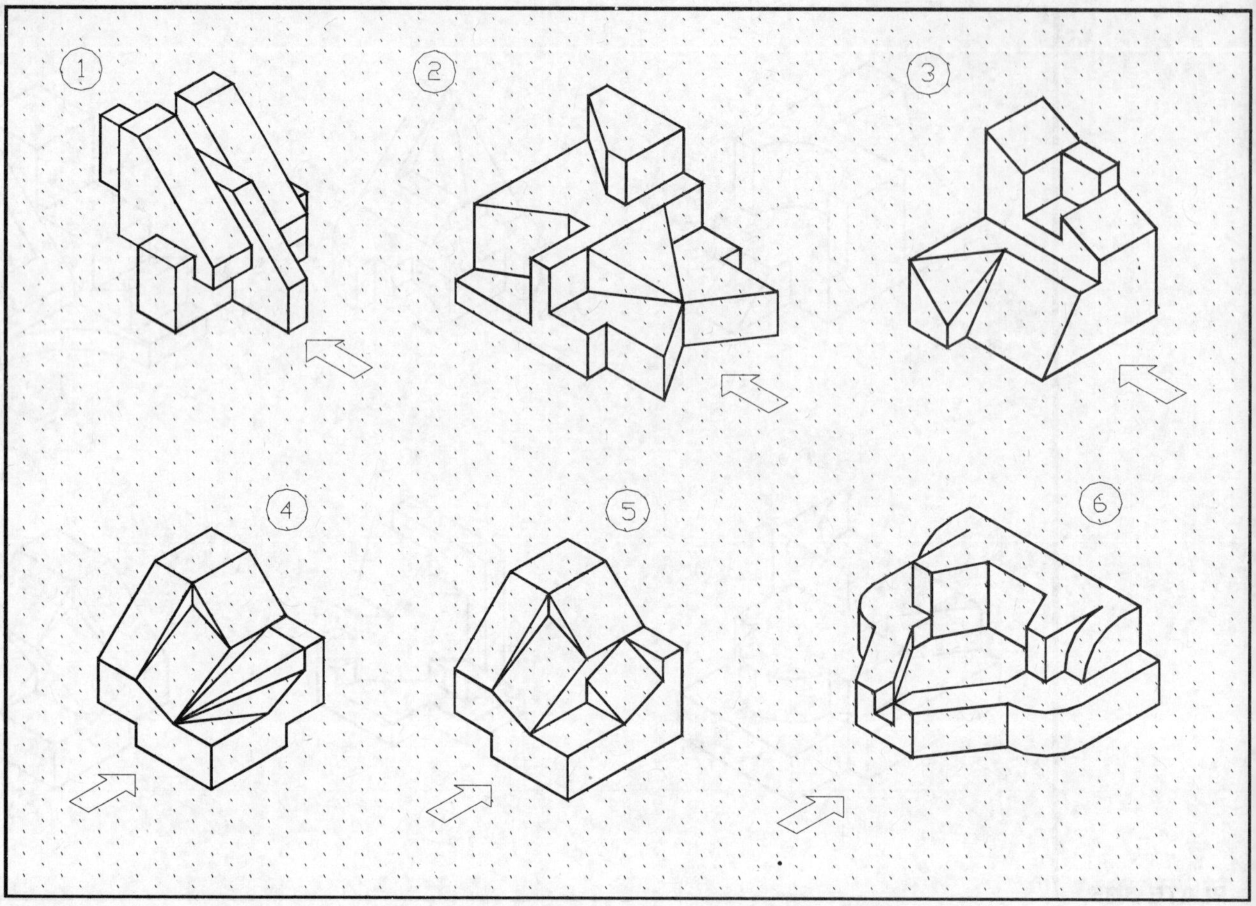


PLATE 5.37

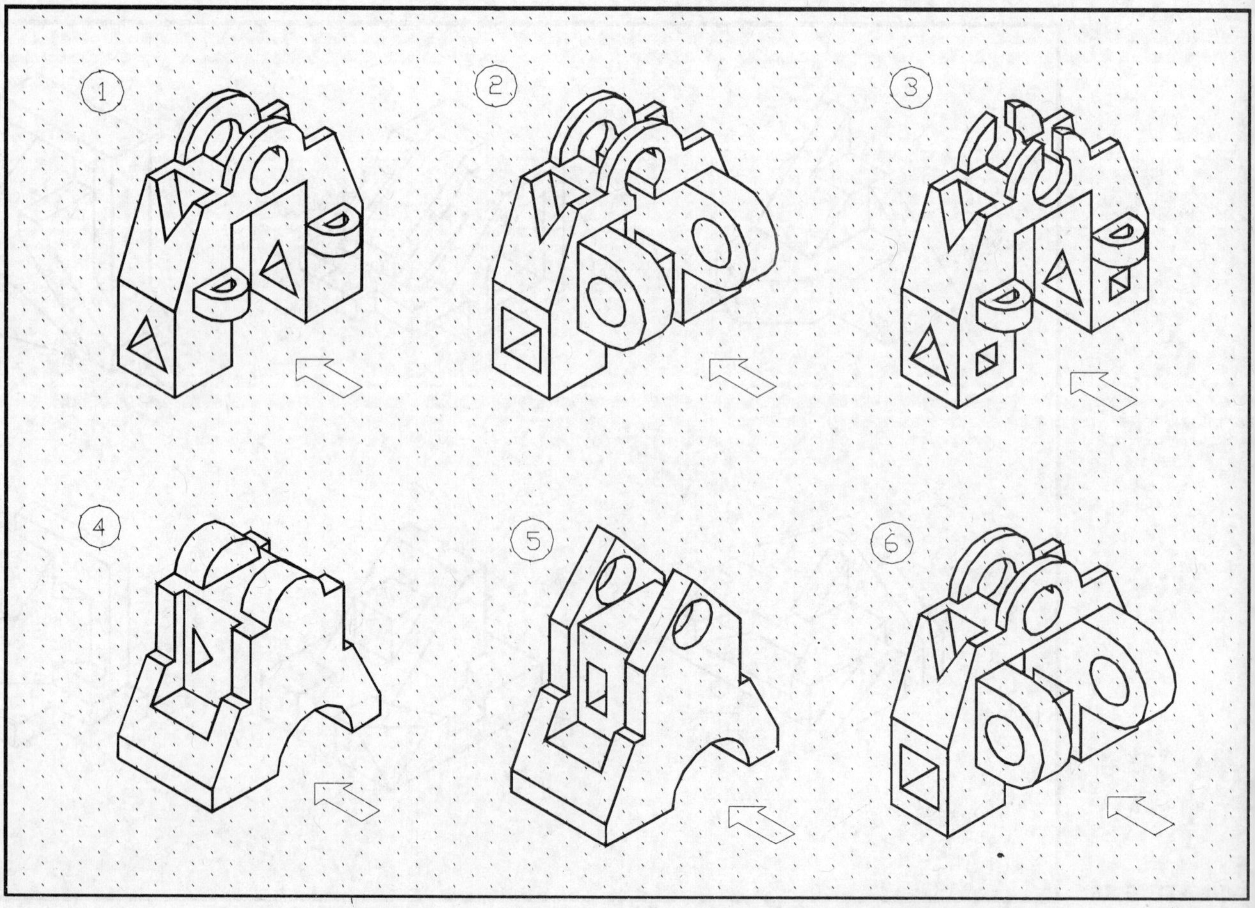


PLATE 5.38

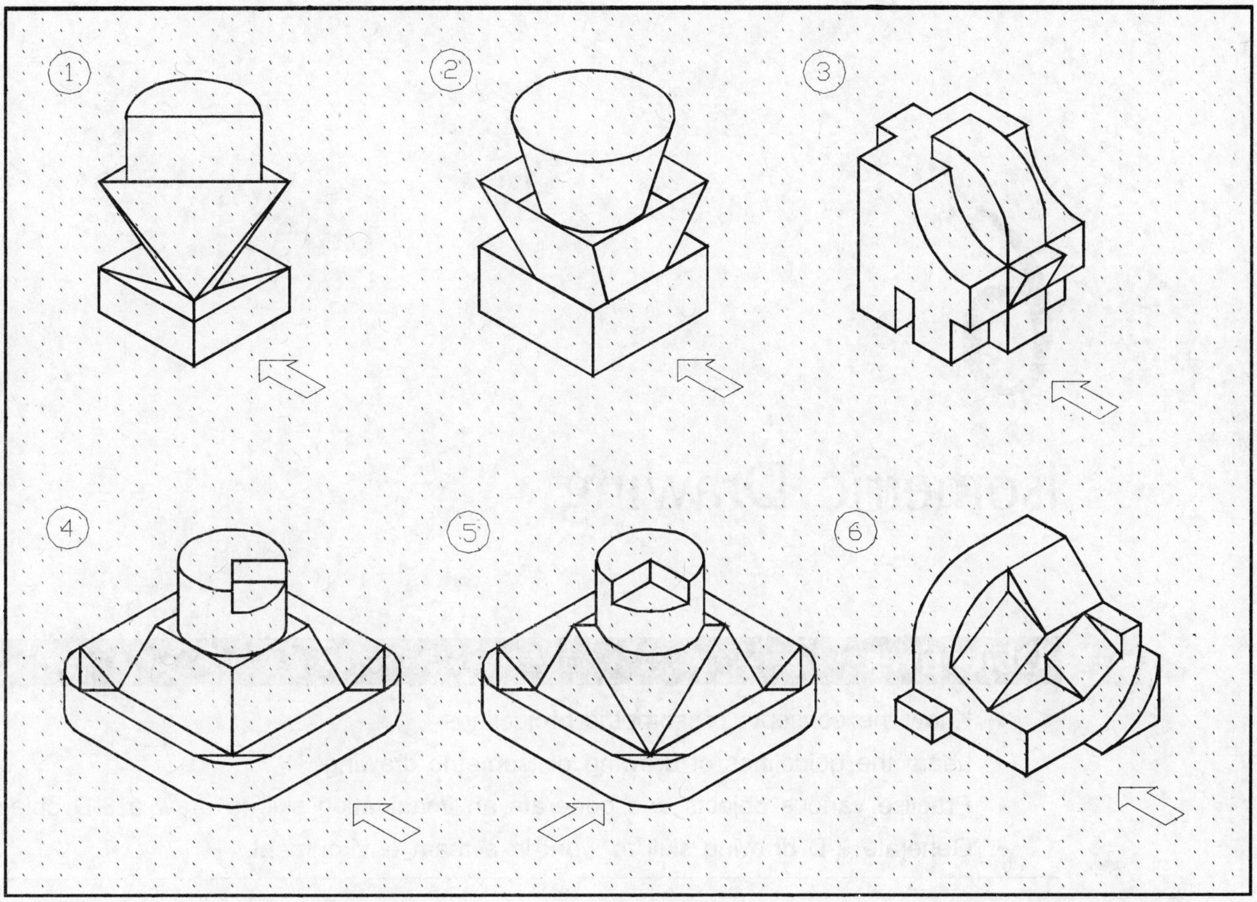


PLATE 5.39

# 6

# Isometric Drawing

## Objectives

- Know the concepts of isometric projections.
- Learn the guidelines of drawing of isometric drawing.
- Practise various objects and generate an imagination skill to draw a 3-D object.
- Generate 3-D drawing skill to work in a team environment.

## Introduction

An orthographic drawing of two or more views describes an object accurately in form and size. Because each of the views shows only two dimensions, however, such a drawing can convey information only to those who are trained in reading the engineering multiview drawings. For this reason, multiview drawings are used mainly by engineers, designers and technologists.

Engineers and technologists often find that they must use pictorial 3-D drawings to convey technical information to persons who do not possess the training necessary to construct an object mentally from multiple views. To make such drawings, several specific schemes of pictorial drawing have been devised that combine the pictorial effect of perspective with the advantage of having the principal dimensions to scale.

## 6.1 Divisions of Pictorial Projection

Pictorial projection is classified in the following three general divisions:

1. Axonometric projection
2. Oblique projection
3. Perspective projection

Perspective methods produce the most realistic drawings, but the necessary construction is more difficult and tedious than are oblique or axonometric. For this reason, engineers customarily use some form of axonometric or oblique projection. Modified axonometric or oblique methods are often used to produce the desired effects.

## 6.2 Divisions of Axonometric Projection

Axonometric projection is a form of orthographic projection. The distinguishing difference is that only one projection plane (the axonometric plane) is used to display the drawing. This type of projection is also called one plane projection. Axonometric projection is an orthographic projection, the only difference is that the object is projected on an axonometric plane, which is inclined to HP, VP and PP.

In multiview drawing, horizontal, vertical, and profile planes are used. The object is turned from its customary position so that the three faces are displayed. Because an object can be placed in an unlimited number of positions relative to the axonometric plane of projection, an infinite number of axonometric views may be drawn. These views will vary in general proportions, lengths and sizes of angles. For practical reasons, a few of these positions have been classified into the following recognised divisions or axonometric projection:

1. Isometric
2. Dimetric
3. Trimetric

## 6.3 Theory of Isometric Projection

In an isometric projection, three dimensions of a solid are shown in one view. This is explained with the help of projections of a cube resting on one corner.

Figure 6.1 shows a TV and FV of a cube placed on the ground with two of its faces (PSWT and SRVW) equally inclined to VP in Stage-I. Corners S and W are nearest to the observer and corners q and u are farthest from the observer. Corners p, t and v, r are at equal distance from the observer. In Stage-II, it is tilted about one corner v on the ground such that the body-diagonal tr becomes parallel to the ground and corresponding projections are drawn. In Stage-III, the object is further tilted about the corner v such that the body-diagonal tr made perpendicular to vp, however, would remain parallel to the ground. The body-diagonal tr becomes a point view in the FV of Stage-III. The resulting FV is an isometric projection (view) of a cube.
Following are the observations from the FV of the Stage-III:

- The top square face of the cube appears to be a rhombus pqrs. Similarly, the remaining five faces also appear as rhombuses.
- Three faces pqrs, qrvu and srvw are equally inclined to VP. These faces are called isometric planes.
- The edges rs, rv and rq are equally inclined to VP and equally foreshortened. These lines which make an angle of 120° with each other are called isometric axes. The lines which are parallel to these axes are called isometric lines.

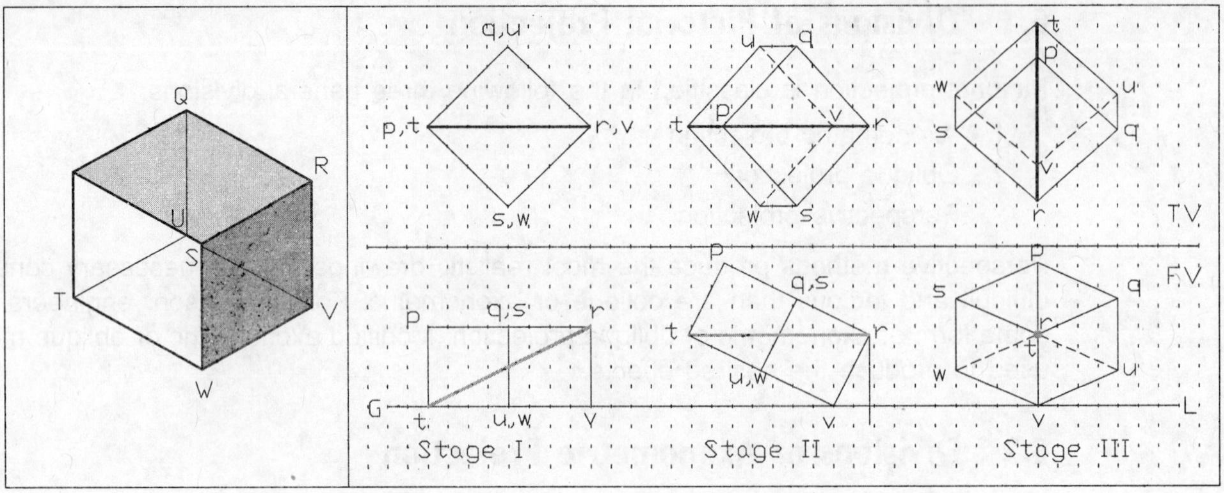


**Fig. 6.1** Theory of isometric projection.

## 6.4 Isometric View and Isometric Drawing

Figure 6.2 shows the rhombus PQRS. The angle <psr is equal to <pqr and equal to 60°. The same figure shows the true view of top face PQRS as PqRs. The intersection of two face diagonals PR and sq is marked as x. The sP is a true length whereas sp is an isometric length.

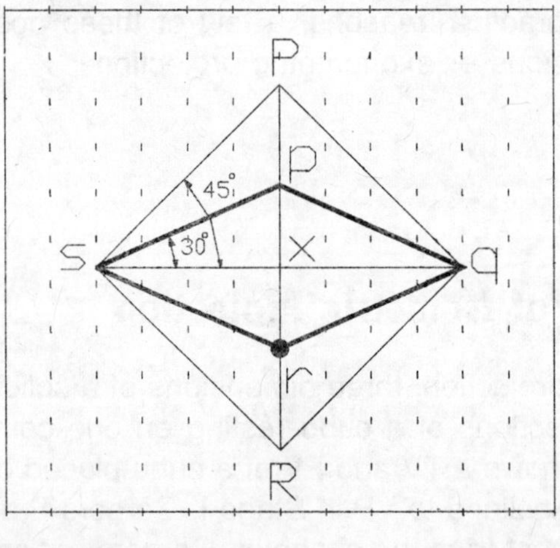


**Fig. 6.2** Isometric projection of a cube.

In triangle psx,

$$sp \cos 30 = sx \tag{6.1}$$

In triangle Psx,

$$sP \cos 45 = sx \tag{6.2}$$

From Eqs. (6.1) and (6.2),

$$sp \cos 30 = sP \cos 45$$

therefore,

$$\frac{sp}{sP} = \frac{\cos 45}{\cos 30} = 0.815$$

The true length is reduced by 81.5% in an isometric projection. Therefore, while drawing an isometric projection, it is necessary to convert the true lengths into isometric lengths for measuring and marking the sizes. This is conventionally done by constructing and making use of an isometric scale.

Objects are seldom drawn in true isometric projection, because the use of an isometric scale (81.5%) is inconvenient and impractical. Instead, a conventional method is used in which all foreshortening is ignored, and the actual true lengths are laid off along isometric axes and isometric lines. To avoid confusion and to set this method apart from true isometric projection, it is called isometric drawing. Figure 6.3 shows TV and FV of a rectangular block. Figure 6.3(a) shows isometric projection of the same obtained by following the procedure shown in Fig. 6.1. Figure 6.3(b) shows isometric drawing with true lengths laid off along isometric lines.

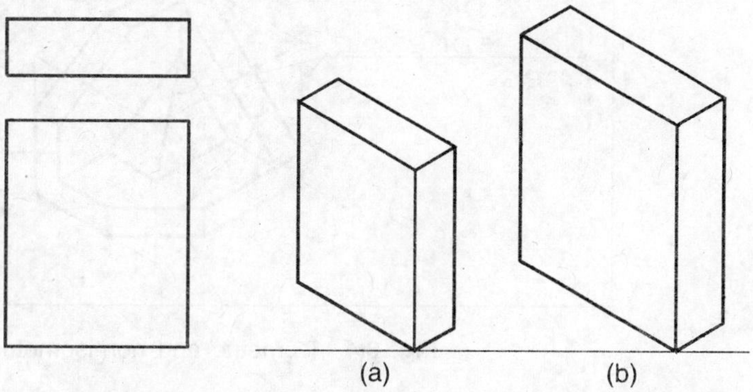

**Fig. 6.3** (a) Isometric projection (b) Isometric drawing.

## 6.5 Advantages of Isometric Projection

Isometric projection is the simplest of these because the principal axes make equal angles ($\alpha = \beta = \gamma$) with the plane of projection and are, therefore, foreshortened equally. Isometric projection is widely used in the industry because of its ease of construction, the ability to measure and compare the dimensions parallel to the isometric axes directly and the ability to combine isometrics from different sources into composite projections.

## 6.6 Disadvantages of Pictorial Drawing

Pictorial drawings, despite certain advantages, however, have disadvantages that limit their use. A few of these are as follows:

- Some pictorial drawings frequently have a distorted and unreal appearance that is disagreeable.
- The time required for execution is, greater than for multiview drawings.
- Pictorial drawings are difficult to dimension.
- Some lines on a pictorial drawing cannot be directly measured.
- Hidden portion cannot be easily imagined unless an object is symmetric.

Even with these limitations, pictorial drawings are used extensively for technical publications, patent office records, piping diagrams and furniture designs. Occasionally they are used, in one form or another, to supplement and clarify machine and structural details that would be difficult to visualise through multiview drawing.

## 6.7 Non-isometric Lines

The lines which are parallel to the isometric axes ($x$, $y$ and $z$) are called isometric lines. Isometric lines that are parallel to the isometric axes can be directly measured.

Those lines that are inclined and are not parallel to the isometric axes are called non-isometric lines. A line of this type cannot be measured directly. Its position and projected length must be established by locating its endpoints. The lines that are parallel on an object are parallel in the pictorial view. Conversely, lines that are not parallel on the object are not parallel in the pictorial view. Figure 6.4 shows the isometric and non-isometric lines in an object along with *x*-, *y*- and *z*-directions.

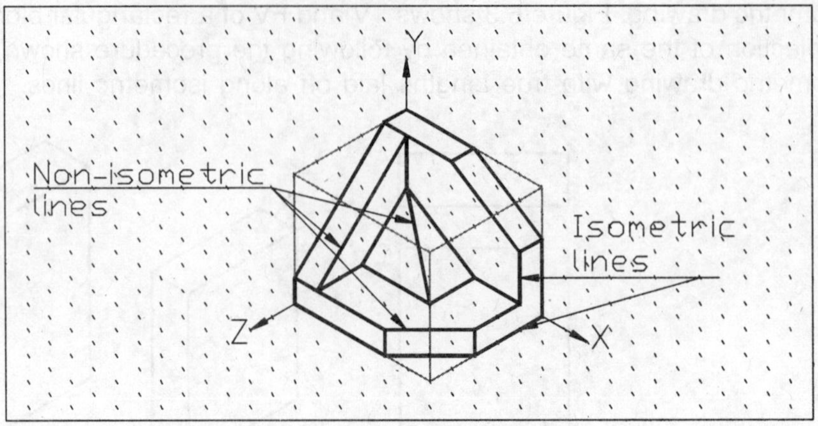


**Fig. 6.4** Isometric and non-isometric lines.

## 6.8 How to Draw an Isometric Drawing?

The following sections discuss on setting of AutoCAD screen for isometric drawing, methods to generate an isometric drawing and few examples.

### 6.8.1 Setting of AutoCAD Screen

First you have to set the AutoCAD screen suitable to draw isometric drawing. This can be done by using a SNAP command.

**Command: SNAP** [Enter]
Snap spacing or ON/OFF/Rotate/Style <0.250>: S (for style) [Enter]
Standard/Isometric <S>: I (for Isometric) [Enter]
Vertical spacing <0.50>: 0.25 [Enter]

Figure 6.5 shows a suitable grid screen with grids running along the principal axes: length L is taken along principal axis X, depth D is taken along principal axis Z, and height h is taken along principal axis Y. The position of X and Z can be interchanged depending upon the FV direction. The normal orthographic cursor mode is converted into an isometric cursor mode as shown in Fig. 6.6.

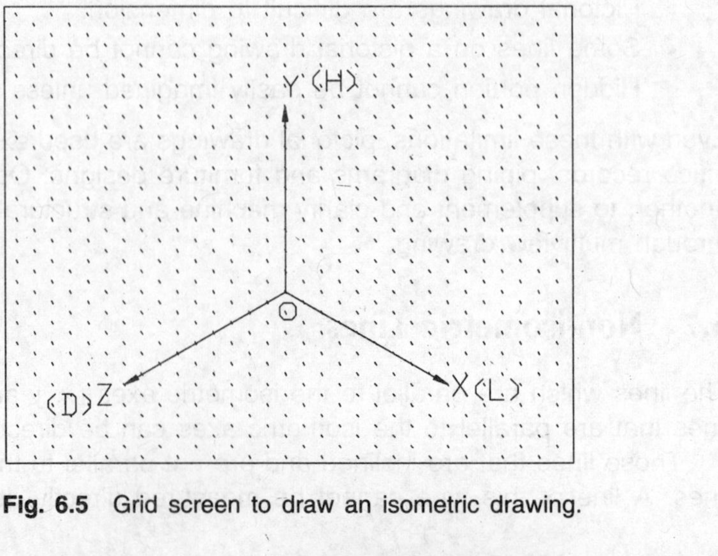


**Fig. 6.5** Grid screen to draw an isometric drawing.

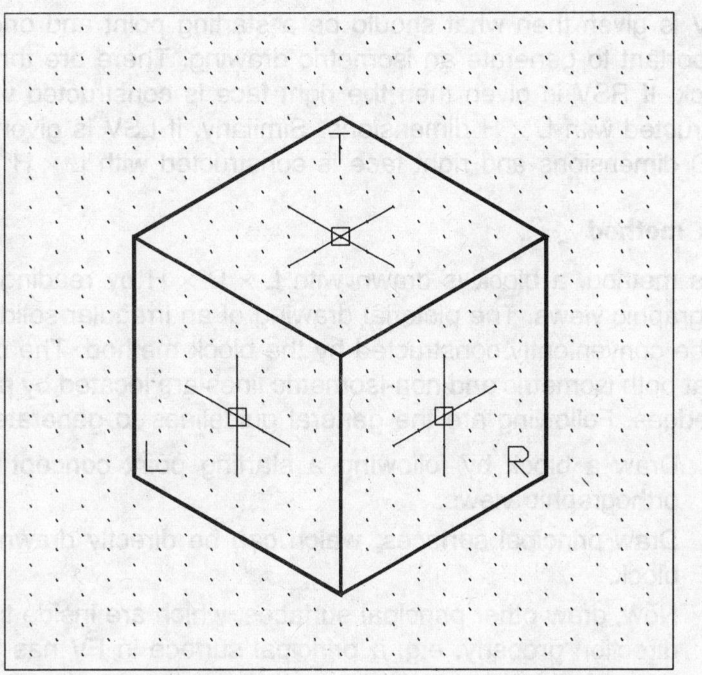


**Fig. 6.6** Three cursor mode for drawing on three isometric planes.

Figure 6.6 shows a box with three faces: left L, right R and top T, which are called isometric planes. The isometric cursor has three orientations: left cursor mode, right cursor mode and top cursor mode to work on respective planes and along the normal to that plane. This cursor mode can be changed by pressing Ctrl and E keys at once.

### 6.8.2 Methods to Generate an Isometric Drawing

There are two methods to generate an isometric drawing:

(a) Block method and

(b) Coordinate axes method.

Basic requirement to generate an isometric drawing is a skill to read the orthographic views. Before we discuss these methods, let us clear a 'starting point' concept. Figure 6.7 (a) shows: if RSV is given then what should be a starting point and orientation of block and similarly Fig. 6.7(b) shows:

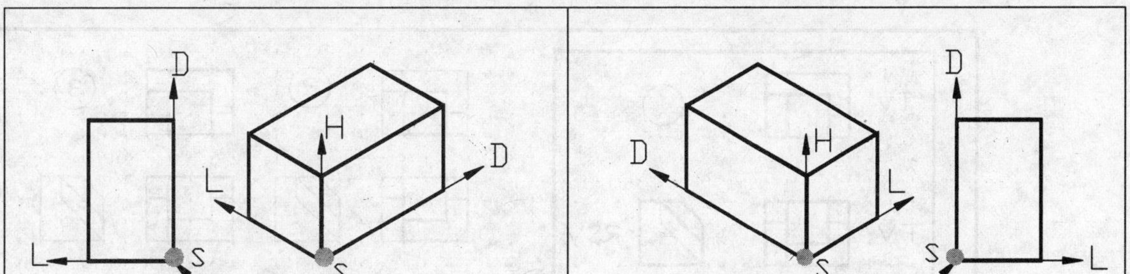


(a) RSV is given (b) LSV is given

**Fig. 6.7** Starting point concept.

if LSV is given then what should be a starting point and orientation of block. The starting point S is important to generate an isometric drawing. There are three visible faces (left, top and right) of a block. If RSV is given then the right face is constructed with H × D dimensions and left face is constructed with L × H dimensions. Similarly, if LSV is given then the left face is constructed with H × D dimensions and right face is constructed with L × H dimensions.

### Block method

In this method, a block is drawn with L × D × H by reading length L, depth D and height H from orthographic views. The pictorial drawing of an irregular solid containing several non-isometric lines may be conveniently constructed by the block method. The object is enclosed in a rectangular box so that both isometric and non-isometric lines are located by points of contact with the block surfaces and edges. Following are the general guidelines to generate an isometric drawing:

1. Draw a block by following a starting point concept and with L × D × H measured from orthographic views.
2. Draw principal surfaces, which can be directly drawn on top, left and right surfaces of the block.
3. Now, draw other principal surfaces, which are inside the block by locating them along a third direction properly, e.g. a principal surface in FV has length L and height H dimensions. To locate inside the block, its depth is required to be read. Similarly, a principal surface in TV has length L and depth D dimensions. To locate inside the block, its height is required to be read. Similarly, a principal surface in SV has height H and depth D dimensions. To locate inside the block, its length is required to be read. This type of reading skill is required to be generated.
4. The inclined surfaces and skew surfaces cannot be drawn directly. Draw the non-isometric lines and connect them to draw inclined and skew surfaces.
(These guidelines generate 80% of isometric view).
5. While doing this make a partly drawn 3-D figure as a question and orthographic view as the answer and satisfy the orthographic view simultaneously.
6. Check for: (a) satisfaction of all the surfaces and invisible line(s) if any and (b) at least three lines must run through a visible point.

The students are advised to follow this method as it takes less time than the coordinate axes method.

**EXAMPLE 6.1** Figure 6.8(1) shows orthographic views of a general object. Draw an isometric drawing.

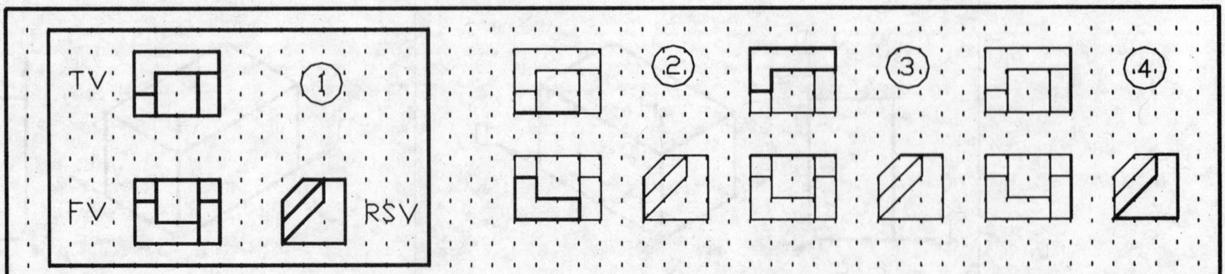


Fig. 6.8

***Solution***

**Steps:** The following steps are correlated with the corresponding steps of Figs. 6.8 and 6.9, simultaneously.

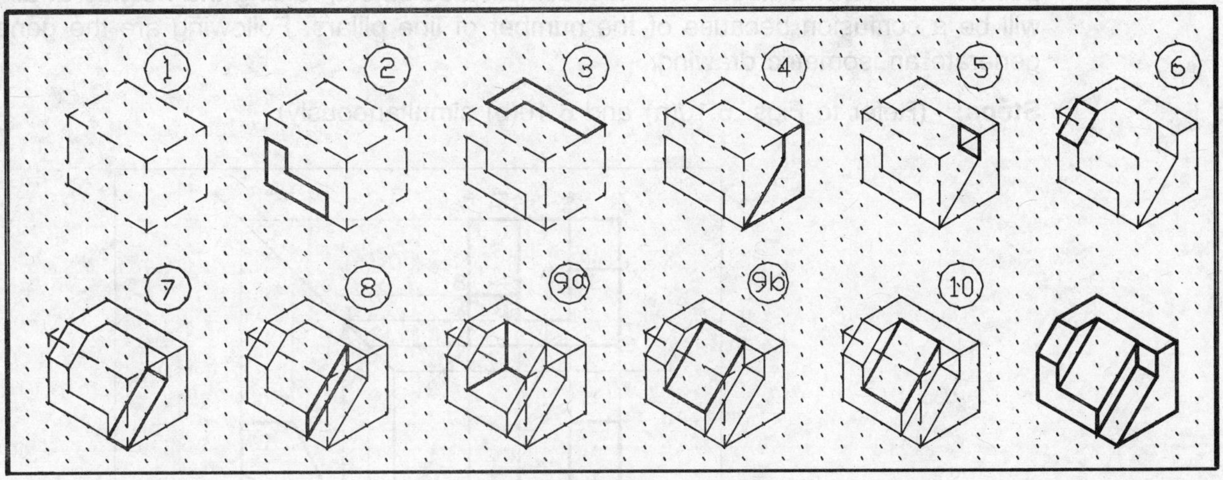


**Fig. 6.9**

1. Read length and height from FV and depth from RSV or TV. The dimensions L × D × H are 4 × 3 × 3 grid spacing. Draw a block of 4 × 3 × 3 grid spacing with proper starting point (i.e., the FV on left face of 4 × 3 grid spacing and RSV on right face of 3 × 3 grid spacing. This is shown in Step (1), Fig. 6.9.
2. In the FV, there is an L-shape surface, which represents a line view in the TV and RSV. The position of this principal surface is directly on the left-face of a block as shown in Step (2), Fig. 6.9.
3. In the TV, there is an L-shape surface, which represents a line view in the FV and RSV. The position of this principal surface is directly on the top-face of a block as shown in Step (3), Fig. 6.9.
4. In the RSV, there is one odd shape surface, which represents a line view in the FV and TV. The position of this principal surface is directly on the right-face of a block as shown in Step (4), Fig. 6.9.
5. In the FV, there is a square surface, which represents a line view in the TV and RSV. The line view is two-grid spacing along depth direction. Draw this principal surface as shown in Step (5), Fig. 6.9.
6. In the FV, there is one more square surface, which represents a square surface in the TV and an inclined line view in the RSV. Draw this inclined surface as shown in Step (6), Fig. 6.9.
7. In the FV, there is a rectangular-shape surface, which represents a rectangular surface in the TV and a line view in the RSV. Draw this inclined surface in a box as shown in Step (7), Fig. 6.9.
8. In the RSV, there is a parallelogram-shape surface, which represents a line view in the TV and FV. Draw this principal surface in a box as shown in Step (8), Fig. 6.9.
9. Now, a square surface is left in the TV and FV to incorporate in the block. To satisfy these square surfaces in the TV and FV, three lines can be drawn as shown in Step (9a). However, this object is not a valid object.

   These two square surfaces represent a line view in the RSV, which is inclined. Therefore, draw an inclined surface in the block as shown in Step (9b). The final object is shown in Step (10), Fig. 6.9.

**Coordinate axes**

In this method, heights at all key points of TV are raised, e.g. a classroom has four corners (key points) on a floor (TV or plan). Raise the height (or erect four pillars) at four points. Join the top points. It will form a classroom. In a complicated object, raising the heights at all key points of TV will be a confusion because of the number of line pillars. Following are the general guidelines to generate an isometric drawing:

**Steps:** (Refer to Figs. 6.10(a) and 6.10(b) simultaneously)

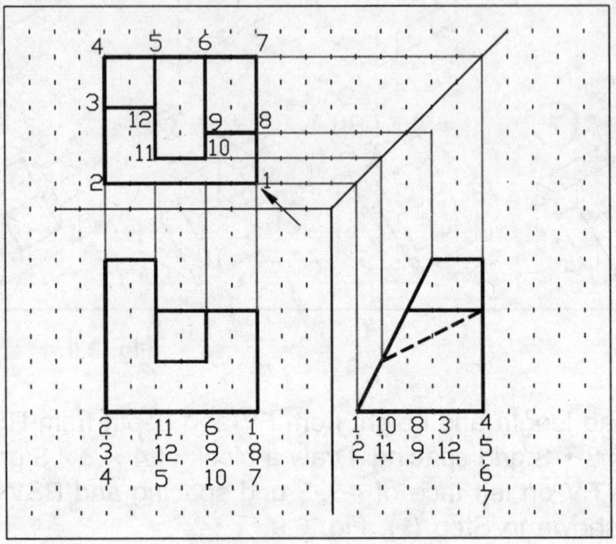

(a) Orthographic views

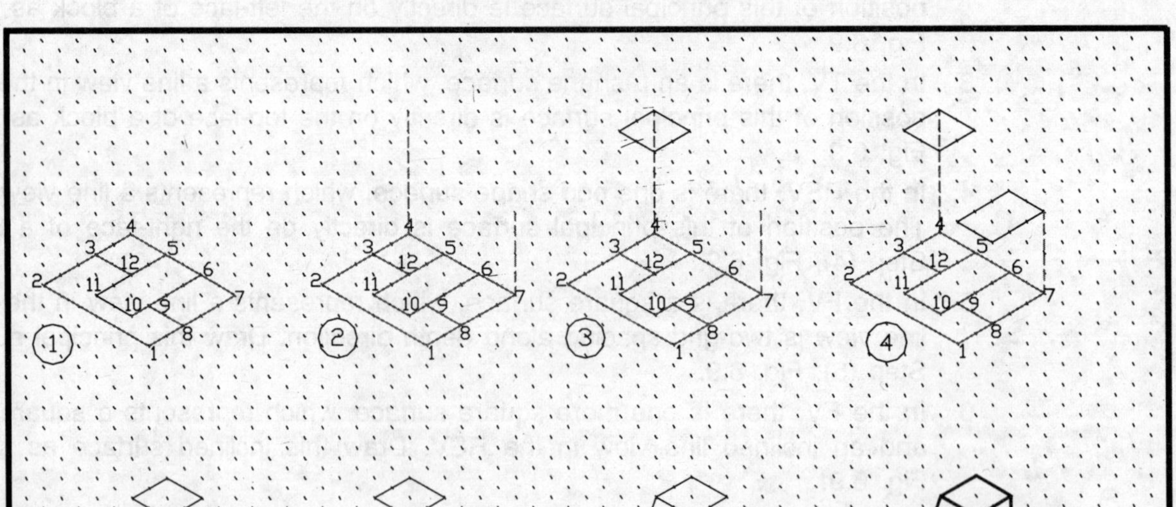

(b) Isometric drawing using a coordinate axes method

**Fig. 6.10** Coordinate axes method.

1. Draw a TV in correct orientation following a starting point concept.
2. Number the key points in TV.
3. First read the heights (in terms of grid spacing) at corner points as the corner points define boundaries of an object. The height is decided by FV and SV. (If there are two possibilities of heights at certain points then a vertical line results joining the two possible heights at the same key points. The raised heights are joined by reading the numbered key points of orthographic TV.
4. Now draw principal surfaces if any. This reduces raising the number of heights.
5. For the remaining points again read and raise the heights.
6. Follow steps (5) and (6) of the block method for final check and complete the isometric drawing.

Figure 6.10(a) shows the orthographic views of an object. Figure 6.10(b) shows the generated isometric drawing by following the steps given for coordinate axes method. Figure 6.10(b) is self-explanatory, which explains the procedure given for the coordinate axes method. The method is simple however laborious and takes more time to generate an isometric drawing.

The authors would like to mention here that the 'Engineering Graphics' is not only an engineering and science but it is also an "art of imagination". However, the engineering and science can be taught in the class, the art of imagination cannot. The skill of imagination can be generated by practising a number of problems.

### 6.8.3 Drawing of Rounded Surface in Isometric Drawing

A circle in isometric view appears as an ellipse, which is called an isocircle. Figure 6.11 shows isocircles on left, right and top of isometric planes. The AutoCAD provides a special command for drawing isocircles.

**Command: ELLIPSE** [Enter]
Arc/Centre/Isocircle/<Axis endpoint 1>: I (for Isocircle) [Enter]
Centre of circle: [select a proper centre]
<Circle radius>/Diameter: [drag the mouse and move along the principal axis to define a radius by clicking the left button].

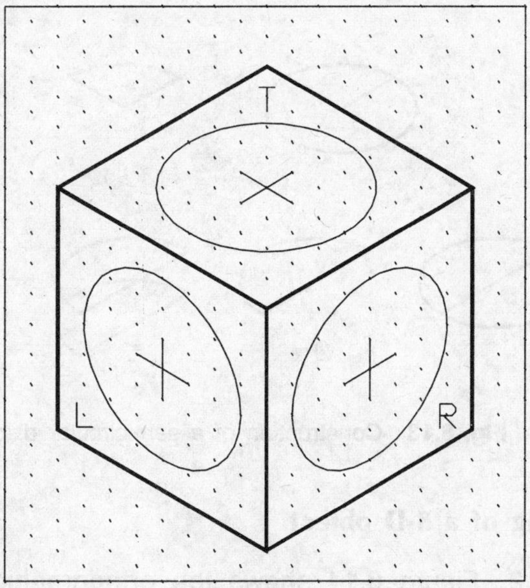


**Fig. 6.11** Isocircles on left, right and top of isometric planes.

### Creation of rounded object

The AutoCAD procedure to create a rounded object is explained with the help of Fig. 6.12. Figure 6.12(a) shows TV and FV of a cylindrical disc with the circle in the TV. The disc has diameter equal to six grid spacing and height equal to two grid spacing. Figure 6.12(b) shows the stepwise AutoCAD procedure to generate a rounded object.

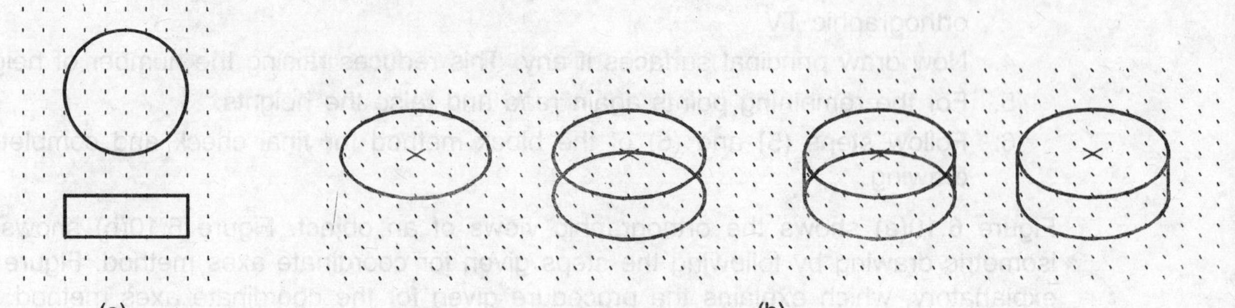

(a) (b)

**Fig. 6.12** Construction of a rounded object.

1. In orthographic view the circle exists in TV, therefore, choose the top cursor mode in order to draw an isocircle.
2. Draw an isocircle using ELLIPSE command with isocircle option.
3. This plane circular surface is required to be protruded in the third (height) direction. Copy the isocircle and shift by two grid spacing with the centre as a base point.
4. Draw the tangents (using the tan option in LINE command) to define the last visible part of the curved surface. While drawing tangents the view should be sufficiently zoomed out with one cursor line along the direction of the tangent.
5. The invisible part of the object can be trimmed out with the help of the TRIM command.

Figure 6.13 shows similar steps followed to generate a semicircular disc and quarter portion of a disc. For complicated object having rounded entities, block method is suggested to follow. The block method restricts us from crossing the boundaries L × D × H. The method also helps in visualisation ability because the reference planes are available. This is explained in EXAMPLE 6.2.

**Fig. 6.13** Construction of a semicircular disc and quarter portion of the disc.

### Understanding of a 3-D object

**EXAMPLE 6.2** Figure 6.14 shows the orthographic views of a mechanical object. Draw an isometric drawing.

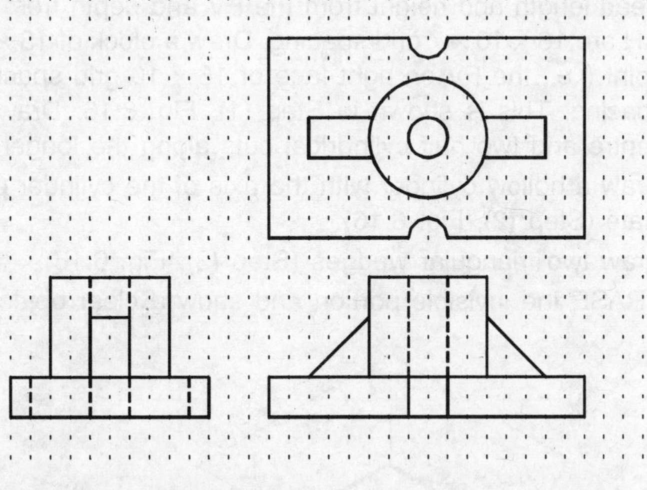

**Fig. 6.14** Orthographic views of a mechanical object.

***Solution*** There are two ways to understand the object:

(a) If we think this object as an assembly point of view, then there are four components as shown in Fig. 6.15(a):

1. One base-plate with one cylindrical hole at the centre and two half cylindrical cut portion along two longer edges;
2. Hollow cylinder; and
3. Two triangular wedges

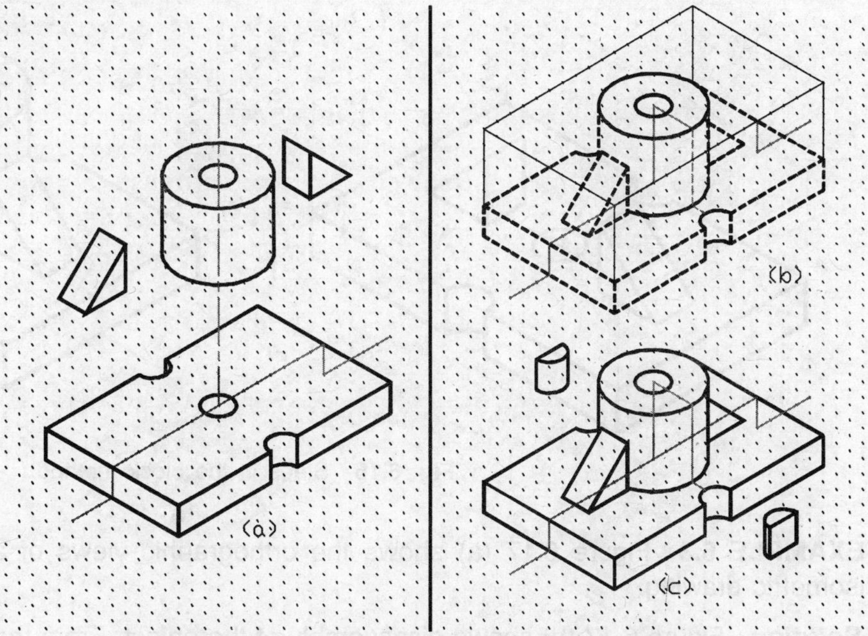


**Fig. 6.15** Components of a mechanical object.

All these four components can be drawn separately and then can be positioned properly. After proper positioning, the invisible part can be trimmed and/or erased. The final object is shown in Fig. 6.15(c).

(b) If we think the object as a single piece, which is our usual practice then a material has been taken out from a block having L × D × H as 16 × 10 × 7 grid spacing measured from orthographic views (Fig. 6.14). This is shown in Fig. 6.15(b). The block method is shown in the following steps:

1. Read length and height from the FV and depth from the RSV or TV. The dimensions L × D × H are 16 × 10 × 7 grid spacing. Draw a block of 16 × 10 × 7 grid spacing with proper starting point (i.e., the FV on right face of 16 × 10 grid spacing and LSV on left face of 10 × 7 grid spacing. This is shown in Step (1), Fig. 6.16. Draw a base-plate with circular hole at the centre and two half cylindrical cuts along the longer edges of the base-plate.
2. Draw a hollow cylinder with the axis of the cylinder passing through the centre of the base-plate (Step (2), Fig. 6.16).
3. Draw two triangular wedges (Step (3), Fig. 6.16).
4. ERASE the invisible portion and show a clear object (Step (4), Fig. 6.16).

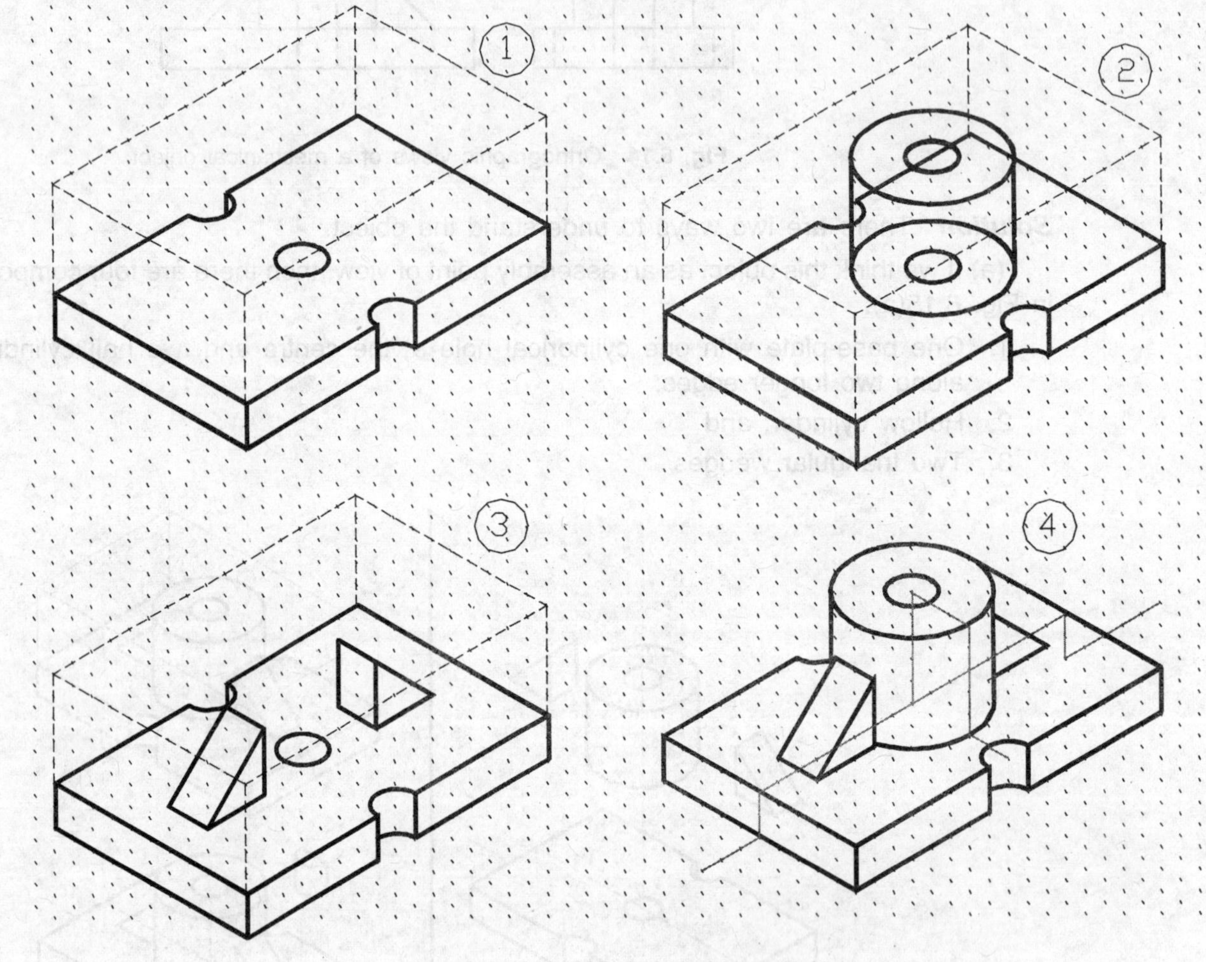


**Fig. 6.16** Steps in the block method.

**EXAMPLE 6.3** Figure 6.17 (a) shows the orthographic views of a mechanical object. Draw an isometric drawing.

***Solution*** Figure 6.17 (b) shows disassembly of the object. From the assembly point of view, there are three components (base plate, vertical plate and extended cylindrical neck).

**Steps:** (Refer to Fig. 6.18)

1. Read the length and height from the FV and depth from the RSV or TV. L × D × H are 10 × 5 × 10 grid spacing. Draw a block of 10 × 5 × 10 grid spacing with proper starting point, (i.e., the FV on left face of 10 × 10 grid spacing and RSV on right face of 5 × 10 grid spacing.
2. Draw a base plate of 10 × 5 × 2 grid spacing.
3. Draw an isocircle of 2 grid spacing radius with ELLIPSE command and ISO option on the left face. The location of the centre is read from orthographic views.

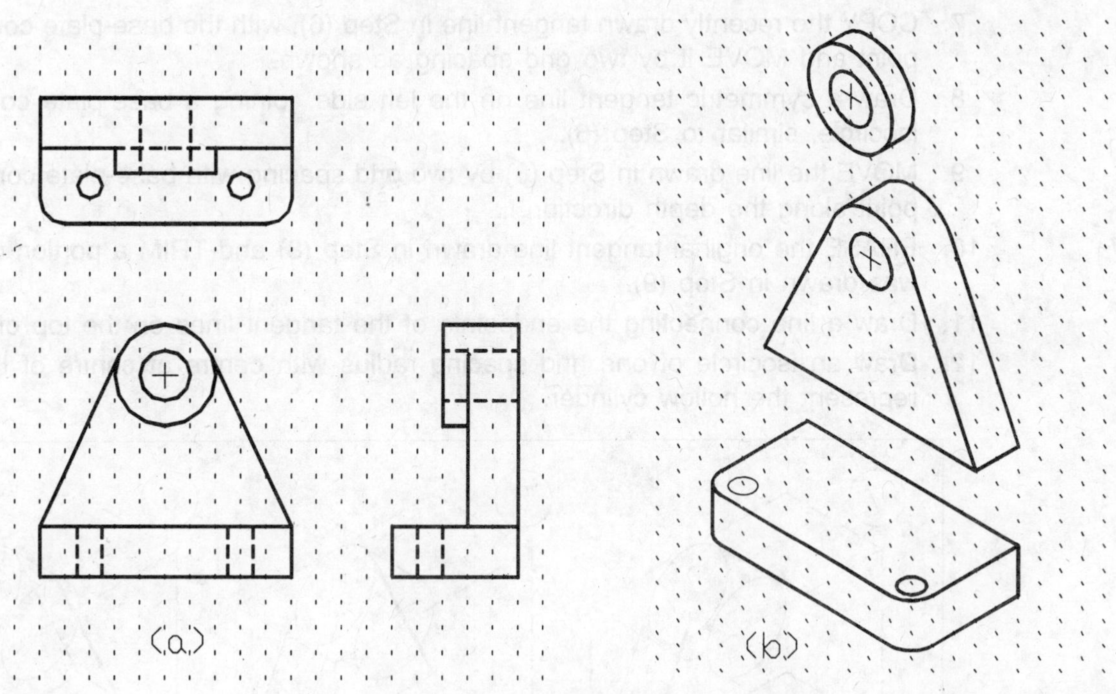


**Fig. 6.17** Disassembly of the object.

4. MOVE this isocircle with the centre as a base point by two grid spacing along the depth direction.
5. Copy the same isocircle and move it with the centre as a base point by three grid spacing along the depth direction.
6. Draw a tangent to two isocircles along the depth direction. Draw a line from a base-plate corner, tangent to the copied isocircle in Step (5).

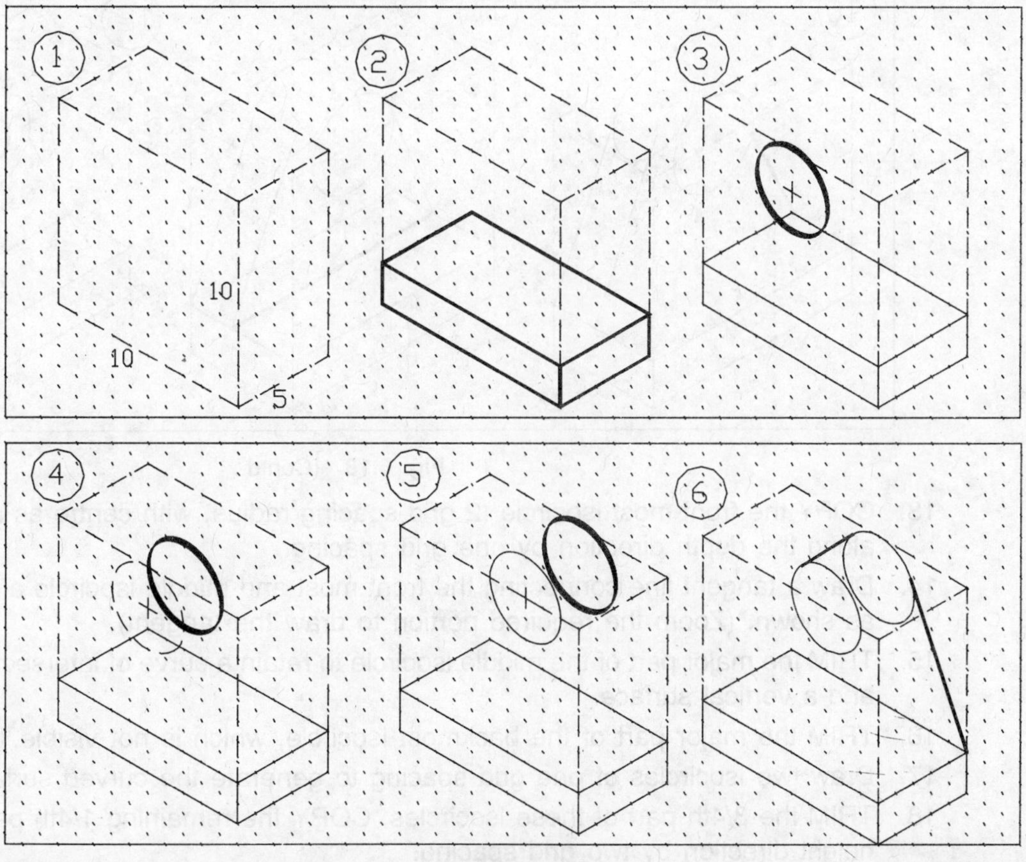


**Fig. 6.18** (Contd.)

7. COPY the recently drawn tangent line in Step (6), with the base-plate corner point as a base point and MOVE it by two grid spacing as shown.
8. Draw a symmetric tangent line on the left side, joining a base-plate corner and tangent to isocircle, similar to Step (6).
9. MOVE the line drawn in Step (8) by two grid spacing with base-plate corner point as a base point along the depth direction.
10. ERASE the original tangent line drawn in Step (8) and TRIM a portion of copied line which was drawn in Step (9).
11. Draw a line connecting the endpoints of the tangent lines on the top of the base-plate.
12. Draw an isocircle of one grid spacing radius with centre at centre of the front isocircle to represent the hollow cylinder.

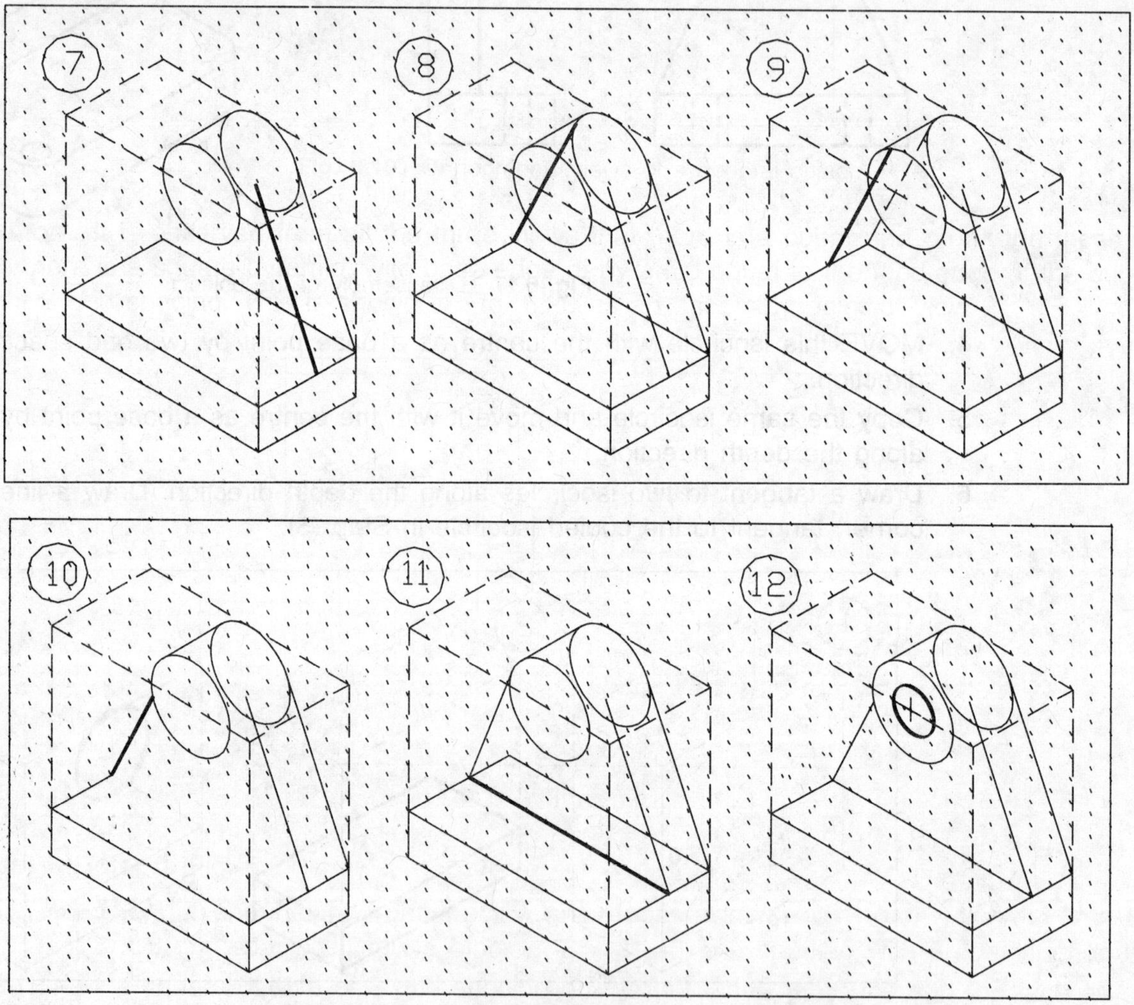


**Fig. 6.18** (Contd.)

13. COPY the front-most isocircle (2 grid spacing radius) with centre as a base point and move along the depth direction by one grid spacing.
14. Draw a tangent line connecting the front most and middle isocircle along the depth direction as shown. (Zoom the required portion to draw the tangent).
15. TRIM the major part of the middle isocircle to retain a curve of intersection of a hollow cylinder and a vertical surface.
16. TRIM the major part of the backmost isocircle, which is not visible.
17. Draw two isocircles of one grid spacing to generate the curved surface of the base-plate.
18. TRIM the 3/4th part of these isocircles. COPY the remaining 1/4th part and move along the height direction by two grid spacing.

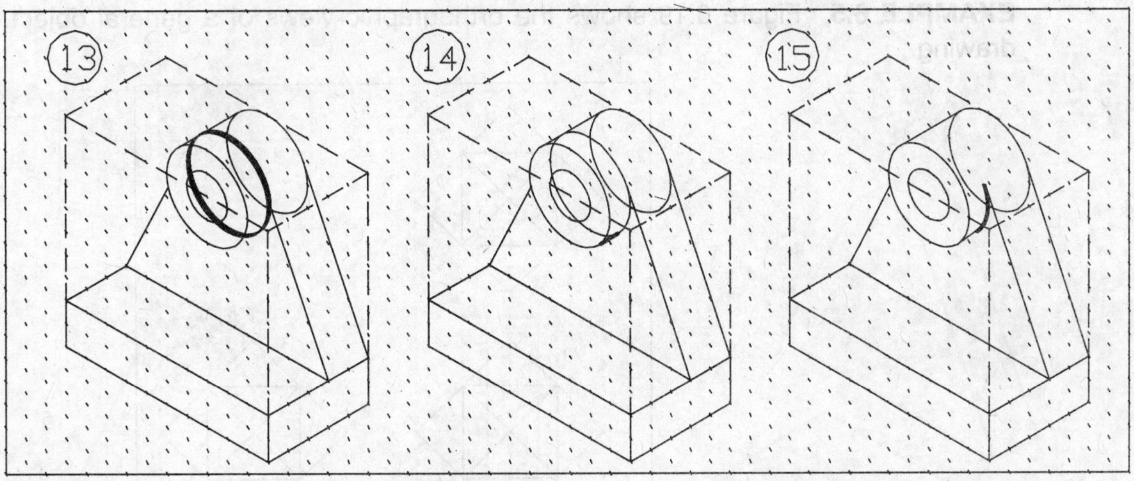

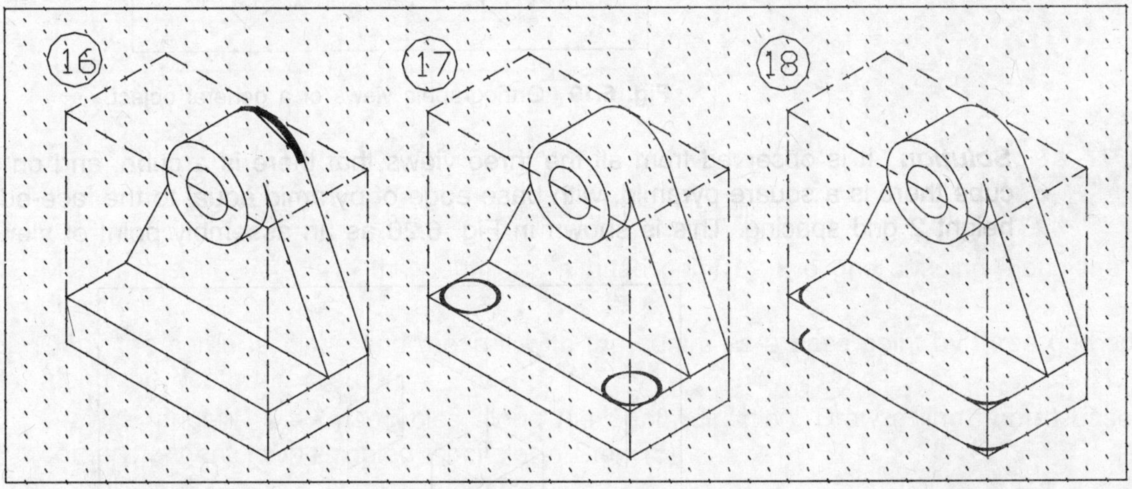

**Fig. 6.18** (Contd.)

19. Draw a tangent line along the height direction connecting two 1/4th segments of isocircles to show the last visible part of curved surface.
20. TRIM the part of the base-edges, edges associated with this 1/4th segment. Draw an isocircle of 1/2 grid spacing radius to represent a hole.
21. ERASE the block lines and show a clean object.

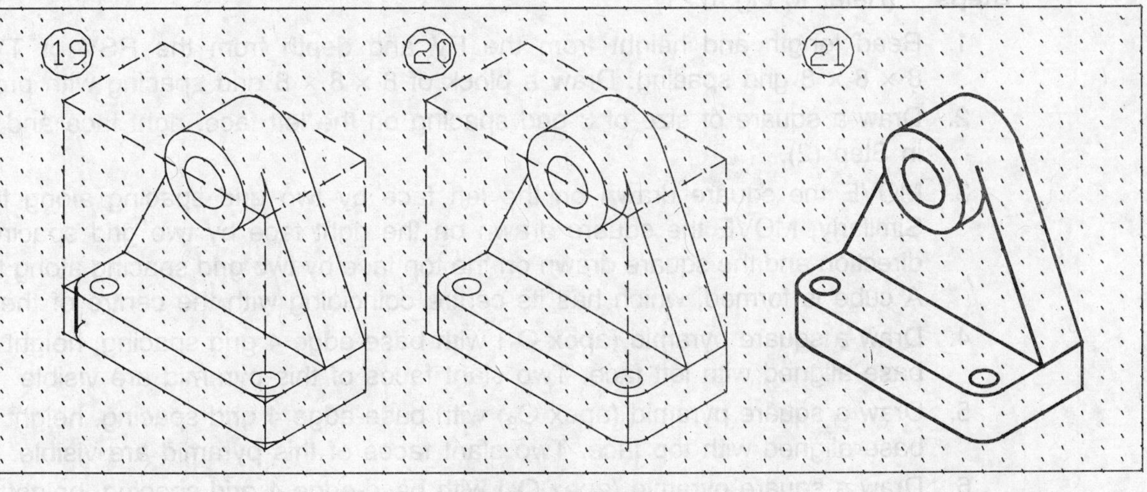

**Fig. 6.18** (Contd.)

**EXAMPLE 6.5** Figure 6.19 shows the orthographic views of a general object. Draw an isometri drawing.

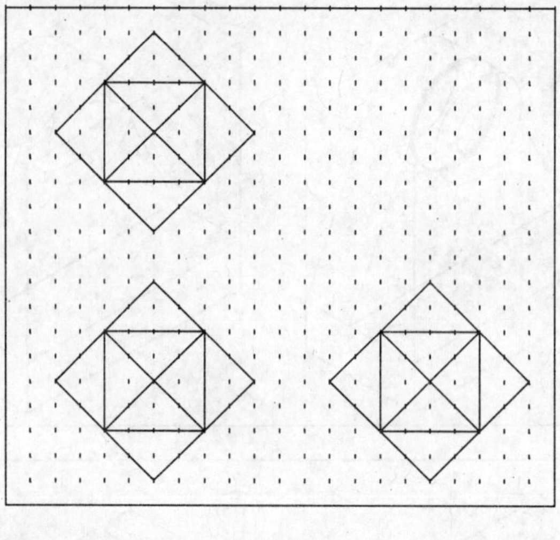

**Fig. 6.19** Orthographic views of a general object.

***Solution*** It is observed from all the three views that there is a cube, and on all the faces of th cube there is a square pyramid, with base-edge of pyramid equal to the face-edge of the cube an height 2 grid spacing. This is shown in Fig. 6.20 as an assembly point of view.

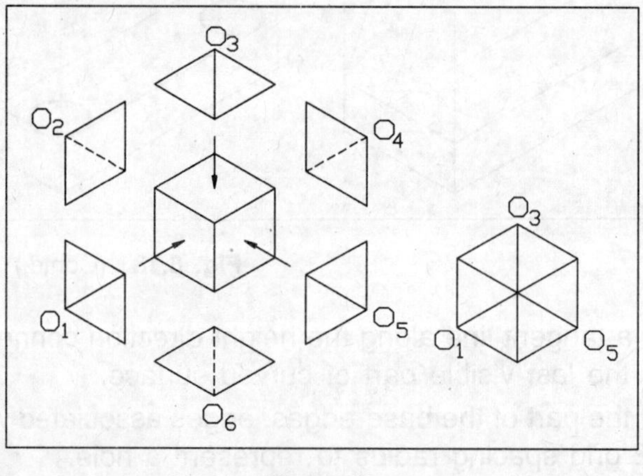


**Fig. 6.20**

**Steps:** (Refer to Fig. 6.21)

1. Read length and height from the FV and depth from the RSV or TV. L × D × H ar 8 × 8 × 8 grid spacing. Draw a block of 8 × 8 × 8 grid spacing with proper starting point
2. Draw a square of size of 3 grid spacing on the left face, right face and top face as show in Step (2).
3. MOVE the square drawn on the left face by two grid spacing along the depth directior Similarly, MOVE the square drawn on the right face by two grid spacing along the lengt direction and the square drawn on the top face by two grid spacing along the height directior A cube is formed, which has its centre coinciding with the centre of the block.
4. Draw a square pyramid (apex $O_1$) with base-edge 4 grid spacing, height 2 grid spacing an base aligned with left face. Two slant faces of this pyramid are visible.
5. Draw a square pyramid (apex $O_3$) with base-edge 4 grid spacing, height 2 grid spacing an base aligned with top face. Two slant faces of this pyramid are visible.
6. Draw a square pyramid (apex $O_5$) with base-edge 4 grid spacing, height 2 grid spacing an base aligned with right face. Two slant faces of this pyramid are visible.

7. Erase the block lines and show a clear object. Object is a cube with six pyramids. The base of the square pyramid is aligned with the face of the cube.

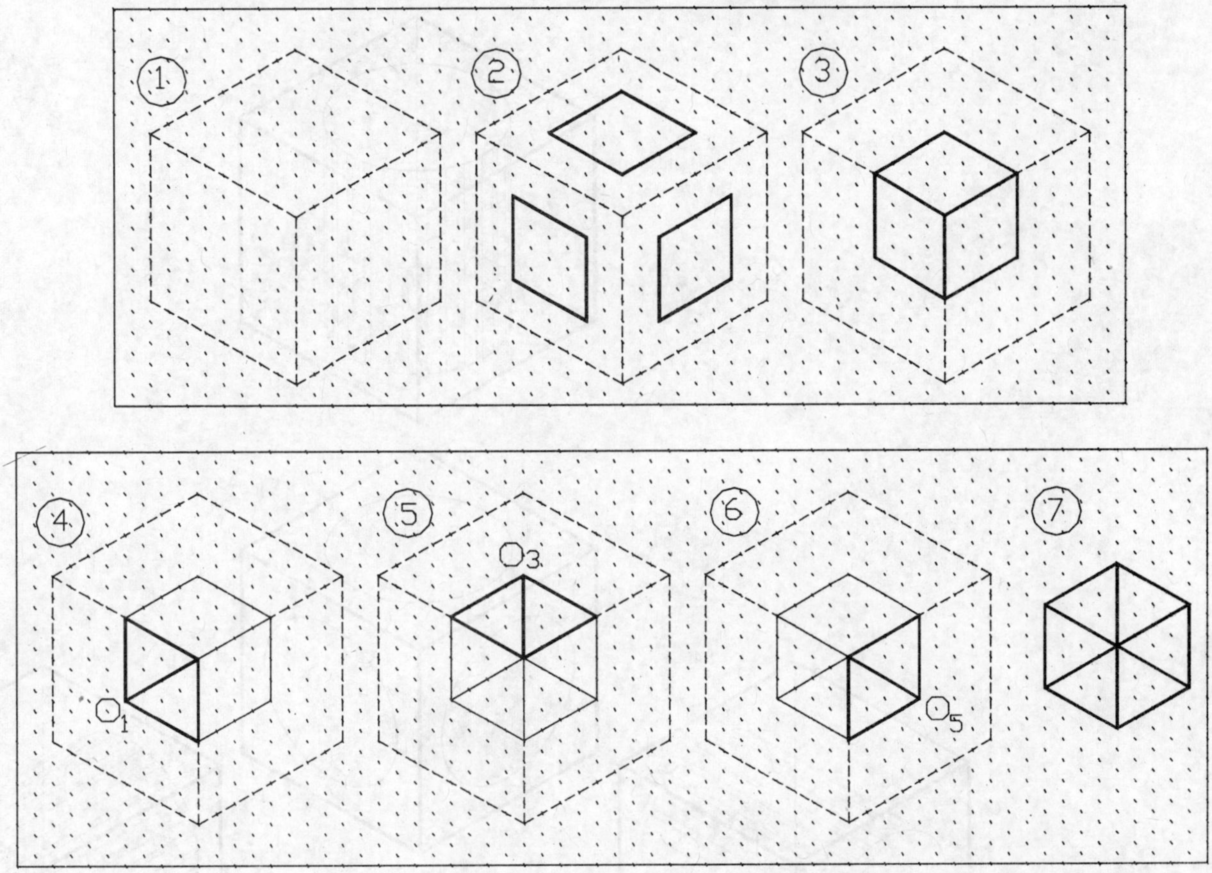


**Fig. 6.21**

**EXAMPLE 6.4** Figure 6.22 shows the orthographic views of a general object. Draw an isometric drawing.

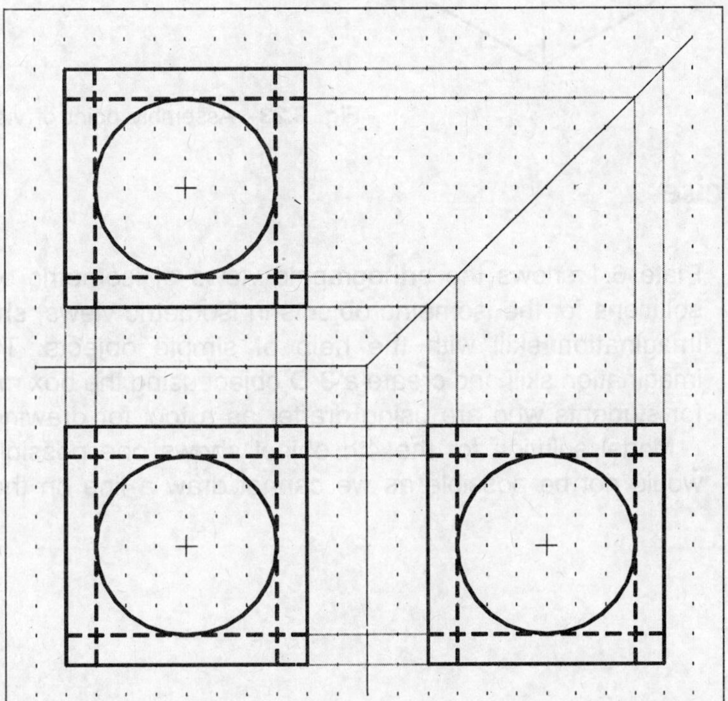

**Fig. 6.22**

***Solution*** It is observed from all the three views that there is a cube with cylindrical hole along th length, depth and height directions. This is shown in Fig. 6.23 as an assembly point of view.

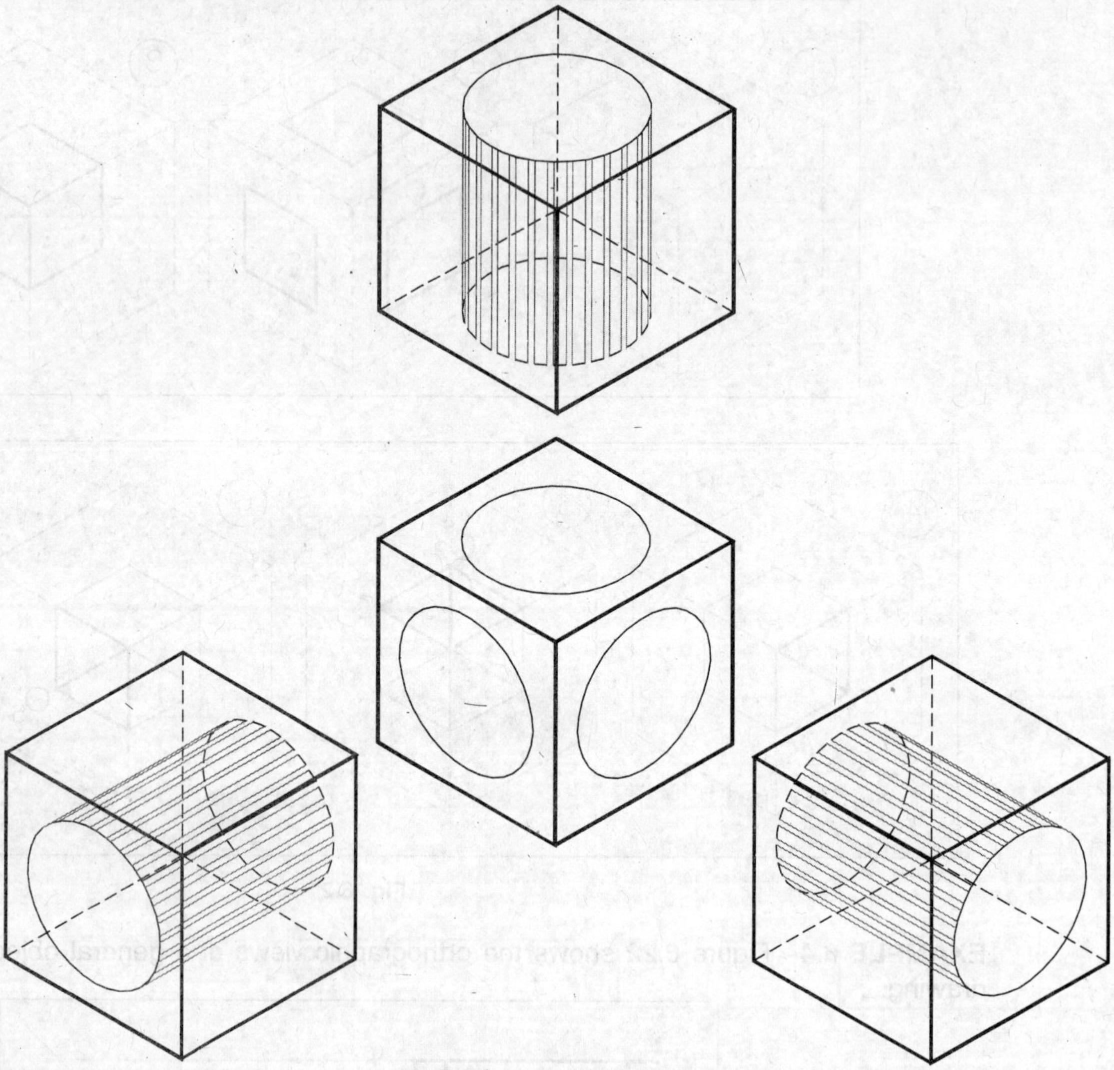

**Fig. 6.23** Assembly point of view.

## Exercise

**6.1** Plate 6.1 shows the orthographic views of isometric objects. Read the given step-by-ste solutions for the isometric objects in isometric views, shown in the same plate. Generate a imagination skill with the help of simple objects. This activity helps you to generat imagination skill and create a 3-D object using the box method. The box method is also usef for students who are using drafter as a tool for drawing.

Model solution for the 4th object shows one possibility at step 4, however, such obje would not be possible as we cannot draw a line on the surface.

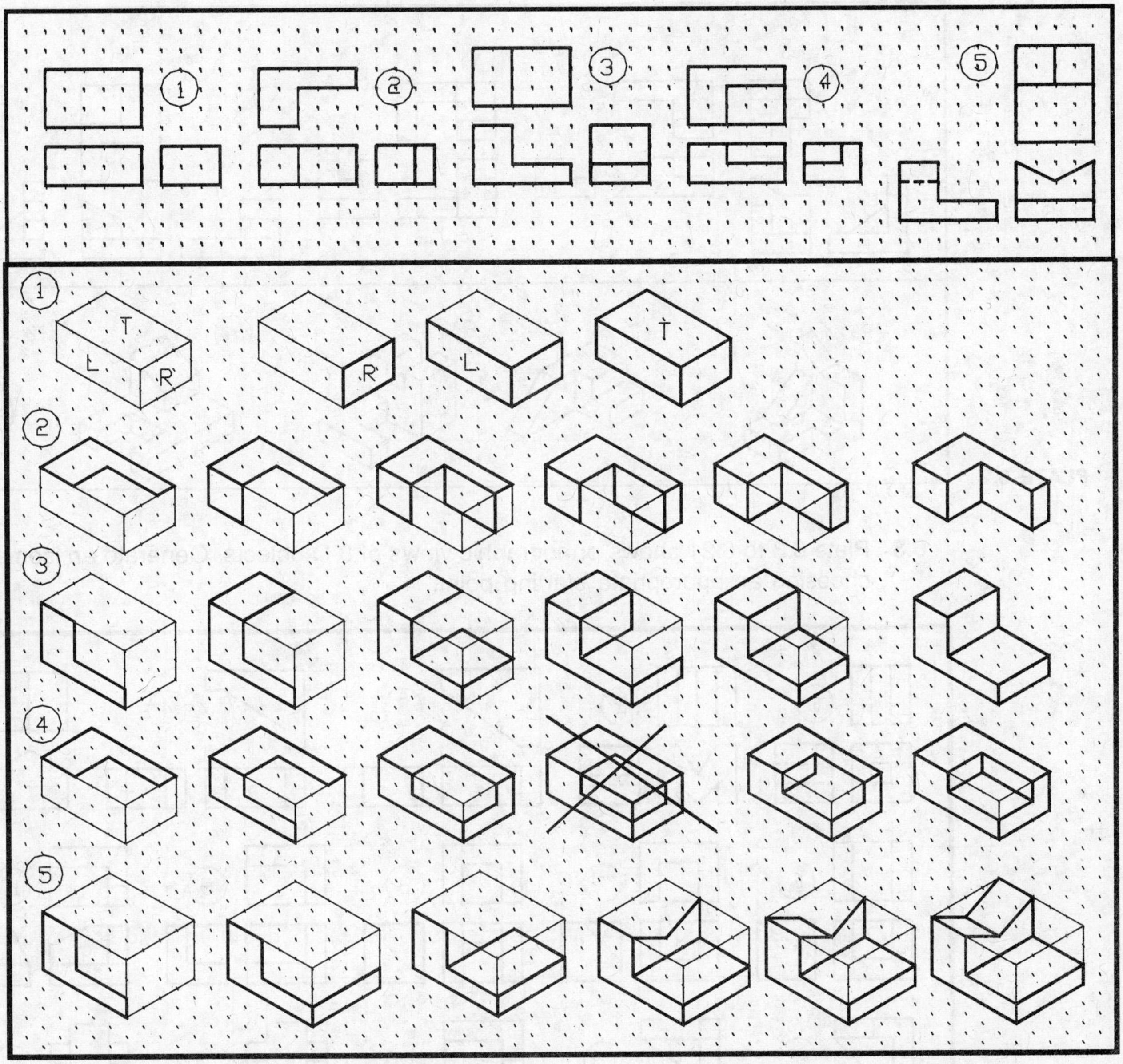

PLATE 6.1

**6.2** Plate 6.2 shows the orthographic views of isometric objects. The same plate shows the solution in isometric views. Practise for isometric views and generate an imagination skill with the help of these simple objects.

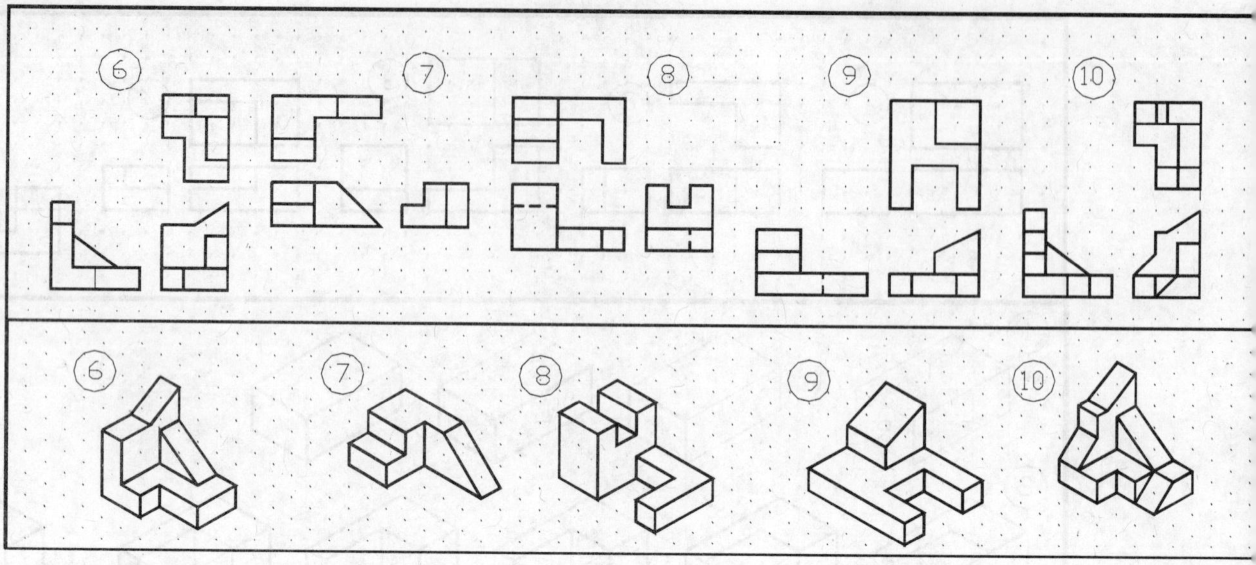

PLATE 6.2

**6.3** Plate 6.3 to 6.21 shows orthographic views of 3-D objects. Generate an isometric drawing by choosing an appropriate starting point.

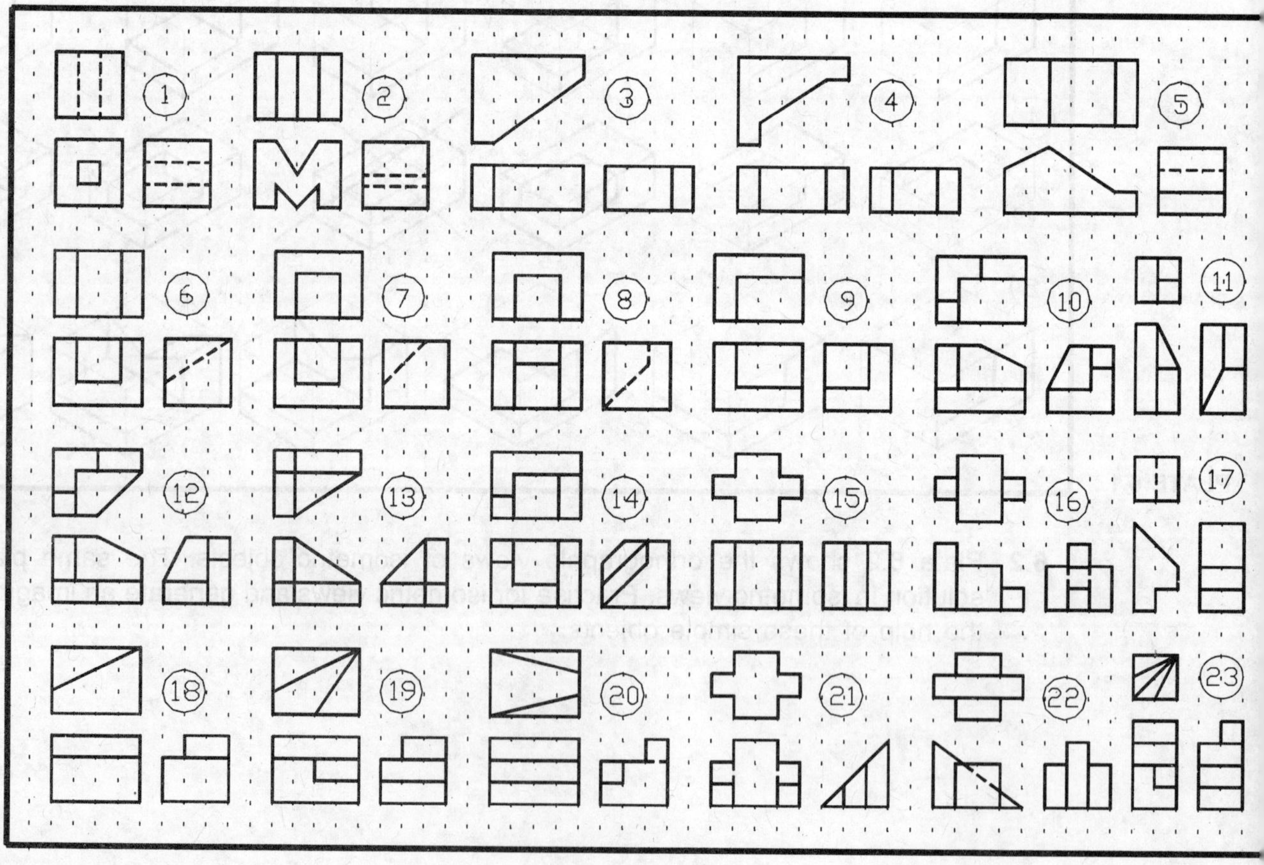

PLATE 6.3

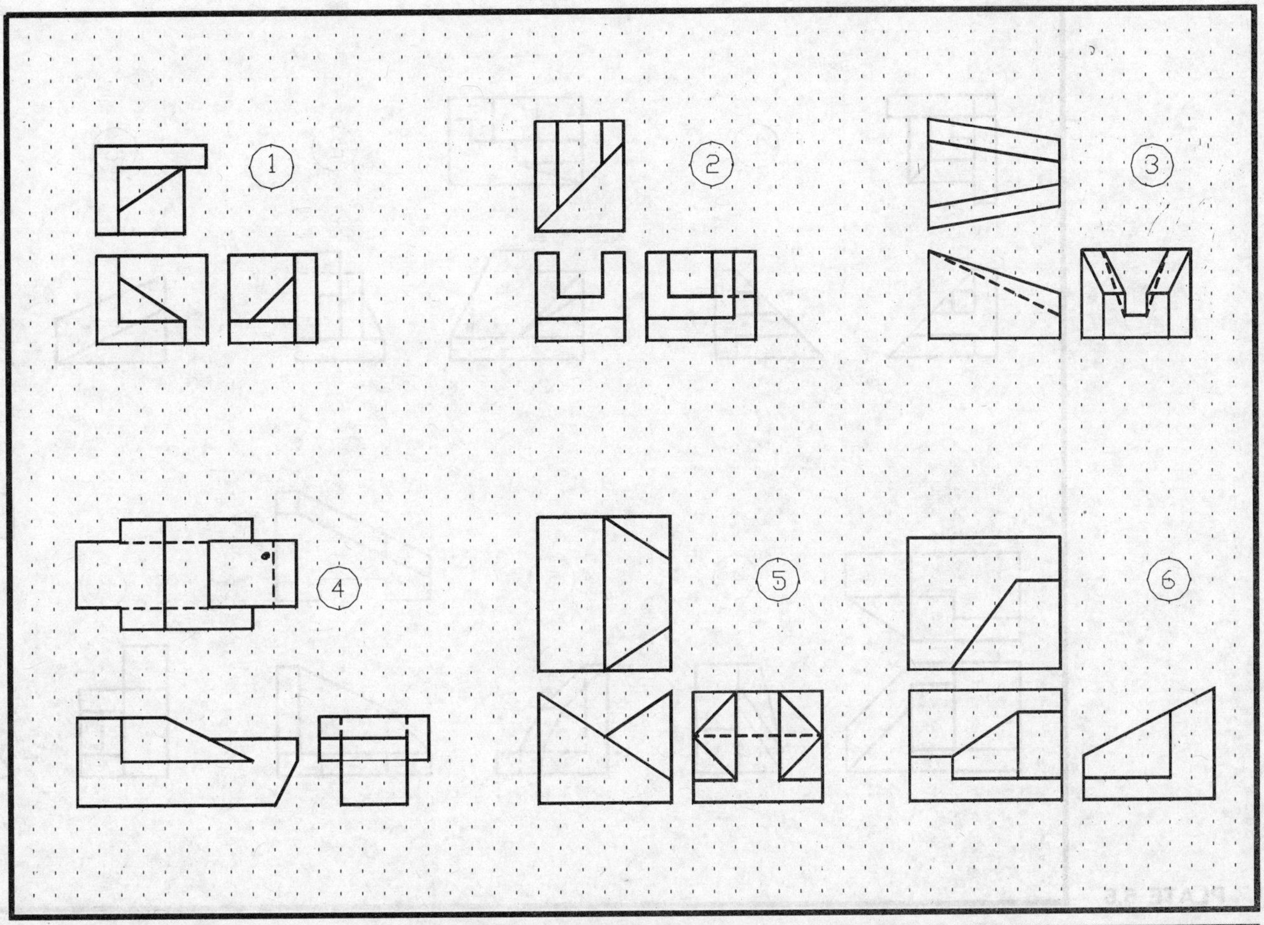


PLATE 6.4

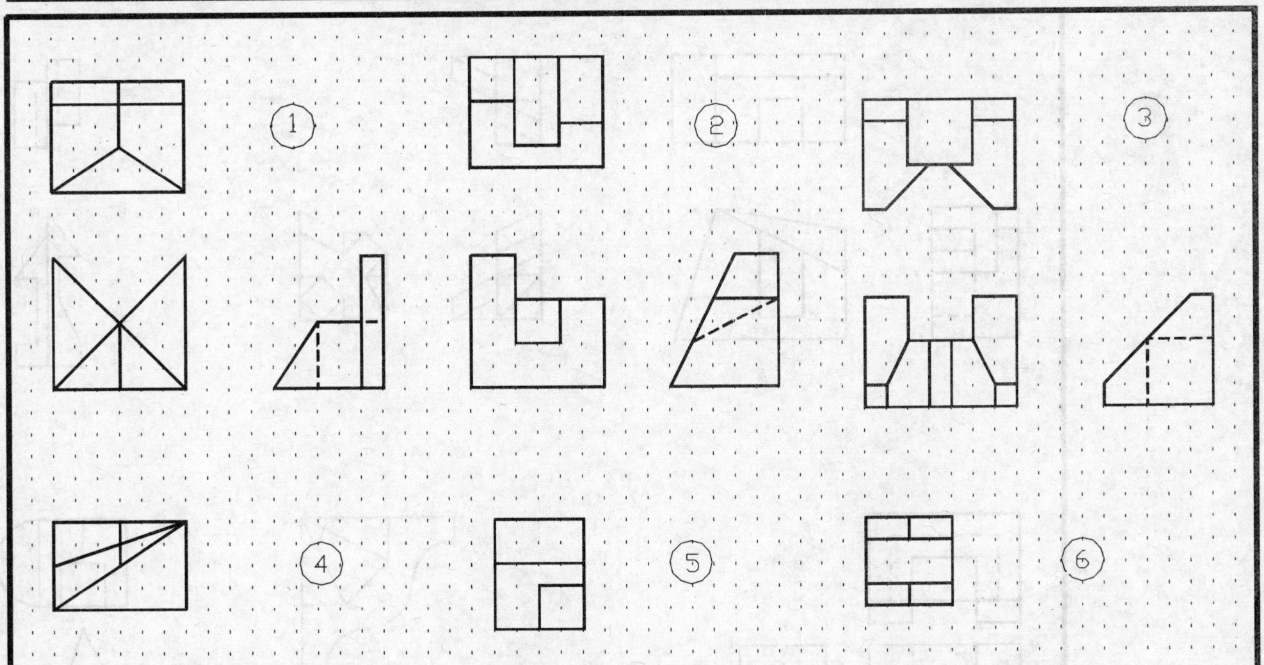


PLATE 6.5

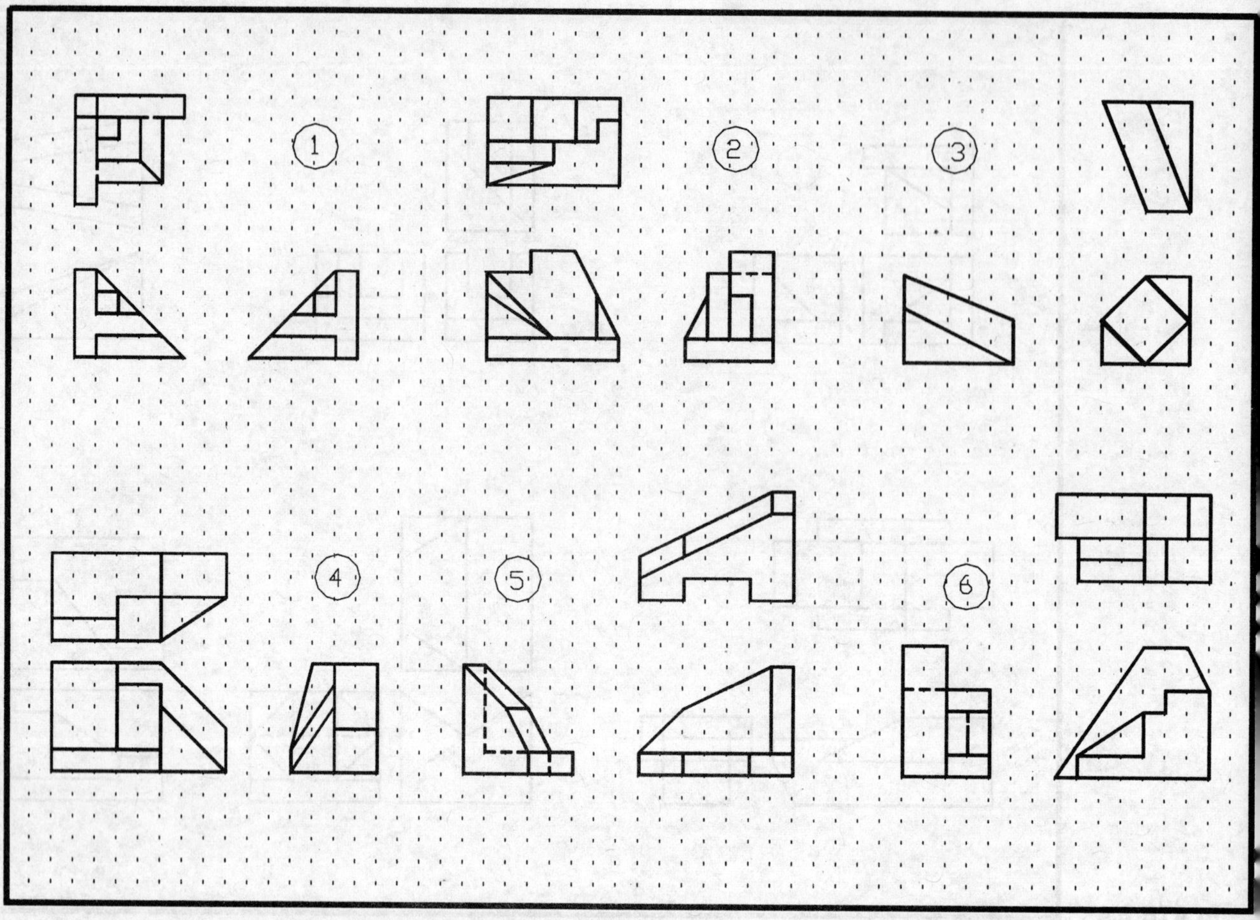


PLATE 6.6

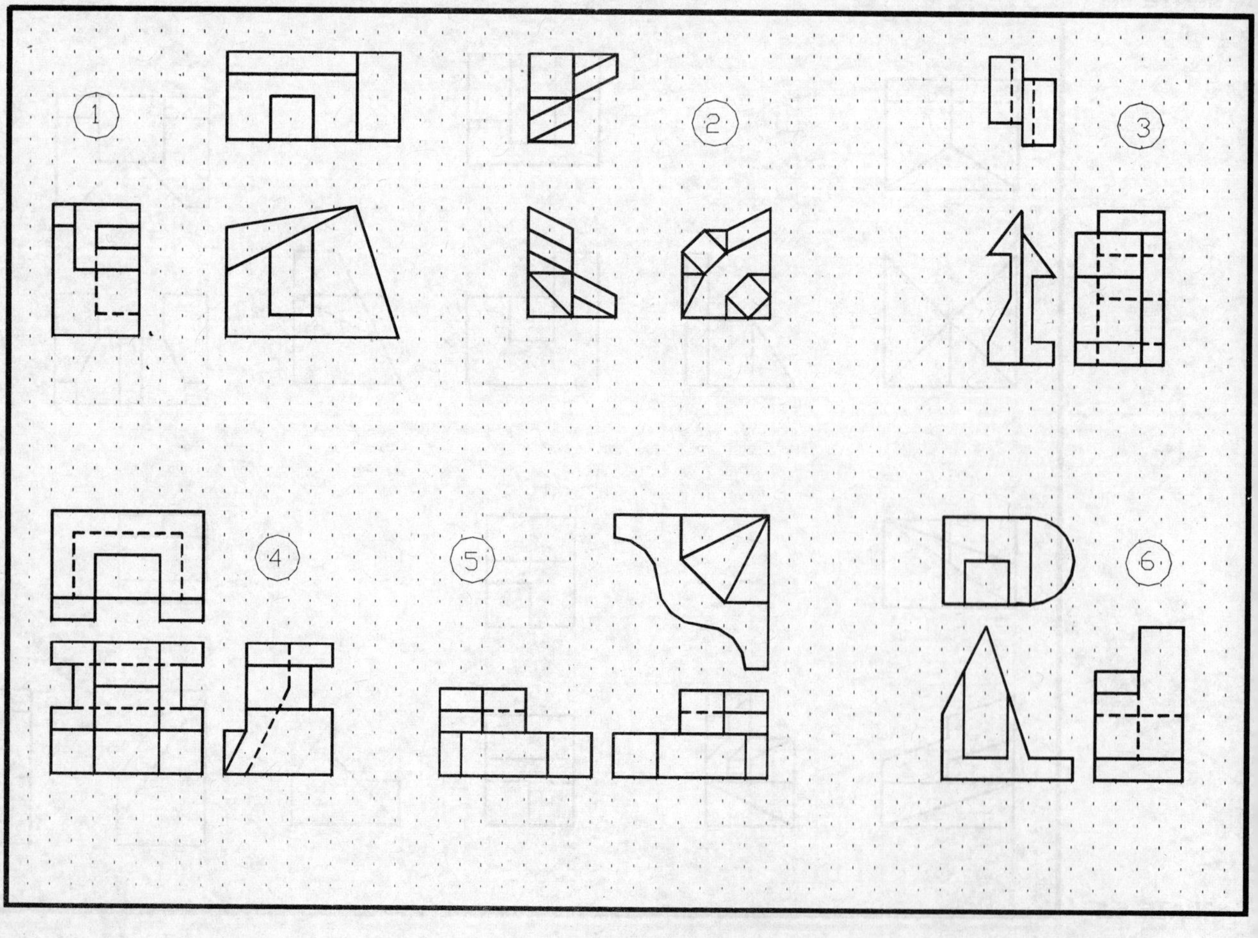


PLATE 6.7

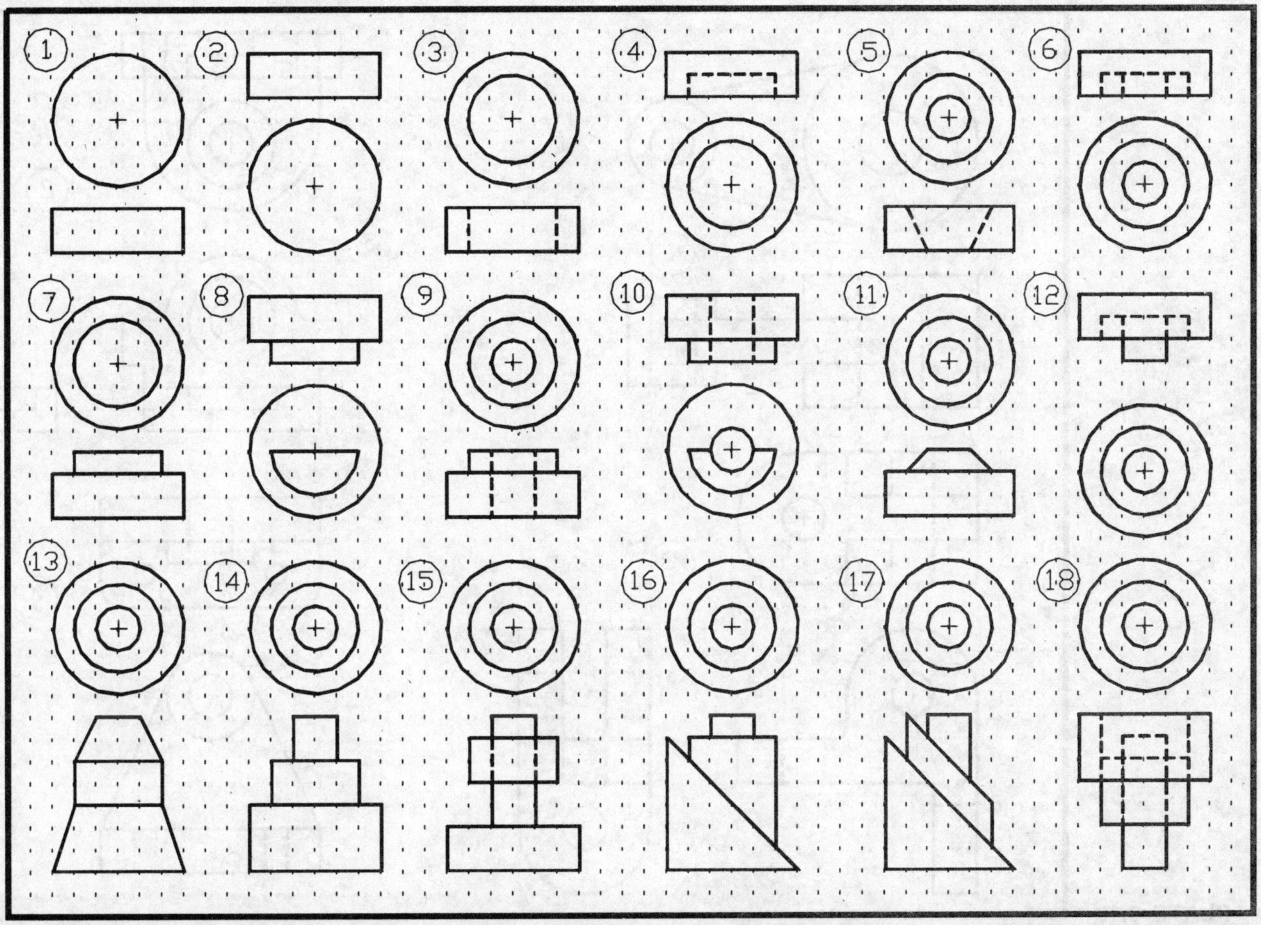


PLATE 6.8

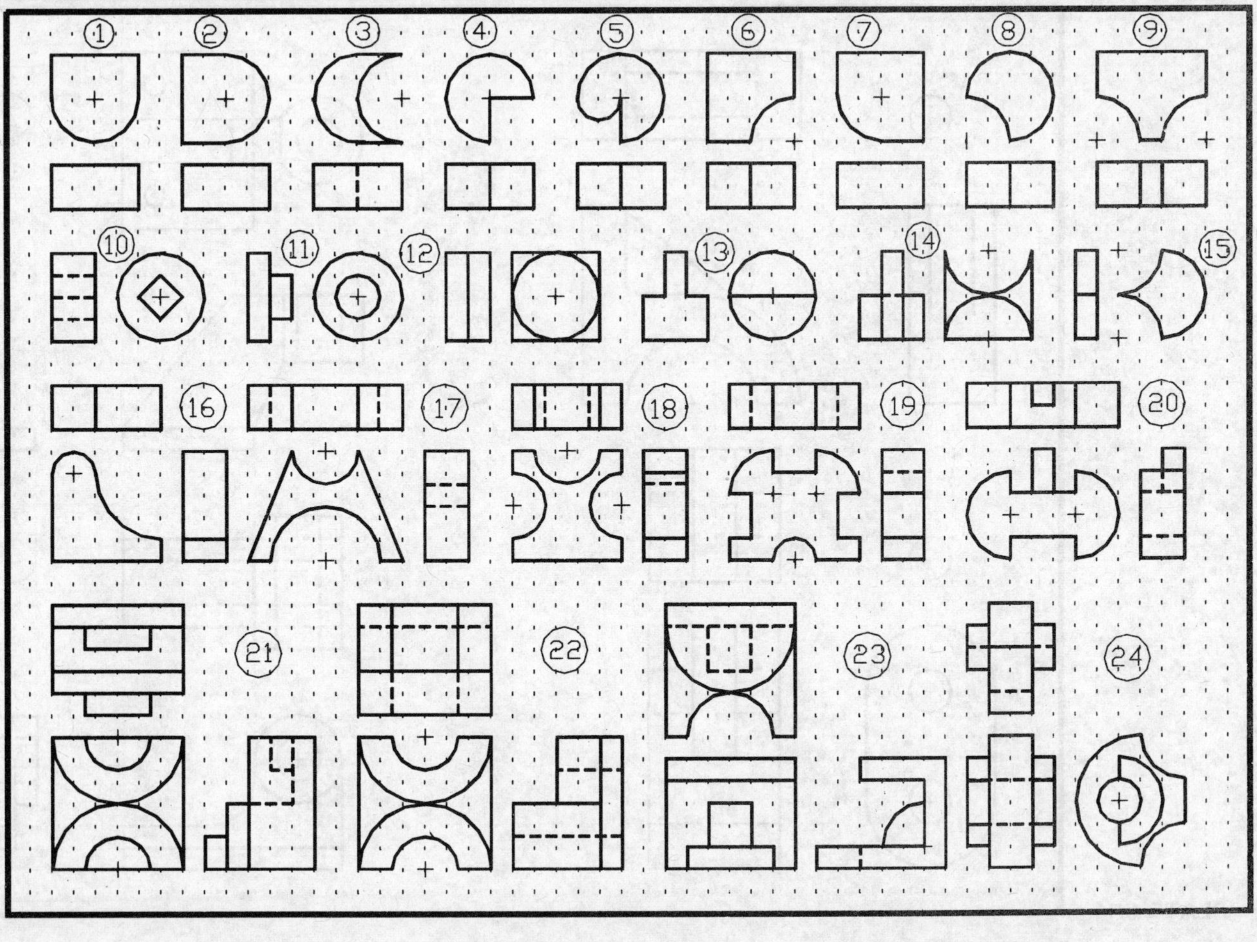


PLATE 6.9

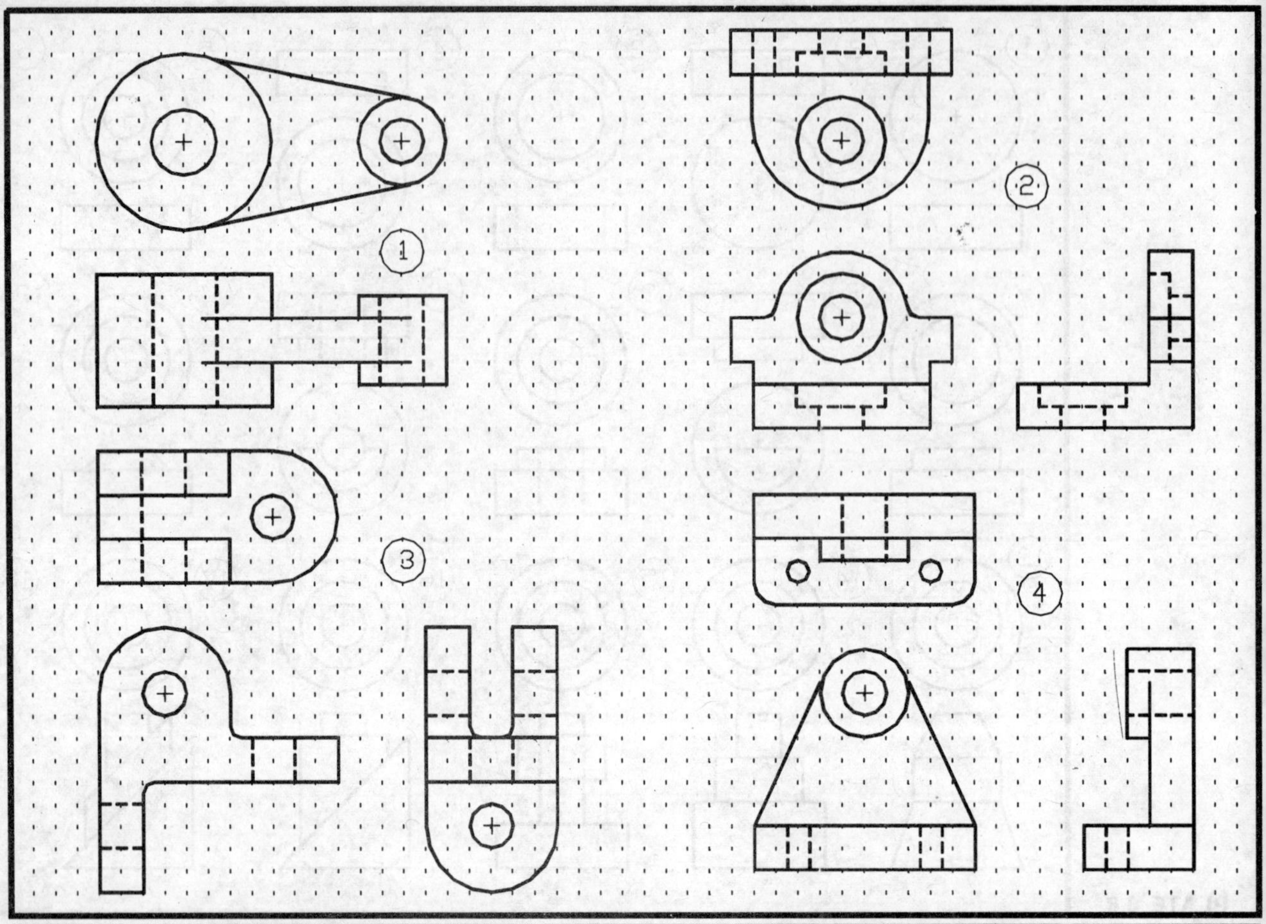


**PLATE 6.10**

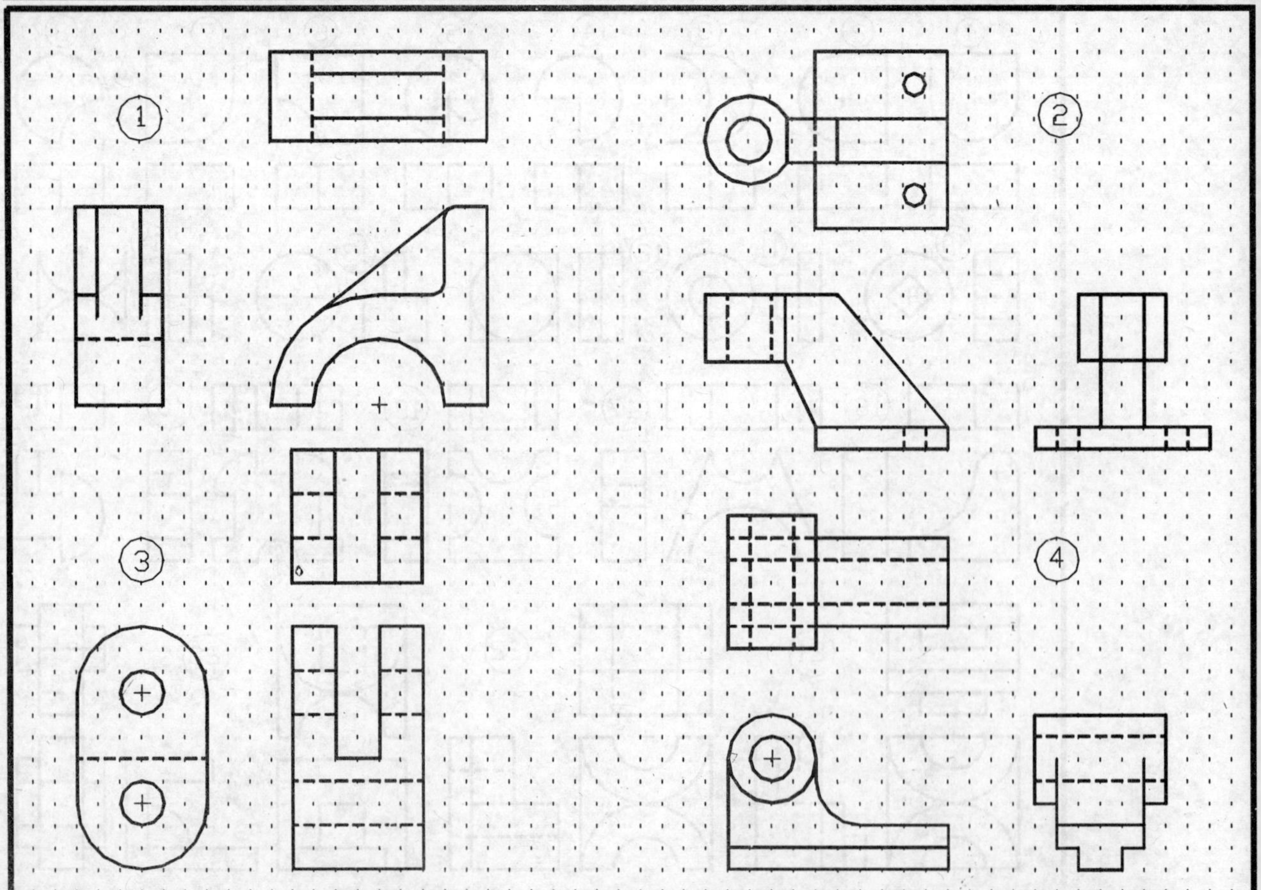


**PLATE 6.11**

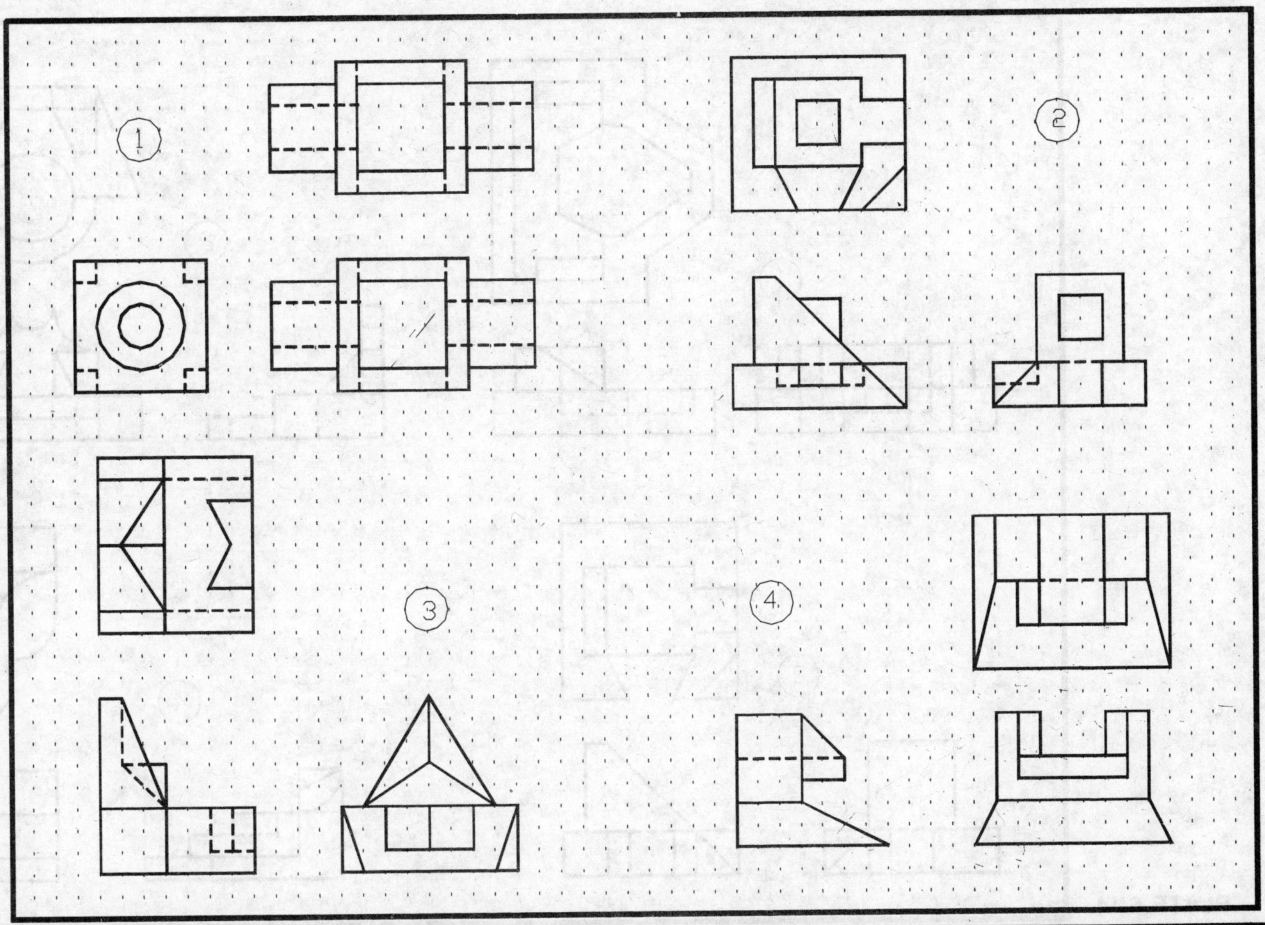


PLATE 6.12

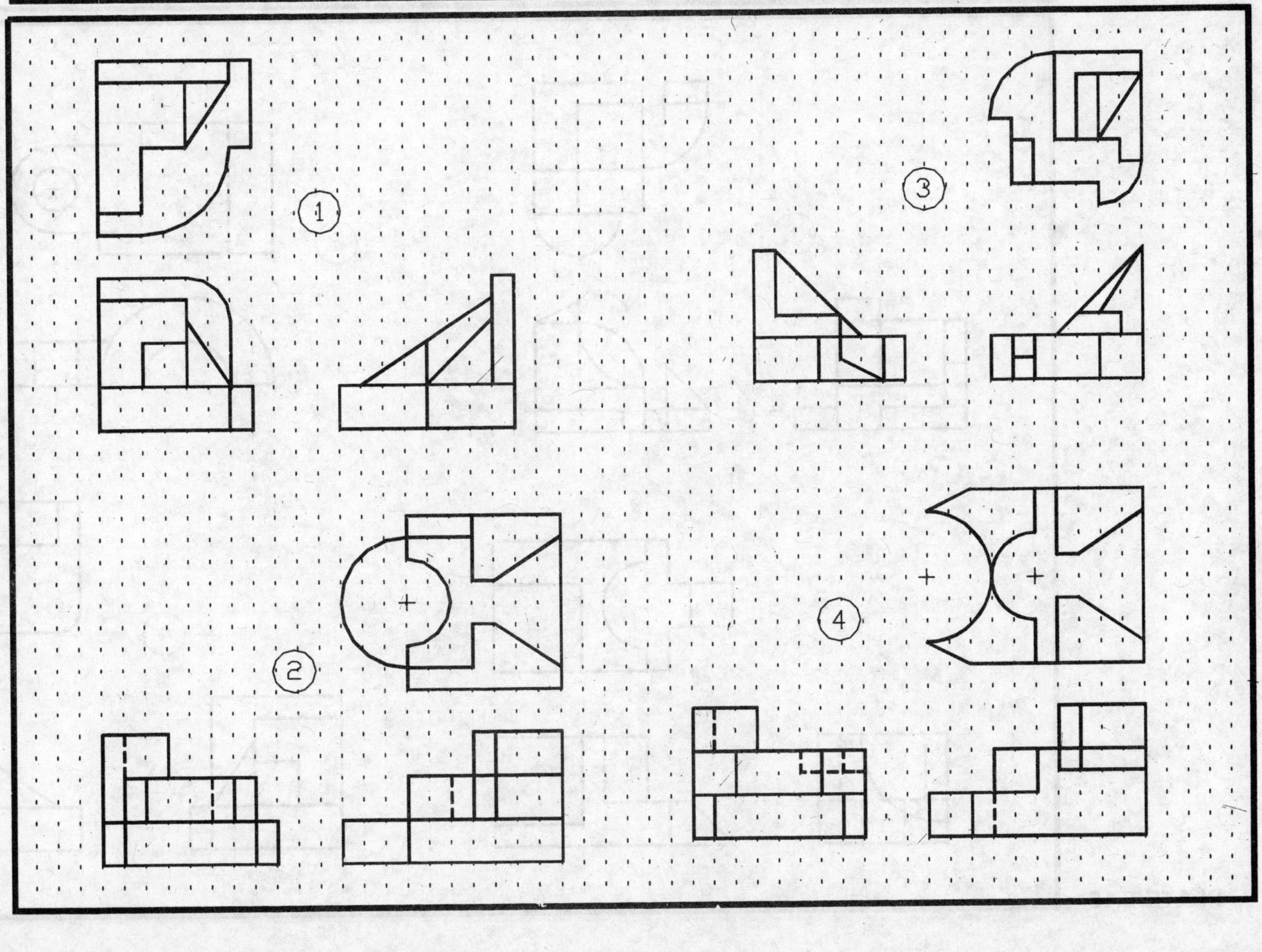


PLATE 6.13

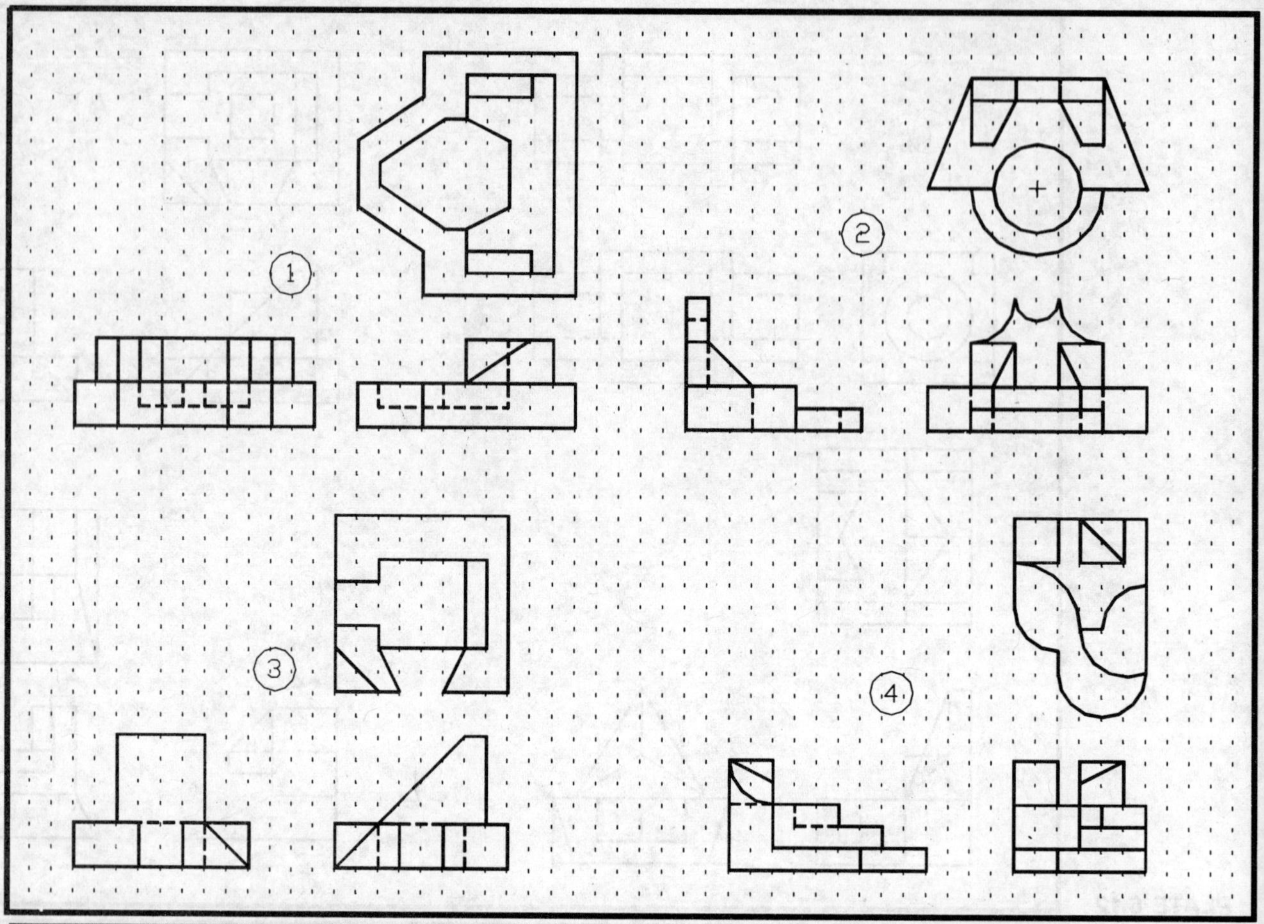


PLATE 6.14

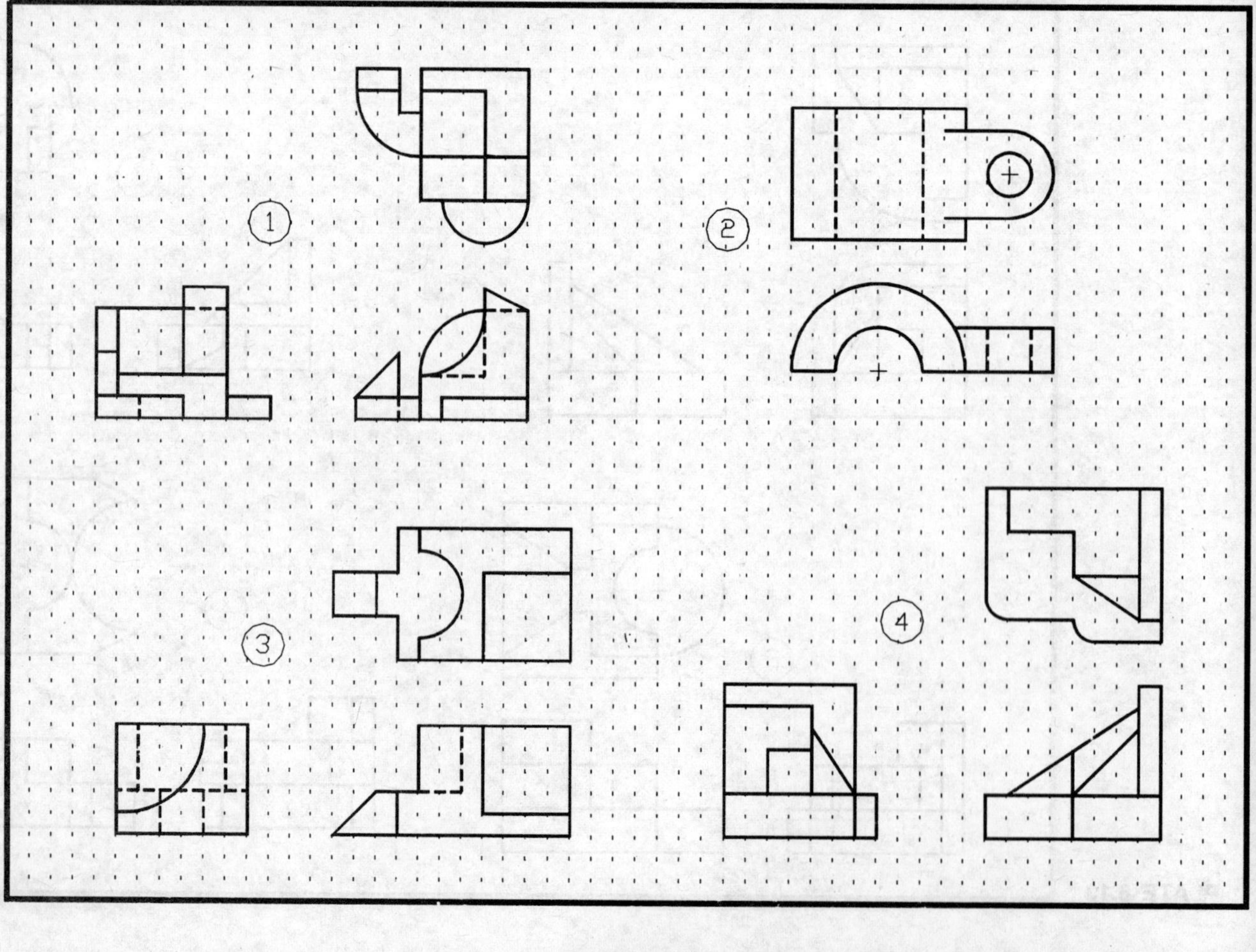


PLATE 6.15

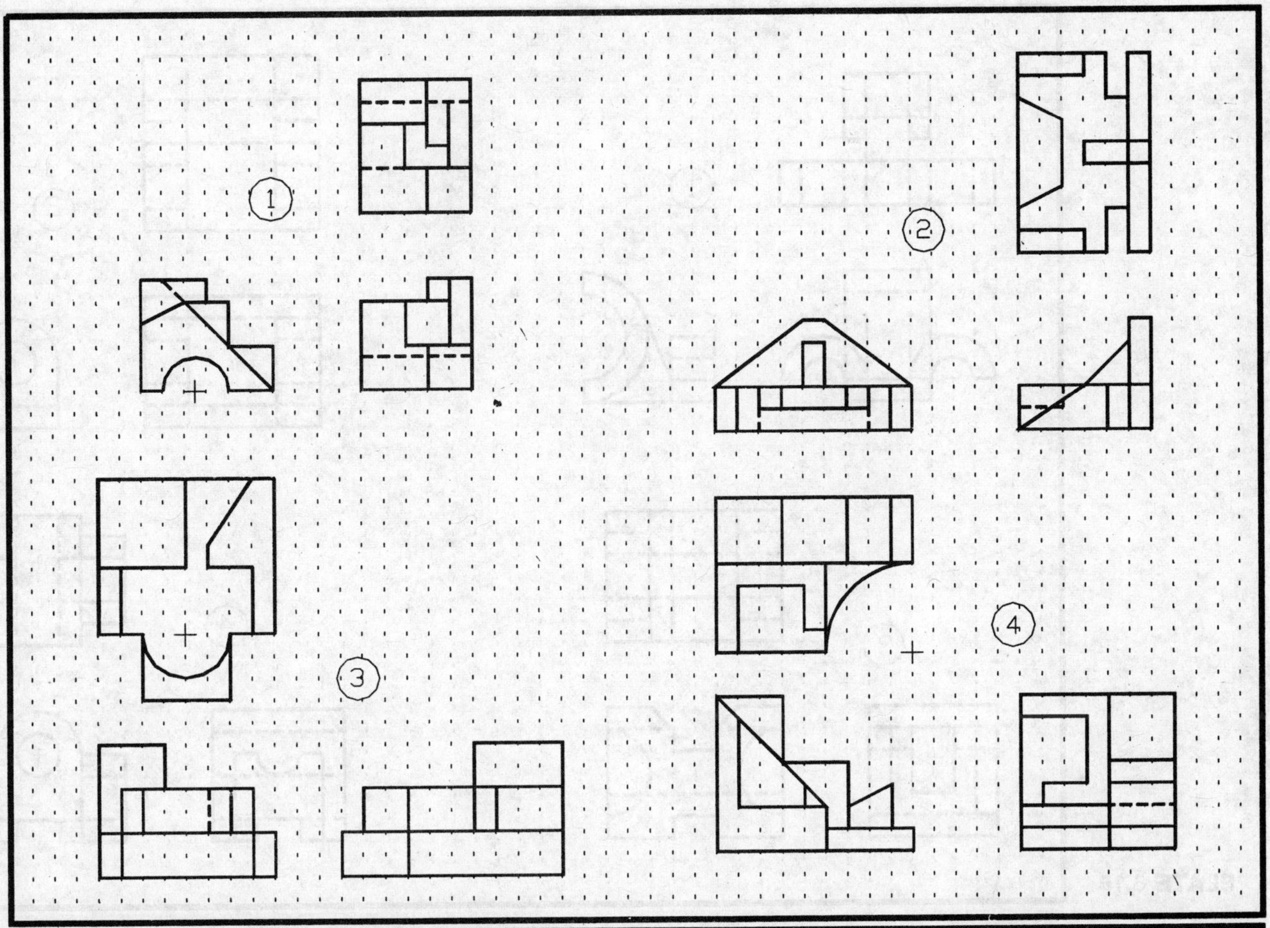


PLATE 6.16

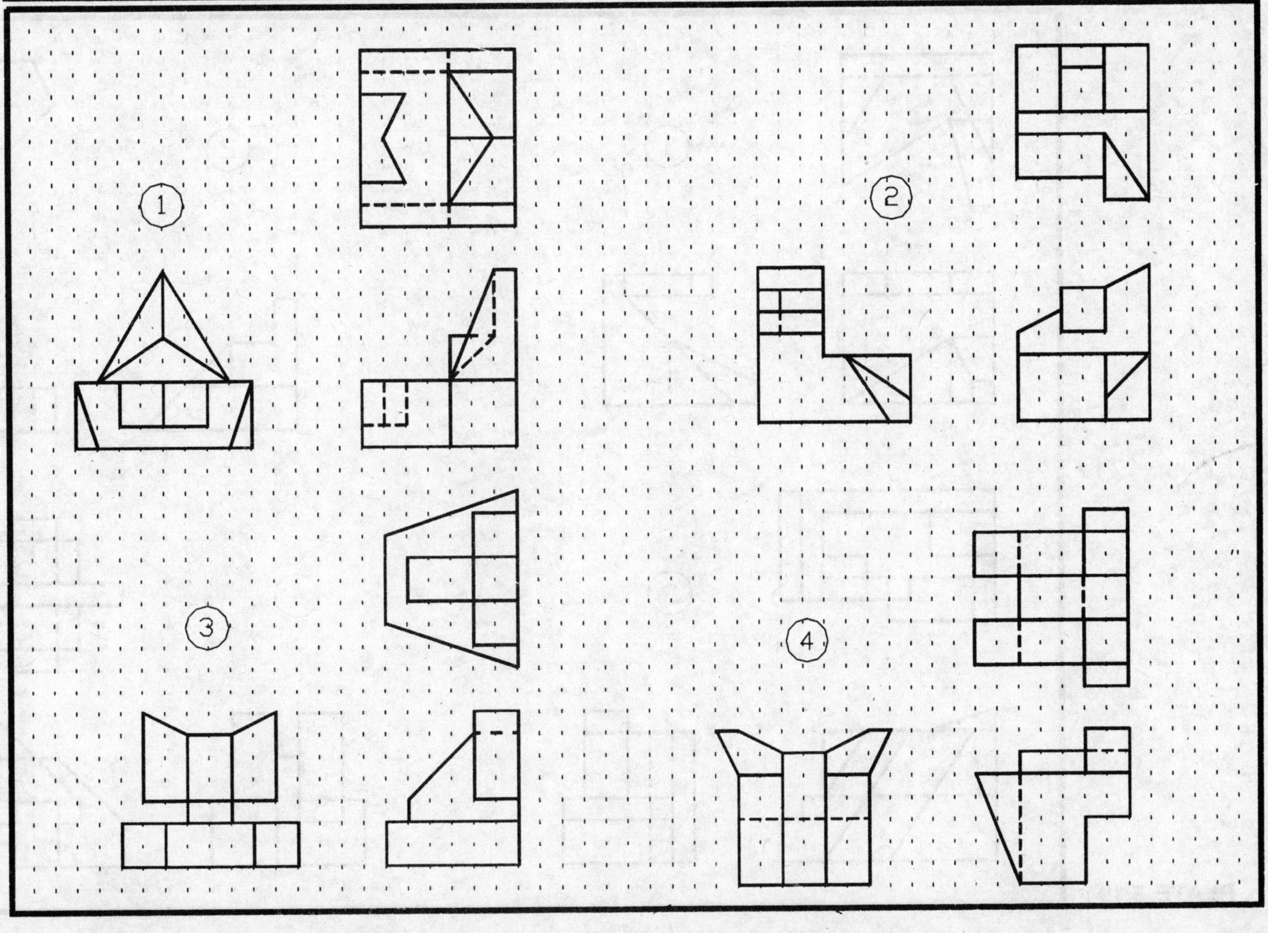


PLATE 6.17

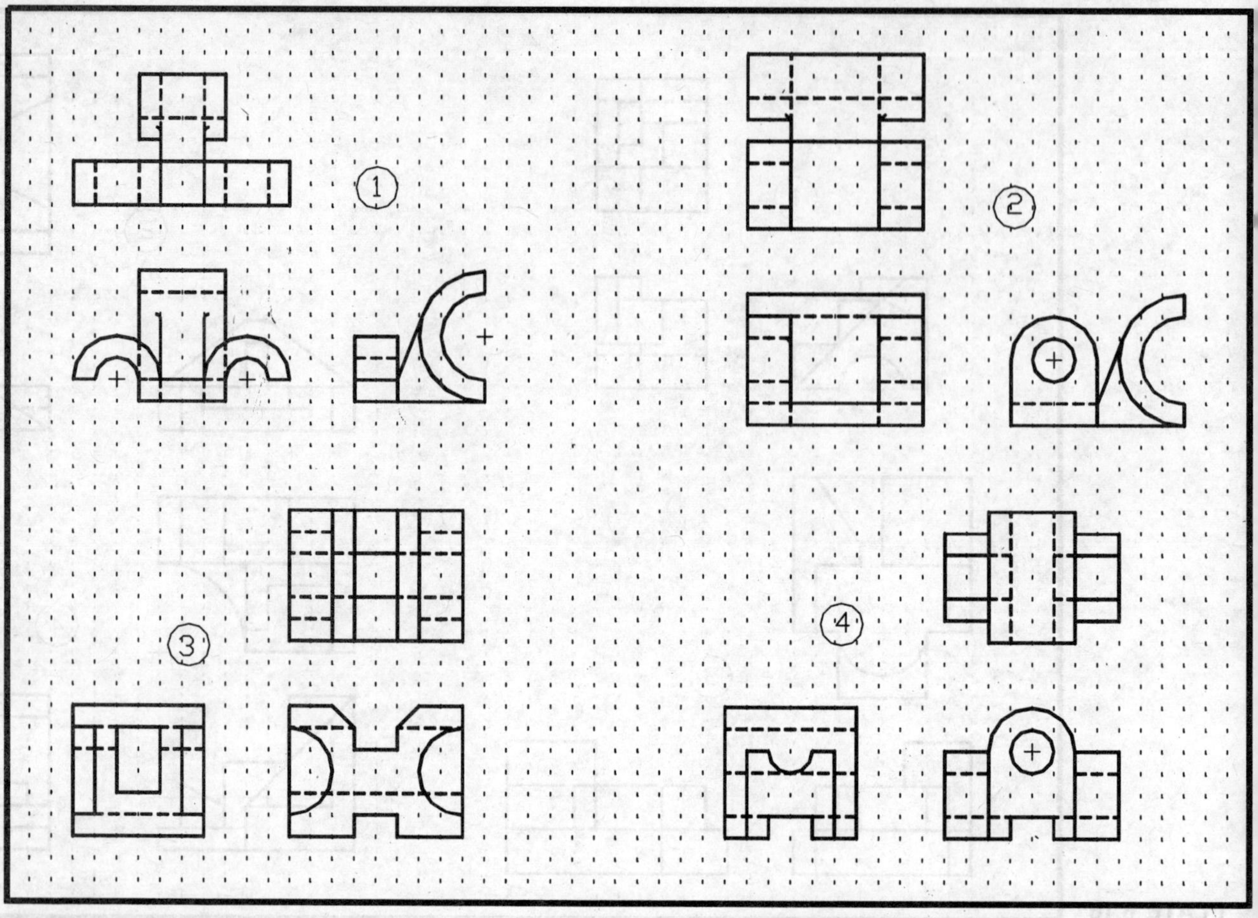


PLATE 6.18

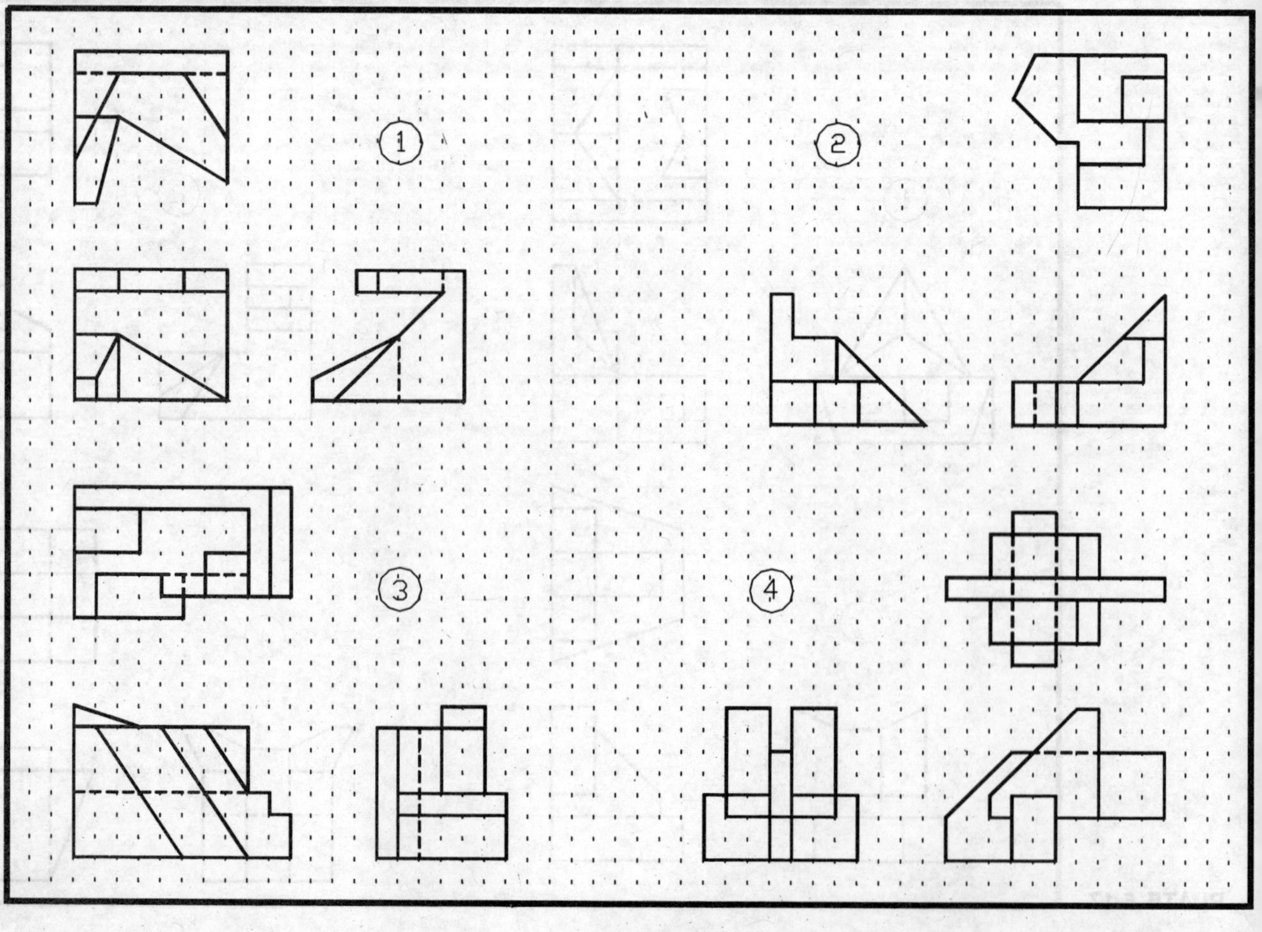


PLATE 6.19

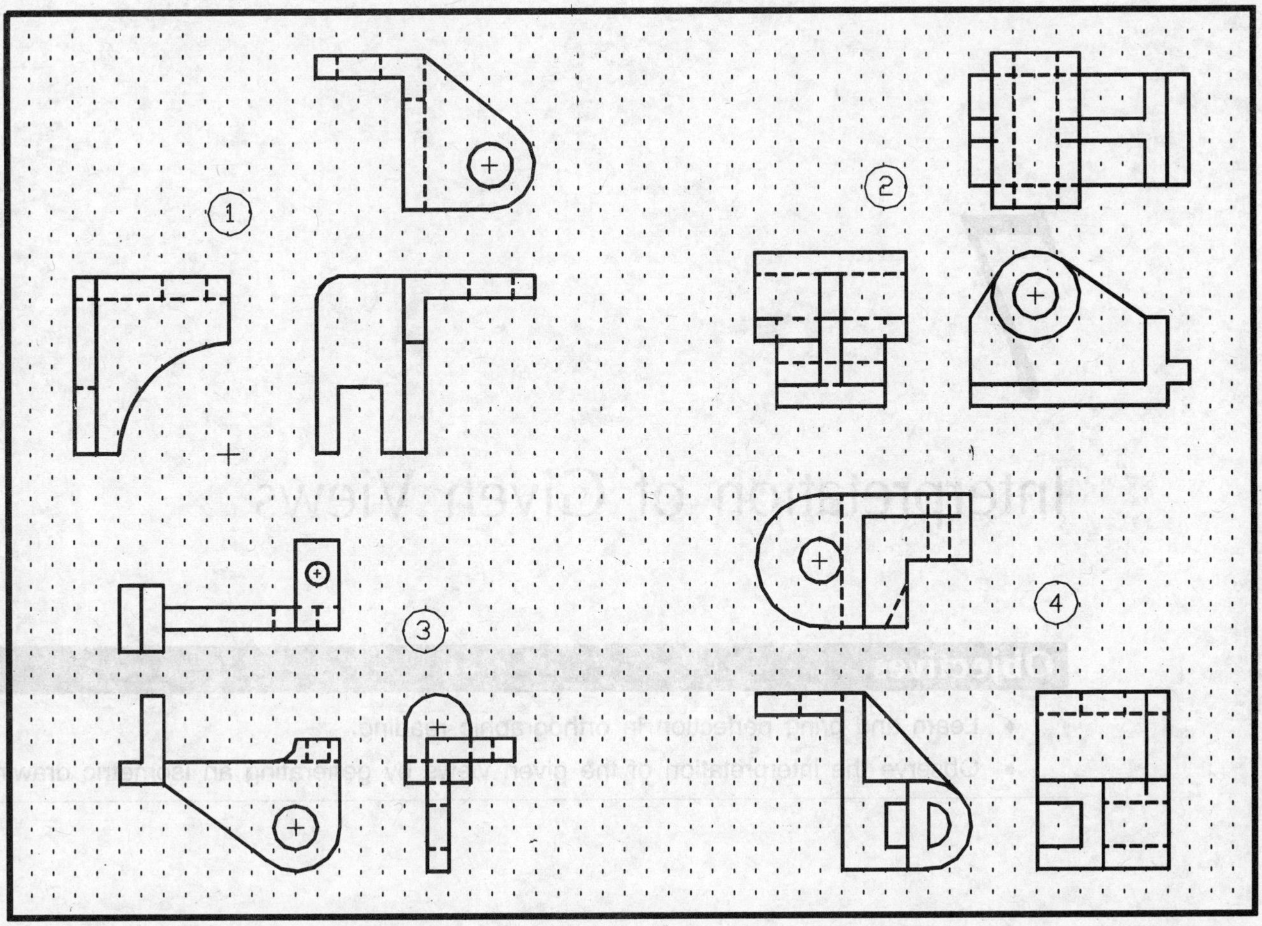


PLATE 6.20

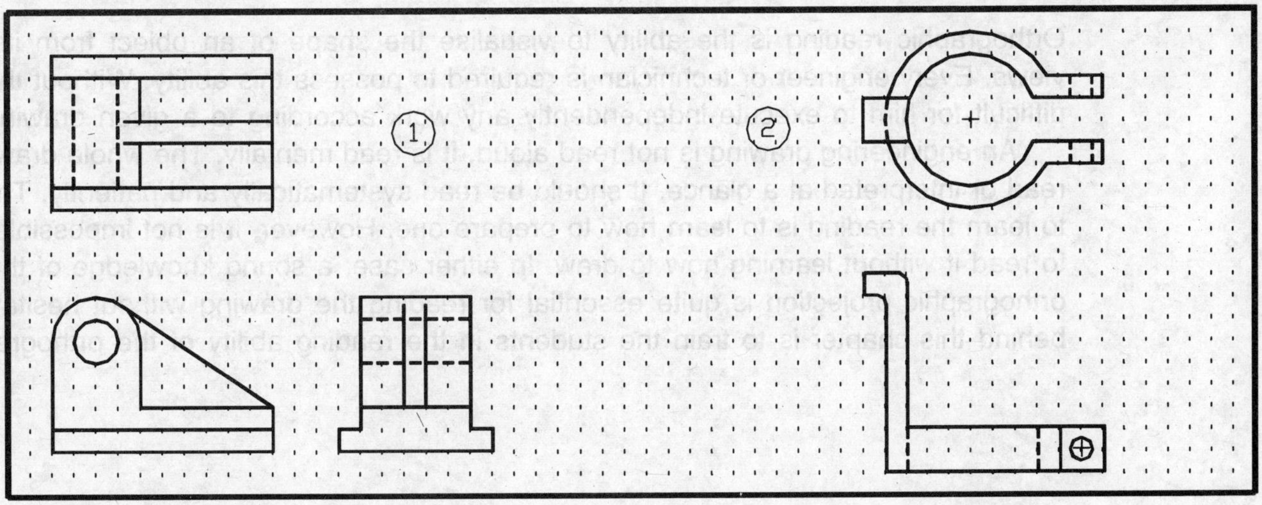


PLATE 6.21

# 7

# Interpretation of Given Views

## Objectives

- Learn and bring perfection in orthographic reading.
- Observe the interpretation of the given views by generating an isometric drawing.

## Introduction

Orthographic reading is the ability to visualise the shape of an object from its orthographic views. Every engineer or technician is required to possess this ability. Without this, it would be difficult for him to execute independently any work according to a given drawing.

An engineering drawing is not read aloud. It is read mentally. The whole drawing cannot be read or interpreted at a glance. It should be read systematically and patiently. The easiest way to learn the reading is to learn how to prepare one. However, it is not impossible to know how to read it without learning how to draw. In either case, a sound knowledge of the principles of orthographic projection is quite essential for reading the drawing without hesitation. The goal behind this chapter is to train the students in the reading ability of the orthographic views.

## 7.1 Visualisation of an Object with Partial Information

Most students in elementary graphics courses find it difficult to imagine an object from partial information of views. In this textbook, an attempt is made to create the reading ability and generate a 3-D imagination skill in students with the help of a number of simple 3-D objects and a number of simple orthographic views discussed in Chapters 5 and 6, respectively. In this Chapter, with the help of partial information, a student is asked to generate a 3-D object and, thereby, complete the given partial information. The given partial information could be of the following two types.

(a) **Missing line(s)**

All three views are given. At least one out of three views is complete. However, out of the remaining two views, either one view is a missing line(s) or both the views are missing line(s).

(b) **Missing view**

Two views are given. Both the views are complete. The third view is missing and is required to be completed.

In both the cases, we have to draw the missing view by creating a 3-D object, i.e., an isometric drawing. Instead of guessing a missing line or a missing view, one should start generating an isometric drawing from the given partial information. This approach is followed in this chapter. With this approach, both, orthographic reading ability and 3-D imagination skill can be achieved together. Creation of an isometric drawing from the given partial information is discussed in the following sections 7.1.1 and 7.1.2.

The block method (L × D × H) is used to create an isometric drawing. This method is already described in Chapter 6.

### 7.1.1 Correlation of Surfaces

From a given partial information, surfaces are correlated with each other and the views are completed without the creation of a 3-D object. Figure 7.1(a) shows TV and FV of a 3-D object. We are asked to complete an RSV. The projection lines are drawn from TV and FV to complete the RSV as shown in (a). Both the views (TV and FV) have got two rectangles, one smaller and one bigger. Bigger rectangle of TV corresponds to bigger rectangle in FV. Within the limits of its projection lines a rectangle is drawn as shown in (b). Similarly, smaller rectangle of TV corresponds to smaller rectangle in FV. Within the limits of its projection lines a square is drawn as shown in (c). The missing view (RSV) is completed without drawing an isometric drawing, which is shown in (d).

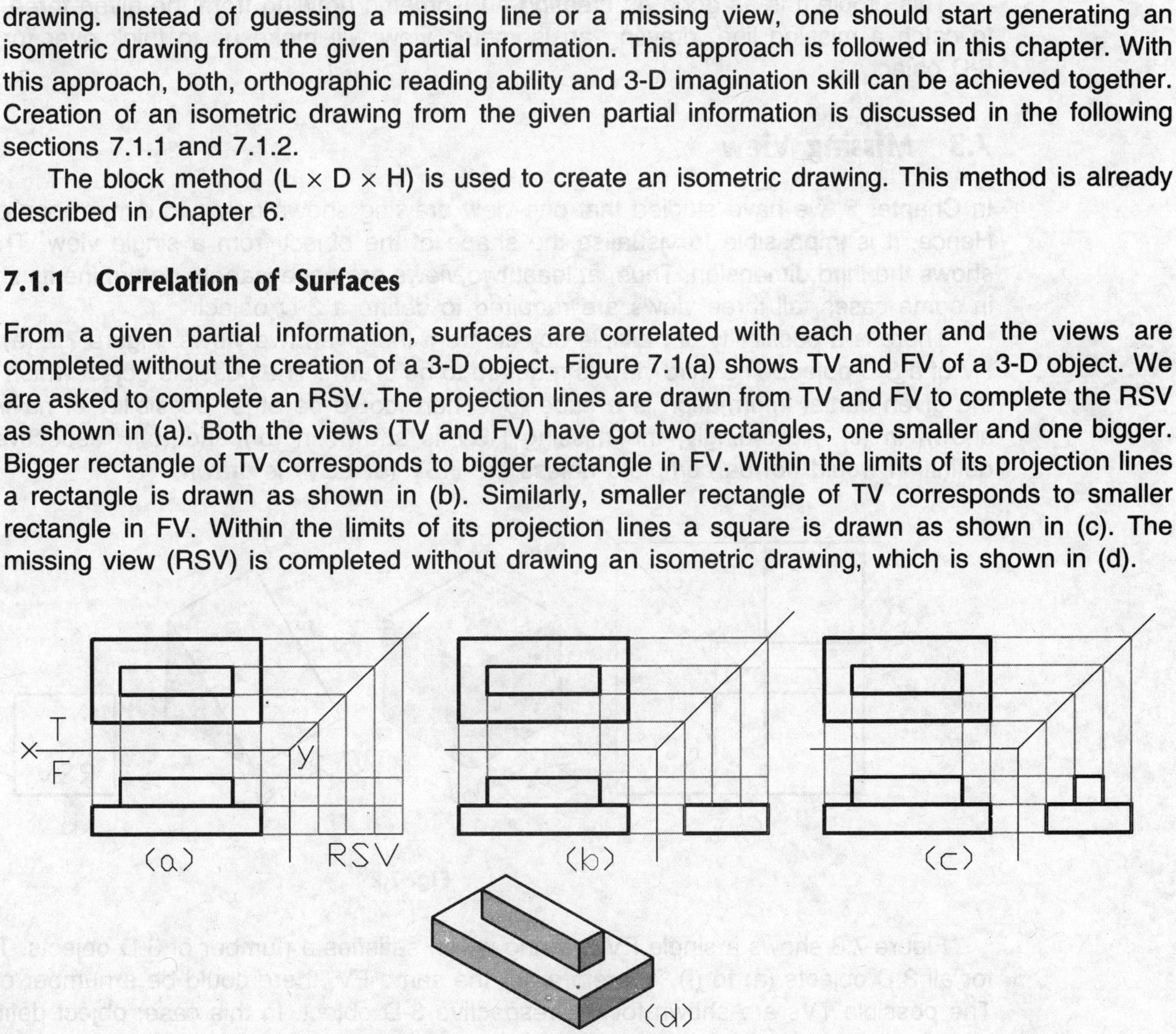


Fig. 7.1

### 7.1.2 3-D Object Imagination with Assembly or Disassembly Approach

The assembly or disassembly is another approach proposed to complete the given partia information. All objects may be considered as composed of elementary geometrical solids (primitives) such as cube, rectangular cuboid, prisms, pyramids, cylinders, cones and spheres o parts of such forms. Primitives may be either positive in the form of solid parts or negative in the form of cavities or holes. The information is discussed in details in Section 5.5.1.

## 7.2 Missing Lines

Sometimes, we come across one view or two views, which miss a line(s). If all three views out o three views miss lines then it may mislead to different object(s) or may waste our time in thinking process. By this time a reading knowledge has been developed through a practice of drawing ar orthographic views and isometric drawing in Chapters 5 and 6, respectively. One should be able to correlate the surfaces and lines in the given views. This correlation is to decide whether a particula line is a principal surface or an inclined surface. It could be a skew surface if there are three area views. The examples are set in such a fashion that, either one or two views out of three, however may miss a line but not all three.

This problem is as good as creating an isometric drawing from the given three views. In orde to catch a missing line, drawing an isometric view will make us to think over for a complicatec 3-D object.

## 7.3 Missing View

In Chapter 5, we have studied that one-view drawing shows only two dimensions of a 3-D object Hence, it is impossible to visualise the shape of the object from a single view. The second view shows the third dimension. Thus, at least two views are necessary to determine its shape. However in some cases, all three views are required to define a 3-D object.

There is a possibility of multiple objects from the given two views. Figure 7.2 (a) shows TV anc FV of a 3-D object. The third view is required to be drawn. The possible object which will not destro the given partial information is a cube (b). There could be other possibility of having a wedge as shown in (c). Accordingly, the missing RSV is shown in (b′) and (c′), respectively. The objec definition would not be complete unless the RSV (or LSV) is drawn.

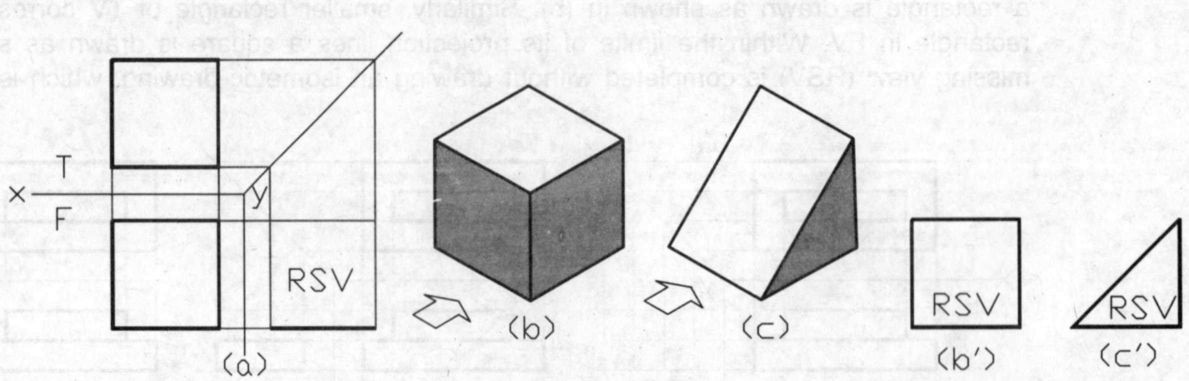


**Fig. 7.2**

Figure 7.3 shows a single FV drawing which satisfies a number of 3-D objects. The FV is same for all 3-D objects (a) to (f). Therefore, for the same FV, there could be a number of possible TVs The possible TVs are shown for the respective 3-D object. In this case, object definition would be complete with the help of TV and FV. If required, RSV or LSV can be drawn from TV and FV information.

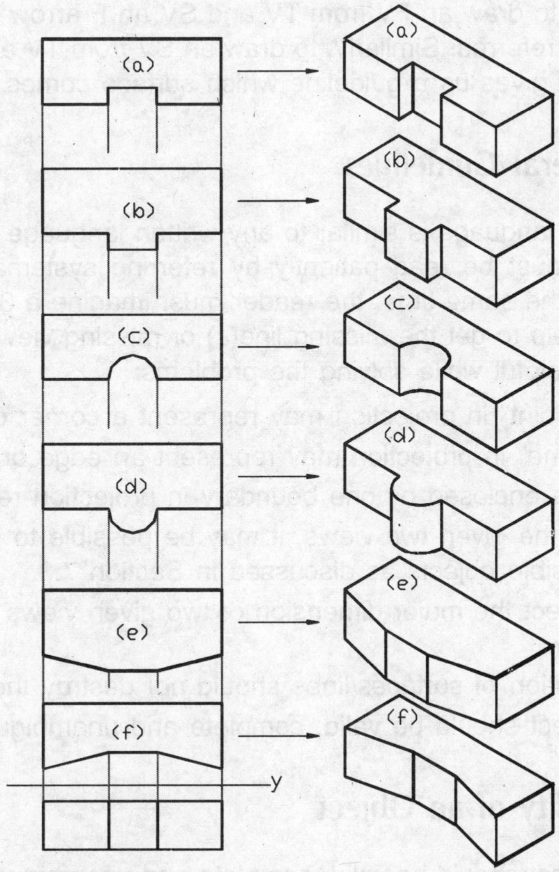


**Fig. 7.3** Multiple possibilities of object from single FV.

## 7.4 Direction of the Observer for the Precedence of Entities

In interpretation of given views, with partial information, it is essential to understand the precedence of entities (lines or surfaces). Figure 7.4 shows TV, FV, LSV and RSV of a rectangular cuboid. The direction of the observer is shown with the help of an arrow to draw one view from the other two.

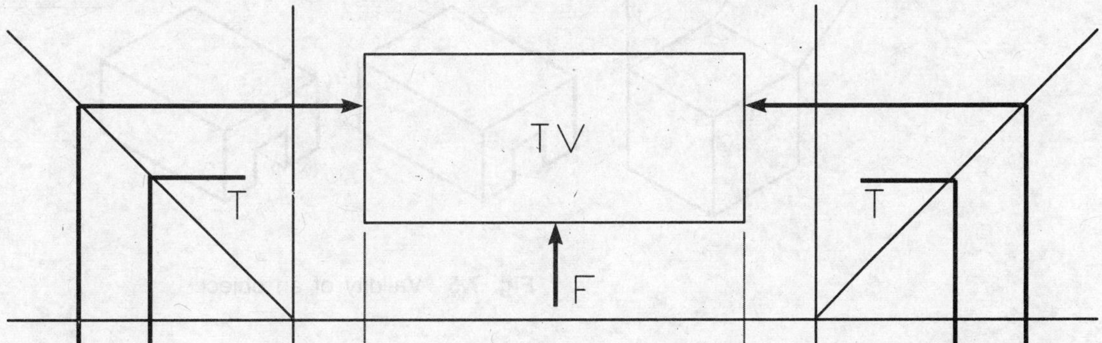


**Fig. 7.4** Direction of the observer for precedence of entities.

In order to draw an FV from TV and SV an F arrow is referred. To draw a TV from FV and SV a T arrow is referred. Similarly, to draw an SV from TV and FV an S arrow is referred. The direction of the arrow gives us a guideline which surface comes first, which is next and which is last.

## 7.5 General Guidelines

The graphic language is similar to any written language and no language can be read at a glance. A drawing must be read patiently by referring systematically, back and forth, from one view to another. At the same time, the reader must imagine a 3-D object. There are no fixed sets of rules which will help to get the missing line(s) or missing view rapidly. However, the following guidelines should be helpful while solving the problems:

(i) A "point" in projection may represent a corner or an edge of an object.
(ii) A "line" in projection may represent an edge or a surface of an object.
(iii) Area enclosed by one boundary in projection represents either a flat or a curved surface.
(iv) For the given two views, it may be possible to imagine a different third view i.e., different possible objects as discussed in Section 7.3.
(v) Project the major dimension of two given views to obtain the limiting boundary of the third view.
(vi) Addition of surfaces/lines should not destroy the given partial information.
(vii) Object should be valid, complete and unambiguous.

## 7.6 Validity of an Object

The 3-D object should be valid, complete and unambiguous. A valid solid is one that does not have dangling edge or dangling surface as shown in Fig. 7.5(a). The solid shown in Fig. 7.5(b) is incomplete and appears to be ambiguous. The reason is that the slot feature is not complete and clear. The correct representation is shown at adjacent location. The line on a surface has no meaning as shown in Fig. 7.5(c).

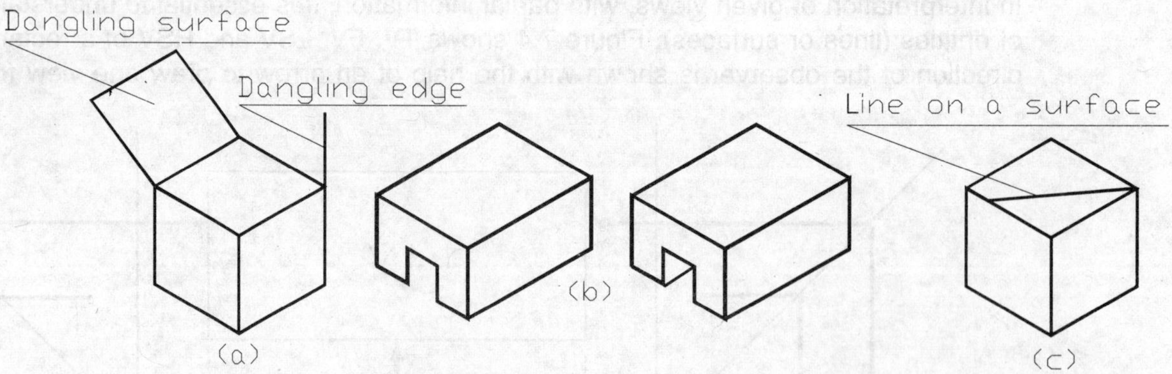


**Fig. 7.5** Validity of an object.

**EXAMPLE 7.1** Figure 7.6(a) shows three views of an object. Out of three views, one view is complete. Add missing line(s) in other view(s).

***Solution***

1. Project the TV, FV and RSV as shown in (a).
2. In TV and FV there is one projection line which is not covered. Therefore, both the views are missing lines.
3. With the help of major dimensions, (L × D × H) draw a box.
4. Add a key feature of RSV and imagine the 3-D object, simultaneously, as shown in (b).

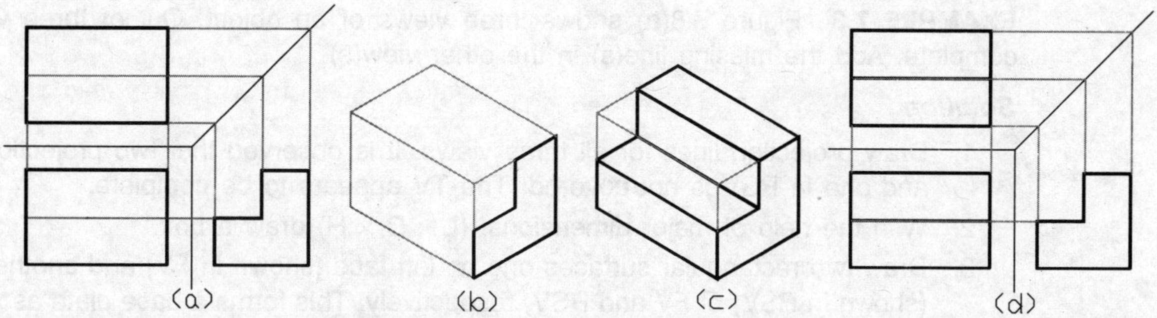


Fig. 7.6

5. Add rectangular surfaces associated with every smaller edge of the key feature in a box. This is shown in (c).
6. Make the generated 3-D object (c) as a question and the given partial information (orthographic views) as the answer. With this, one can check that the addition of imagined surfaces has not destroyed the given partial information.
7. Add missing line in TV and FV as shown in (d).
8. Check finally, for the 3-D object and the solution you proposed.

**EXAMPLE 7.2** Figure 7.7(a) shows three views of an object. Out of three views, one view is complete. Add the missing line(s) in the other view(s).

***Solution***

1. Project the TV, FV and RSV as shown in (a).
2. FV has one projection line which is not covered. Therefore, FV view is a missing line.
3. With the help of major dimensions, (L × D × H) draw a box.
4. Add a key feature of RSV on the right face of a box and imagine the 3-D object, simultaneously. This is shown in (b).
5. Add rectangular surfaces associated with every edge of the key feature in a box as shown in (c).
6. Make the generated 3-D object (c) as a question and the given partial information (orthographic views) as answer. With this, one can check that the addition of imagined surfaces has not destroyed the given partial information.
7. Add a missing line, which is a hidden line in FV as shown in (d). This hidden line represents an invisible surface from the front (F) direction.
8. Check finally, for the 3-D object and the solution you proposed.

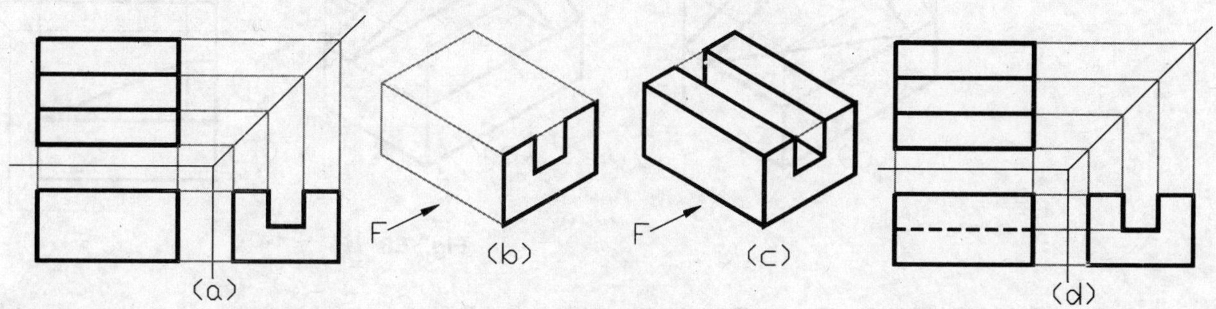


Fig. 7.7

**EXAMPLE 7.3** Figure 7.8(a) shows three views of an object. Out of three views, one view is complete. Add the missing line(s) in the other view(s).

***Solution***

1. Draw projection lines for all three views. It is observed that two projection lines, one in FV and one in RSV is not covered. The TV appears to be complete.
2. With the help of major dimensions, (L × D × H) draw a box.
3. Draw two rectangular surfaces one on left face (shown in FV) and another on the right face (shown in RSV) of FV and RSV, respectively. This forms a base plate as shown in (b). Draw a square surface on the top face of the box as shown in (b).
4. Draw two rectangular slant surfaces connecting square surface with base plate as shown in the TV. This is shown in (e).
5. Make the generated 3-D object (c) as a question and the partial information as the answer. With this, one can check that the addition of imagined surfaces has not destroyed the given partial information. However, the object is not complete.
6. As shown in (d), we can draw a line on a base plate connecting two slant rectangular surfaces. The generated object is valid, complete. However, it will destroy the given partial TV (which is complete).
7. In order to satisfy the uncovered projection lines, two vertical triangular surfaces are constructed by drawing three lines.
8. Draw missing lines in TV and FV as shown in (f).
9. Check finally, for the 3-D object validity and the solution you proposed.

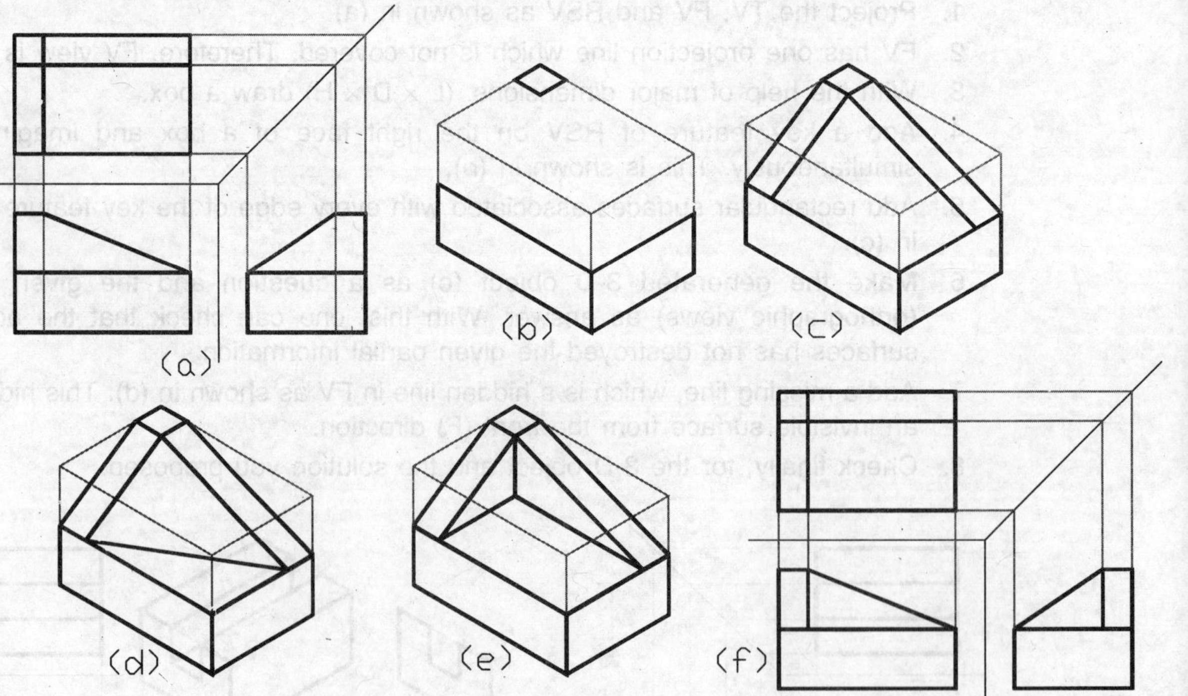


**Fig. 7.8**

**EXAMPLE 7.4** Figure 7.9(a) shows TV and RSV of an object. Imagine the possible front view and draw the same.

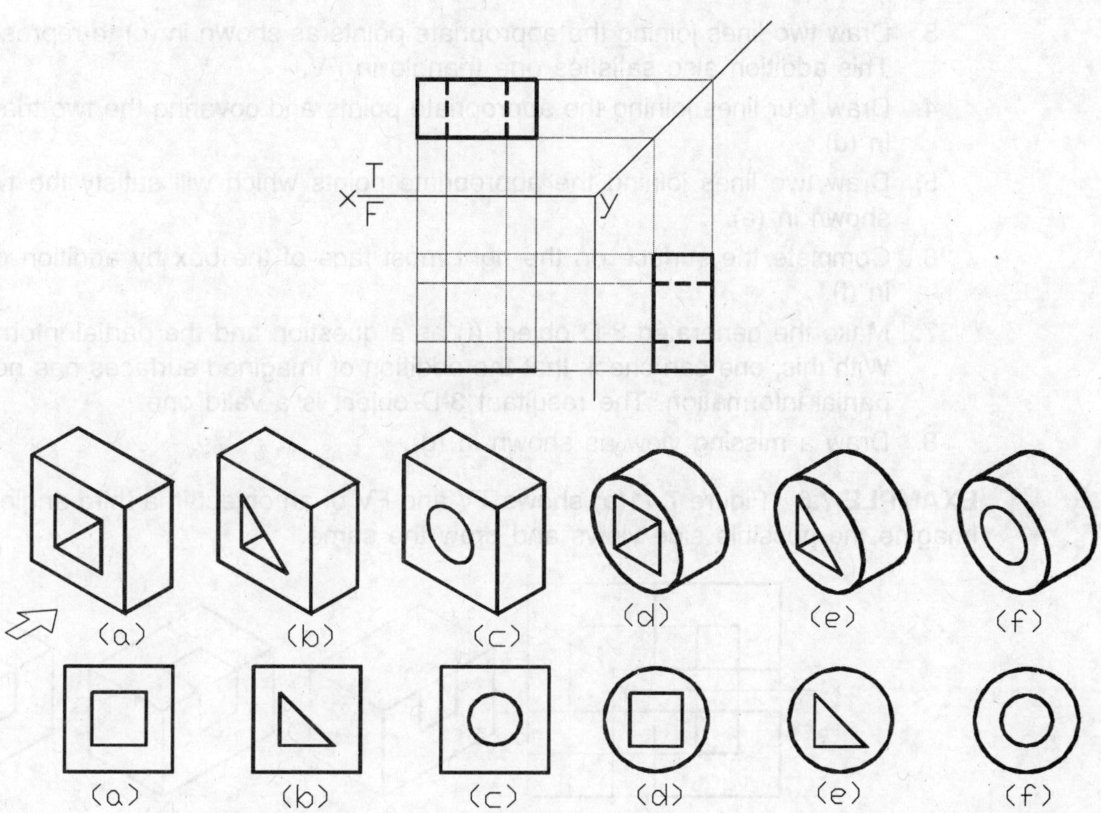


**Fig. 7.9**

***Solution*** Figure (a), (b), (c), (d), (e) and (f) show possible 3-D objects which can satisfy the given two views. Accordingly, the FVs are drawn for all six possibilities as shown in the figure.

**EXAMPLE 7.5** Figure 7.10(a) shows TV and FV of an object. Imagine the possible right side view and draw the same.

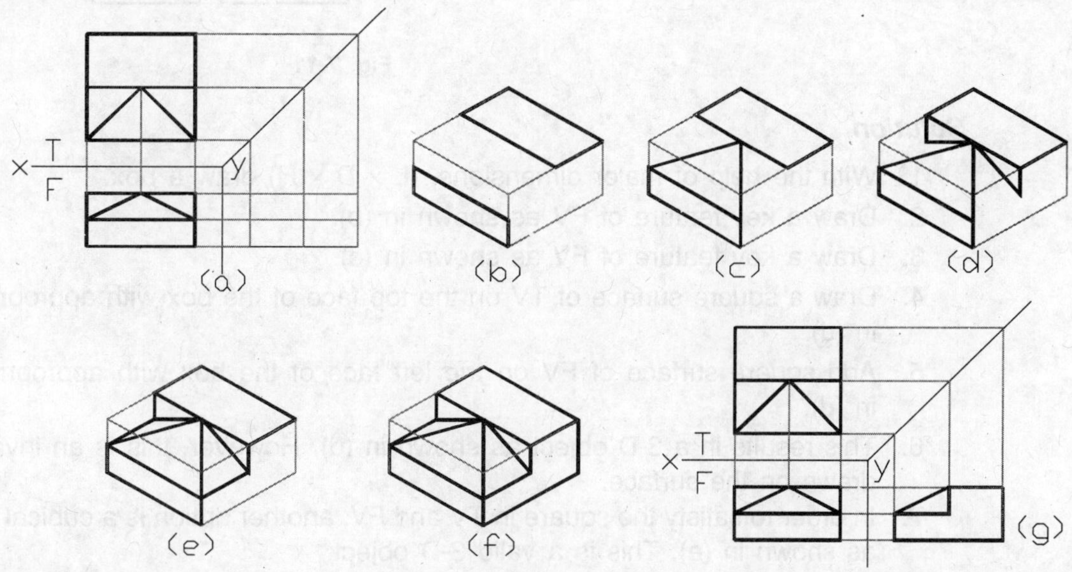


**Fig. 7.10**

***Solution***

1. With the help of major dimensions, (L × D × H) draw a box.
2. Draw two rectangular surfaces of TV and FV at appropriate location in the box as shown in (b).

3. Draw two lines joining the appropriate points as shown in (c) to represent the triangle of TV. This addition also satisfies one triangle in FV.
4. Draw four lines joining the appropriate points and covering the two triangles of FV as shown in (d).
5. Draw two lines joining the appropriate points which will satisfy the two triangles of TV as shown in (e).
6. Complete the surface on the right most face of the box by addition of two lines as shown in (f).
7. Make the generated 3-D object (f) as a question and the partial information as the answer. With this, one can check that the addition of imagined surfaces has not destroyed the given partial information. The resultant 3-D object is a valid one.
8. Draw a missing view as shown in (g).

**EXAMPLE 7.6** Figure 7.11(a) shows TV and FV of an object in a third-angle method of projection. Imagine the possible side views and draw the same.

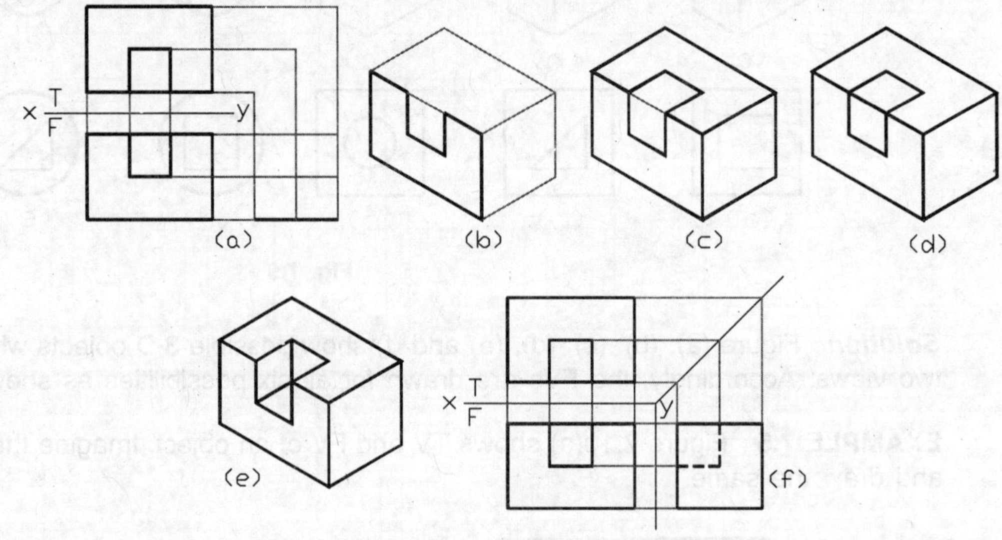


**Fig. 7.11**

***Solution***

1. With the help of major dimensions, (L × D × H) draw a box.
2. Draw a key feature of FV as shown in (b).
3. Draw a key feature of FV as shown in (c).
4. Draw a square surface of TV on the top face of the box with appropriate location as shown in (d).
5. Add square surface of FV on the left face of the box with appropriate location as shown in (d).
6. This results in a 3-D object as shown in (d). However, this is an invalid object as lines are drawn on the surface.
7. In order to satisfy the square in TV and FV, another option is a cubical cavity inside the object as shown in (e). This is a valid 3-D object.
8. Draw a missing view as shown in (f).

## Exercise A (Missing Lines)

**7.1** Copy the given views and add the missing LINE(S) or VIEW(S) for the problems given in Plate Nos. 7.1, 7.2, 7.3, 7.4, 7.5, 7.6, 7.7 and 7.8. Change the colour of the missing line.

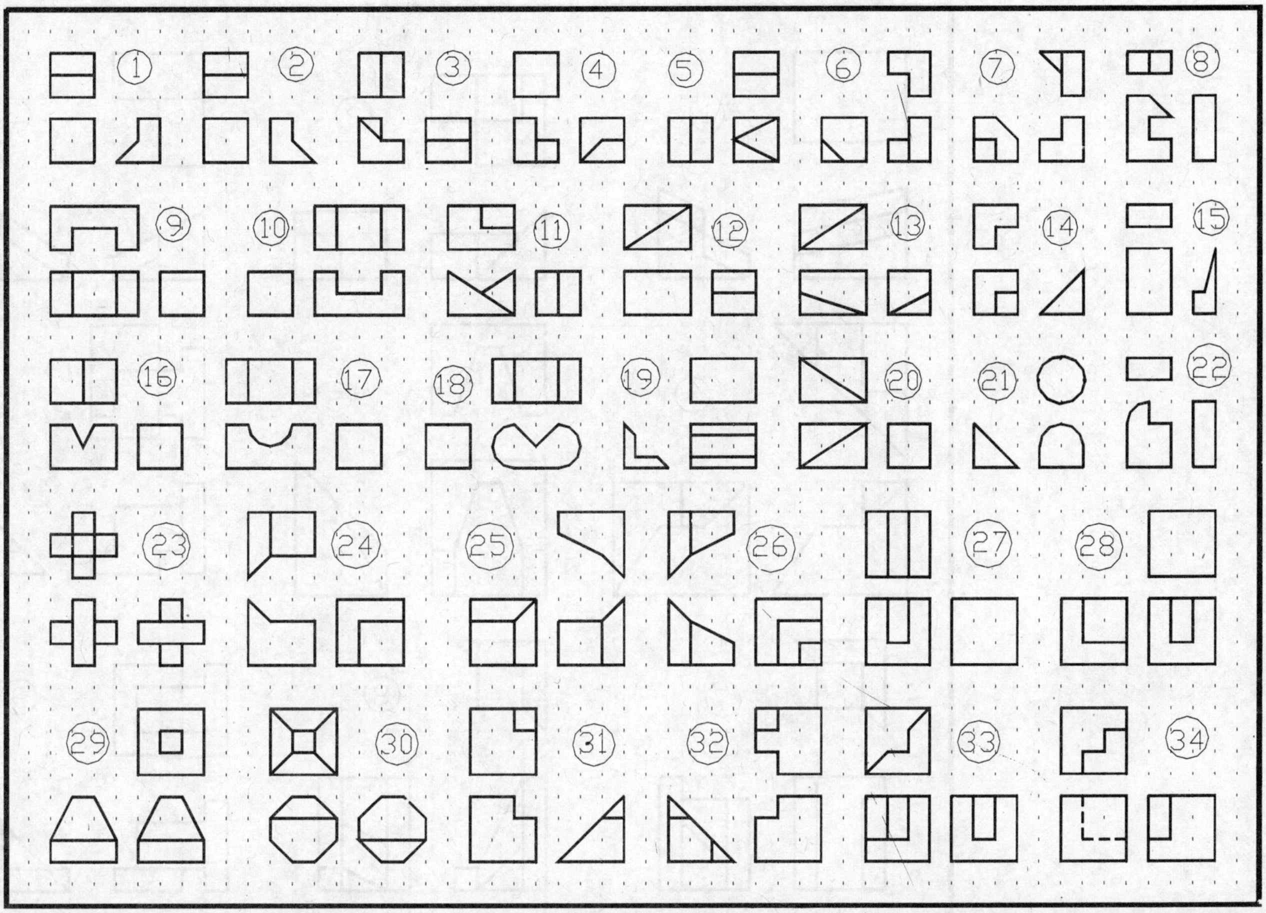


PLATE 7.1

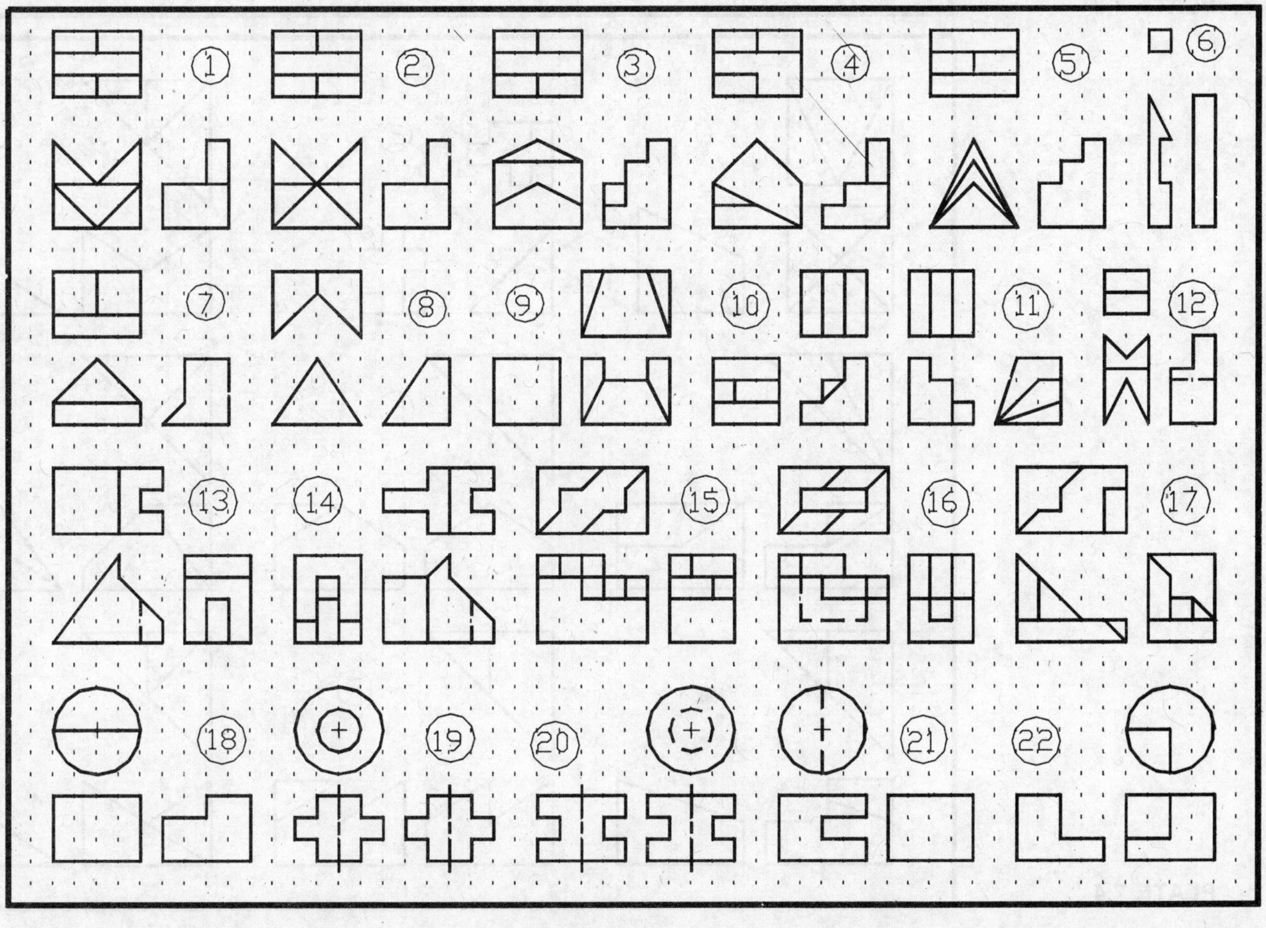


PLATE 7.2

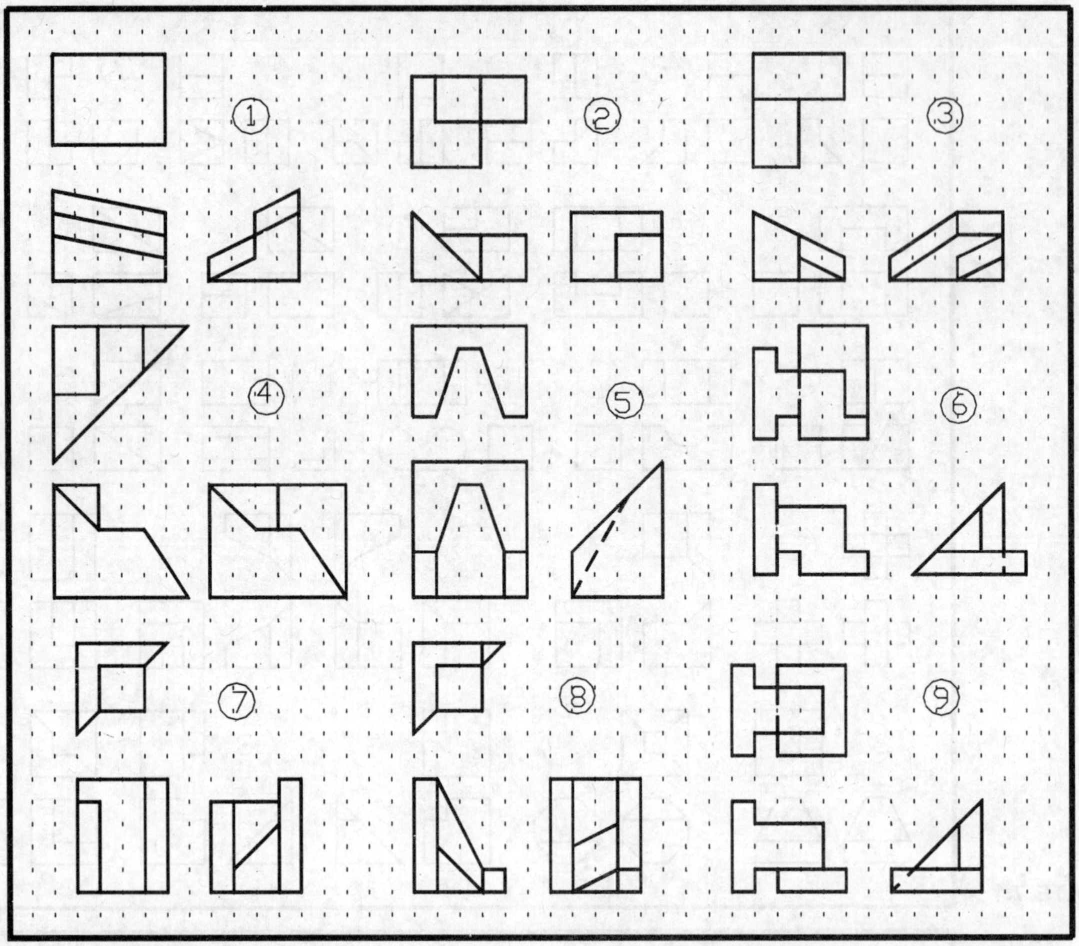


**PLATE 7.3**

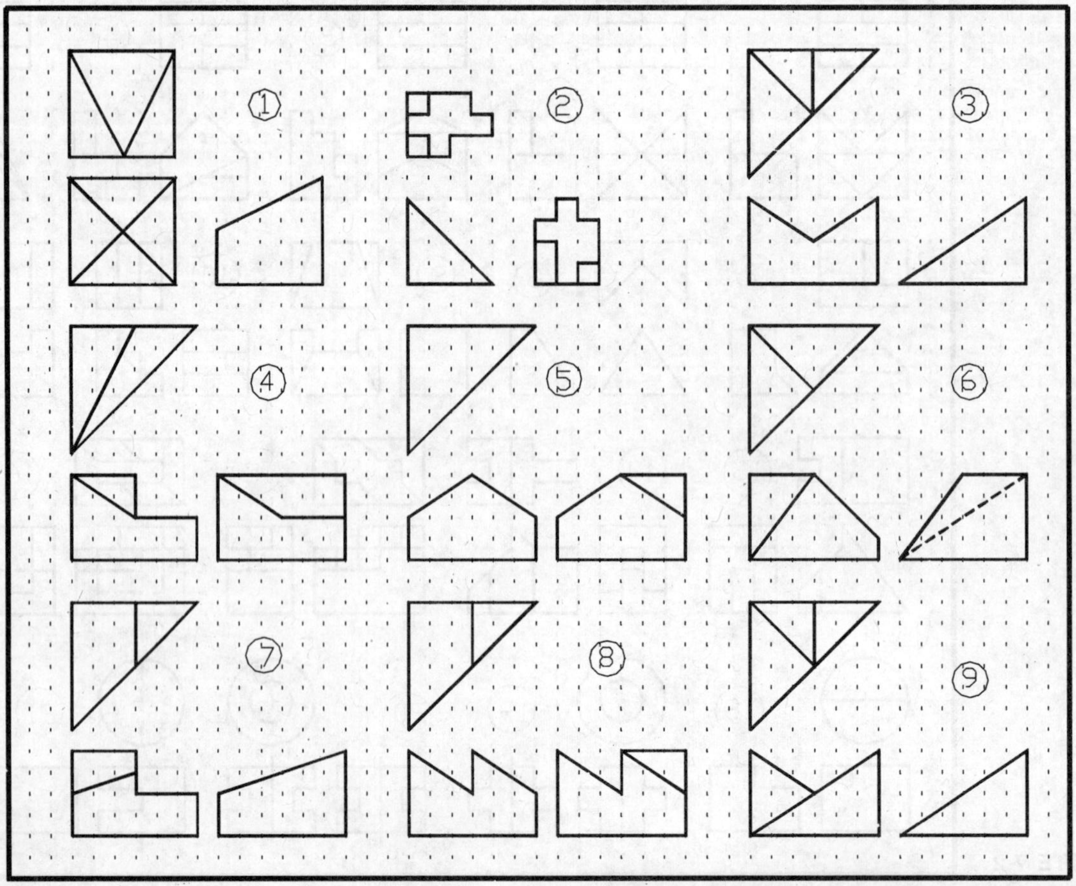


**PLATE 7.4**

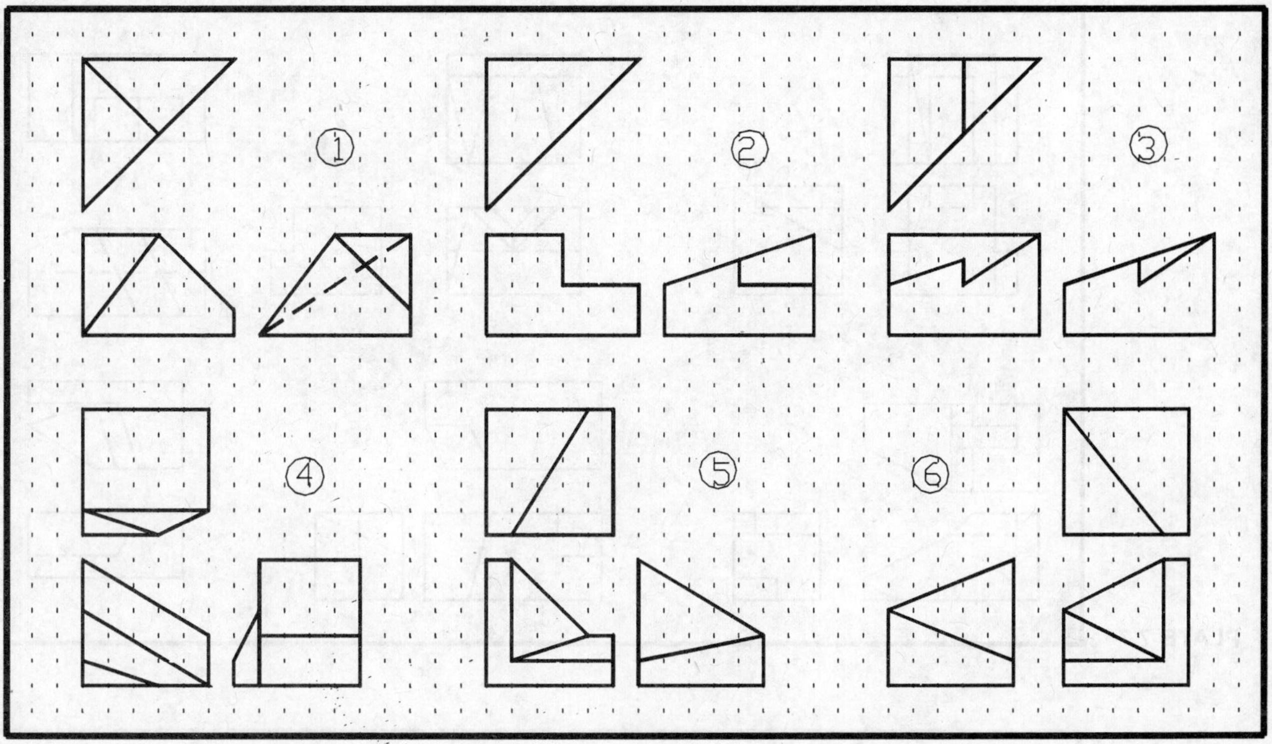


PLATE 7.5

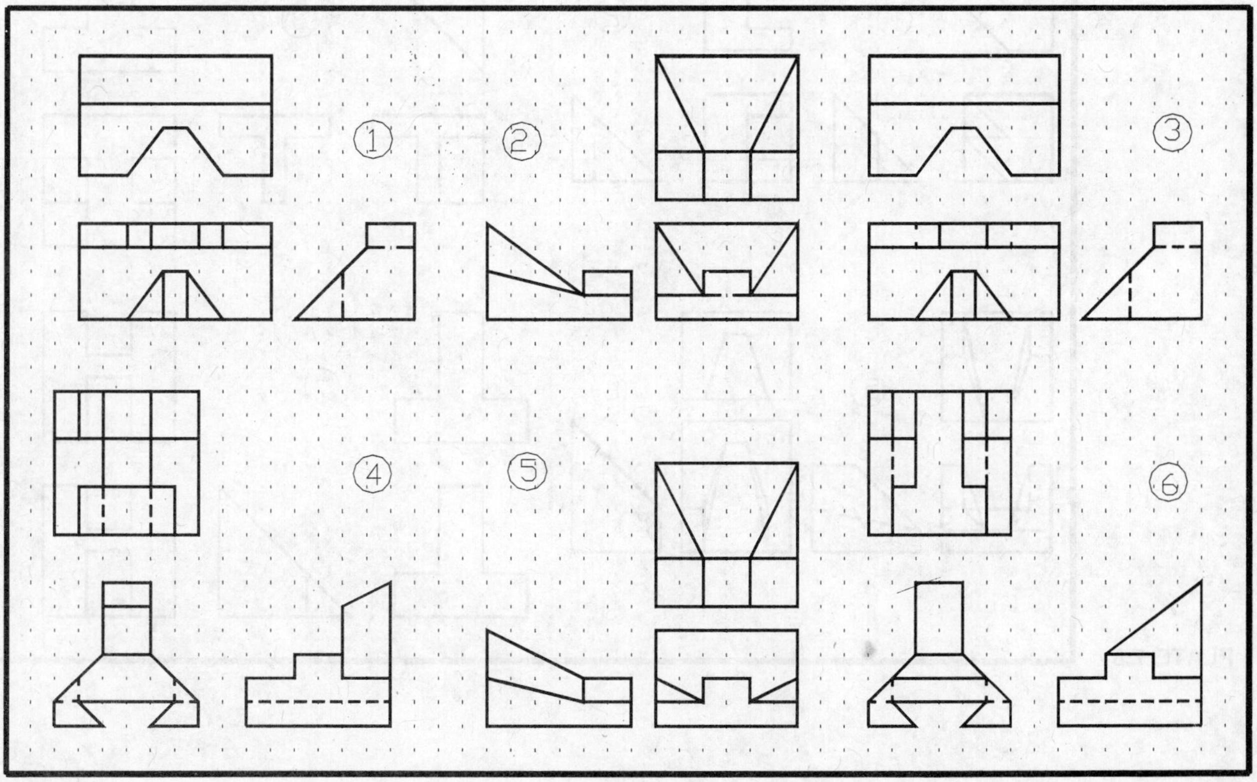


PLATE 7.6

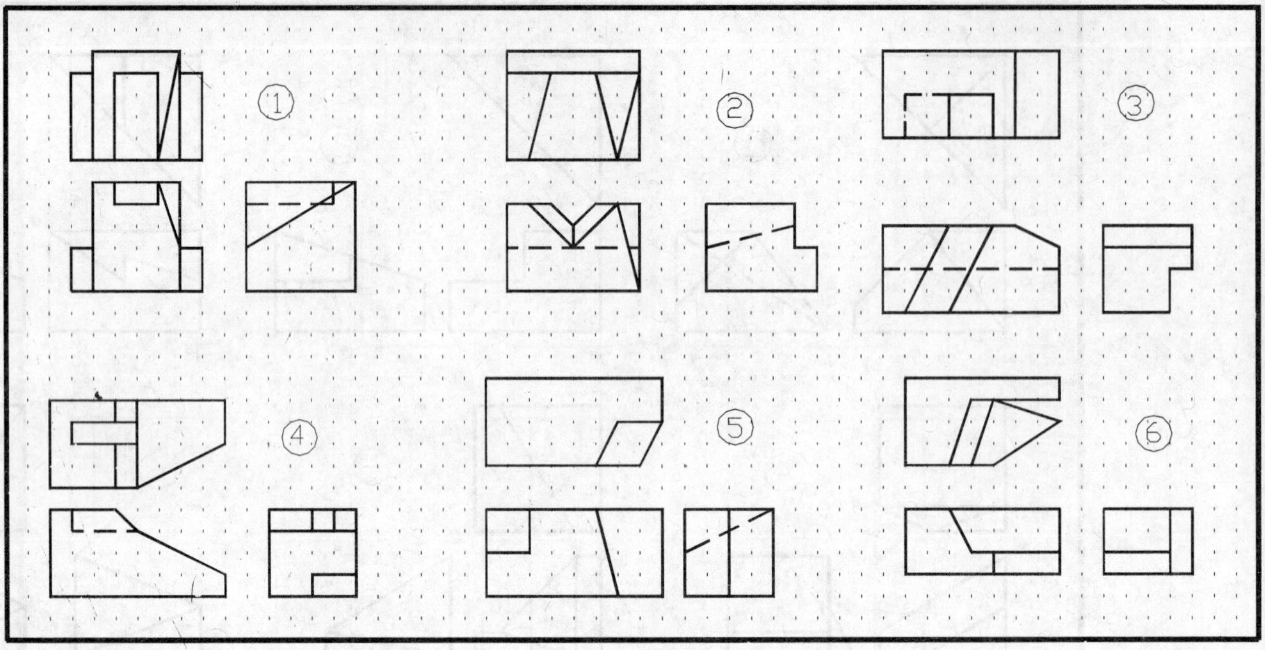


PLATE 7.7

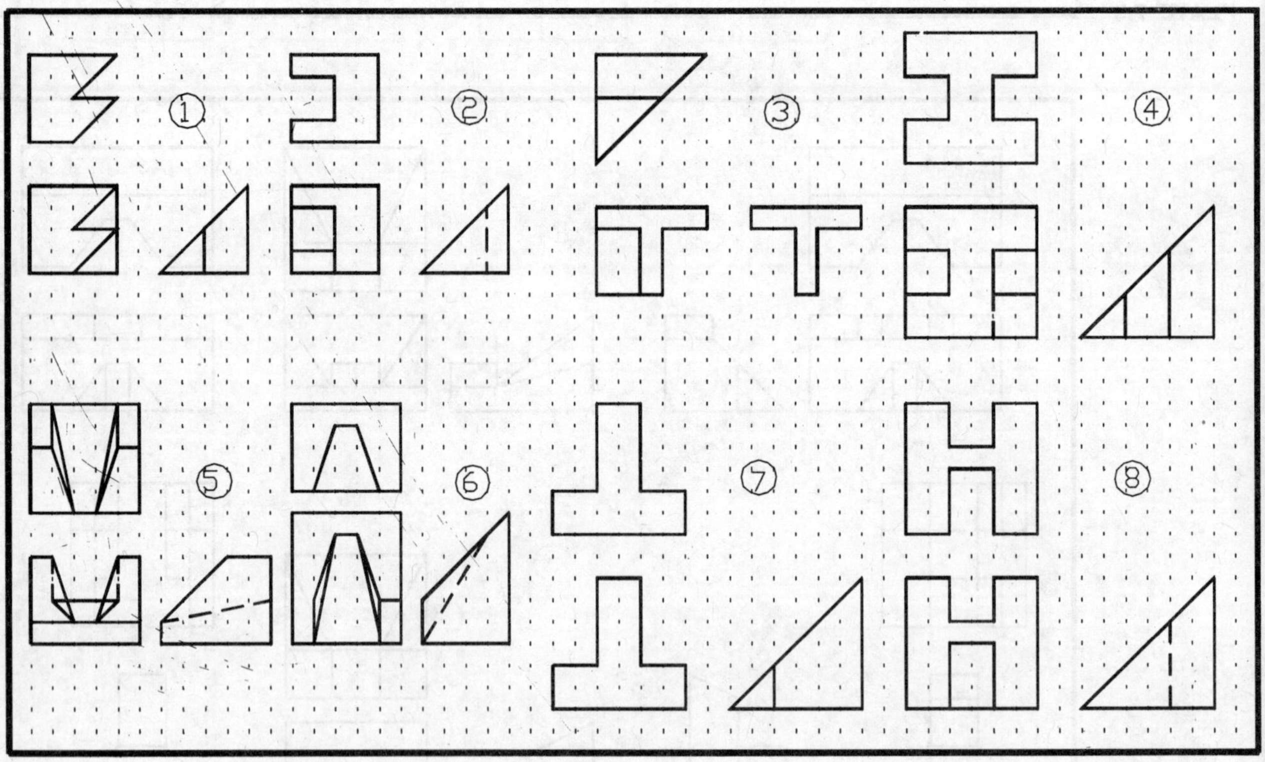


PLATE 7.8

## Exercise B (Missing Views)

**7.2** Copy the given views and add the missing VIEW(S) for the problems given in Plate Nos. 7.9, 7.10 and 7.11 and draw an isometric view.

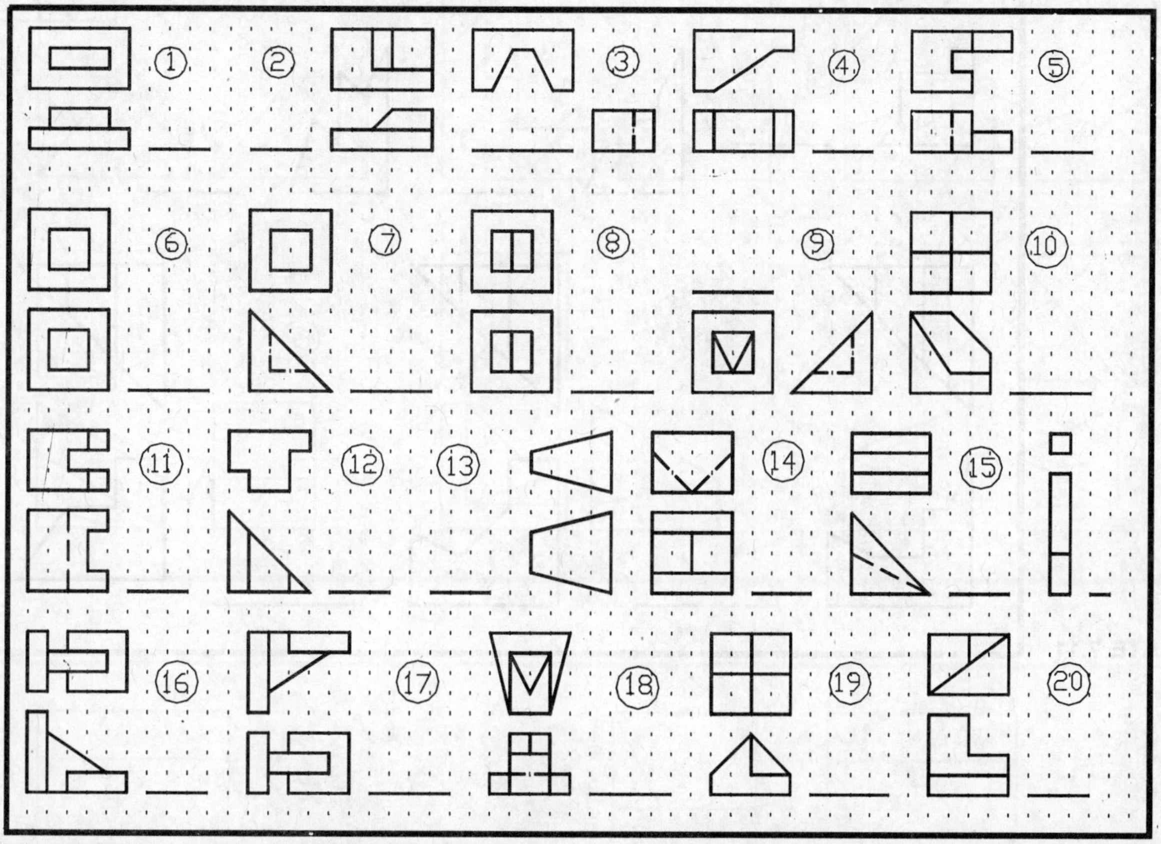


**PLATE 7.9**

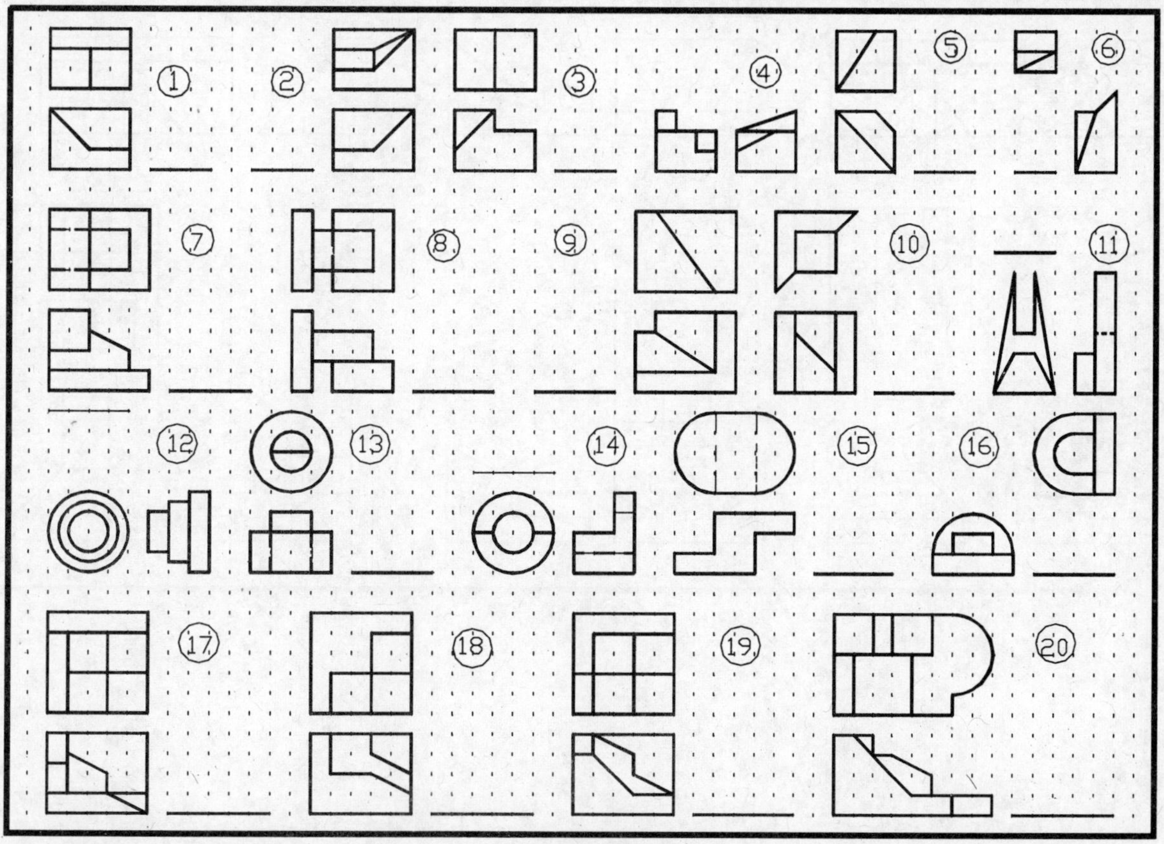


**PLATE 7.10**

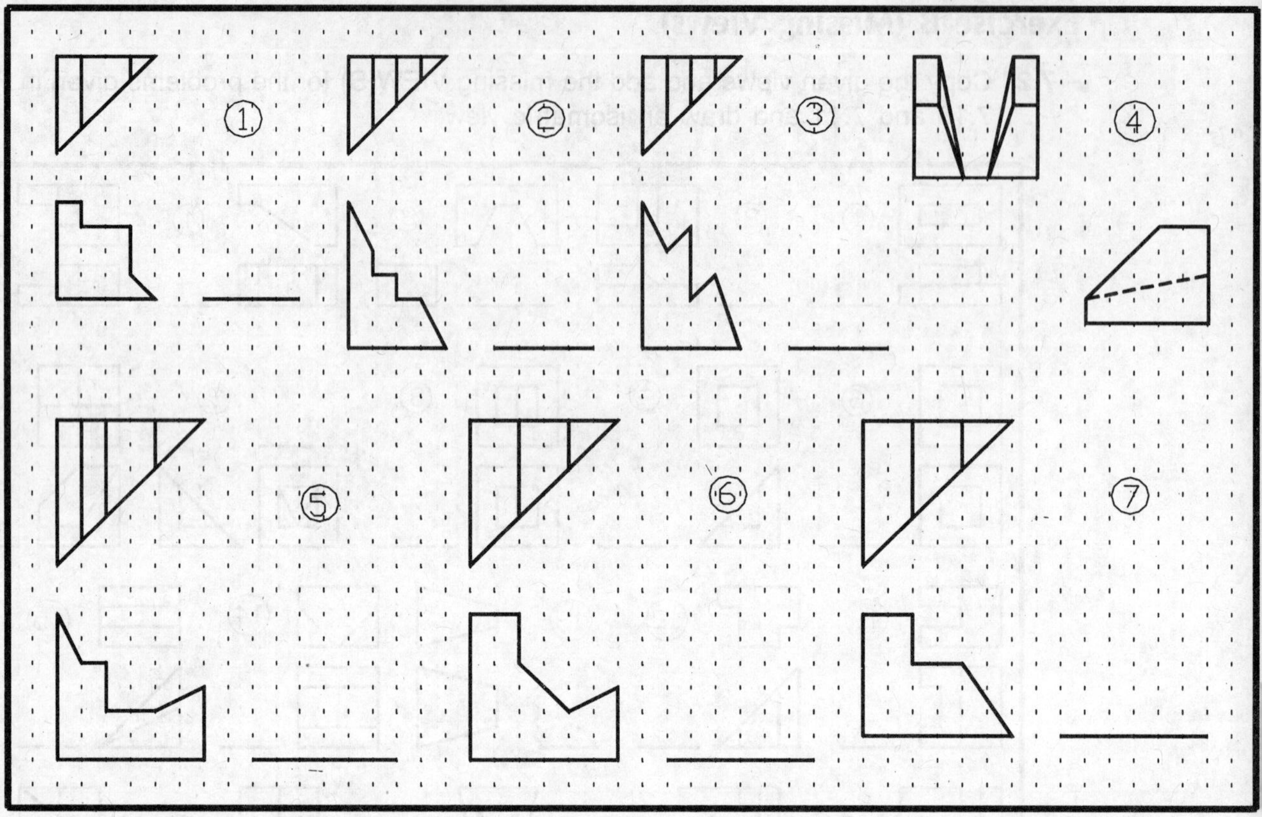


PLATE 7.11

# 8

# Auxiliary Projections

## Objectives

- Studying auxiliary projection planes.
- Learning AutoCAD technique to obtain the true shape and desired features of an inclined surface in an object using an auxiliary view.
- Knowing AutoCAD technique to obtain the true shape and desired features of an oblique surface in an object using auxiliary views.

## Introduction

When the surface of an object are at right angles to one another and parallel or perpendicular to either of the three principal planes (HP, VP and PP), regular views (TV, FV and SV) are adequate for their representation. However, if there is an inclined or an oblique surface, the regular views will not show the true shape of the inclined or oblique surface. To show the true shape, an auxiliary (additional) view is used. The additional view obtained on an auxiliary projection plane is called an auxiliary view.

## 8.1 Need of Auxiliary Views

The theories underlying the principal views apply to the auxiliary views as well. In other words, a auxiliary view shows the object as it would appear to an observer stationed at an infinite distanc away.

In commercial drafting, an auxiliary view is usually a partial view showing only the desire information. The reason for this is that a projection showing the entire object adds very little to th description of the shape. Showing the entire object in the auxiliary view is likely to defeat th intended purpose of the view. For example, a complete drawing must include an auxiliary view c the inclined surface to show the true shape of the surface, profile and the location of the holes.

Figure 8.1 shows an object with an inclined surface. A square hole is cut through the objec which is perpendicular to the inclined surface as shown in the figure.

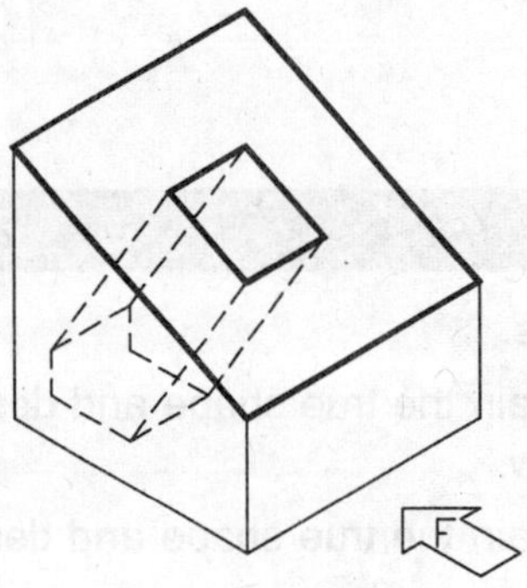


(a) Object

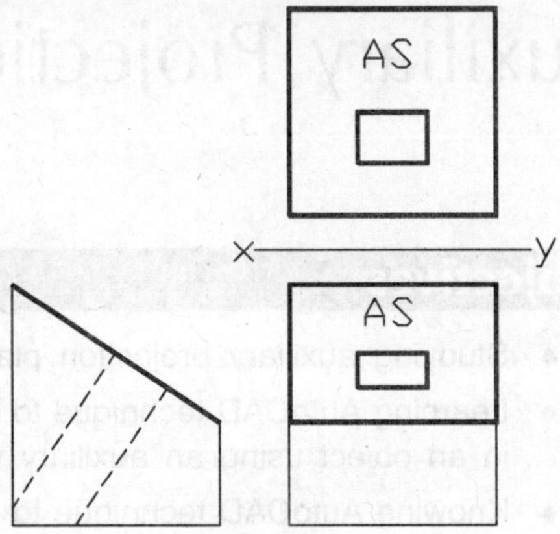


(b) Views

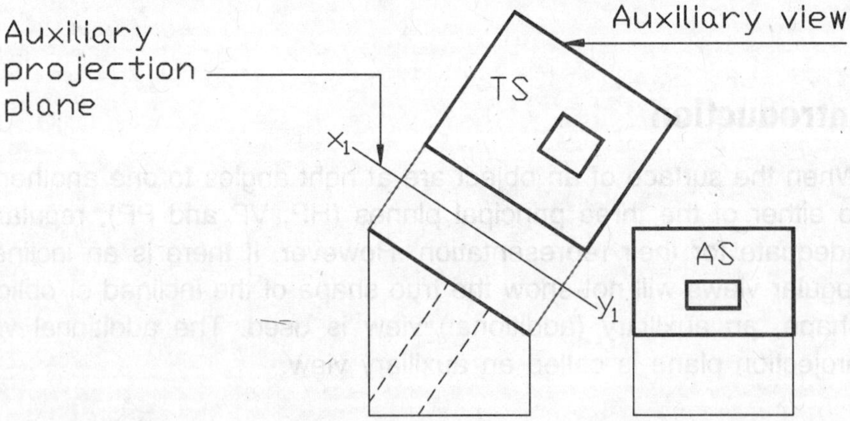


(c) Use of auxiliary plane to obtain the true shape with an auxiliary view

**Fig. 8.1** Inclined surface.

Figure 8.1(b) shows three views TV, FV and LSV. In all three primary views, the details of th holes are not clear because the surface is inclined and the hole is perpendicular to the incline surface. In order to view a true shape of the inclined surface and true shape of the hole-profile, eith the object or the observer should be adjusted in such a way that the visual rays are perpendicula to the inclined surface and parallel to the interior hole surface.

Figure 8.1(c) shows an additional (auxiliary) view with the help of an additional (auxiliary) projection plane ($x_1 - y_1$) parallel to the inclined surface. This view shows a true shape of the inclined surface, the true shape of a square hole as well as location of square hole with respect to the inclined surface.

Therefore, a partial auxiliary view is often needed to complete the description of a foreshortened feature in a principal view. Another important function of auxiliary view is to complete a principal view.

## 8.2 Types of Auxiliary Projection Planes

Auxiliary projection planes may have an infinite number of positions in relation to the three principal planes of projections (HP, VP and PP). Based on the orientation of the projection plane relative to the three principal planes, the auxiliary planes are classified.

### Projection plane perpendicular to HP and inclined to VP and PP

In addition to the three primary projection planes, an additional projection plane which is perpendicular to HP and inclined to VP and PP is used to obtain a true shape of an inclined surface which is also perpendicular to HP and inclined to VP and PP. This plane of projection is called auxiliary vertical plane (AVP). Figure 8.2 shows the orientation of AVP with reference to the primary projection planes.

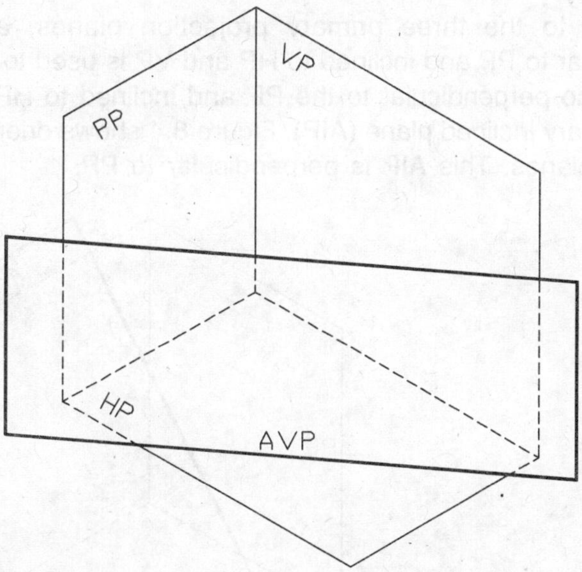


**Fig. 8.2** Auxiliary plane perpendicular to HP.

### Projection plane perpendicular to VP and inclined to HP and PP

In addition to the three primary projection planes, an additional projection plane which is perpendicular to VP and inclined to HP and PP is used to obtain a true shape of an inclined surface which is also perpendicular to VP and inclined to HP and PP. This plane of projection is called auxiliary inclined plane (AIP). Figure 8.3 shows orientation AIP with reference to the primary projection planes. This AIP is perpendicular to VP.

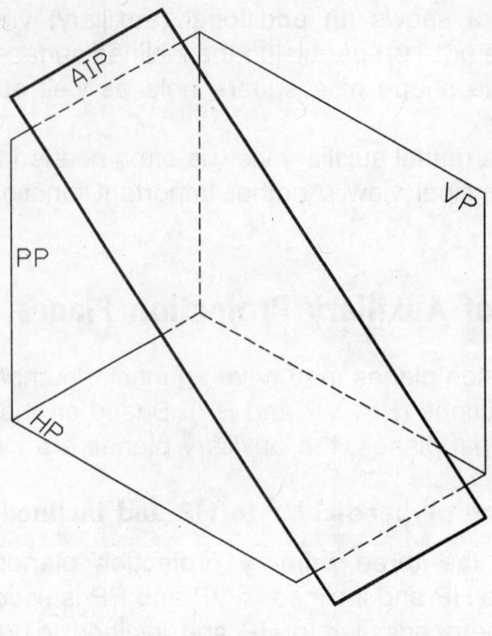


**Fig. 8.3** Auxiliary plane perpendicular to VP.

### Projection plane perpendicular to PP and inclined to HP and VP

In addition to the three primary projection planes, an additional projection plane which is perpendicular to PP and inclined to HP and VP is used to obtain a true shape of an inclined surface which is also perpendicular to the PP and inclined to HP and VP. This plane of projection is also called auxiliary inclined plane (AIP). Figure 8.4 shows orientation of AIP with reference to the primary projection planes. This AIP is perpendicular to PP.

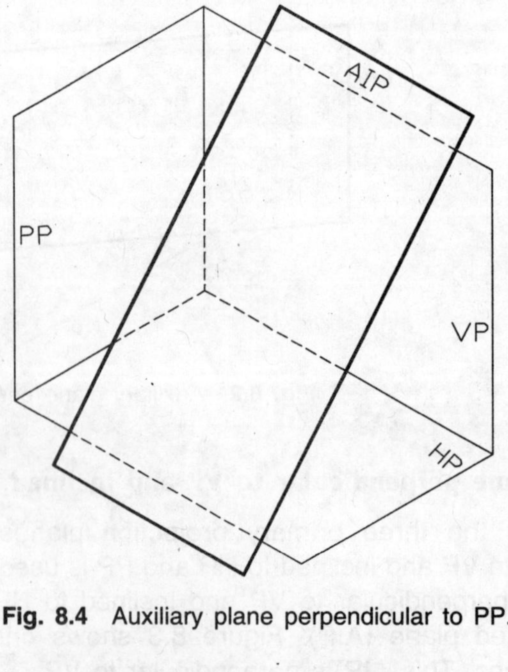


**Fig. 8.4** Auxiliary plane perpendicular to PP.

### Projection plane inclined to HP, VP and PP

In addition to the three primary projection planes, an additional projection plane which is inclined to HP, VP and PP is used to obtain a true shape of an inclined surface which is also inclined to HP VP and PP. This plane of projection is called auxiliary oblique plane (AOP). Figure 8.5 shows orientation of AOP with reference to the primary projection planes.

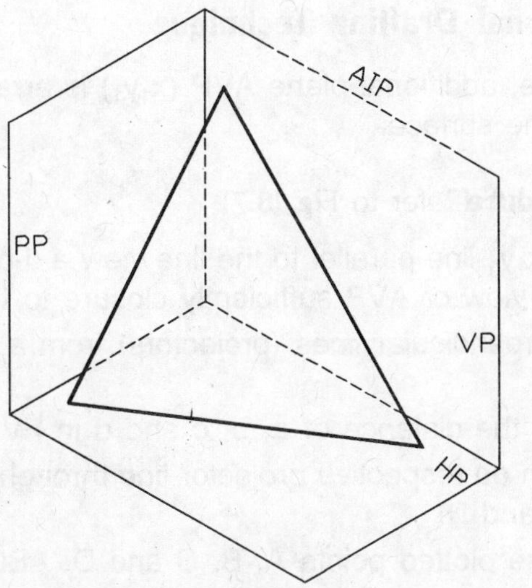


**Fig. 8.5** Auxiliary plane inclined to HP, VP and PP.

## 8.3 Techniques to Obtain Auxiliary Views

There are two techniques to obtain the true shape of the surface ABCD.

(a) Traditional drafting technique
(b) AutoCAD drafting technique

In order to understand the use of four types of auxiliary projection planes four cases are discussed. First three cases are to obtain the true shape of an inclined surface. The fourth case is to obtain the true shape of an oblique surface.

**Case I: Surface is perpendicular to HP and inclined to VP and PP**

Figure 8.6 shows an object with shaded surface ABCD perpendicular to HP and inclined to VP and PP. On left side, projections are drawn to show the TV and FV of the object. The FV shows an apparent shape (AS) of the surface. The TV shows a line view for the inclined surface. Therefore, we require an AVP parallel to the surface ABCD to obtain a true shape of the surface ABCD.

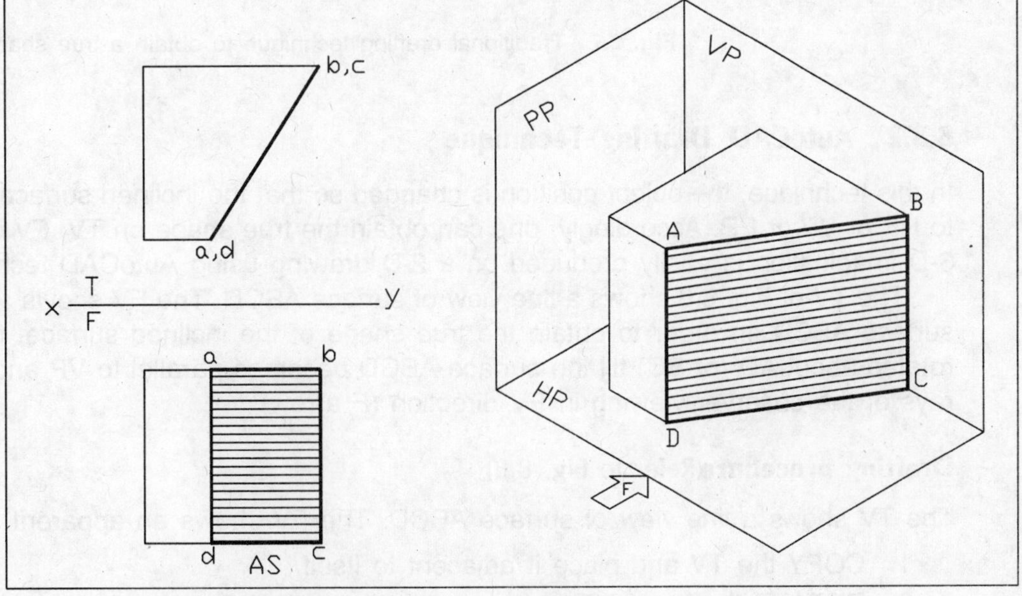


**Fig. 8.6** Inclined surface perpendicular to HP.

### 8.3.1 Traditional Drafting Technique

In this technique, additional plane AVP ($x_1y_1$) is arranged parallel to the surface ABCD to obtain a true shape of the surface.

**Drafting procedure**(Refer to Fig. 8.7)

1. Draw a $x_1y_1$ line parallel to the line view a,d-b,c of surface ABCD in the TV region. Line $x_1y_1$ is a line view of AVP sufficiently closure to the line view of the inclined surface (a,d-b,c).
2. Draw perpendicular lines (projectors) from a,d and b,c intersecting $x_1y_1$ line and projecting further.
3. Measure the distance of a, b, c and d in FV from xy line.
4. Plot them on respective projector line through a,d and b,c from the $x_1y_1$ line. Name these as A, B, C and D.
5. Join these plotted points A, B, C and D. ABCD represents the true shape of the surface.

The traditional drafting technique is laborious in relation to measuring and plotting the distances to get the key points for the true shape.

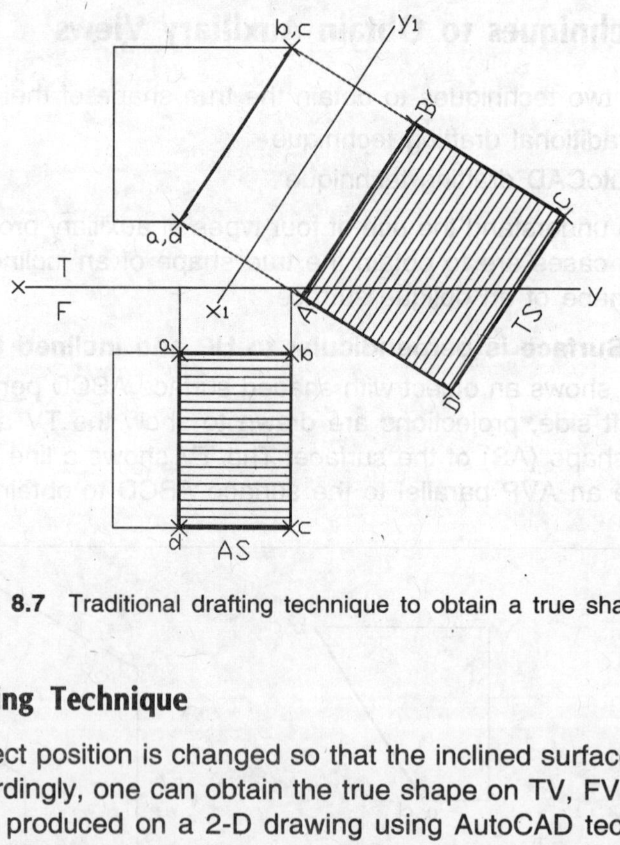


**Fig. 8.7** Traditional drafting technique to obtain a true shape.

### 8.3.2 AutoCAD Drafting Technique

In this technique, the object position is changed so that the inclined surface becomes either parallel to HP or VP or PP. Accordingly, one can obtain the true shape on TV, FV or SV, respectively. This 3-D imagination is easily produced on a 2-D drawing using AutoCAD technique.

The TV of Fig. 8.6 shows a line view of surface ABCD. The FV shows an apparent shape of the surface ABCD. In order to obtain the true shape of the inclined surface, the object in Fig. 8.6, is rotated about AD (or BC) till the surface ABCD becomes parallel to VP and perpendicular to visual rays of the observer viewing in FV direction (F arrow).

**Drafting procedure**(Refer to Fig. 8.8)

The TV shows a line view of surface ABCD. The FV shows an apparent shape of the same.

1. COPY the TV and place it adjacent to itself.
2. ROTATE the copied TV with respect to 'a, d' till the line view (a,d-b,c) becomes parallel to the xy line.

3. Project the marked cross points from the second stage TV and the first stage FV. Intersection of projection lines give us the key points to draw a true shape (TS) of the inclined surface ABCD.

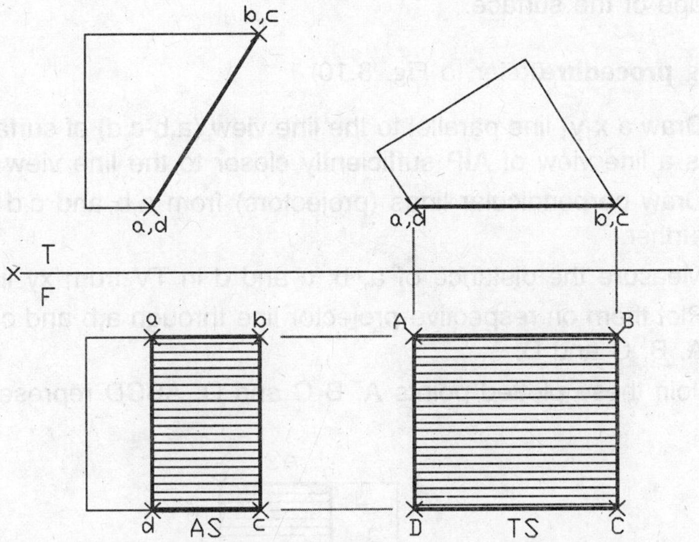


**Fig. 8.8** AutoCAD drafting technique to obtain a true shape.

### Case II: Surface is perpendicular to VP and inclined to HP and PP

Figure 8.9 shows an object with shaded surface ABCD perpendicular to VP and inclined to HP and PP. On the left side, projections are drawn to show the TV and FV of the object. The TV shows an apparent shape (AS) of the surface. The FV shows a line view for the inclined surface. Therefore, we require an AIP parallel to the surface ABCD to obtain a true shape of the surface. This AIP would be perpendicular to VP.

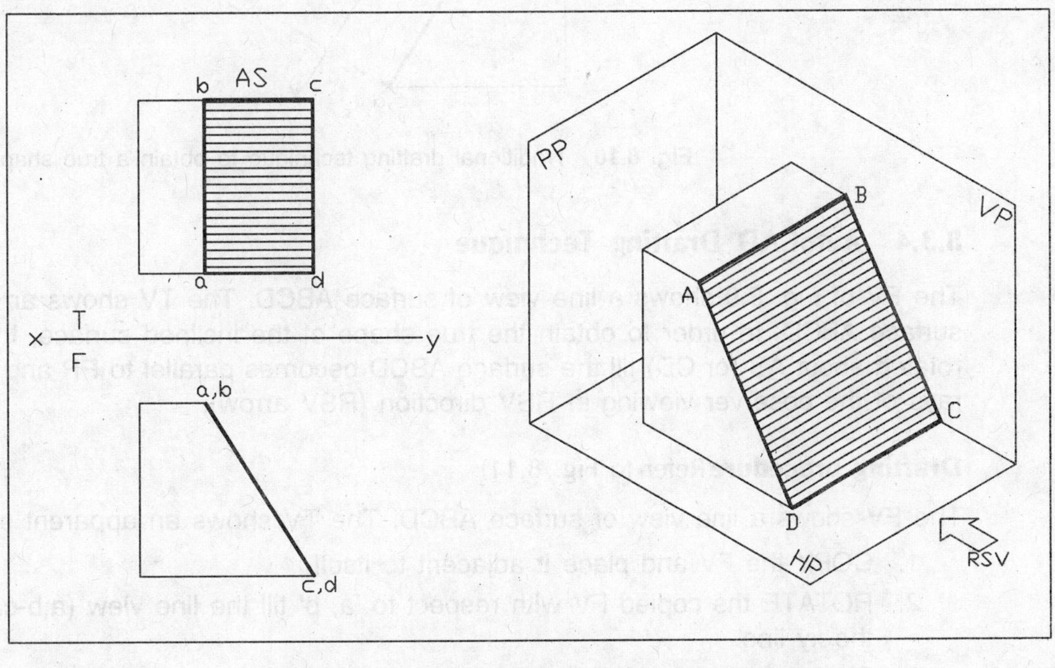


**Fig. 8.9** Inclined surface perpendicular to VP.

### 8.3.3 Traditional Drafting Technique

In this technique, additional plane AIP ($x_1y_1$) is arranged parallel to the surface ABCD to obtain a true shape of the surface.

**Drafting procedure**(Refer to Fig. 8.10)

1. Draw a $x_1y_1$ line parallel to the line view (a,b-c,d) of surface ABCD in the FV region. Line $x_1y_1$ is a line view of AIP sufficiently closer to the line view of the inclined surface (a,b-c,d).
2. Draw perpendicular lines (projectors) from a,b and c,d intersecting $x_1y_1$ line and projecting further.
3. Measure the distance of a, b, c and d in TV from xy line.
4. Plot them on respective projector line through a,b and c,d from the $x_1y_1$ line. Name these as A, B, C and D.
5. Join these plotted points A, B C and D. ABCD represents the true shape of the surface.

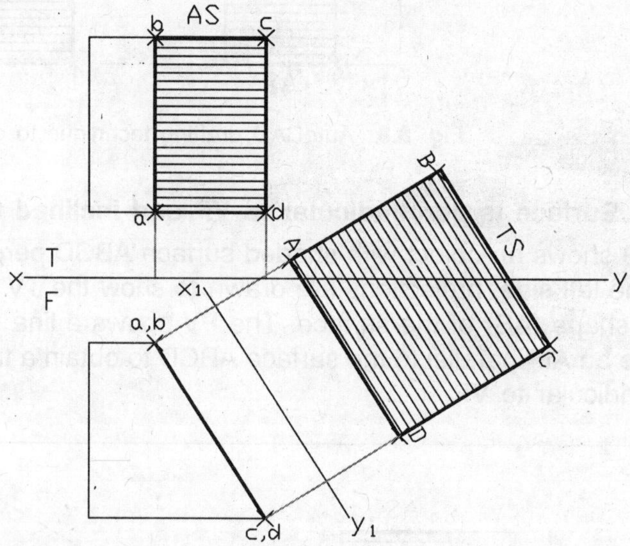


**Fig. 8.10** Traditional drafting technique to obtain a true shape.

### 8.3.4 AutoCAD Drafting Technique

The FV of Fig. 8.9 shows a line view of surface ABCD. The TV shows an apparent shape of the surface ABCD. In order to obtain the true shape of the inclined surface, the object in Fig. 8.9, is rotated about AB (or CD) till the surface ABCD becomes parallel to PP and perpendicular to visual rays of the observer viewing in RSV direction (RSV arrow).

**Drafting procedure**(Refer to Fig. 8.11)

The FV shows a line view of surface ABCD. The TV shows an apparent shape of the same.

1. COPY the FV and place it adjacent to itself.
2. ROTATE the copied FV with respect to 'a, b' till the line view (a,b-c,d) becomes parallel to the xy line.
3. Project the marked cross points from the second stage FV and the first stage TV. Intersection of projection lines give us the key points to draw a true shape (TS) of the inclined surface ABCD.

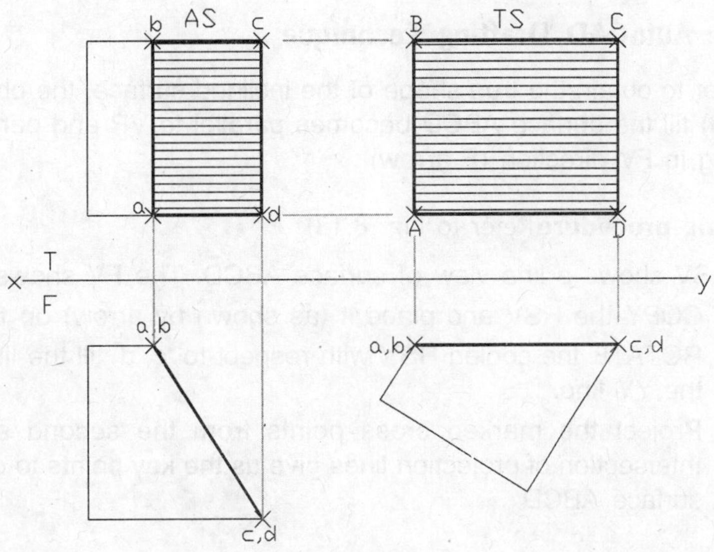


**Fig. 8.11** AutoCAD drafting technique to obtain a true shape.

In all three cases, one auxiliary projection plane is required to obtain the true shape of the inclined surface, therefore, it is also called single auxiliary projection and the obtained true shape, single auxiliary view.

**Case III: Surface is perpendicular to PP and inclined to HP and VP**

Figure 8.12 shows an object with shaded surface ABCD perpendicular to the PP and inclined to HP and VP. On left side, projections are drawn to show TV, FV and SV of the object. The TV and FV show an apparent shape of the surface. The RSV shows a line view (a,b-c,d) of the inclined surface. Therefore, we require an AIP parallel to the surface ABCD to obtain a true shape of the surface ABCD. The AIP would be perpendicular to PP. The traditional drafting technique is not discussed for this case.

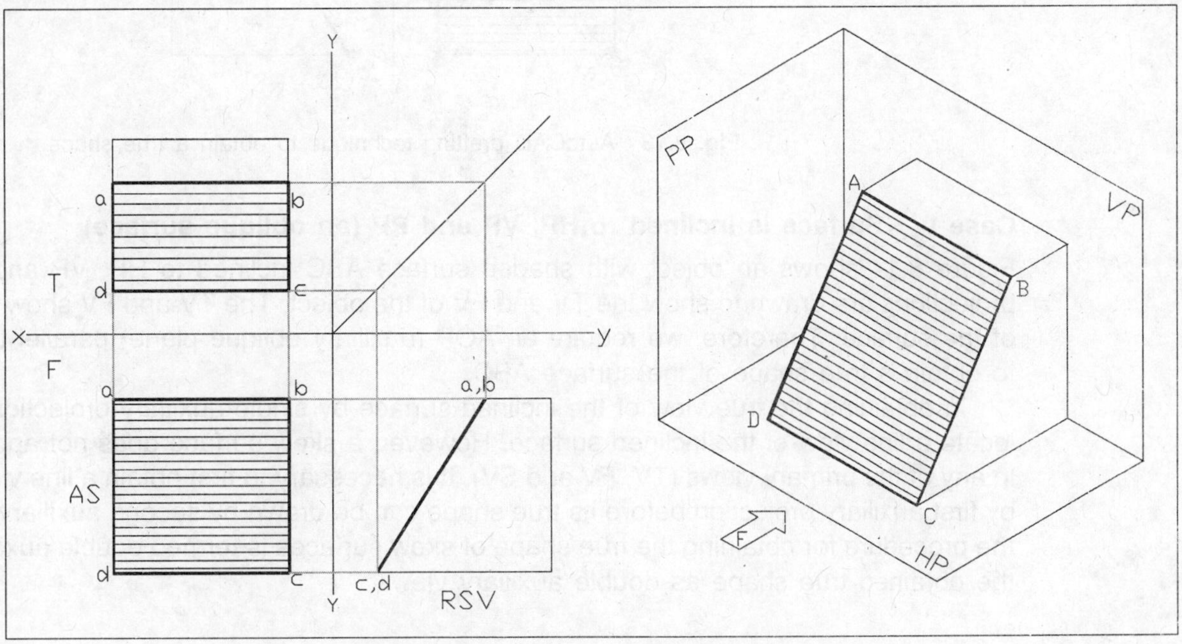


**Fig. 8.12** Inclined surface perpendicular to PP.

### 8.3.5 AutoCAD Drafting Technique

In order to obtain the true shape of the inclined surface, the object in Fig. 8.12, is rotated about CD (or AB) till the surface ABCD becomes parallel to VP and perpendicular to visual rays of observer viewing in FV direction (F arrow).

**Drafting procedure**(Refer to Fig. 8.13)

The RSV shows a line view of surface ABCD. The FV shows an apparent shape of the same.

1. COPY the RSV and place it (as shown by arrow) on the top of itself.
2. ROTATE the copied RSV with respect to 'c, d' till the line view (c,d-a,b) becomes parallel to the YY line.
3. Project the marked cross points from the second stage RSV and the first stage FV. Intersection of projection lines give us the key points to draw a true shape (TS) of the inclined surface ABCD.

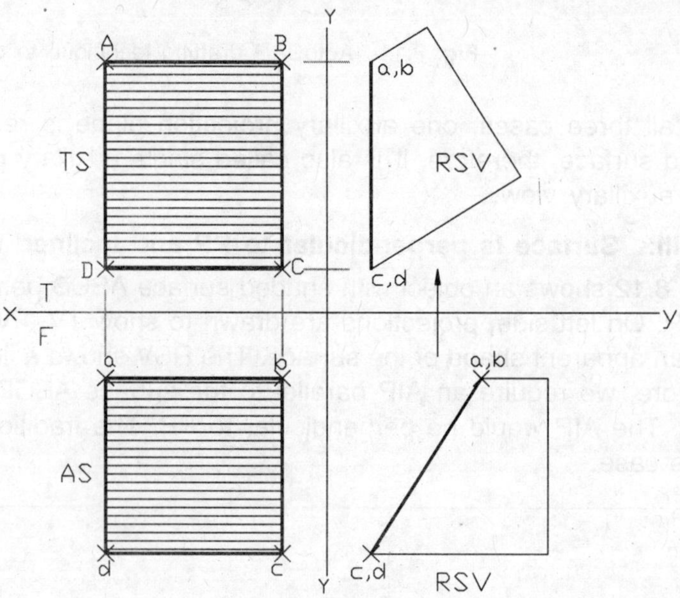


**Fig. 8.13** AutoCAD drafting technique to obtain a true shape.

**Case IV  Surface is inclined to HP, VP and PP (an oblique surface)**

Figure 8.14 shows an object with shaded surface ABC inclined to HP, VP and PP. On left side projections are drawn to show the TV and FV of the object. The TV and FV show an apparent shape of the surface. Therefore, we require an AOP (auxiliary oblique plane) parallel to the surface ABC to obtain a true shape of the surface ABC.

In obtaining the true view of the inclined surface by single auxiliary projection it is necessary to locate a line view of the inclined surface. However, a skew surface does not appear in its line view in any of the primary views (TV, FV and SV). It is necessary to first obtain a line view of skew surface by first auxiliary projection before its true shape can be drawn by second auxiliary projection. Hence the procedure for obtaining the true shape of skew surfaces is termed double auxiliary projection and the obtained true shape as double auxiliary view.

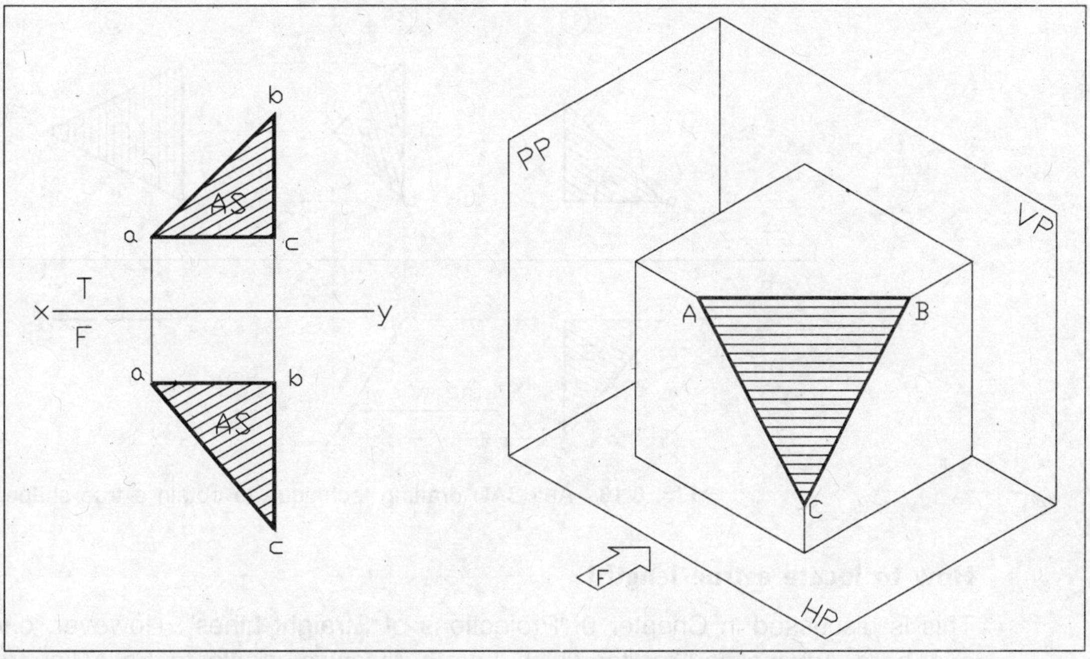


**Fig. 8.14** AutoCAD drafting technique to obtain a true shape.

### 8.3.6 AutoCAD Drafting Technique

In order to obtain a line view of skew surface, first an auxiliary plane is required which is perpendicular to one of the edges of the skew surface. This is shown in Fig. 8.15. The edge AB, being perpendicular to the projection plane, falls into a point view (a,b). Therefore, the skew surface ABC falls into a line view (a,b-c).

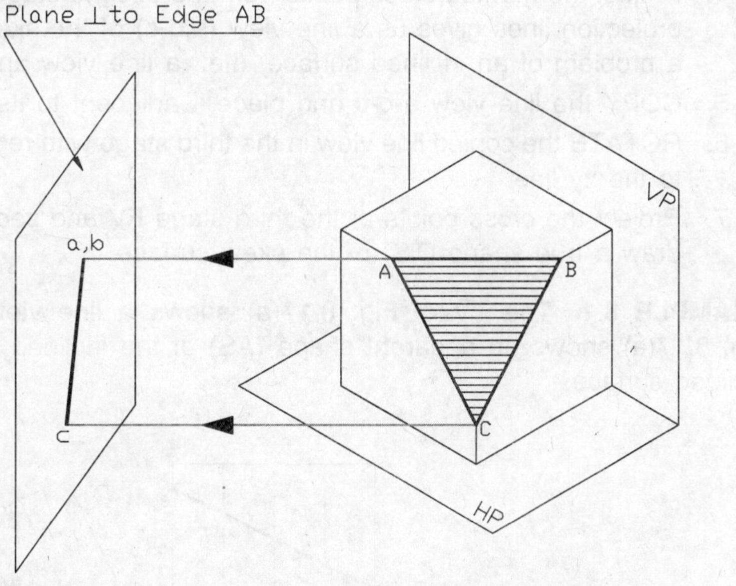


**Fig. 8.15** Projection plane to obtain a line view of skew surface.

#### Drafting procedure

The TV and FV of Fig. 8.16 shows an apparent shape of the skew surface ABC.

1. Locate a true length.

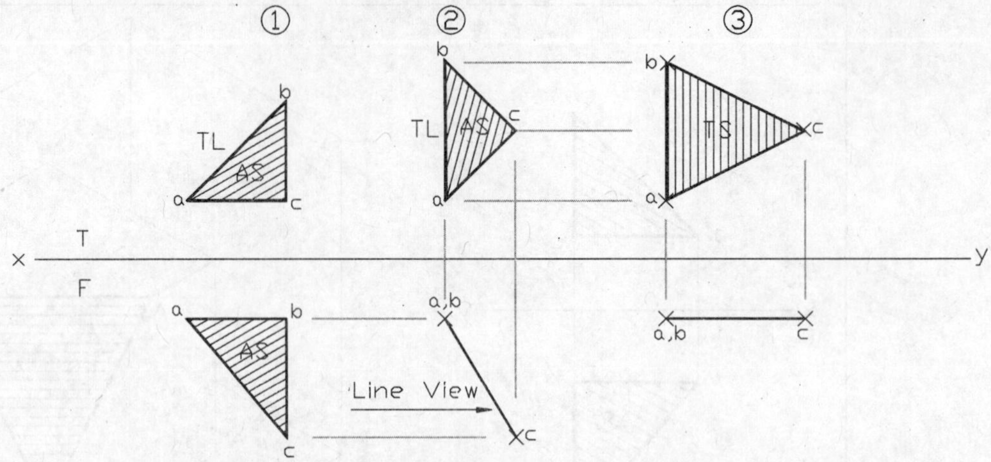


**Fig. 8.16** AutoCAD drafting technique to obtain a true shape.

### How to locate a true length?

This is discussed in Chapter 9 "Projections of Straight Lines". However, one simple clue is give over here. Any inclined line for which the other view is parallel to the xy line, represents a true lengt for the same. In this example, in the TV, ab is inclined. In the FV, ab is parallel to the xy line Therefore, ab shows a true length. (Similarly, ac in the FV, also represents a true length as ac i the TV is parallel to the xy line.)

2. COPY the surface a-b-c in the TV, in which we have ab as a true length and place it adjacer to itself.
3. ROTATE the copied surface a-b-c with respect to 'a' such that the true length ab become perpendicular to the xy line.
4. Project the marked cross points from the second stage TV and first stage FV. Intersection c projection lines gives us a line view (a,b-c) of the skew surface a-b-c. The second stage i a problem of an inclined surface, (i.e., a line view and an apparent shape).
5. COPY the line view a-b-c and place it adjacent to itself.
6. ROTATE the copied line view in the third stage with respect to a,b point till it becomes paralle to the xy line.
7. Project the cross points of the third stage FV and second stage TV to get the key points t draw a true shape (TS) of the skew surface.

**EXAMPLE 8.1** The TV of Fig. 8.17(a) shows a line view of an inclined surface. The FV c Fig. 8.17(a) shows an apparent shape (AS) of the inclined surface. Obtain the true shape of th inclined surface.

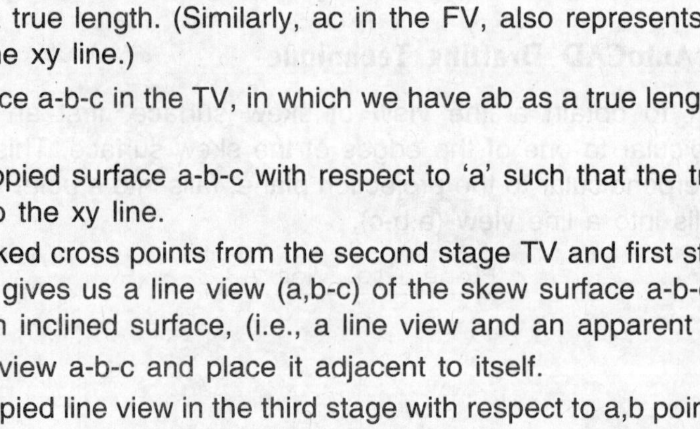


**Fig. 8.17**

### *Solution*

1. COPY the TV and place it adjacent to itself.
2. ROTATE the copied TV with respect to 'O' till the line view becomes parallel to the x-y line as shown in (b).
3. Project the marked cross points from the second stage TV and the first stage FV. Intersection of projection lines give us the key points to draw a true shape (TS) of the inclined surface. This is shown in (b).

**EXAMPLE 8.2** The FV of Fig. 8.18(a) shows a line view of an inclined surface. The TV of Fig. 8.18(a) shows an apparent shape (AS) of the inclined surface. Obtain the true shape of the inclined surface.

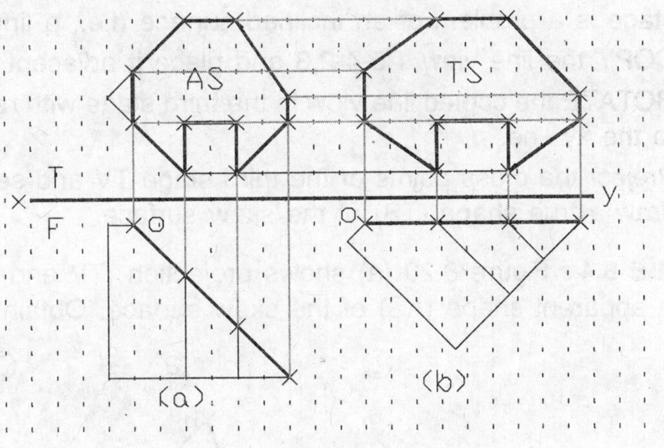


**Fig. 8.18**

### *Solution*

1. COPY the FV and place it adjacent to itself.
2. ROTATE the copied FV with respect to 'O' till the line view becomes parallel to the x-y line as shown in (b).
3. Project the marked cross points from the second stage FV and the first stage TV. Intersection of projection lines give us the key points to draw a true shape (TS) of the inclined surface. This is shown in (b).

**EXAMPLE 8.3** Figure 8.19 (a) shows projections (TV and FV) of a skew surface. Both the views show an apparent shape (AS) of the skew surface. Obtain the true shape of the skew surface.

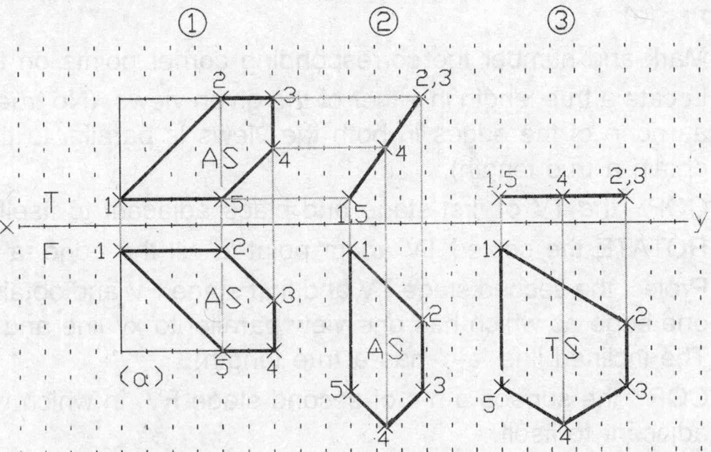


**Fig. 8.19**

***Solution***

1. Mark and number the corresponding corner points on the skew surface in both the views.
2. Locate a true length in either of the given views. (In this case, 1-5 in the FV can be taken as a true length. One can choose 1-2 or 4-5 as true lengths in the TV. One can also choose 2-3 as a true length in the FV.)
3. COPY the surface 1-2-3-4-5 in the FV, in which we have one true length (1-5) and place it adjacent to itself.
4. ROTATE the surface 1-2-3-4-5 with respect to '1' such that the true length, 1-5, becomes perpendicular to the xy line.
5. Project the marked cross points from the second stage FV and first stage TV. Intersection of projection lines gives us a line view (1,5-4-2,3) of the skew surface 1-2-3-4-5. The second stage is a problem of an inclined surface (i.e., a line view and an apparent shape).
6. COPY the line view 1,5-4-2,3 and place it adjacent to itself.
7. ROTATE the copied line view in the third stage with respect to 1,5 point till it becomes parallel to the xy line.
8. Project the cross points of the third stage TV and second stage FV to get the key points to draw a true shape (TS) of the skew surface.

**EXAMPLE 8.4** Figure 8.20 (a) shows projection (TV and FV) of a skew surface. Both the views show an apparent shape (AS) of the skew surface. Obtain the true shape of the skew surface.

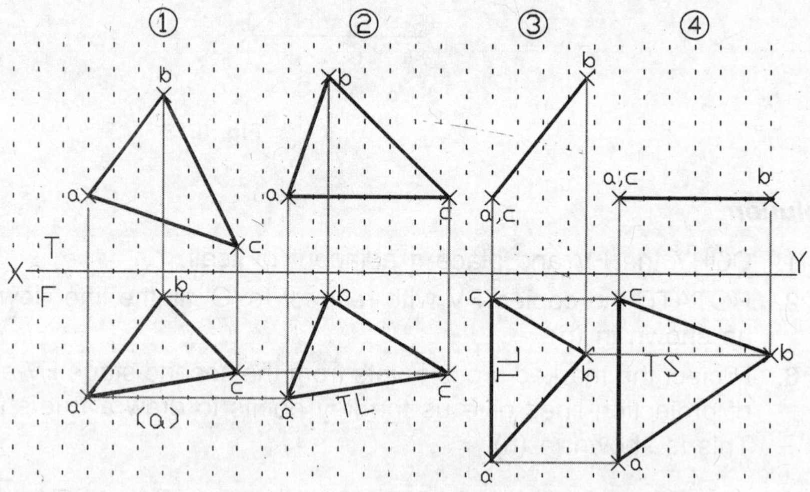


**Fig. 8.20**

***Solution*** (a)

1. Mark and number the corresponding corner points on the skew surface in both the views.
2. Locate a true length in either of the given views. (No true length is available in both the views as none of the edges in both the views is parallel to the xy line. Now, need has arised to obtain a true length).
3. COPY the TV of first stage and place adjacent to itself.
4. ROTATE the copied TV about point 'a' till the edge 'a-c' becomes parallel to xy.
5. Project the second stage TV and first stage FV and obtain a second stage FV. (Now, we have one edge ac which has one view parallel to xy line and another view inclined to the xy line. The inclined line 'a-c' has a true length).
6. COPY the surface a-b-c of second stage FV, in which we have one true length and place it adjacent to itself.
7. ROTATE the copied surface a-b-c with respect to 'a' such that the true length a-c becomes perpendicular to the xy line.

8. Project the marked cross points from the third stage FV and second stage TV. Intersection of the projection lines give us a line view (a,c-b) of the skew surface a-b-c. (The third stage is a problem of an inclined surface i.e., a line view and an apparent shape).
9. COPY the line view a,c-b and place it adjacent to itself.
10. ROTATE the line view in the fourth stage with respect to a,c point till it becomes parallel to the xy line.
11. Project the cross points of the fourth stage TV and third stage FV to get the key points to draw a true shape (TS) of the skew surface.

***Solution*** (b) [Alternate solution for EXAMPLE 8.4; (Refer to Fig. 8.21)]

1. Mark and name the corresponding corner points on the skew surface in both the views.
2. Draw a guest LINE from 'a' parallel to the xy line intersecting bc at d.
3. Project point 'd' to the FV on bc and join with 'a'. Line 'ad' represents a true length as the TV for the same is parallel to the xy line.
4. COPY the surface a-b-d-c of first stage FV, in which we have one true length and place it adjacent to itself.
5. ROTATE the surface a-b-d-c with respect to 'a' such that the true length a-d becomes perpendicular to the xy line.
6. Project the marked cross points from the second stage FV and first stage TV. Intersection of projection lines gives us a line view (c-a,d-b) of the skew surface c-d-b-a. (The second stage is a problem of an inclined surface i.e., a line view and apparent shape).
7. COPY the line view c-a,d-b and place it adjacent to itself.
8. ROTATE the copied line view in the third stage with respect to c till it becomes parallel to the xy line.
9. Project the cross points of the third stage TV and second stage FV to get the key points to draw a true shape (TS) of the skew surface.

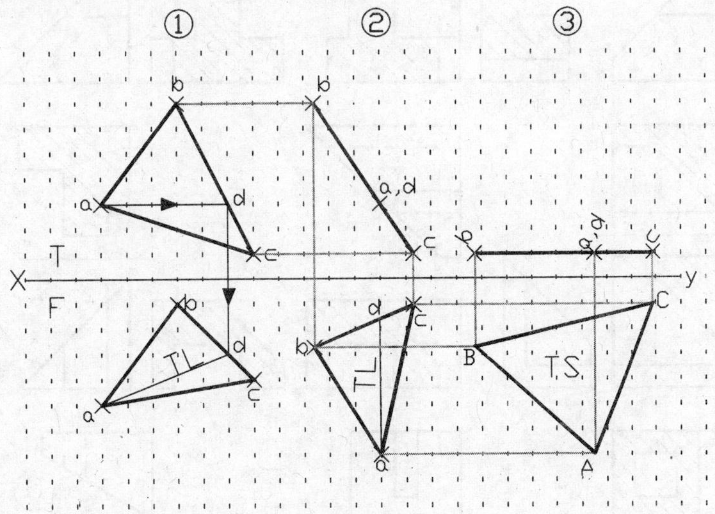


**Fig. 8.21**

## Exercise A (Single Auxiliary View)

**8.1** (a) P1 and P2 of Plate 8.1 show a single auxiliary technique to obtain a true shape of the inclined surface from a line view. Use the same technique to solve P3, P4, P5, P6, P7 and P8 to obtain their respective true shapes.

(b) Copy two necessary views in P1 to P12 problems of Plate 8.2 and obtain the true shape of the inclined surface. The projections are drawn in the III-quadrant.

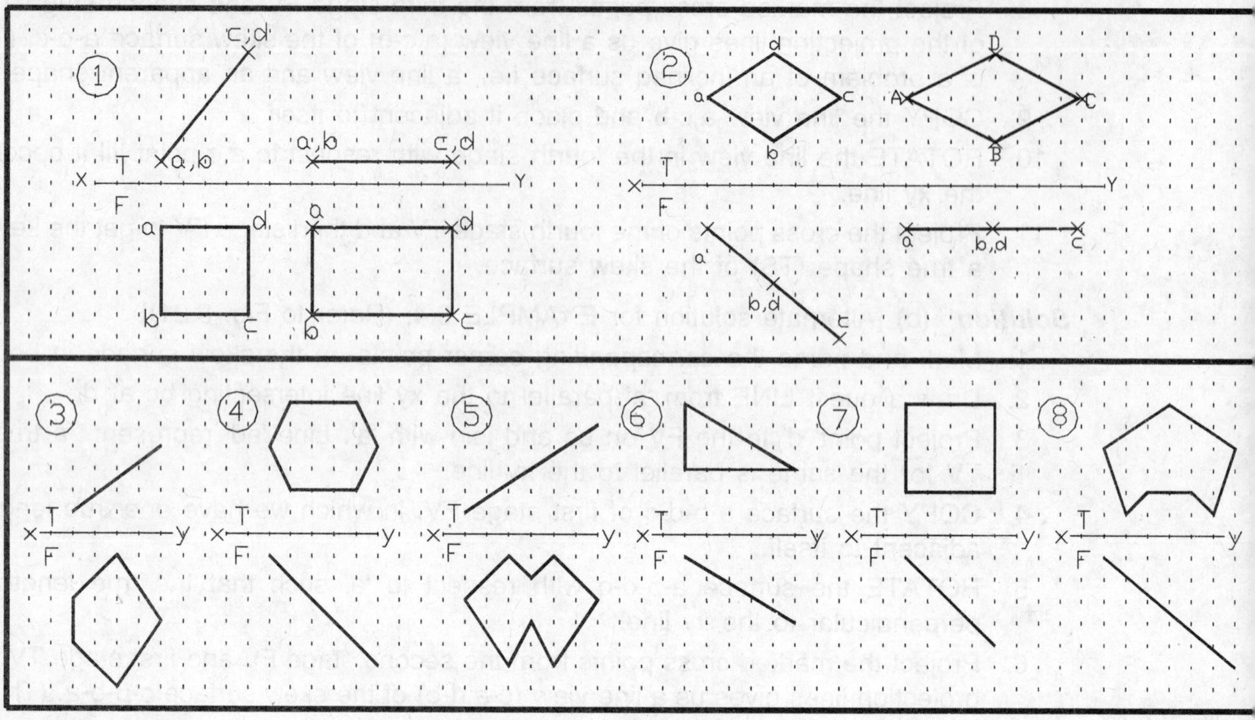

PLATE 8.1

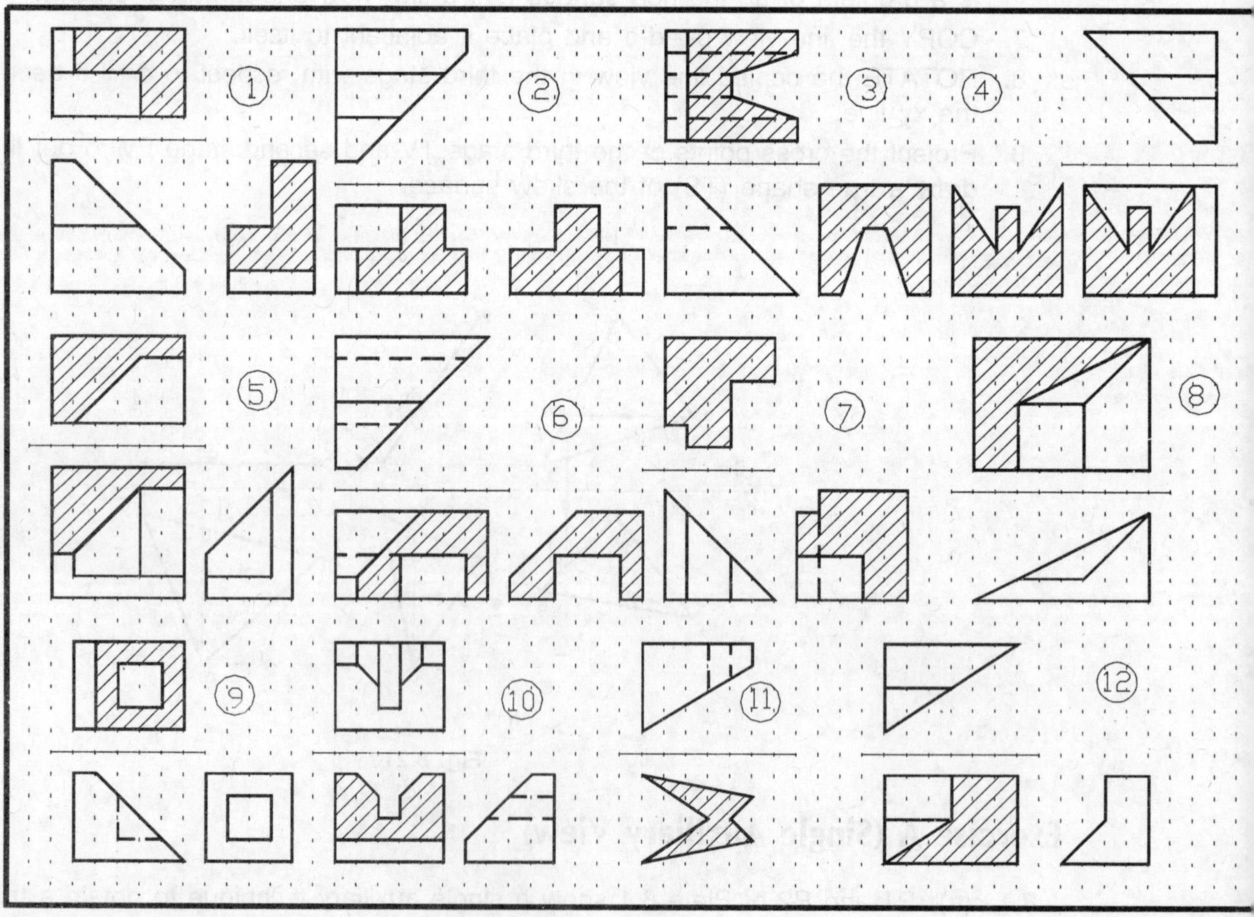

PLATE 8.2

## Exercise B (Double Auxiliary View)

**8.1** (a) P1 of Plate 8.3 shows a double auxiliary technique to obtain a true shape of an oblique surface from a line view. Use the same technique to solve P2, P3, P4, P5, P6, P7 and P8 to obtain their respective true shapes.

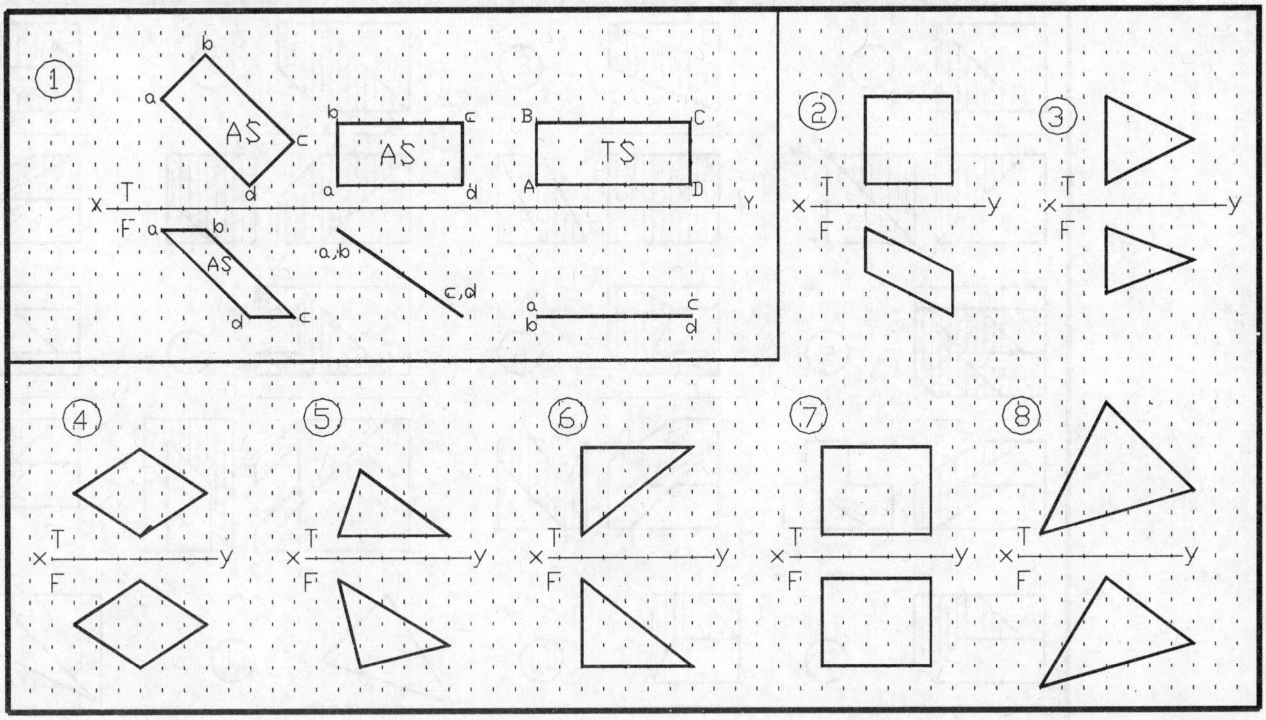


PLATE 8.3

(b) Copy two necessary views in Plate 8.4 and 8.5 problems and obtain the true shape of the inclined or skew surface.

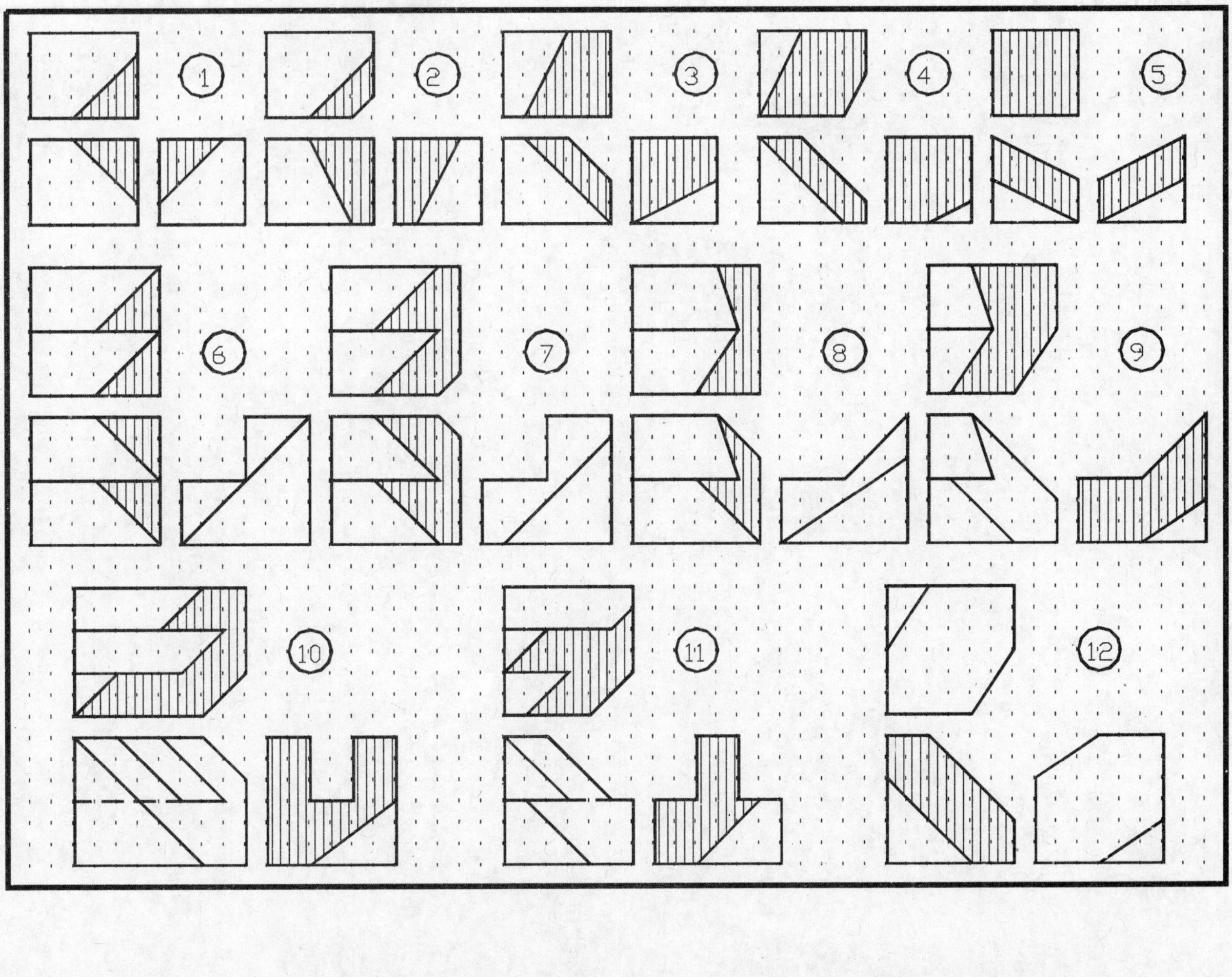


PLATE 8.4

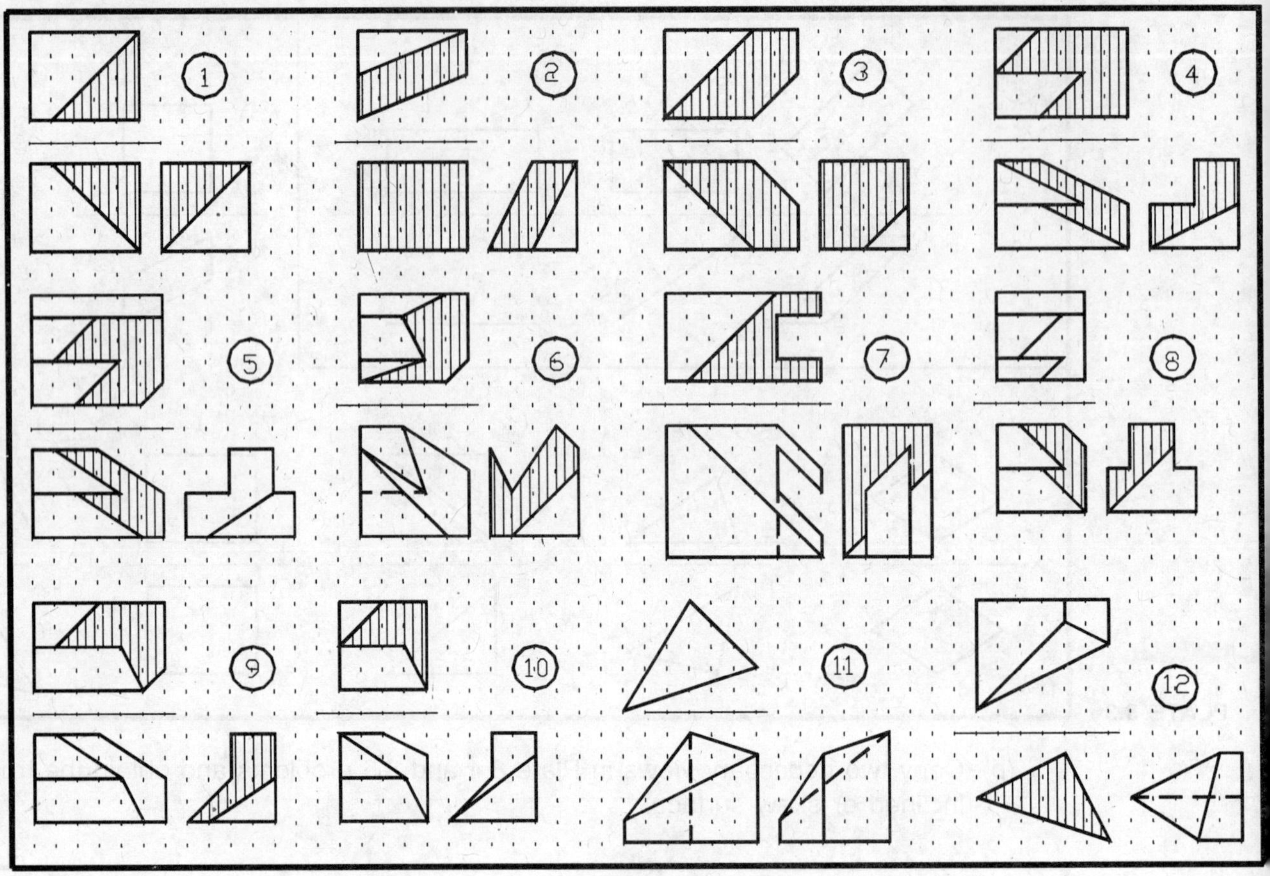


PLATE 8.5

# PART III

## Descriptive Geometry

# 9

# Projection of Straight Lines

## Objectives

- Learn the description of quadrant on a plane of paper.
- Observe the projection of point.
- Study the projection of straight lines in a 3-D space.
- Know the projection of intersection of planes and straight lines.

## Introduction

A straight line is the shortest distance between two points. Hence, the projections of a straight line may be drawn by joining the respective projections of its ends which are points. The knowledge of projection of straight lines is very useful in the field of underground pipelines and air conditioning ducts in order to find out the shortest distance, inclinations as well as location of their intersection point with walls and ground. The knowledge of projection of straight lines is also useful for the management of air traffic.

Before we look into the projection of straight lines, it is necessary to understand the description of quadrant on a plane of paper with respect to the principal planes and projection of points.

## 9.1 Object in Four Quadrant

With the help of the principal planes (HP, VP) four quadrants can be formed as shown in Fig. 9.1. The object is located in any one of the quadrant and the position of the object is described with reference to the principal planes (HP and VP). The object (entity) can be a point, a straight line, a plane or a solid. The position of any object with reference to a particular quadrant can be described as:

1. Above HP and in front of VP, if it is in I-quadrant.
2. Above HP and behind VP, if it is in II-quadrant.
3. Below HP and behind VP, if it is in III-quadrant.
4. Below HP and in front of VP, if it is in IV-quadrant.

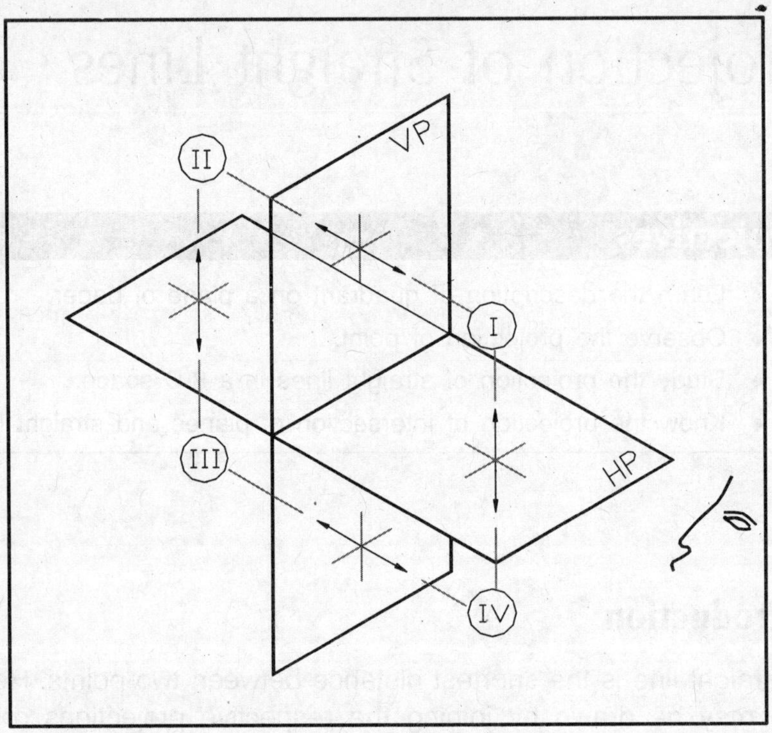


**Fig. 9.1** Four quadrants.

Figure 9.2 shows point as an object and its position is described with the help of the principal planes HP and VP. Point 'A' is above HP and in front of VP in I-quadrant. Point 'B' is above HP and behind VP in II-quadrant. Point 'C' is below HP and behind VP in III-quadrant. Point 'D' is below HP and in front of VP in IV-quadrant.

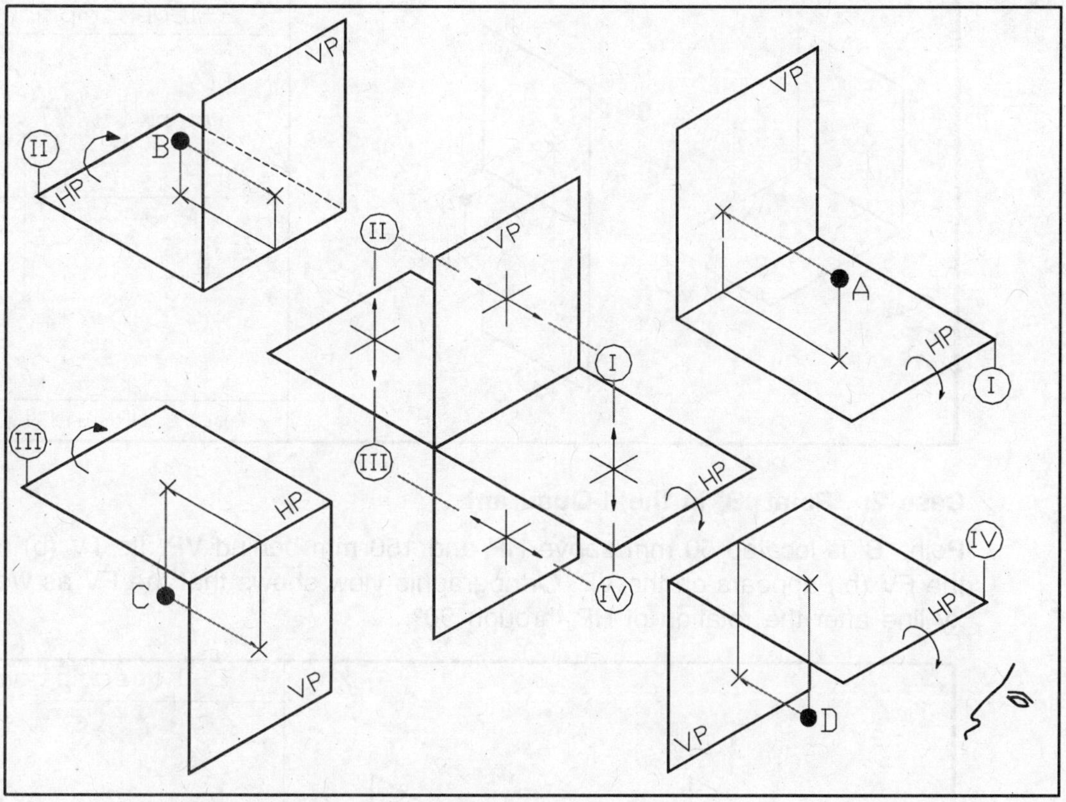


**Fig. 9.2** Projection of points in I, II, III and IV-quadrant (3-D view).

## 9.2 2-D Description of Quadrants

In order to communicate the design and drawing ideas it is always necessary to know the 2-D description of a quadrant on a plane of paper. The knowledge of 2-D description helps to avoid the imagination of the team members involved in the discussion of design and development of the product. This knowledge will be useful to understand the projection of straight lines, planes, solids, section of solids, development of surfaces of solids and intersection of surfaces of solids. The following topic focusses on the 2-D description of an individual quadrant with reference to a point as an object.

## 9.3 Projections of Point

In order to understand the 2-D description of a quadrant, individual quadrants are considered and a point is chosen as an entity. The reason to choose a point as an entity, is solid, surface and straight line all represent a set of points. While discussing the description of a quadrant on a plane of paper, few important observations are summarised below:

- The xy line represents a FV of HP and TV of VP for the given position of the observer.
- Distance measured with respect to HP gives the FV location of a point and distance measured with respect to VP gives the TV location of a point.
- In order to show the projection on a drawing sheet, the HP is rotated through 90° towards the feet of the observer.

**Case 1: Point 'A' in the I-Quadrant**

Point 'A' is located 75 mm above HP and 125 mm in front of VP. Its TV (a) appears on the HP and the FV (a′) appears on the VP. Orthographic view shows the FV above xy line and the TV below xy line after the rotation of HP through 90°.

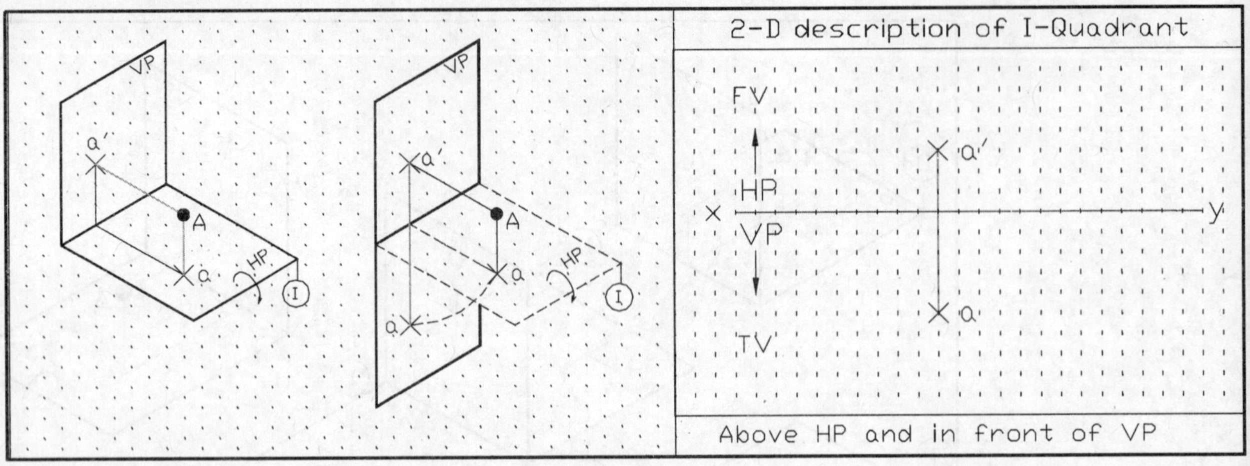


**Case 2: Point 'B' in the II-Quadrant**

Point 'B' is located 50 mm above HP and 150 mm behind VP. Its TV (b) appears on the HP and the FV (b′) appears on the VP. Orthographic view shows that the FV as well as the TV are above xy line after the rotation of HP through 90°.

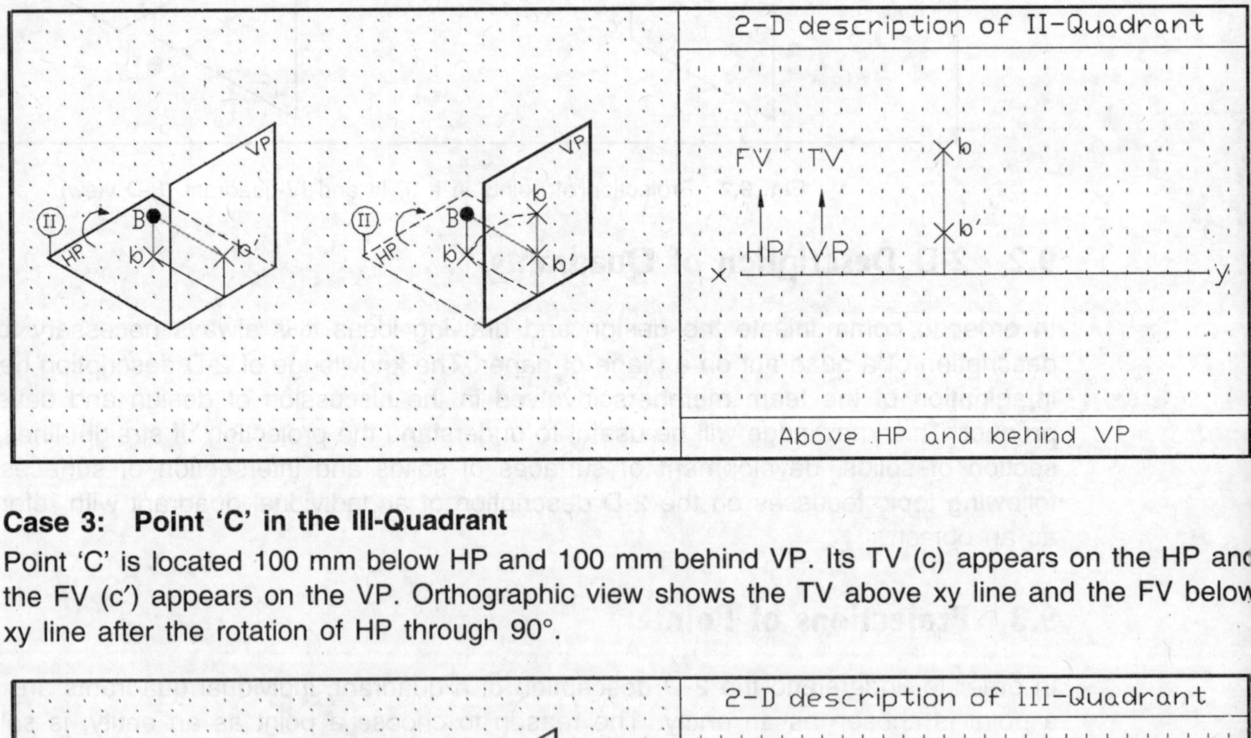


**Case 3: Point 'C' in the III-Quadrant**

Point 'C' is located 100 mm below HP and 100 mm behind VP. Its TV (c) appears on the HP and the FV (c′) appears on the VP. Orthographic view shows the TV above xy line and the FV below xy line after the rotation of HP through 90°.

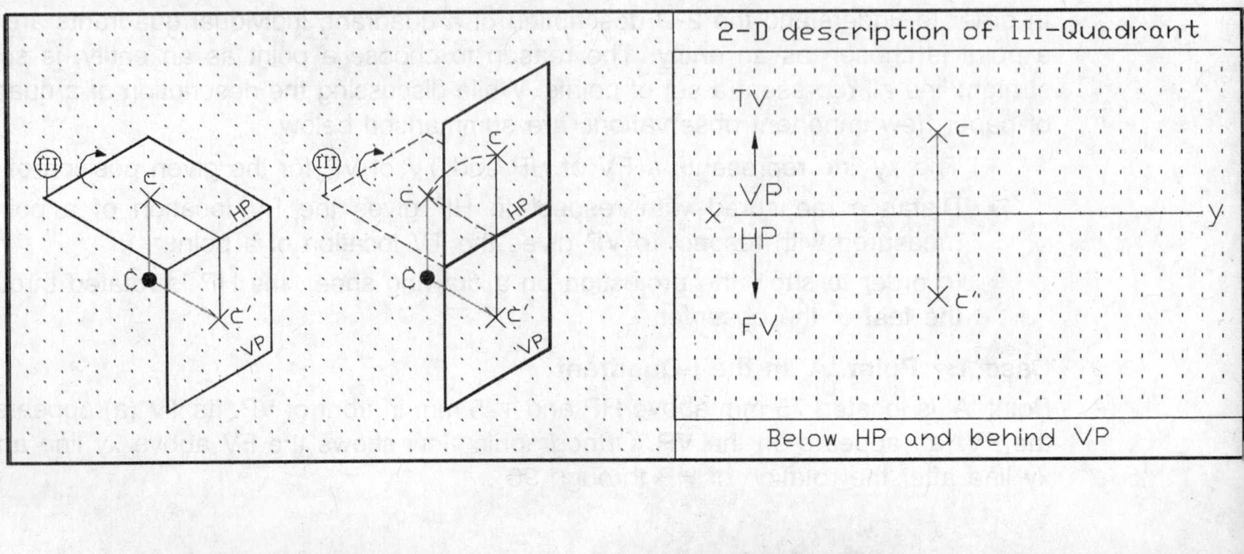

**Case 4: Point 'D' in the IV-Quadrant**

Point 'D' is located 125 mm below HP and 75 mm in front of VP. Its TV (d) appears on the HP and the FV (d′) appears on the VP. Orthographic view shows that the FV as well as the TV are below xy line after the rotation of HP through 90°.

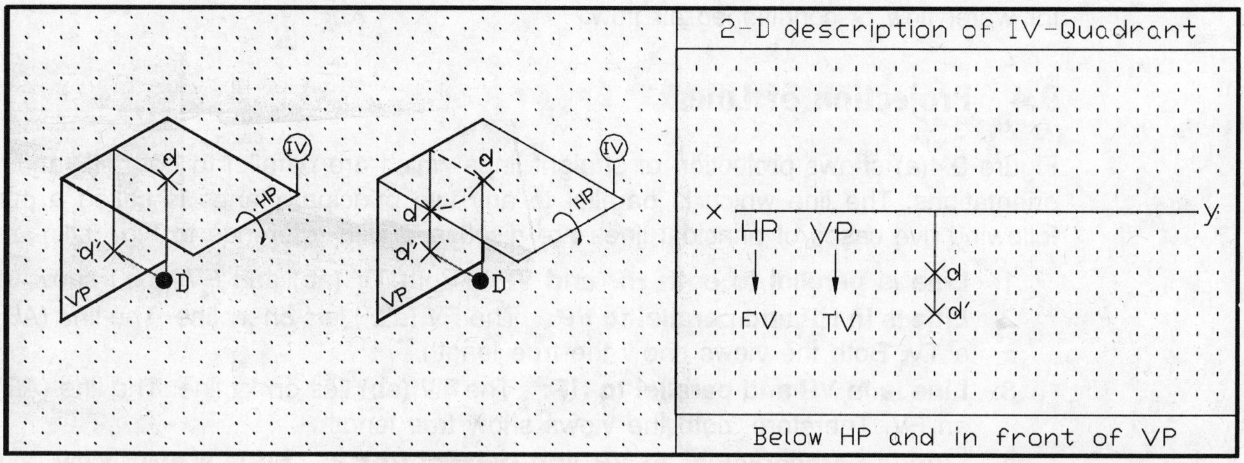


Figure 9.3(a) shows the 2-D description of I- and III-quadrants together for immediate comparison. This information is very useful for the projection of points, straight lines, plane surfaces and solids.

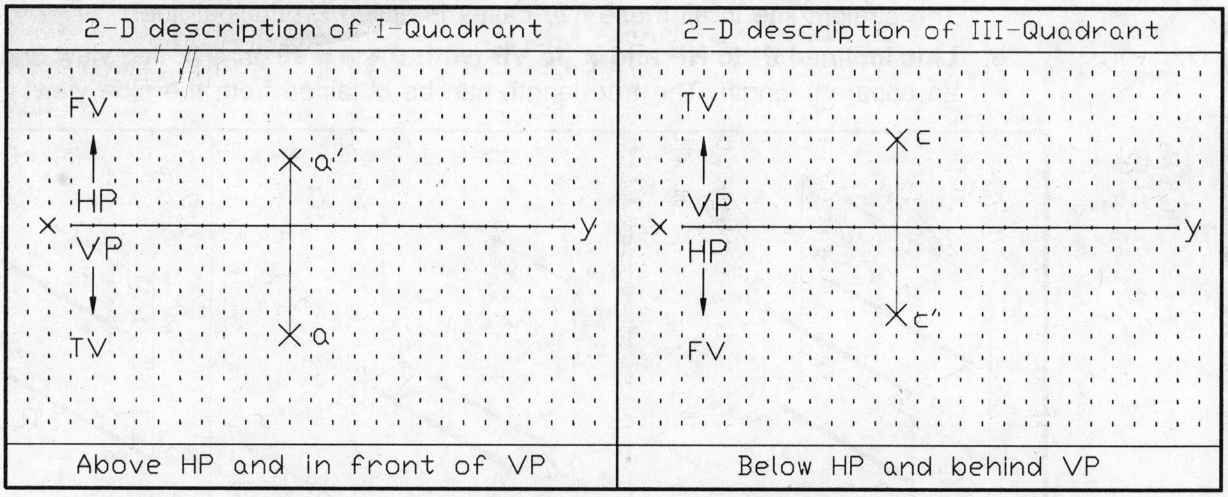


**Fig. 9.3**(a) 2-D description for I and III quadrant (comparison).

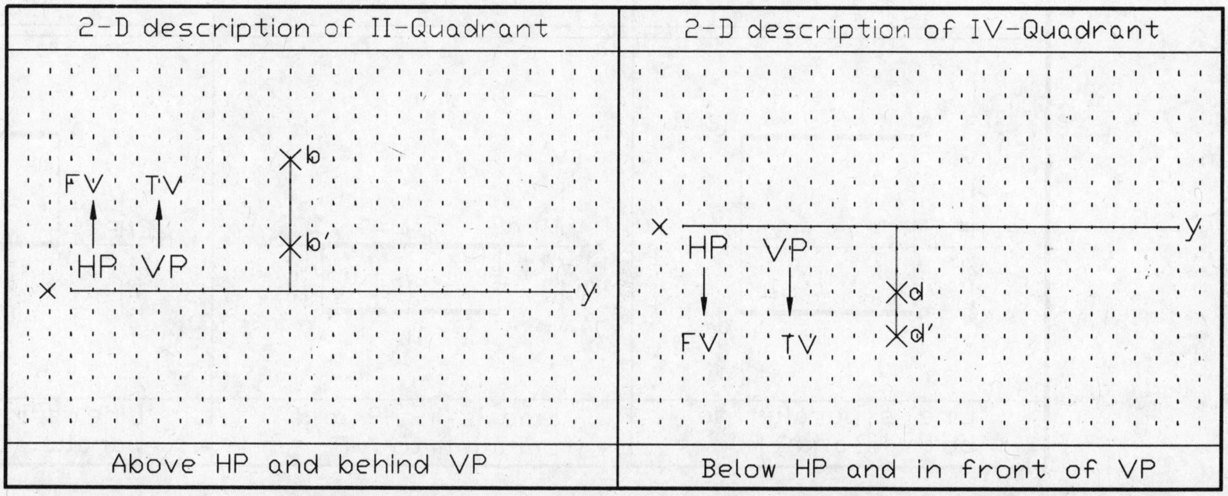


**Fig. 9.3**(b) 2-D description for II- and IV-quadrant (comparison).

Figure 9.3(b) shows the 2-D description of II- and IV-quadrants together for immediate comparison. This information is very useful for the projection of points, and straight lines. Straight line problems are required to be solved in any quadrant. In some cases of straight lines, one end point is in one quadrant and another endpoint is in the other quadrant. From the application point of view, the pipelines in a complicated building structure runs from one room to another room either for water flow or conditioned air flow.

## 9.4 Projection of Lines

Figure 9.4(a) shows projection of straight lines which are parallel to both HP and VP with various orientations. The line which is parallel to any two principal planes is called a principal line. The following five cases of principal lines are discussed with reference to Fig. 9.4.

1. **Line is parallel to both HP and VP:** Both TV (ab) and FV (a′b′) show true length.
2. **Line is in HP and parallel to VP:** The FV (a′b′) lies on xy line. The line (AB) itself becomes a TV. Both the views show the true length.
3. **Line is in VP and parallel to HP:** The TV (ab) lies on xy line. The line (AB) itself becomes an FV. Therefore, both the views show true length.
4. **Line is perpendicular to VP and parallel to HP:** FV is a point view (a′,b′) and TV (ab) shows a true length.
5. **Line is perpendicular to HP and parallel to VP:** TV is a point view (a,b) and FV (a′b′) shows a true length.
   The straight line in all these five cases is called a principal line.
6. **Line inclined $\theta$° to HP and $\phi$° to VP (with $\theta + \phi = 90°$):** The top view and front view show an apparent length. The true length can be obtained from the side view.

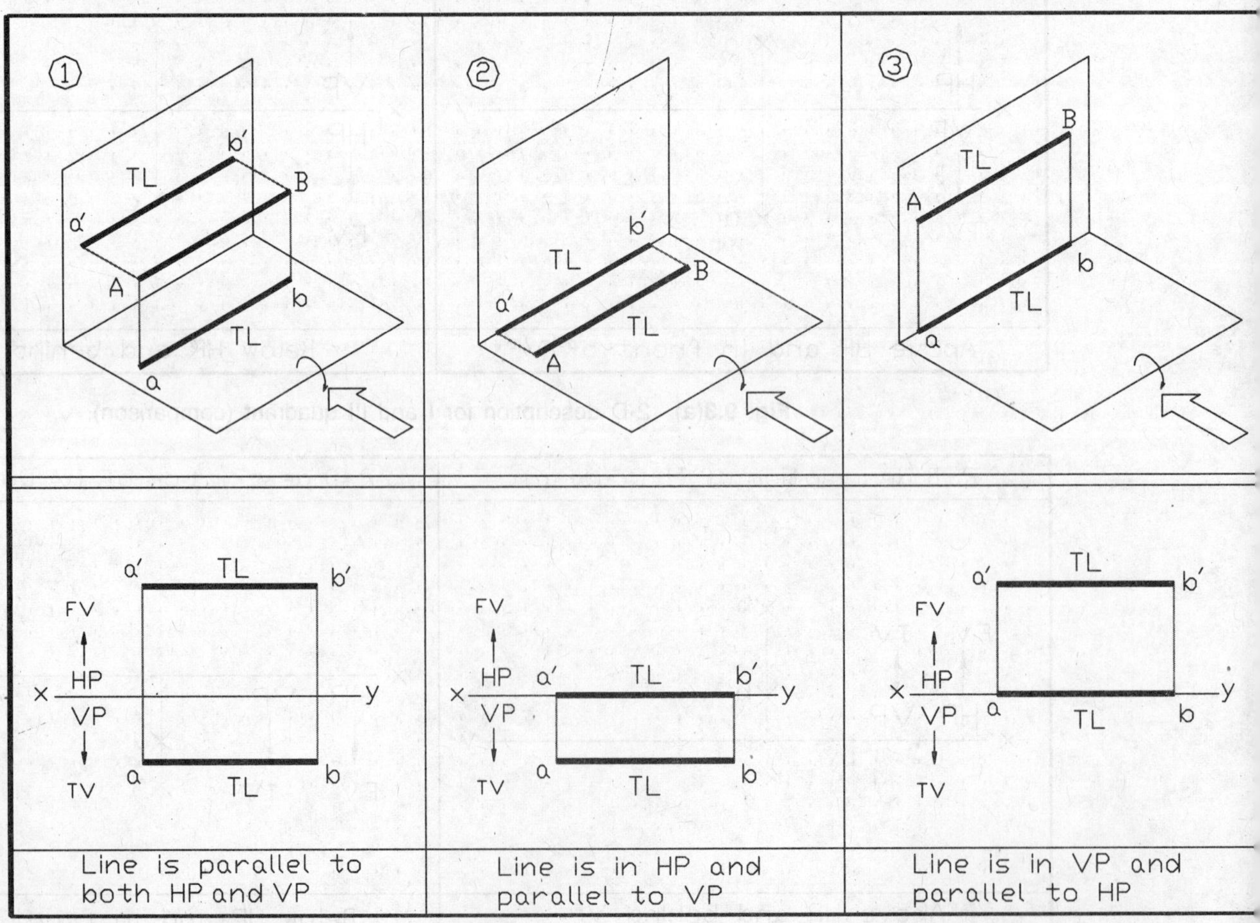


(a)

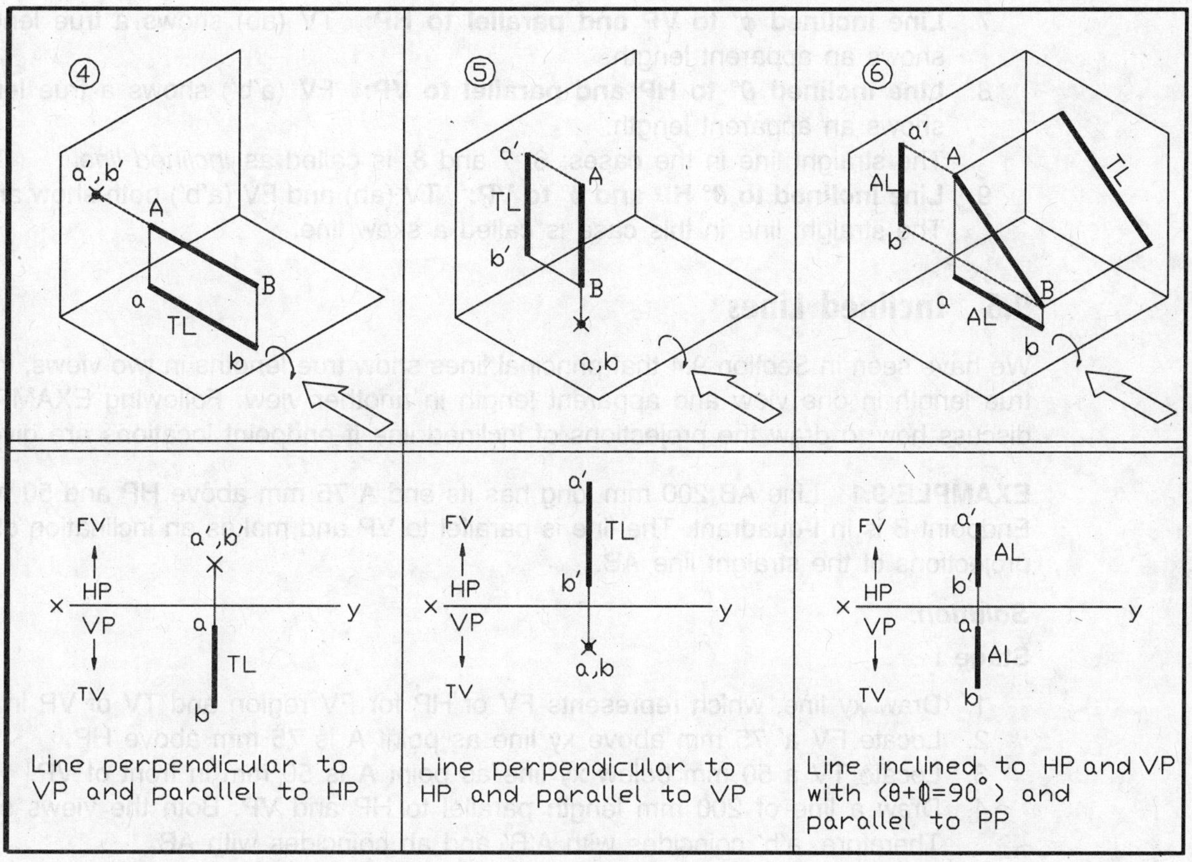


(b)

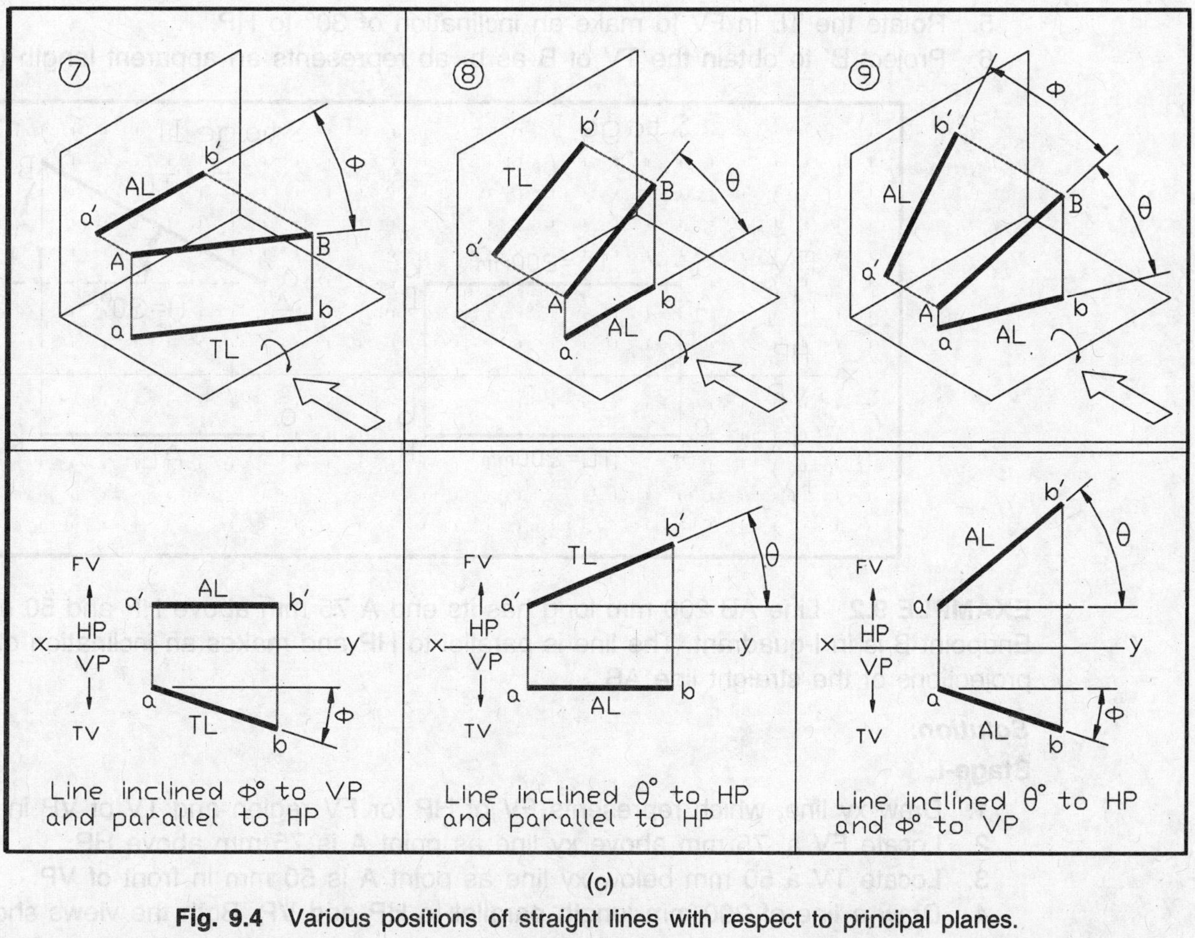


(c)

**Fig. 9.4** Various positions of straight lines with respect to principal planes.

7. **Line inclined $\phi$° to VP and parallel to HP:** TV (ab) shows a true length and FV (a′b′) shows an apparent length.
8. **Line inclined $\theta$° to HP and parallel to VP:** FV (a′b′) shows a true length and TV (ab) shows an apparent length.
   The straight line in the cases: 6, 7 and 8, is called as *inclined line.*
9. **Line inclined to $\theta$° HP and $\phi$° to VP:** TV (ab) and FV (a′b′) both show an apparent length. The straight line in this case is called a skew line.

## 9.5 Inclined Lines

We have seen in Section 9.4 that principal lines show true lengths in two views, inclined lines show true length in one view and apparent length in another view. Following EXAMPLES 9.1 and 9.2 discuss how to draw the projections of inclined line if endpoint locations are given.

**EXAMPLE 9.1** Line AB 200 mm long has its end A 75 mm above HP and 50 mm in front of VP. Endpoint B is in I-quadrant. The line is parallel to VP and makes an inclination of 30° to HP. Draw projections of the straight line AB.

***Solution:***

**Stage I**

1. Draw xy line, which represents FV of HP for FV region and TV of VP in TV region.
2. Locate FV a′ 75 mm above xy line as point A is 75 mm above HP.
3. Locate TV a 50 mm below xy line as point A is 50 mm in front of VP.
4. Draw a line of 200 mm length parallel to HP and VP. Both the views show true lengths. Therefore, a′b′ coincides with A′B′ and ab coincides with AB.

**Stage II**

5. Rotate the TL in FV to make an inclination of 30° to HP.
6. Project B′ to obtain the TV of B as b. ab represents an apparent length (AL).

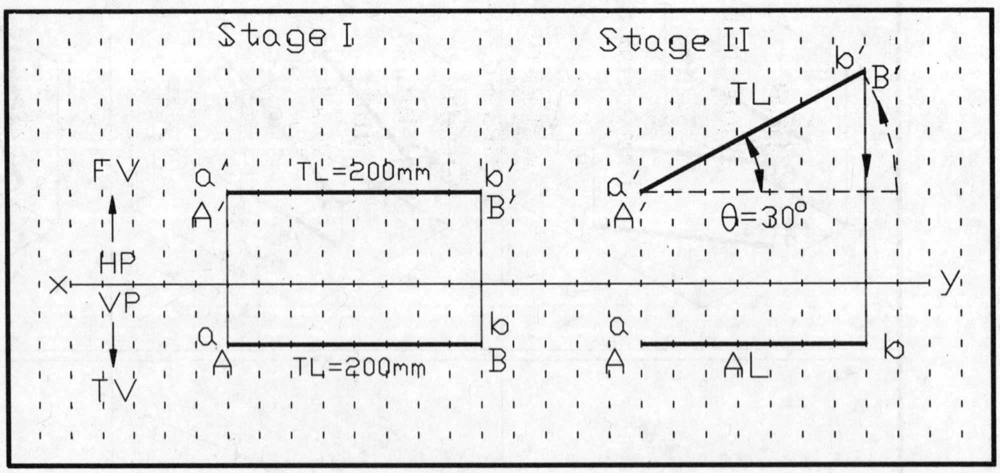


**EXAMPLE 9.2** Line AB 200 mm long has its end A 75 mm above HP and 50 mm in front of VP. Endpoint B is in I-quadrant. The line is parallel to HP and makes an inclination of 40° to VP. Draw projections of the straight line AB.

***Solution:***

**Stage-I**

1. Draw xy line, which represents FV of HP for FV region and TV of VP in TV region.
2. Locate FV a′ 75 mm above xy line as point A is 75 mm above HP.
3. Locate TV a 50 mm below xy line as point A is 50 mm in front of VP.
4. Draw a line of 200 mm length parallel to HP and VP. Both the views show true lengths.

**Stage-II**

5. Rotate the TL in TV to make an inclination of 40° to VP.
6. Project B to obtain the FV of B as b′. a′b′ represents an apparent length (AL).

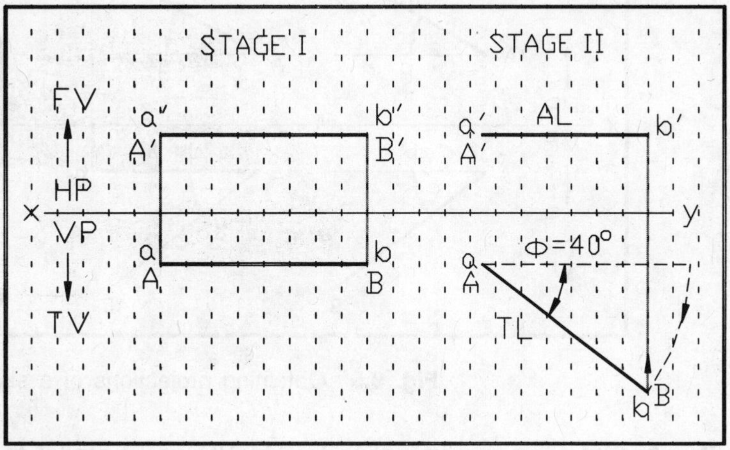


## 9.6 Projection of a Skew Line

If a line is inclined $\theta$° to HP and parallel to VP, its TV length is shorter than its true length (refer to EXAMPLE 9.1). Similarly, if a line is inclined $\phi$° to VP and parallel to HP, its FV length is shorter than its true length (refer to EXAMPLE 9.2). Line inclined to both the principal planes, i.e., $\theta$° to HP and $\phi$° to VP, is called a skew line. If a line is inclined to both the principal planes, then its TV and FV lengths are shorter than its true length. These shorter lengths are called apparent lengths and will be inclined at apparent angles to the xy line in their orthographic view.

The problem of projection of skew line is formulated in two types. The order of solution is just the reverse for the two types: Type (A) and Type (B).

**Type (A):** The true lengths and true inclinations are given. Find out the apparent lengths and apparent inclinations.

**Type (B):** The apparent lengths and apparent inclinations are given. Find out the true lengths and true inclinations.

Before we look into the solution for both the types of problem, notations are defined for true lengths TL, true inclinations TI, apparent lengths AL and apparent inclinations AI. These notations are shown for a line AB in the following table:

| Lengths | TL | AL | Angles | TI | AI |
|---|---|---|---|---|---|
| TV | AB | ab | HP | $\theta$ | $\alpha$ |
| FV | A′B′ | a′b′ | VP | $\phi$ | $\beta$ |

**Type (A)**

True lengths (AB and A′B′) and true inclinations ($\theta$, $\phi$) are seen in Fig. 9.5 (a). We are required to find out the apparent lengths (ab, a′b′) and apparent inclinations ($\alpha$, $\beta$). The endpoints A and B are located in the I-quadrant. The solution is obtained in two stages:

**Stage I:** Assume the line is inclined $\theta$° to HP and parallel to VP. Draw its projections. (Refer to Fig. 9.5)

(a) Fix the endpoint 'A' so that point A and a coincides in TV and point A′ and a′ coincides in FV.
(b) Assume the line is inclined $\theta$° to HP and parallel to VP. Project B′ on the locus line passing through a. $ab_1$ represents an apparent length [Fig. 9.5 (b)].
(c) Draw a locus line for TV of B i.e., b passing through point B. Draw an arc with a, A as centre and $ab_1$ as radius cutting locus line passing through B. The intersection of locus line and arc gives b. The apparent length ab makes inclination $\beta$° with VP [Fig. 9.5 (c)].

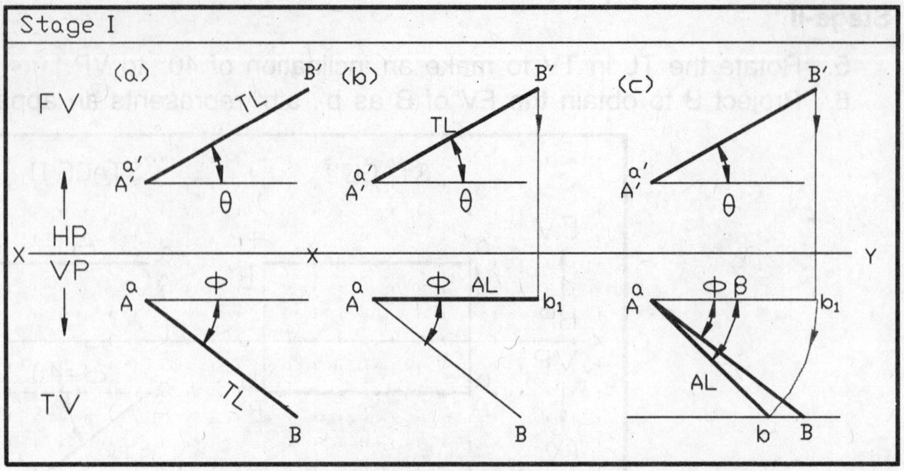


**Fig. 9.5** Obtaining projections of a skew line.

**Stage II:** Assume the line is inclined $\phi°$ to VP and parallel to HP. Draw its projections. (Refer to Fig. 9.6).

(d) Assume the line is inclined $\phi°$ to VP and parallel to HP. Project B on the locus line passing through a′. a′b′$_1$ represents an apparent length [Fig. 9.6 (d)].

(e) Draw a locus line for FV of B i.e., b′ passing through point B′. Draw an arc with a′, A′ as centre and a′b′$_1$ as radius cutting the locus line passing through B′. The intersection of locus line and arc gives b′. The apparent length a′b′ makes inclination $\alpha°$ with HP [Fig. 9.6 (e)].

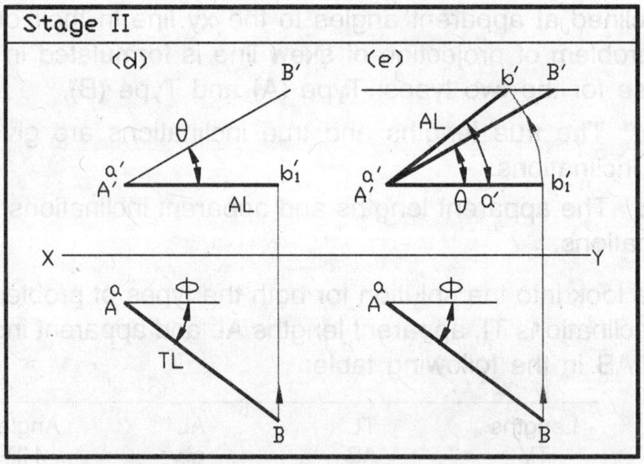


**Fig. 9.6** Obtaining projections of a skew line.

Figure 9.7 shows a solution combining Stage I and Stage II. In this example, both the endpoints are in the same quadrant. In some cases, one endpoint is located in one quadrant and another end point is located in another quadrant. Some observations are summarised in order to solve such complicated problems.

**Observations:**(Refer to Fig. 9.7)

1. The projections (apparent lengths: ab and a′b′) have got common end projectors a′-a and b′-b.
2. The apparent angles are greater than the true angles, i.e., $\alpha > \theta$ and $\beta > \phi$.
3. The path to obtain TV of B, i.e., b is:
   {B′–[Projection up to a locus line passing through a]–b$_1$–[Rotation]–b}
4. The path to obtain FV of B, i.e., b′ is
   {B–[Projection up to a locus line passing through a′]–b′$_1$–[Rotation]–b′}

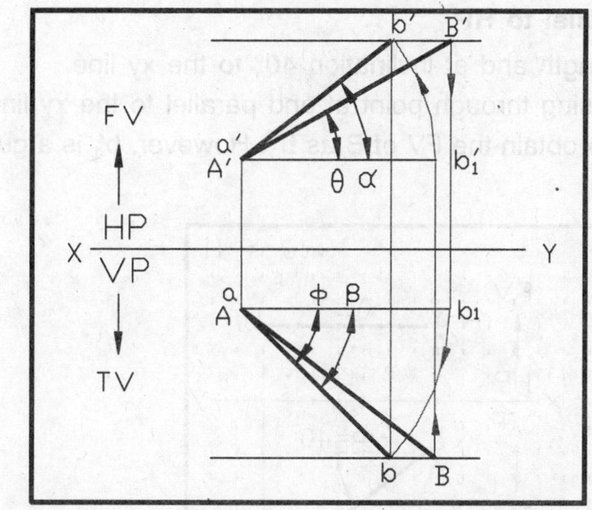


| | *TL* | *Tl* | *AL* | *Al* |
|---|---|---|---|---|
| TV | AB | $\theta$ | ab | $\alpha$ |
| FV | A′B′ | $\phi$ | a′b′ | $\beta$ |

**Fig. 9.7** Combining Stage I and Stage II.

**Type (B)**

Follow a reverse order of solution given for Type (A). In this type, we know b and b′ and we are required to find B and B′. The path to obtain B′ is {b–[Rotation]–$b_1$–[Projection up to a locus line passing through b′]–B′}. The path to obtain B is {b'–[Rotation]–$b'_1$–[Projection up to a locus line passing through b]–B}.

**EXAMPLE 9.3** Line AB 200 mm long has its end A 75 mm above HP and 50 mm in front of VP. Endpoint B is in I-quadrant. The line makes an angle of 30° to HP and 40° to VP. Draw the projections of the straight line AB and find the apparent inclinations.

***Solution*** In EXAMPLE 9.1, the line is inclined 30° to HP and in EXAMPLE 9.2, the line is inclined 40° to VP. In this problem, the line is inclined 30° to HP and 40° to VP. The solution for this problem is a combination of the solutions of EXAMPLES 9.1 and 9.2.

1. Draw xy line, which represents FV of HP for FV region and TV of VP for TV region.
2. Locate a′ 75 mm above xy line, as point A is 75 mm above HP.
3. Locate a 50 mm below xy line, as point A is 50 mm in front of VP.
4. Fix the endpoint A so that a will coincide with A′ and a′ will coincide with A′.

**Stage I Line is $\theta$° to HP and parallel to VP**

5. Draw a line A′B′ of 200 mm length and at inclination 30° to the xy line.
6. Draw a locus for point $b_1$ passing through point a and parallel to the xy line.
7. Project B′ on this locus to obtain the TV of B as $b_1$. However, $b_1$ is a guest TV of point B and not the final one.

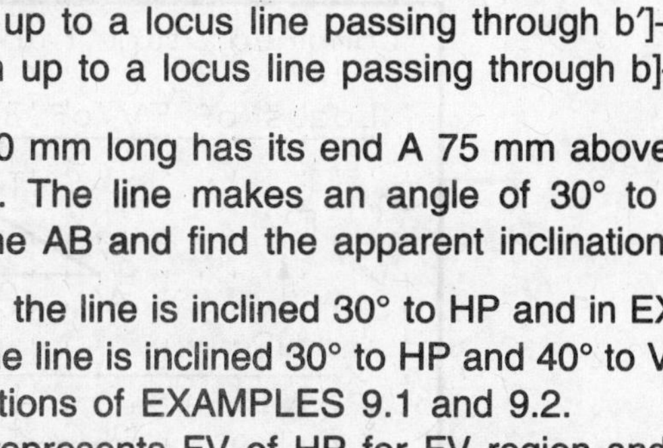

**Stage II Line is $\phi$° to VP and parallel to HP**

8. Draw a line AB of 200 mm length and at inclination 40° to the xy line.
9. Draw a locus for point $b'_1$ passing through point a′ and parallel to the xy line.
10. Project point B on this locus to obtain the FV of B as $b'_1$. However, $b'_1$ is a guest FV of point B and not the final one.

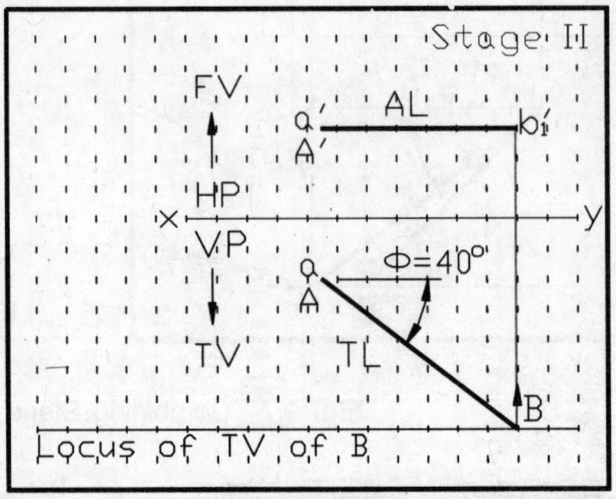


Now, two solutions are combined and shown below.

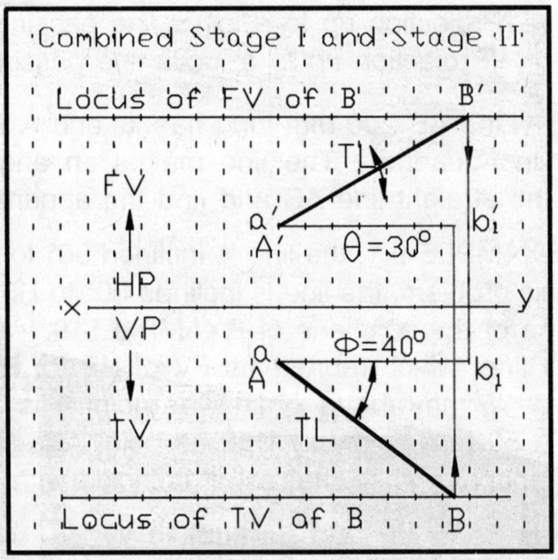


**Stage III Line is $\theta$° to HP and $\phi$° to VP**

Now the target is to obtain projections of point B (b and b′) and apparent inclinations, which ab makes with VP as $\beta$ and a′b′ makes with HP as $\alpha$.

11. Draw an arc with centre a′ and radius $a'b'_1$ till it cuts the locus of FV of point B and obtain point view b′.
12. Join a′b′ and determine inclination $\alpha$° to HP, ($\alpha$ =39°).
13. Draw an arc with centre a and radius $ab_1$ till it cuts the locus of TV of point B and obtain the point view b.
14. Join ab and determine the inclination $\beta$° to VP, ($\beta$ = 45°).

At this stage, a student is required to imagine and follow the procedure given below in order to understand the solution.

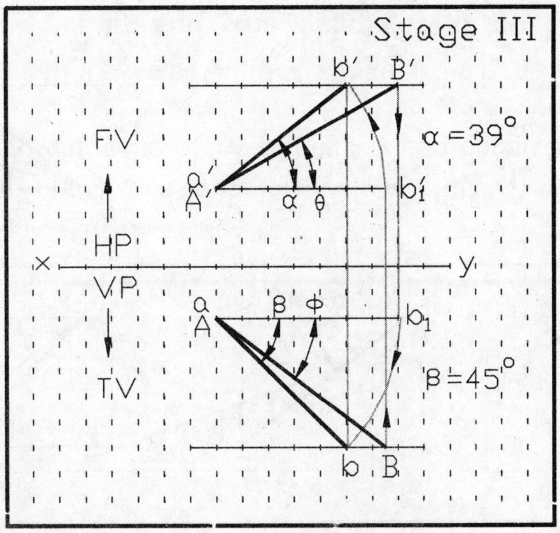


**First imagine the line is $\theta°$ to HP and parallel to VP (Fig. 9.8)**

Fix the endpoint A and move the end point B in a horizontal plane so that the inclination $\theta°$ to HP will be maintained. Move the point B in the horizontal plane till the line AB makes an inclination $\phi°$ to VP. The movement in the horizontal plane appears along a horizontal line in the FV, which is referred as locus of FV of B to obtain point b′. At this stage, the line is a skew line because it is inclined $\theta°$ to HP and $\phi°$ to VP.

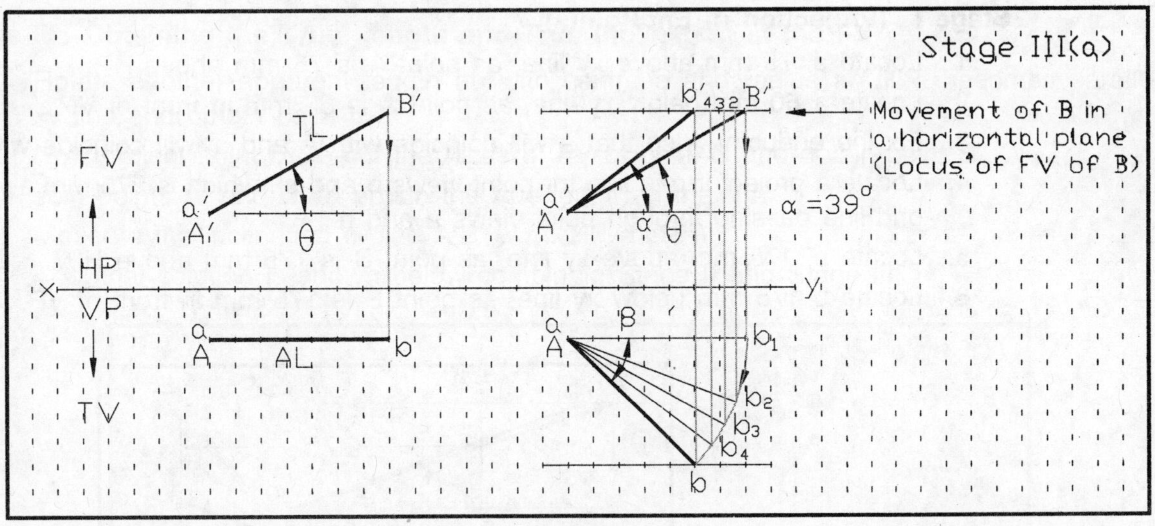


Fig. 9.8

**Now imagine the line is $\phi°$ to VP and parallel to HP (Fig. 9.9)**

Fix the endpoint A and move the endpoint B in a vertical plane so that the inclination $\phi°$ to VP will be maintained. Move the point B in the vertical plane till the line AB makes an inclination $\theta°$ to HP. The movement in the vertical plane appears along a horizontal line in the TV, which is referred as the locus of TV of B to obtain point b. At this stage, the line is a skew line because it is inclined $\phi°$ to VP and $\theta°$ to HP.

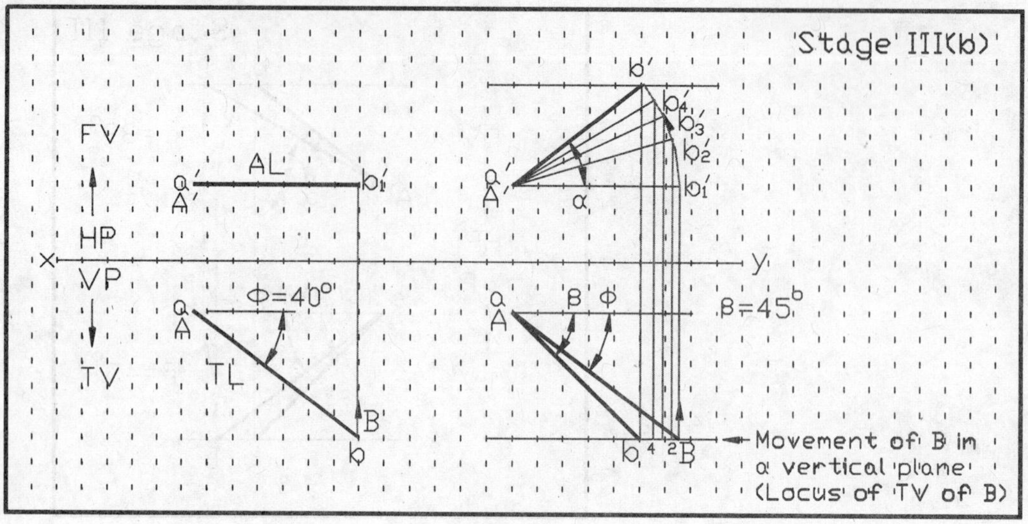


**Fig. 9.9**

**EXAMPLE 9.4** Line AB has its end A 75 mm above HP and 50 mm in front of VP. The other end B is 175 mm above HP and 175 mm in front of VP. The projector distance is 275 mm. Draw projections of line AB. Obtain the true length and true inclinations.

***Solution*** Both the endpoints (A and B) are above HP and in front of VP, therefore, the line AB is in the I-quadrant.

**Stage I (Projection of EndPoints)**

1. Locate a′ 75 mm above xy line, as point A is 75 mm above HP.
2. Locate a 50 mm below xy line, as point A is 50 mm in front of VP.
3. Fix the endpoint A so that a will coincide with A and a′ will coincide with A′.
4. Locate a projector grid line for point views b and b′, which is 275 mm away from a projector grid line passing through point views a and a′.
5. Locate b′ 175 mm above xy line, as point B is 175 mm above HP.
6. Locate b 175 mm below xy line, as point B is 175 mm in front of VP.

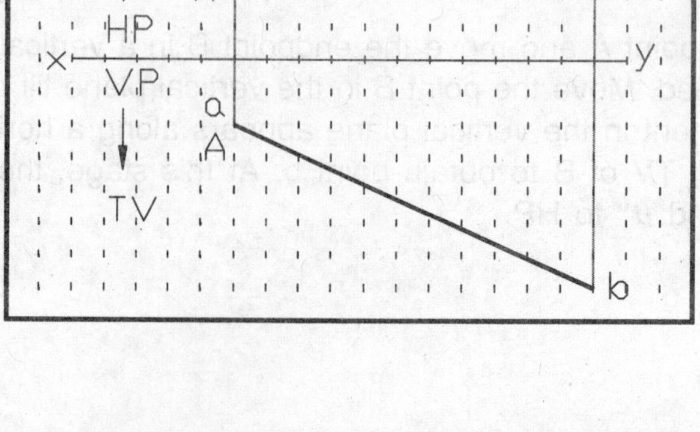

**Stage II Obtaining the True Length and True Inclination in the TV**

7. Rotate the apparent length a′b′ with centre a′ and radius a′b′ till it becomes parallel to xy line and obtain point $b'_1$.
8. Now project point $b'_1$ to cut locus of TV of point B and obtain point B.
9. Join points A and B to obtain true length in the TV.
10. Measure inclination of line AB w.r.t. VP ($\phi$°).

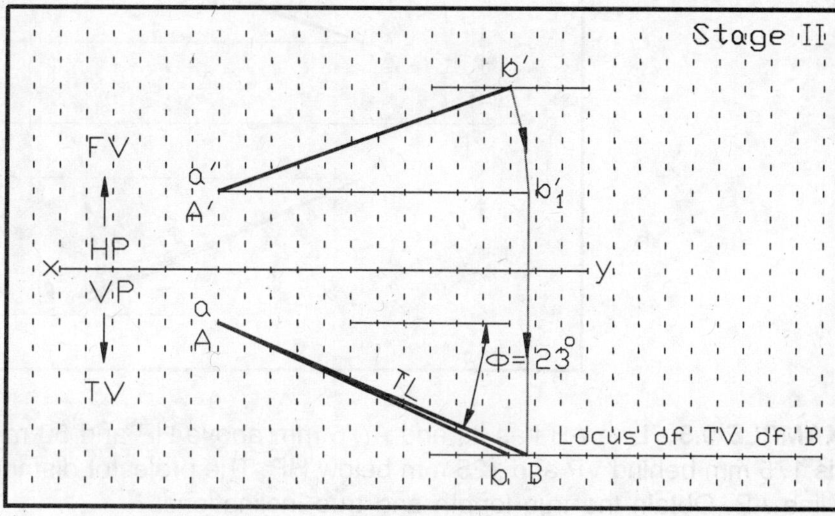


**Stage III Obtaining the True Length and True Inclination in the FV**

11. Rotate another apparent length ab with centre a and radius ab till it becomes parallel to xy line and obtain point $b_1$.
12. Now project point $b_1$ to cut the locus of FV of point B and obtain point B′.
13. Join points A′ and B′ to obtain the true length in the FV.
14. Measure the inclination of line A′B′ w.r.t. HP as ($\theta$°).
15. The final solution is shown below combining Stage II and Stage III.

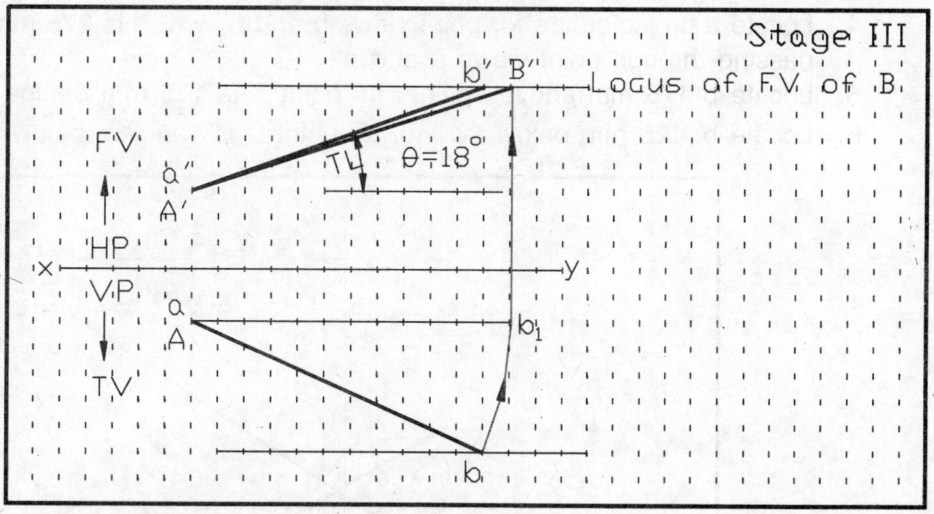

***Note:*** Students are required to draw the solution in one stage as shown below.

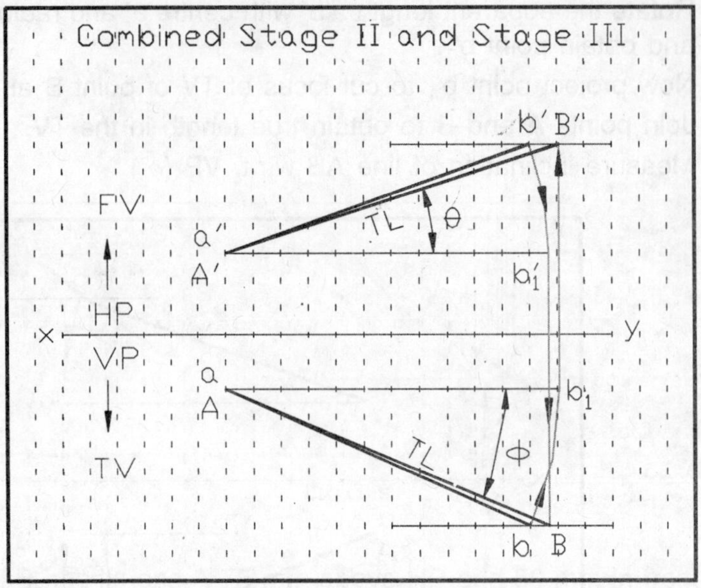


**EXAMPLE 9.5** Line AB has its end A 75 mm above HP and 50 mm in front of VP. The other er B is 175 mm behind VP and 125 mm below HP. The projector distance is 275 mm. Draw projectior of line AB. Obtain the true length and true inclinations.

***Solution*** The point A is above HP and in front of VP, therefore, it is in the I-quadrant and the poi B is behind VP and below HP, therefore, it is in the III-quadrant. For point A, the quadrant descriptic is written to the left side. Therefore, for point A, the FV region is above xy line and TV region is belo xy line. Similarly, for point B, the quadrant description is written to the right side. Therefore, for poi B, the TV region is above xy line and FV region is below xy line.

**Stage I (Projection of End Points)**

1. Locate a′ 75 mm above xy line, as point A is 75 mm above HP.
2. Locate a 50 mm below xy line, as point A is 50 mm in front of VP.
3. Fix the endpoint A so that a will coincide with A and a′ will coincide with A′.
4. Locate a projector line for point views b and b′, which is 275 mm away from a projector lir passing through point views a and a′.
5. Locate b 175 mm above xy line, as point A is 175 mm behind VP.
6. Locate b′ 125 mm below xy line, as point B is 125 mm below HP.

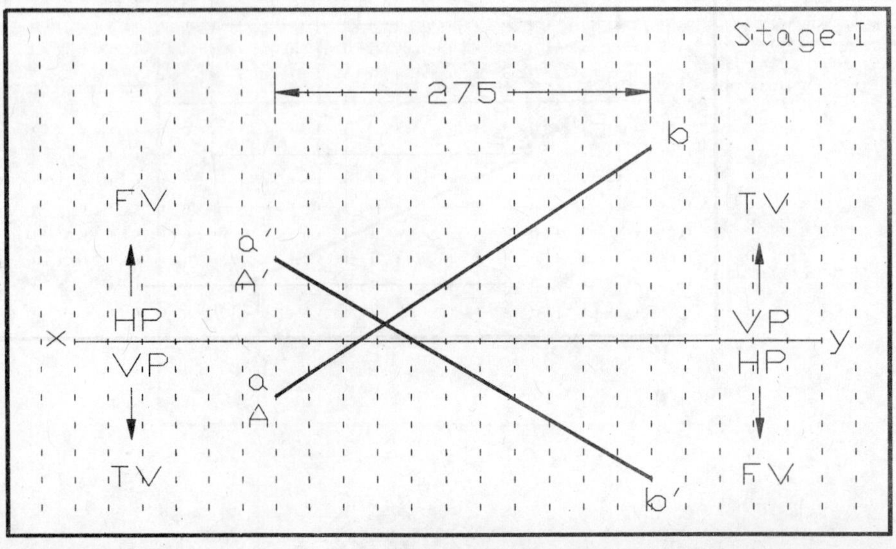

**Stage II Obtaining the true length and true inclination in the FV**

7. Rotate apparent length ab with centre a and radius ab till it becomes parallel to the xy line and obtain projection $b_1$.
8. Now project $b_1$ to cut locus of FV of point B and obtain B′.
9. Join points A′ and B′ to obtain the true length in the FV.
10. Measure inclination of A′B′ w.r.t. HP as $\theta$ referring to the right side description of quadrant. (FV makes an angle with HP, $\theta$ = 29°).

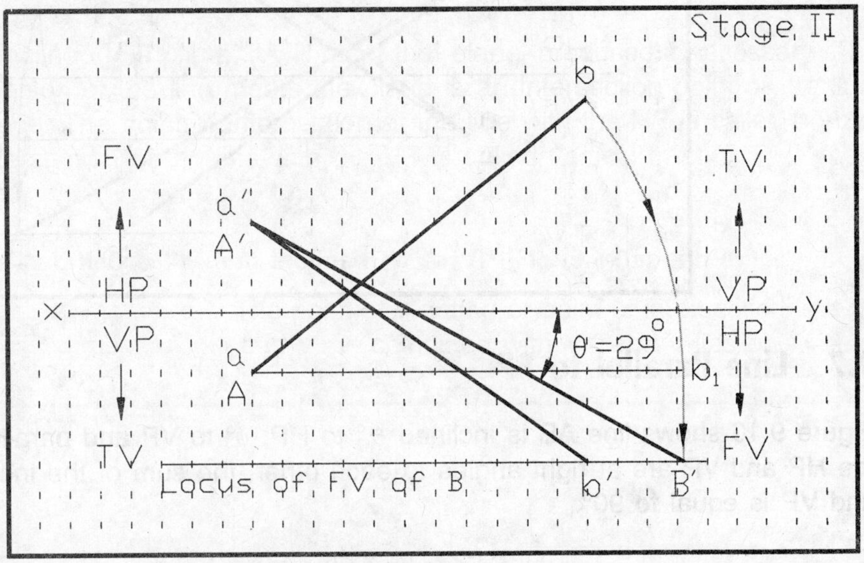


**Stage III (Obtaining the true length and true inclination in the TV region)**

11. Rotate another apparent length a′b′ with centre a′ and radius a′b′ till it becomes parallel to xy line and obtain projection $b'_1$.
12. Now project $b'_1$ to cut locus of TV of point B and obtain B.
13. Join points A and B to obtain the true length in the TV.
14. Measure inclination of line AB w.r.t. VP as $\phi$ referring to the right side description of quadrant. (TV makes an angle with VP, $\phi$ = 33°).

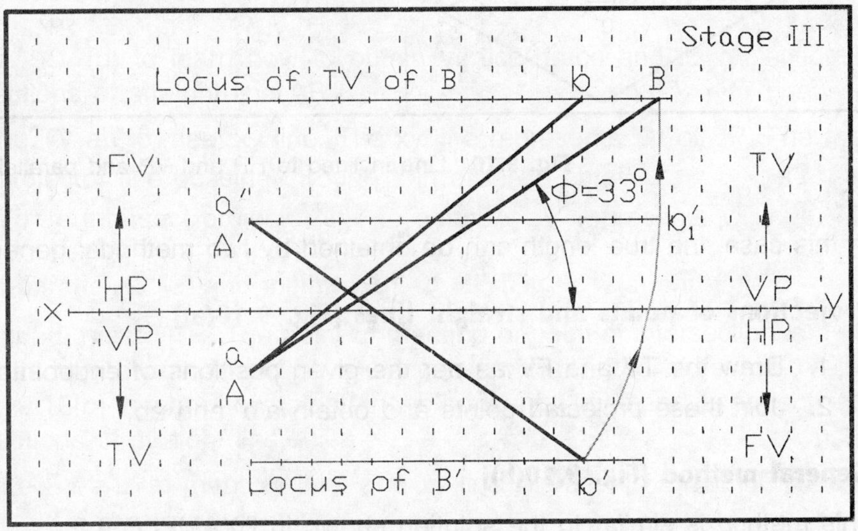

15. The final solution is shown below, in which, Stage I, Stage II and Stage III are combined

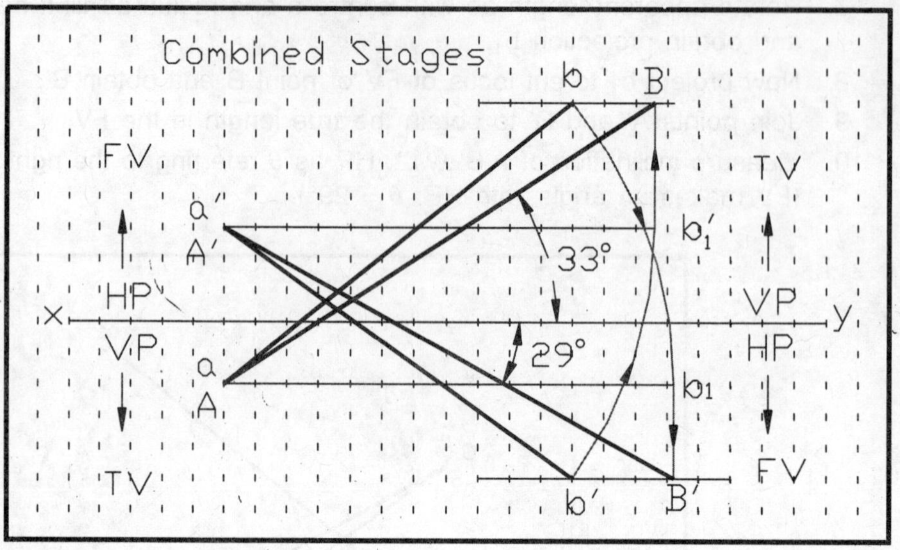


## 9.7 Line Parallel to PP

Figure 9.10 shows line AB is inclined $\theta°$ to HP, $\phi°$ to VP and parallel to a profile plane (PP). As the HP and VP are at right angles to each other, the sum of the inclinations of a line with the HP and VP is equal to 90°.

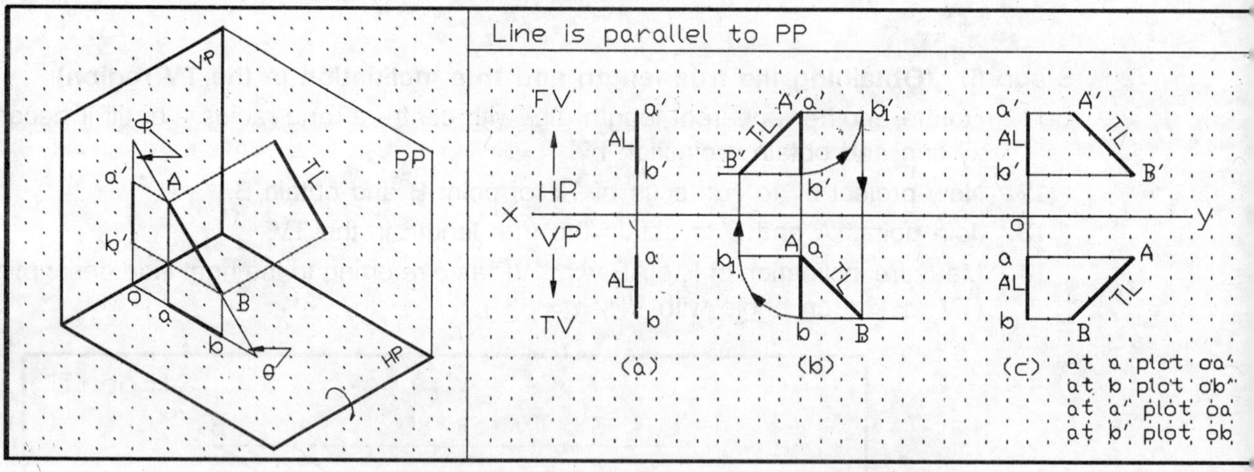


**Fig. 9.10** Line inclined to HP and VP and parallel to PP.

In this case, the true length can be obtained by two methods: general method and offset method

**Projections of points and straight lines [Fig. 9.10(a)]**

1. Draw the TV and FV as per the given positions of endpoints 'A' and 'B'.
2. Join these projected points and obtain a′b′ and ab.

**General method [Fig. 9.10(b)]**

This method is similar to the solution shown in EXAMPLE 9.3.

**Offset method [Fig. 9.10(c)]**

In this method, with reference to a 3-D representation shown in Fig. 9.10, offset distances are plotted to obtain true length in FV and TV region. This is shown in Fig. 9.10 (c). Referring to a 3-D representation, the distance of A from a′ is oa, the distance of B from b′ is ob, the distance of A from a is oa′, the distance of B from b is ob′. The AB represents the true length in TV and A′B′ represents the true length in FV.

## 9.8 Traces of Line

When a line is inclined to a plane, it will meet that plane, produced if necessary. The point in which the line or extension of the line meets the plane is an intersection point of the line and the plane and is called trace. The point of intersection of the line with the HP is called horizontal trace (HT). The point of intersection of the line with the VP is called vertical trace (VT). Principal lines do not have any traces.

**Case 1: Line is parallel to HP and inclined $\phi°$ to VP [Fig. 9.11(a) and (b)]**

Line AB, if extetnded meets, VP. The point of meeting or point of intersection is a VT. The VT itself is a point and has projections: v and v′. v′ coincides with VT and v lies on xy line.

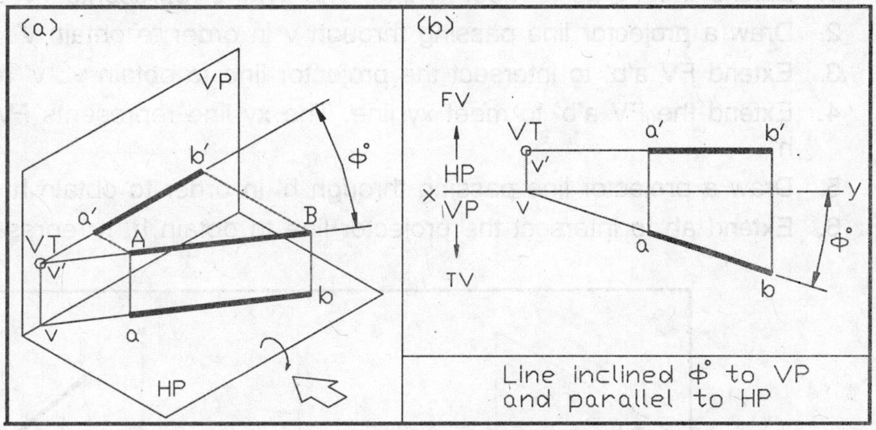


**Fig. 9.11** Obtaining vertical trace (VT).

Refer to Fig. 9.11(b) to learn how to obtain vertical trace and its projections. Figure 9.11(b) shows the projections of straight line AB.

1. Extend the TV ab to meet xy line. The xy line represents TV of VP. The meeting point is v.
2. Draw a projector line passing through v in order to obtain v′.
3. Extend a′b′ to intersect projector line to obtain v′. v′ represents VT.

**Case 2: Line is parallel to VP and inclined $\theta°$ to HP [Fig. 9.12(a) and (b)]**

Line AB, if extended, meets HP. The point of meeting or point of intersection is a HT. The HT itself is a point and has projections: h and h′. h coincides with HT and h′ lies on xy line.

Refer to Fig. 9.12(b) to learn how to obtain the horizontal trace and its projections. Figure 9.12(b) shows the projections of straight line AB.

1. Extend the FV a′b′ to meet xy line. The xy line represents FV of HP. The meeting point is h′.
2. Draw a projector line passing through h′ in order to obtain h.
3. Extend ab to intersect the projector line to obtain h. h represents HT.

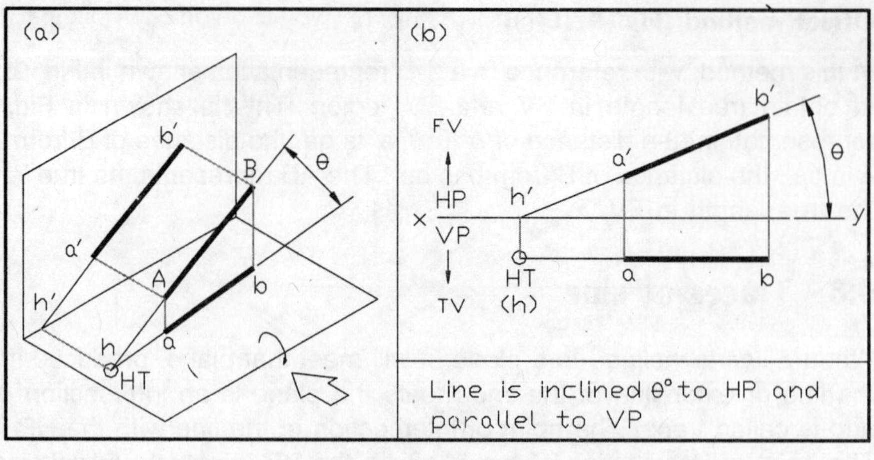


**Fig. 9.12** Obtaining horizontal trace (HT).

**Case 3: Line is inclined $\theta°$ to HP and $\phi°$ to VP [Fig. 9.13(a) and (b)]**

Refer to Fig. 9.13(b) to learn how to obtain vertical trace and its projections. Figure 9.13(b) shows the projections of straight line AB.

1. Extend the TV ab to meet xy line. The xy line represents TV of VP. The meeting point is v
2. Draw a projector line passing through v in order to obtain v′.
3. Extend FV a′b′ to intersect the projector line to obtain v′. v′ represents VT.
4. Extend the FV a′b′ to meet xy line. The xy line represents FV of HP. The meeting point is h′.
5. Draw a projector line passing through h′ in order to obtain h.
6. Extend ab to intersect the projector line to obtain h. h represents HT.

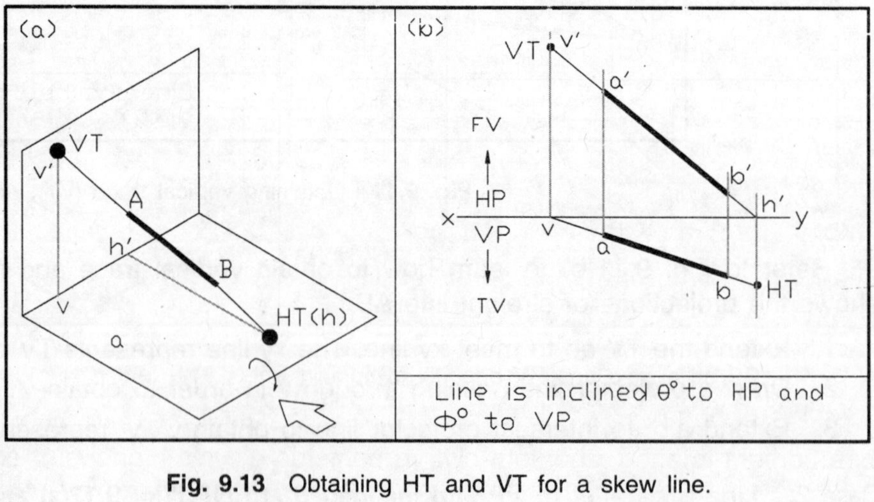


**Fig. 9.13** Obtaining HT and VT for a skew line.

EXAMPLES 9.1, 9.2 and 9.4 are taken to learn how to obtain the traces of lines: (a) when the line is $\theta°$ to HP, (b) when the line is $\phi°$ to VP and (c) when the line is $\phi°$ to HP and to $\phi°$ VP respectively.

**EXAMPLE 9.6** Line AB 200 mm long has its end A 75 mm above HP and 50 mm in front of VP. Endpoint B is in I-quadrant. The line is parallel to VP and makes an inclination of 30° to HP. Draw the projections of straight line AB. Obtain HT and VT.

***Solution*** This exercise is similar to EXAMPLE 9.1 to draw the projections of line. The procedure to obtain HT and VT is explained here.

1. EXTEND the FV a′b′ to meet xy line. (FV of line intersects the HP).
2. Note the intersection point of extended line and xy line as h′ as h′ lies on xy line.
3. Draw a projector line passing through point h′.
4. EXTEND the TV ab to meet the projector line passing through h′.
5. Note the intersection point of extended line and projector line as HT (or h).
6. There is no VT as the line is parallel to VP.

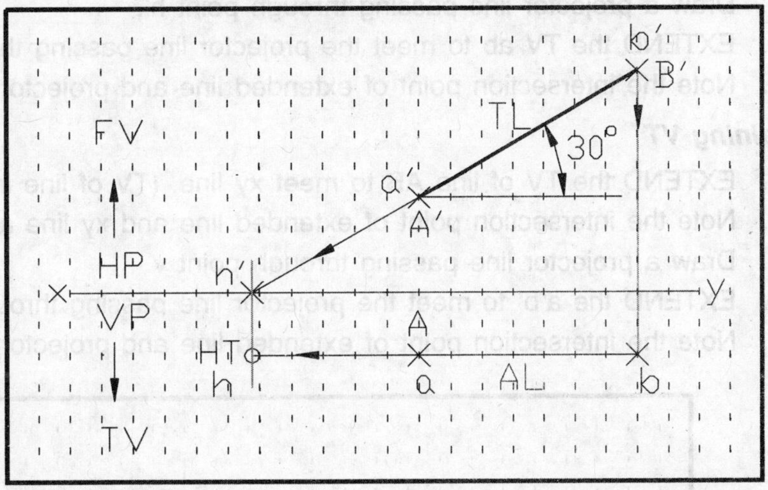


**EXAMPLE 9.7** Line AB 200 mm long has its end A 75 mm above HP and 50 mm in front of VP. Endpoint B is in I-quadrant. The line is parallel to HP and makes an inclination of 40° to VP. Draw the projections of straight line AB. Obtain HT and VT.

***Solution*** This exercise is similar to EXAMPLE 8.2 to draw the projections of line. The procedure to obtain HT and VT is explained here.

1. Extend the TV of line AB to meet xy line. (TV of line intersects the VP).
2. Note the intersection point of extended line and xy line as v as v lies on xy line.
3. Draw a projector line passing through point v.
4. Extend the FV of line AB (a′b′) to meet the projector line passing through v.
5. Note the intersection point of extended line and projector line as VT (or v′).
6. There is no HT as the line is parallel to HP.

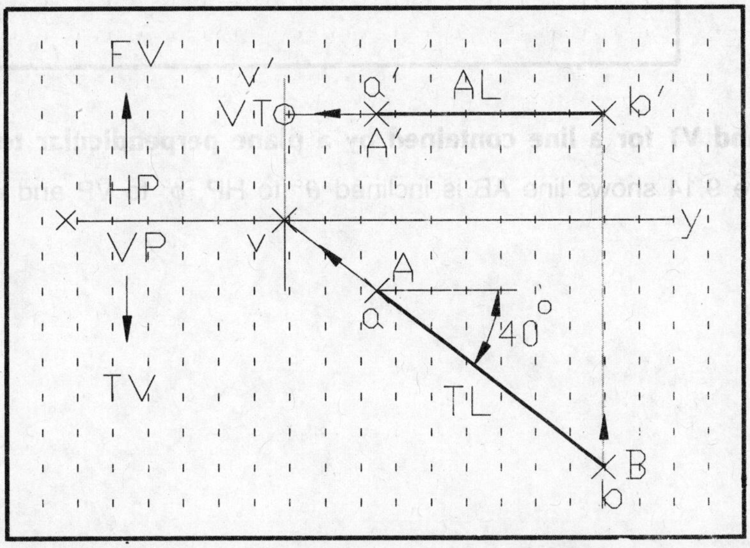

**EXAMPLE 9.8** Line AB has its end A 75 mm above HP and 50 mm in front of VP. The other en B is 175 mm above HP and 175 mm in front of VP. The projector distance is 275 mm. Dra projections of line AB. Obtain HT and VT.

***Solution*** This exercise is similar to EXAMPLE 9.4 to draw the projections of line. The procedur to obtain HT and VT is explained here.

***Obtaining HT***

1. EXTEND the FV a′b′ to meet xy line. (FV of line intersects the HP).
2. Note the intersection point of extended line and xy line as h′ as h′ lies on xy line.
3. Draw a projector line passing through point h′.
4. EXTEND the TV ab to meet the projector line passing through h′.
5. Note the intersection point of extended line and projector line as HT (or h).

***Obtaining VT***

6. EXTEND the TV of line AB to meet xy line. (TV of line intersects the VP).
7. Note the intersection point of extended line and xy line as v as v lies on xy line.
8. Draw a projector line passing through point v.
9. EXTEND the a′b′ to meet the projector line passing through v.
10. Note the intersection point of extended line and projector line as VT (or v′).

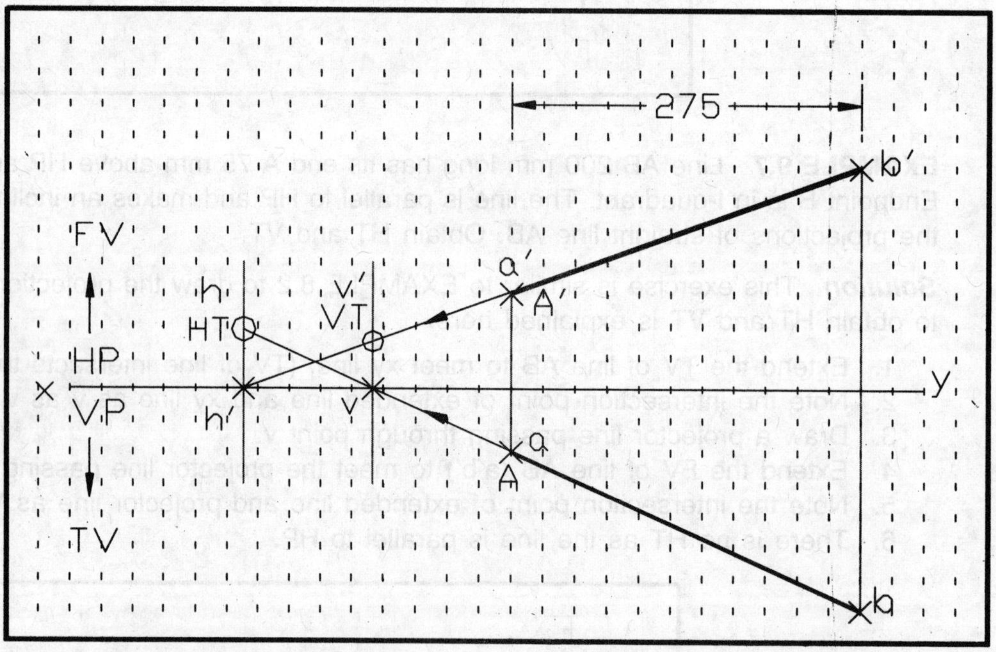


**HT and VT for a line contained by a plane perpendicular to HP and VP**

Figure 9.14 shows line AB is inclined $\theta°$ to HP, $\phi°$ to VP and parallel to a profile plane (PP).

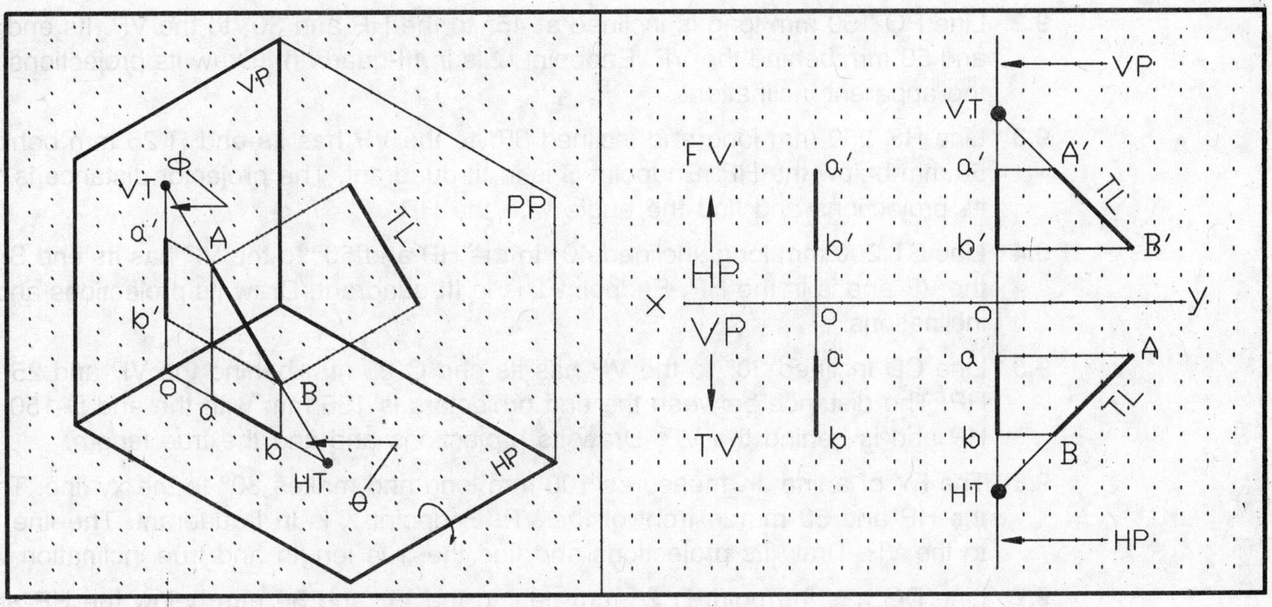


**Fig. 9.14** Obtaining HT and VT for a line parallel to PP.

The steps to determine the true lengths in TV and FV are already discussed in Section 9.7. In this case, if we follow a procedure mentioned in EXAMPLES 9.6, 9.7 and 9.8 to obtain HT and VT, we will not be able to get HT and VT. The reason is TV, FV, HT and VT lie on a single projection line. In order to obtain HT and VT a specific procedure is mentioned below with reference to Fig. 9.14 (b). VT is a FV and HT is a TV. HT and VT lie on the same projector line.

***Obtaining VT***

In FV region, xy line represents FV of HP. Therefore, any line perpendicular to xy line (HP) represents a VP. Therefore, single projection line on which FV is lying represents VP for FV region. EXTEND the true length A′B′. The point at which the extended line intersects the projection line is VT.

***Obtaining HT***

In TV region, xy line represents TV of VP. Therefore, any line perpendicular to xy line (VP) represents a HP. Therefore, single projection line on which TV is lying represents HP for TV region. EXTEND the true length AB. The point at which the extended line intersects the projection line is HT.

## Exercise A (Projection of Straight Lines)

**Instructions for solving the problems**

1. PDMODE : 3
2. PDSIZE : 0.25
3. Text Size : 0.2
4. Unit : 4 grid spacing = 100 mm.
5. Use EXTEND command and perpendicular (perp) option wherever necessary.
6. Change colour of projection line to red.

**9.1** Line AB has its end A 25 mm below the HP and 50 mm behind the VP. The end B is 100 mm below the HP and 100 mm behind the VP. The projector distance between these ends is equal to 150 mm. Draw the projections of AB and show its true length, angle with the HP and VP.

**9.2** Line PQ 150 mm long is inclined at 45° to the HP and 30° to the VP. Its end P is in the HP and 50 mm behind the VP. Endpoint Q is in III-quadrant. Draw its projections and determine the apparent inclinations.

**9.3** Line RS 200 mm long and inclined 30° to the VP has its end R 25 mm behind the VP and 50 mm below the HP. Endpoint S is in III-quadrant. The projector distance is 100 mm. Draw its projections and find the angle with the HP.

**9.4** Line ST 200 mm long, inclined 40° to the HP and 50° to the VP has its end S 25 mm behind the VP and is in the HP. Endpoint T is in III-quadrant. Draw its projections and find apparent inclinations.

**9.5** Line CD inclined 30° to the VP has its end C 25 mm behind the VP and 25 mm below the HP. The distance between the end projectors is 150 mm with the end D 150 mm below the HP and is behind the VP. Draw its projections and find the true length.

**9.6** The FV of a line JK measures 100 mm long and makes 30° to the xy line. The point J is in the HP and 50 mm in front of the VP. Endpoint K is in I-quadrant. The line is inclined 40° to the VP. Draw its projections and find the true length and true inclination with the HP.

**9.7** Line DG has the point D 25 mm behind the VP and 50 mm below the HP and the point G 100 mm above the HP and 75 mm in front of the VP. The end projector distance is 125 mm. Draw its projections and find the true length and true inclinations.

**9.8** Line AC has its TV and FV on the single projector line with the FV 100 mm long and TV 75 mm long. Endpoint A is 25 mm behind the VP and 50 mm below the HP. Draw its projections and find the true length and true inclinations. (Consider the line in III-quadrant)

**9.9** Two fan motors P and Q are hung from a ceiling of a hall 12 m length L × 5 m depth D × 8 m height H at a height of 4 m and 6 m, respectively from the floor. The motors are 5 m and 9 m from the left end wall, 3 m and 4 m from the front wall, respectively. (a) Draw the projections of these fan motors with reference to their location from the end wall and front wall. (b) Determine graphically, the distance between the two motors. [Solve using the III-quadrant. Scale: 0.25 AutoCAD unit (1 grid spacing) = 1 m].

**9.10** Two mangoes A and B are on a tree, 2 m and 4 m above the ground, respectively. P and Q are two compound walls at right angles with Q at the front and P at the left. Mango 'A' is 1 m from the wall P and 3 m from the wall Q. Mango 'B' is 6 m from the wall P and 7 m from the wall Q. Find out the true distance between the two mangoes. [Solve using the III-quadrant. Scale: 0.25 AutoCAD unit (1 grid spacing) = 1 m].

**9.11** Three vertical poles: AA (4 m), BB (6 m) and CC (5 m) are placed at the corners of an equilateral triangle ABC (side = 5 m) lying on the ground with the AC perpendicular to the direction of the observer. The top ends of poles AA and CC are required to be joined to point 'O', centre of the equilateral triangle ABC, which is on the ground. Draw the projections of these poles. Determine the true lengths of the lines joining centre point O and top of AA and CC. [Solve using the III-quadrant. Scale: 0.25 AutoCAD unit (1 grid spacing) = 1 m].

**9.12** Three vertical poles: AA (4 m), BB (6 m) and CC (5 m) are placed at the corners of an isosceles triangle ABC (with base: AC = 5 m and height 5 m) lying on the ground with the AC perpendicular to the direction of the observer. One rope connects the top of AA and BB and the other rope connects the top of CC and BB. Find out the true lengths of these two ropes. [Solve using the III-quadrant. Scale: 0.25 AutoCAD unit (1 grid spacing) = 1 m].

**9.13** A bulb is located at the centre of a ceiling of a hall of 8 m length (along x-direction), 6 m depth (along z-direction) and height (along y-direction). A switch is located on either of the two shorter walls at a height of 2 m from the floor and at a depth of 5 m from either of the two bigger walls. The shortest distance between the bulb and the switch is 5.4 m.

Using the III-quadrant, draw the projections of the room with shorter walls at left and right hand sides of the observer. Locate the bulb and switch in projections. Find the height of the ceiling. Find the angle which the line joining the bulb and the switch makes with the floor. (Unit: 1 grid spacing = 1 m).

**9.14** A picture frame ABCD (length, AB = 8 m and height, BC = 6 m) is to be fixed on a wall. The frame is railing about its bottom edge CD on a wall with the help of two straight chains attached to the top corners A and B of the frame. The other ends of the two chains are to be fixed to a hook H on the wall along the centre line of length of the frame and 7 m above the railing edge CD. The frame is to make an angle of 45° with the wall and chains. Using I-quadrant, draw the projections of the frame along with the chains and hook. Find the true length of the chains. Find the true inclinations in between the two chains. [Scale: 0.25 AutoCAD unit (1 grid spacing) = 1 m].

**9.15** Two wire ropes are attached to the top corner C of a 6 m high wall. The other end is attached to the top of a vertical pole A, 4 m high. Similarly, the other end of the second wire rope is attached to the top of a vertical pole B, 3 m high. These two poles are 4 m apart and aligned in a straight line parallel to and at a distance of 5 m from the wall. The bottom ends of these two poles are at equidistance from the corner C. Using III-quadrant draw the projections of the two wire ropes and poles w.r.t. corner C. Find the required length of the two wire-ropes. Find the true inclination made by the two wire-ropes with each other. [Scale: 0.25 AutoCAD unit (1 grid spacing) = 1 m].

**9.16** Determine graphically, the length of a body diagonal of a cube having edge length 100 mm.

**9.17** Determine graphically, the height of a triangular pyramid having length of the base edges equal to the length of the slant edges, which is equal to 150 mm.

**9.18** Two square pyramids having equal dimensions are meeting at their bases with base edges perfectly aligned. The combined solid is standing on one of the apexes on the ground with common axis of both the pyramids perpendicular to the ground and two base edges perpendicular to the VP. Determine the distance between the two apexes. (Base edge = slant edge = 150 mm).

## Exercise B (Traces of Straight Lines)

### Instructions:

Mark the traces by encircling the point along with a text adjacent to it (e.g. O).

**9.1** Line AB makes an inclination 30° with the HP and parallel to the VP. The TV measures 150 mm. Its endpoint A is at an equidistance, 50 mm from the reference planes and both the endpoints are in I-quadrant. Draw the projections and find the traces (if any). (Use entry III-quadrant method).

**9.2** Line AB 175 mm long is inclined to the VP and parallel to the HP. The FV is 125 mm long. Its end A is 50 mm in front of the VP and 75 mm above the HP. End point B is in I-quadrant. Draw its projections and find the traces (if any).

**9.3** The TV and FV of a line AB measure 150 mm and 100 mm, respectively. The line is inclined 30° to the HP. The end A is 50 mm above the HP and 25 mm in front of the VP. The other end B is in the I-quadrant. Draw the projections and locate the traces.

**9.4** The distance between two end projectors for a line AB is 200 mm. End point A is in the HP and in front of the VP. Endpoint B is in the VP and above the HP. The HT lies on the projector through A and 100 mm in front of the VP. The VT lies on the projector through B and 150 mm above the HP. Draw the projections and find the true lengths and true inclinations.

**9.5** The distance between two end projectors for a line AB is 100 mm. Endpoint A is 25 mm behind the VP and below the HP. Endpoint B is 150 mm behind the VP and 75 mm above the HP. VT lies 75 mm below the HP. Draw the projections and find the true length and true inclinations.

**9.6** The distance between two end projectors for a line AB is 150 mm. Endpoint A is 100 mm above the HP and 50 mm in front of the VP. Endpoint B is 25 mm above the HP and 75 mm behind the VP. Draw its projections and find the true length and true inclinations. Find traces (if any).

**9.7** Endpoint A of line AB is 50 mm above the HP and in front of the VP. Endpoint B is 150 mm above the HP and in front of the VP. The FV makes 45° with the HP. VT is 175 mm above the HP and HT is 125 mm in front of the VP. Draw its projections and find the true lengths and true inclinations.

**9.8** Line AB is parallel to the PP with an endpoint A 175 mm above the HP and 50 mm in front of the VP. Endpoint B is 50 mm above the HP and 150 mm in front of the VP. Draw its projections. Find the true length and true inclinations. Also locate HT and VT.

**9.9** Line AB is parallel to the PP with an endpoint A 150 mm behind the VP and at the same distance below the HP. Endpoint B is in the HP and in the VP. Draw its projections. Find the true length and true inclinations. Also locate HT and VT.

**9.10** Line AB has endpoint A 25 mm behind the VP and 50 mm below the HP and endpoint B 100 mm above the HP and 75 in front of the VP. End projector distance is 125 mm. Draw its projections and find the true length and true inclinations and locate HT and VT.

**9.11** Line AB has its endpoint A 50 mm above the HP and 125 mm behind the VP. Endpoint B is 50 mm in front of the VP and 125 mm below the HP. Draw its projections and true lengths if HT and VT lie on the same projector line. Specify the TL, $\theta$ and $\phi$, the distance of HT and VT from the xy line.

**9.12** Line AB, 150 mm long has its endpoint A 50 mm below the HP and 50 mm behind the VP. The line is equally inclined to the HP and VP. Draw its projections when it is parallel to the PP. Specify the apparent length and the distance of HT and VT from the xy line.

**9.13** Line AB has its endpoint A 75 mm behind the VP and 75 mm below the HP. Endpoint B is 100 mm above the HP and 100 mm in front of the VP. Draw its projections and obtain the true lengths if the end projectors are 200 mm apart. Specify the TL, $\theta$, $\phi$ and the distance of HT and VT from the xy line.

**9.14** Mid point of a line PQ lies on HP and VP. FV and TV length of the line are equal to 250 mm. The front view of the line is inclined 45° with the xy line. Draw the projections of the line. Find its true lengths and true inclinations with HP and VP. Locate HT and VT.

**9.15** Endpoint A of a line AB is 200 mm above the HP and 100 mm in front of the VP and endpoint B is on the VP. A mid point C of the line AB is 100 mm above the HP. The distance between the projectors passing through point A and C is 175 mm. Draw the projections of the line AB. Find its true length and true inclinations with the reference planes.

# 10

# Projection of Planar Surface

## Objectives

- Learn the projection of planes in a 3-D space.
- Know the projection of intersection of planes with the principal planes.

## Introduction

A plane surface has only two dimensions: length and breadth. The surface does not have thickness. A plane surface may be assumed to be contained by a plane, and its projections can be drawn, if the position of that plane with respect to the principal planes of projection is known. The knowledge of projection of planes is useful to solve problems of section of solids and intersection of surfaces of solids.

## 10.1 Planes Under Study

Various types of regular planar surfaces are shown in Fig. 10.1. Various plane surfaces under stud are: (1) equilateral triangle (2) isosceles triangle (3) right angled triangle (4) trapezium (5) squar (6) rectangle (7) rhombus (8) parallelogram (9) pentagon (10) hexagon (11) circle (12) semicircle

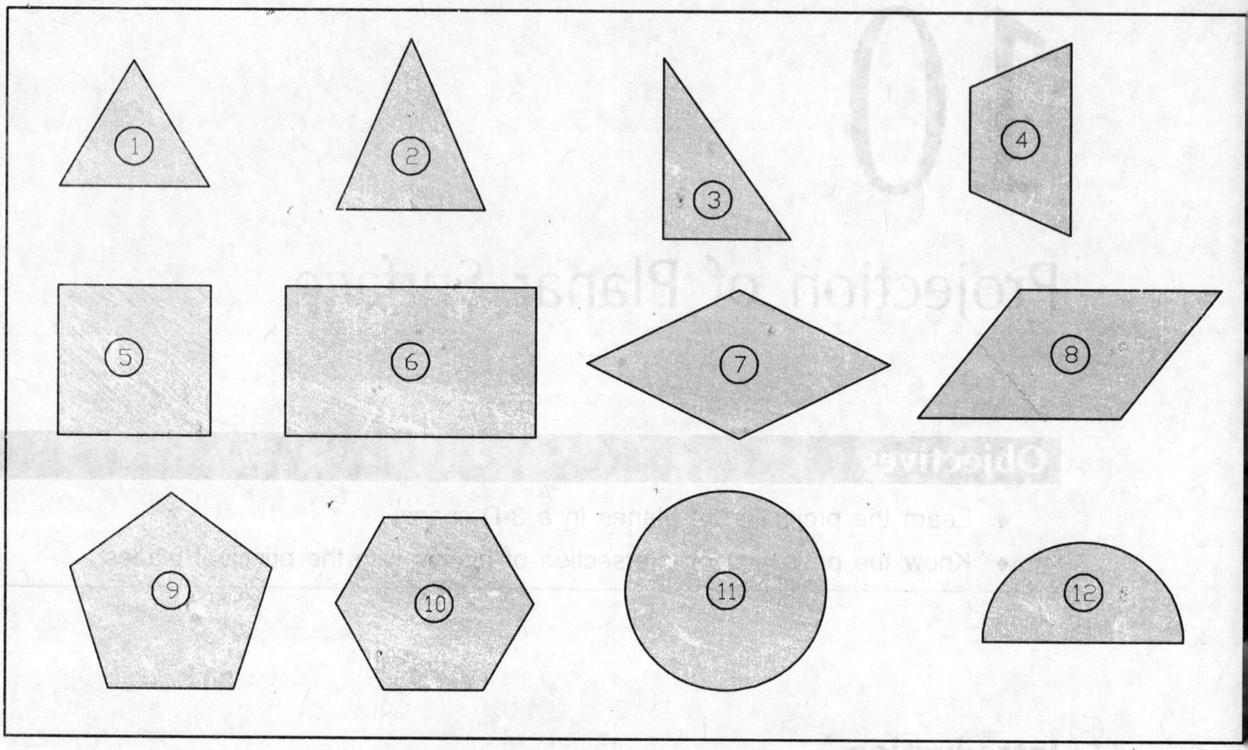


**Fig. 10.1** Various types of plane surfaces.

## 10.2 Classification of Planar Surface

In Engineering Graphics, planar surfaces are divided into two types: (i) perpendicular planes anc (ii) oblique planes.

### Perpendicular Planes

Perpendicular planes can be further divided into the following sub-types with reference to the principal planes: HP and VP.

***Perpendicular to one reference plane and parallel to the other:*** In this case, the true shape car be obtained on a plane to which the given surface is parallel as shown in Fig. 10.2(a) and (b).

***Perpendicular to one plane and inclined to the other:*** In this case, the true shape is no available in either TV or FV. One view shows the line view and another view shows the apparen shape (AS) as shown in Fig. 10.2(c).

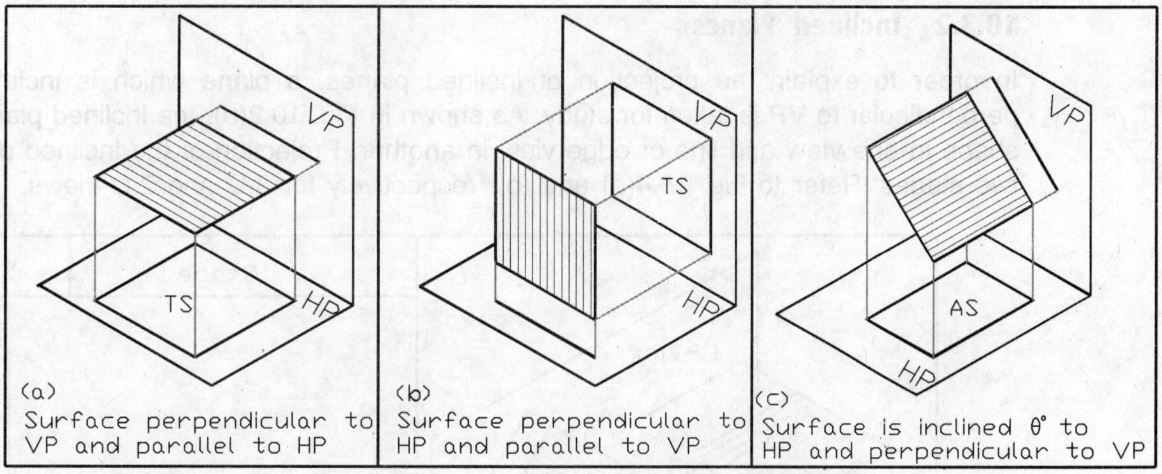


**Fig. 10.2** Perpendicular planes.

### Oblique Planes

Planes which are inclined to both the reference planes are called oblique planes. In this case, the true shape is not available in either TV or FV. Both the views show apparent shapes. This is shown in Fig. 10.3.

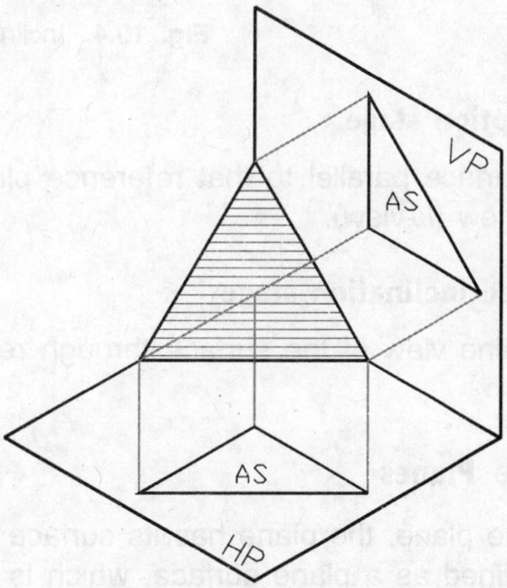


**Fig. 10.3** Oblique planes.

## 10.3 Projection of Planar Surface

Projection of planar surface on principal planes (HP and VP) can be obtained in various stages. This depends upon the position of planar surface with respect to the reference planes (HP and VP). Following are the discussions on projection of principal plane surface, inclined plane surface and oblique (skew) surface.

### 10.3.1 Principal Planes

If it is parallel to HP, the true shape of the surface will be in the TV and if it is parallel to VP, the true shape of the surface will be in the FV. This is shown in Fig. 10.2(a) for both the cases.

### 10.3.2 Inclined Planes

In order to explain the projection of inclined planes, a plane which is inclined $\theta°$ to HP and perpendicular to VP is taken for study. As shown in Fig. 10.2(b), the inclined planes show apparent shape in one view and line or edge view in another. Projection of the inclined plane is obtained in two stages: Refer to Fig. 10.4(a) and (b), respectively for 3-D and 2-D views.

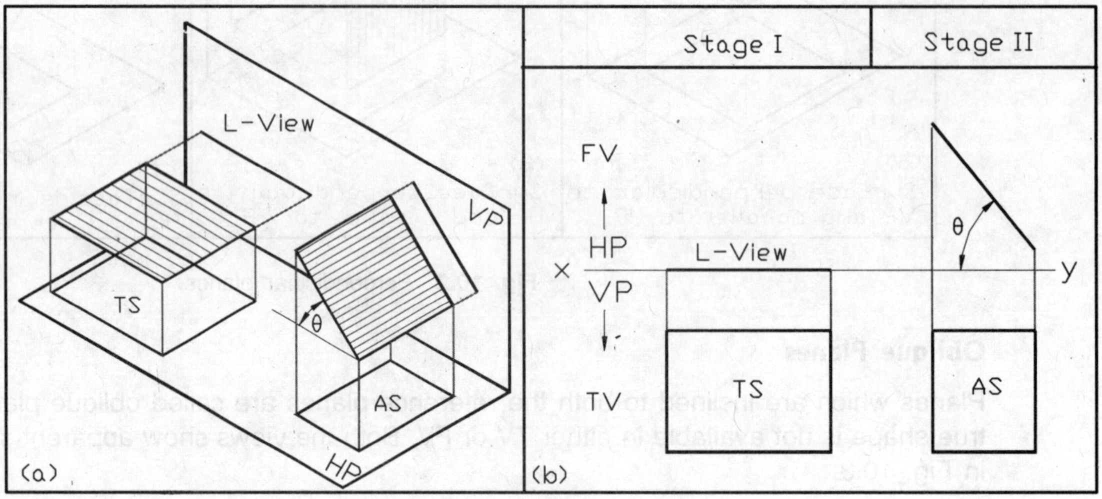


**Fig. 10.4** Inclined planes.

**Stage I: Assumption stage**

Assume plane surface parallel to that reference plane to which it is inclined and obtain the true shape and line view (L-view).

**Stage II: Surface inclination stage**

Now, rotate the line view of the surface through required inclination ($\theta°$) and obtain an apparent shape.

### 10.3.3 Oblique Planes

In case of oblique plane, the plane has its surface inclined to both the reference planes. Oblique plane is also defined as a plane surface, which is inclined to one reference plane and an edge/ diameter/diagonal is inclined to another reference plane. The projections of this plane are obtained in three stages: Refer to Fig. 10.5(a) and (b), respectively for 2-D and 3-D views.

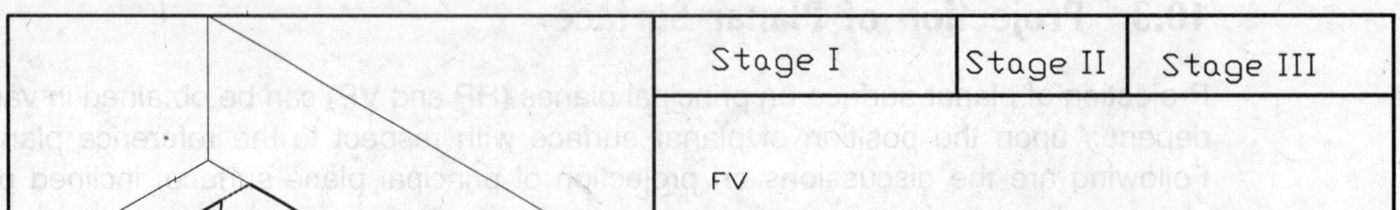


**Fig. 10.5** Stage wise projections of oblique planes.

**Stage I: True Shape stage (assumption)**

Assume the plane surface parallel to the reference plane to which it is inclined. There are two options to draw a true shape: (a) If plane is resting on its edge, draw a true shape of the surface with the resting edge perpendicular to the xy line. (b) If the plane surface is resting on one of its corner, then draw the true shape of the plane surface such that the line joining the resting corner and the centre of the plane parallel to the xy line. In this stage, we obtain the true shape in one view and the line view of surface in another view.

**Stage II: Surface inclination stage**

Now, ROTATE the line view of the surface, obtained in Stage I, through required inclination and obtain an apparent shape. In this stage, a surface is called an inclined surface. In this stage, we obtain an apparent shape in one view and line view of surface in another view.

**Stage III: Edge inclination**

In this stage, the edge of the apparent shape of the plane surface is rotated through the required inclination (say $\alpha°$) to obtain the final projections.

**EXAMPLE 10.1** A rectangular plate of negligible thickness having 150 mm length and 100 mm width is resting on one of its smaller side on HP. The surface is perpendicular to VP and makes an inclination of 30° to HP. Draw projections of the plate.

***Solution*** The plate is resting on a smaller side of HP and makes an inclination with HP. Therefore, the plate will be made parallel to HP so that it will lie on HP. The true shape of the plate will be visible in the TV region.

1. In the TV region, draw a rectangle of 150 mm length and 100 mm width with 100 mm side perpendicular to the xy line. Name the corners as a, b, c and d.
2. Project all the four points on the xy line as the plate is assumed to lie on HP. Obtain a line view a′b′-c′d′ of the plate coinciding with xy line.
3. In the Stage II, COPY the line view of the Stage I and place it on the xy line. ROTATE the line view with respect to side a′b′ through a given inclination of 30°.
4. Project all the four points from the FV of the Stage II to the TV region of Stage II. Project all the four points from the TV of Stage I to the TV of Stage II.
5. Join the respective intersecting points of the projector lines to obtain the final projection points a, b, c and d. We get a reduced apparent shape of the rectangular plate.

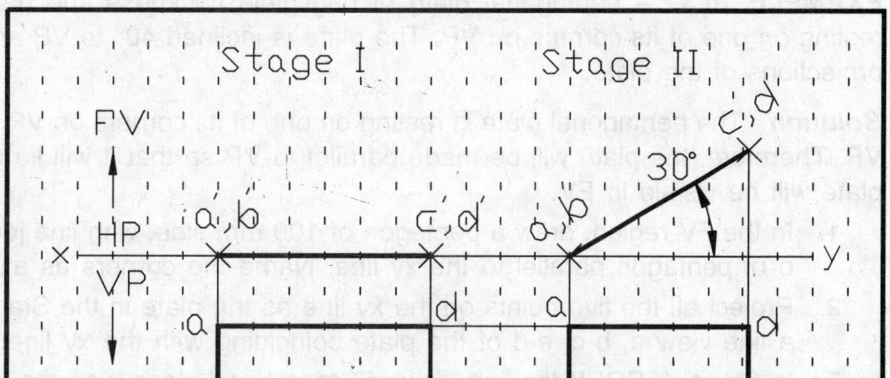

**EXAMPLE 10.2** A parallelogram ABCD has its parallel length AB and CD 100 mm long. The distance between AB and CD is 150 mm. It is resting on side AB on HP with surface inclined 25° to HP and perpendicular to VP. Draw projections of the plate.

***Solution*** The parallelogram is resting on side AB on HP and makes an inclination with HP. The parallelogram will be made parallel to HP so that it will lie on HP. The true shape of the parallelogram will be visible in the TV region.

1. In the TV region, draw a parallelogram of parallel sides 100 mm separated by 150 mm, with side ab perpendicular to the xy line. Name the corners as a, b, c and d.
2. Project all the four points on the xy line as the parallelogram in the Stage I, is lying on HP and obtain a line view a′b′-c′d′ of plate coinciding with the xy line. This completes the Stage I.
3. In the Stage II, COPY the line view of the Stage I and place it on the xy line. ROTATE the line view about side a′b′ through a given inclination of 25°.
4. Project all the four points from the FV of Stage II to the TV region of the Stage II. Project all the four points from the TV of Stage I to the TV of Stage II.
5. Join respective intersecting points of projector lines to obtain the final projection points a, b, c and d. We get a reduced shape of the parallelogram.

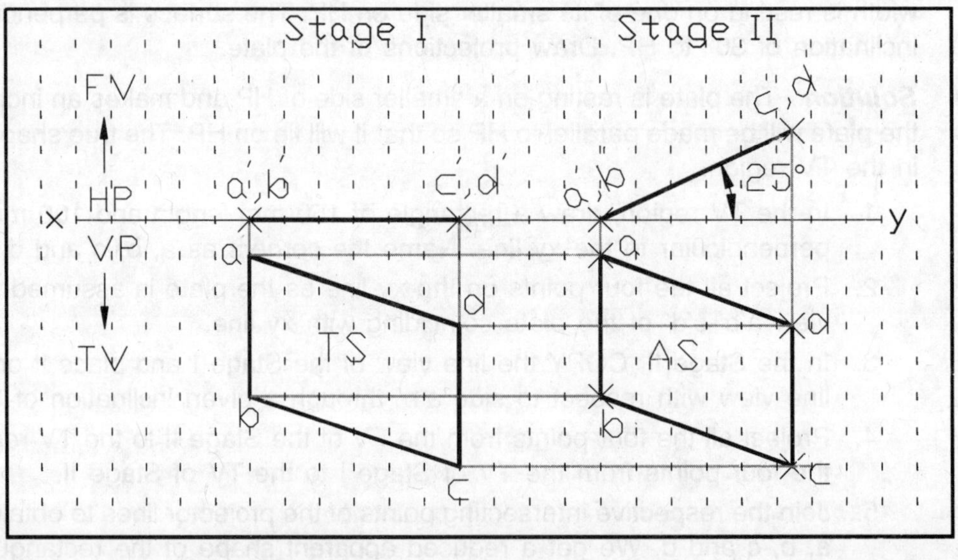


**EXAMPLE 10.3** A pentagonal plate of negligible thickness and having edge length 100 mm is resting on one of its corners on VP. The plate is inclined 40° to VP and perpendicular to HP. Draw projections of the plate.

***Solution*** The pentagonal plate is resting on one of its corners on VP and makes an inclination with VP. Therefore, the plate will be made parallel to VP so that it will lie on VP. The true shape of the plate will be visible in FV.

1. In the FV region, draw a pentagon of 100 mm side, with line joining one corner to the centre o of pentagon parallel to the xy line. Name the corners as a′, b′, c′, d′ and e′.
2. Project all the five points on the xy line as the plate in the Stage I is lying on VP and obtain a line view a, b-c, e-d of the plate coinciding with the xy line.
3. In Stage II, COPY the line view of Stage I and place it on the xy line. ROTATE the line view about the corner point d through a given inclination of 40°.
4. Project all the four points from the TV of Stage II to the FV of Stage II. Project all the four points from the FV of Stage I to the FV of Stage II.
5. Join the respective intersecting points of the projector lines to obtain the final projection points a′, b′, c′, d′ and e′. We get a reduced shape of the pentagon.

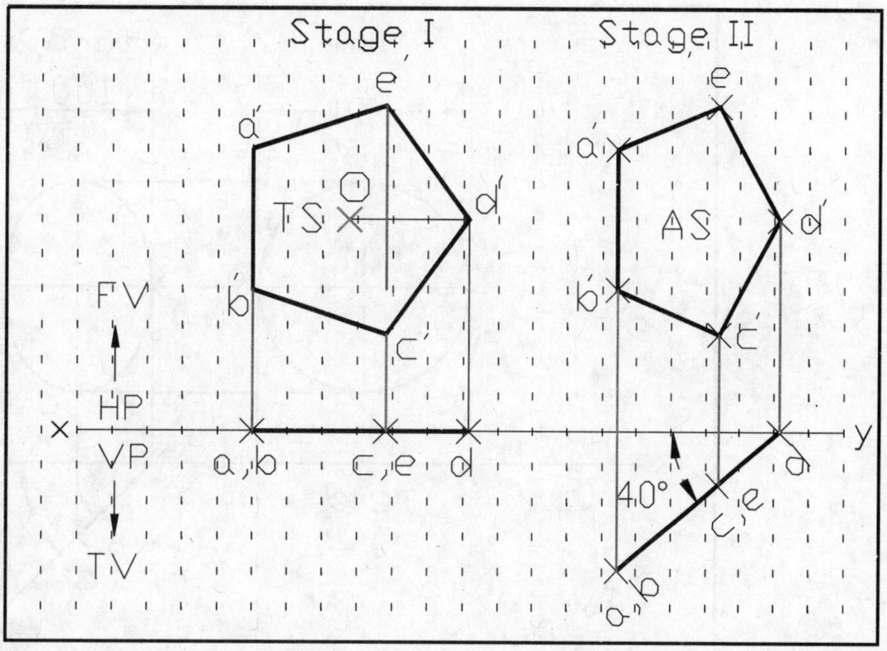


**EXAMPLE 10.4** A circular plate is resting on its circumferential point on VP with surface perpendicular to HP. It is inclined to VP such that the FV appears as an ellipse. Find the surface inclination at which the ellipse has minor axis 100 mm and major axis 150 mm.

***Solution*** If a circular surface is perpendicular to HP and inclined to VP, it appears as an ellipse on the VP (i.e., in the FV). In this situation, a horizontal diameter is inclined to the VP and appears as a minor axis of the ellipse. The vertical diameter which is perpendicular to HP remains unchanged in dimension and appears as a major axis.

The circle appears as an ellipse with major axis 150 mm (equal to the diameter of the circle) and minor axis 100 mm.

As the plate is inclined to VP it is made parallel to VP so that it will lie on VP. The true shape will be visible in the FV.

1. In the FV, draw a circle of 150 mm diameter. Draw two diameters inside the circle, one is parallel and the other is perpendicular to xy line. Name the diameters endpoints as a′b′, and c′d′.
2. Project all the four points on the xy line as the plate in the Stage I is lying on VP and obtain a line view a-c, d-b of plate coinciding with xy line. This completes the Stage I.
3. Draw an ELLIPSE with major axis 'CD' equal to 150 mm and minor axis a′b′ equal to 100 mm in Stage II FV.
4. Project the axes end points a′b′, c′ and d′.
5. COPY the Stage I TV (line view) and place it such that a coincides with the intersection point of projector line through a′ and the xy line.
6. ROTATE the line view a-c, d-b about a till the end b of the line view touches the projector line for b′.
7. Measure the inclination of the line view w.r.t. xy line, (i.e., VP) with DIM command and ANG option. The line view (surface) makes inclination of 45° with VP.

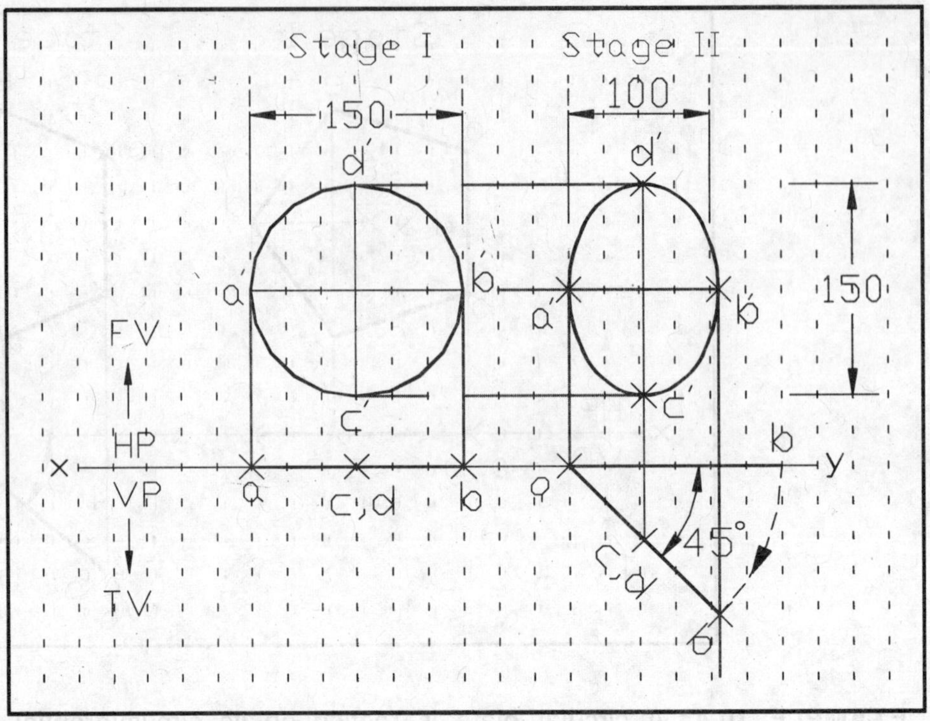


**EXAMPLE 10.5** A rectangular plate of negligible thickness having 150 mm length and 100 mm width is resting on one of its smaller side on HP. The surface makes an inclination of 30° to HP and the smaller side makes an inclination of 60° to VP. Draw projections of the plate.

***Solution*** First five steps are similar to EXAMPLE 10.1.

1. COPY the reduced shape of the rectangular plate from TV region of the Stage II and place it in the TV region of Stage III.
2. ROTATE the copied view so that the smaller side ab makes 60° to VP.
3. Project all the four points from the TV region of Stage III to the FV region of Stage III. Project all the four points from the FV of Stage II to the FV of Stage III.
4. Join the respective intersecting points of the projector lines to obtain the final projection points a′, b′, c′, and d′. The shape of the rectangular plate appears to be as a parallelogram.

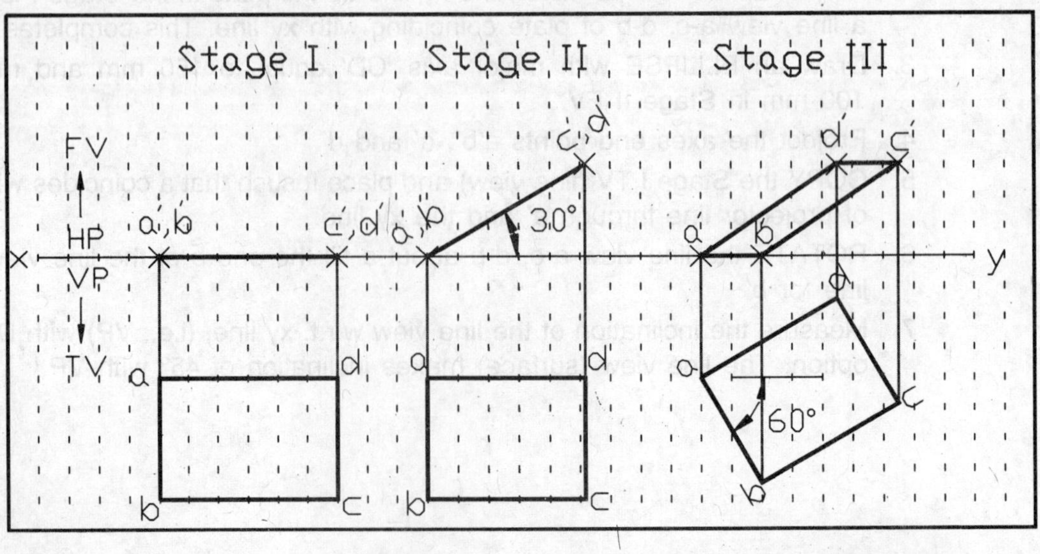

**EXAMPLE 10.6** A pentagonal plate of negligible thickness and having edge length 100 mm is resting on one of its sides on VP. The plate is inclined 40° to VP and the resting side makes an inclination of 50° to HP. Draw projections of the plate.

***Solution*** First five steps (Stage I and Stage II) are similar to EXAMPLE 10.3.

1. COPY the reduced shape of the pentagonal plate from FV region of Stage II and place it in the FV region of Stage III.
2. ROTATE the copied view about b′ so that the resting side a′b′ makes 50° to HP.
3. Project all the five points from the FV of Stage III to the TV of Stage III. Project all the five points from the TV of Stage II to the TV of Stage III.
4. Join the respective intersecting points of the projector lines to obtain the final projection points a, b, c, d and e and obtain an irregular pentagonal shape.

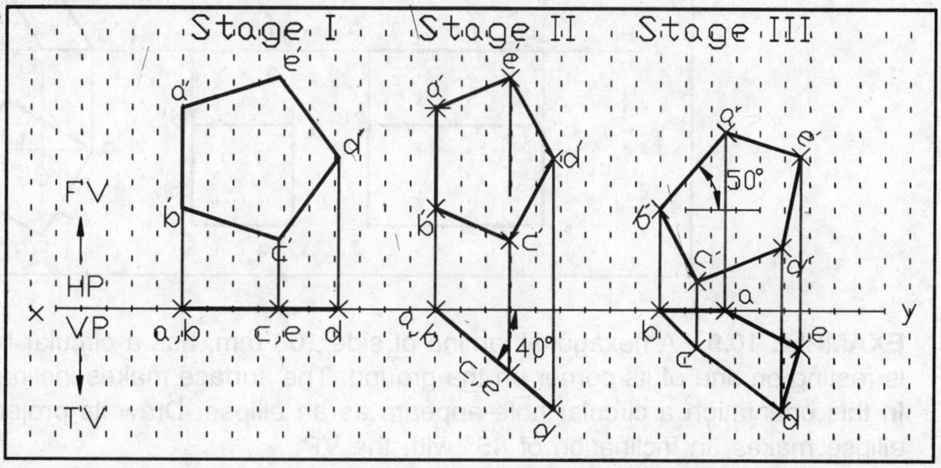


**EXAMPLE 10.7** A square lamina with diagonal 200 mm is resting on one of its corners on the ground. The surface makes inclination of 60° with the ground and the diagonal which is not passing through the resting point makes an inclination of 30° with the VP. Draw projections of the square lamina.

***Solution*** The figure shows the stepwise projections for the given position of the square lamina.

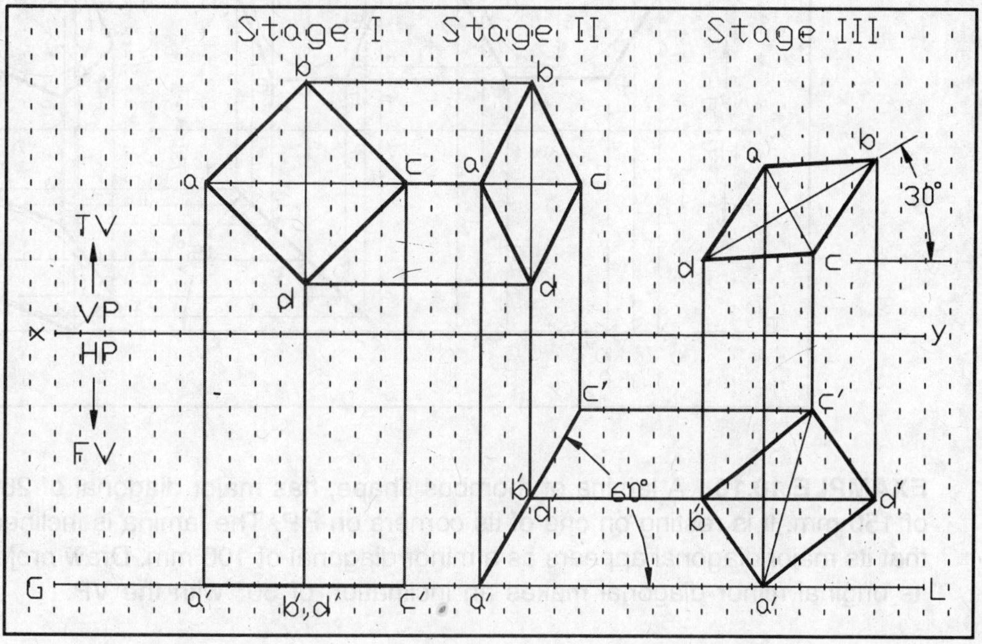

**EXAMPLE 10.8** A rectangular lamina with longer edge 175 mm and smaller edge 100 mm, is resting on one of its smaller edges on the HP. It is inclined with the HP in such a way that its TV appears as a square with maximum dimensions. Draw projections if the smaller edge makes an inclination of 60° with the VP.

***Solution*** The figure shows the stepwise projections for the given position of the rectangular lamina.

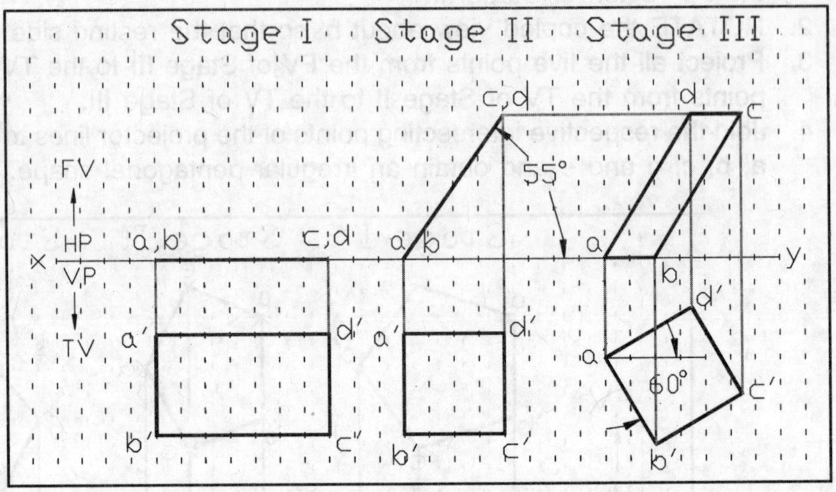


**EXAMPLE 10.9** A hexagonal lamina of side 100 mm, has a circular hole of 150 mm. The lamina is resting on one of its corner on the ground. The surface makes inclination of 45° with the ground. In this orientation a circular hole appears as an ellipse. Draw its projections if a major axis of the ellipse makes an inclination of 45° with the VP.

***Solution*** The figure shows the stepwise projections for the given position of the hexagonal lamina.

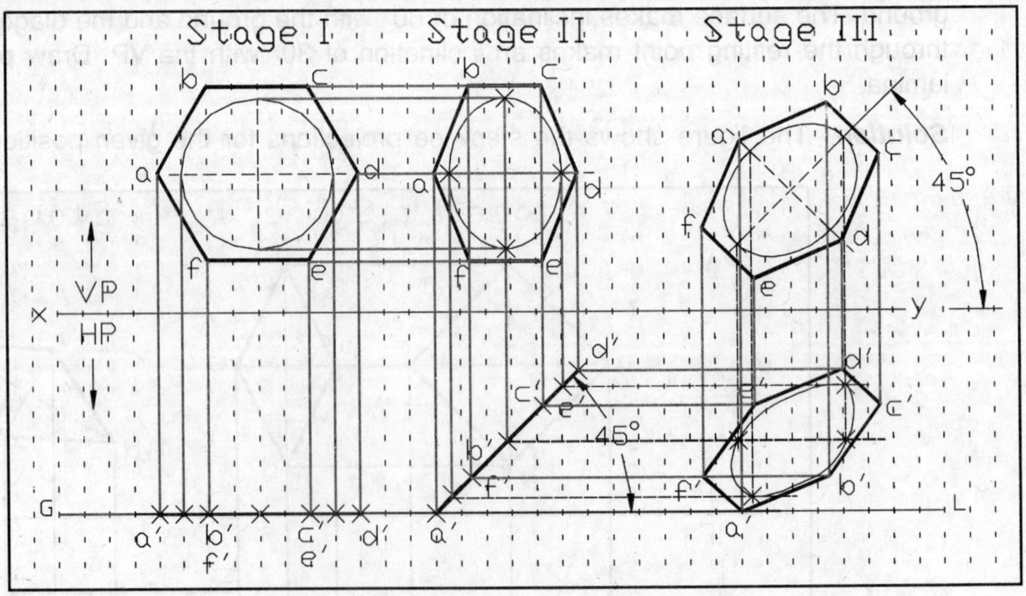


**EXAMPLE 10.10** A lamina of rhombus shape, has major diagonal of 200 mm and minor diagonal of 150 mm. It is resting on one of its corners on HP. The lamina is inclined to the HP in such a way that its major diagonal appears as a minor diagonal of 100 mm. Draw projections of the lamina when its original minor diagonal makes an inclination of 30° with the VP.

***Solution*** The figure shows the stepwise projections for the given position of the hexagonal lamina.

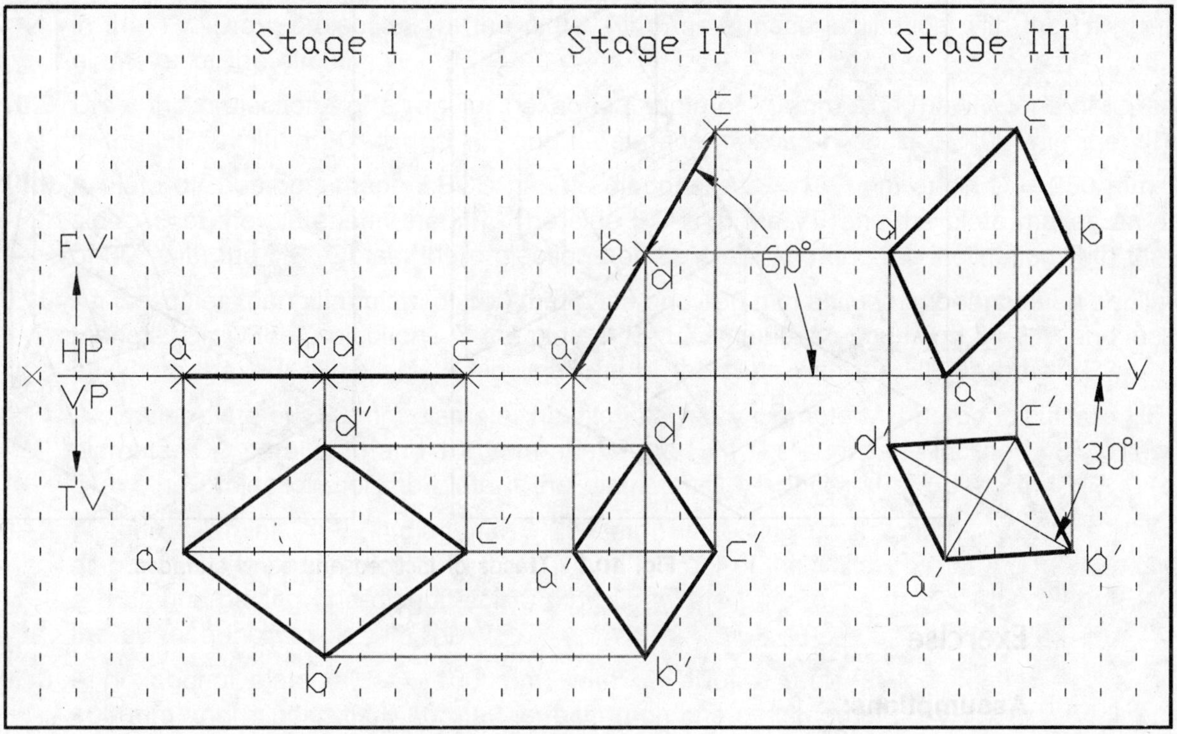


## 10.4 Traces of Planar Surface

The concept of trace of a plane is similar to the one, which we have used in case of traces of a line. Plane surface also have the traces. A plane surface, extended if necessary, will meet the reference planes (HP and VP) in lines, unless it is parallel to any one of them. These lines are called the traces of plane.

The line at which the surface meets the VP is called a vertical trace (VT). The line at which the surface meets the HP is called a horizontal trace (HT). The HT of a plane appears in TV. The VT of a plane appears in FV. This is shown in Fig. 10.6. A principal plane gives either HT or VT depending upon their parallelism. An inclined and skew surface gives both HT and VT. This is shown in Fig. 10.7.

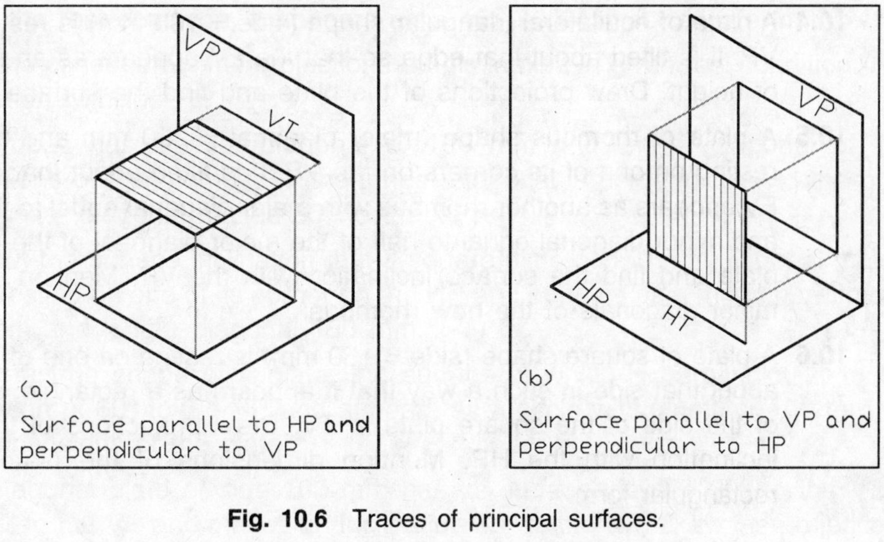


**Fig. 10.6** Traces of principal surfaces.

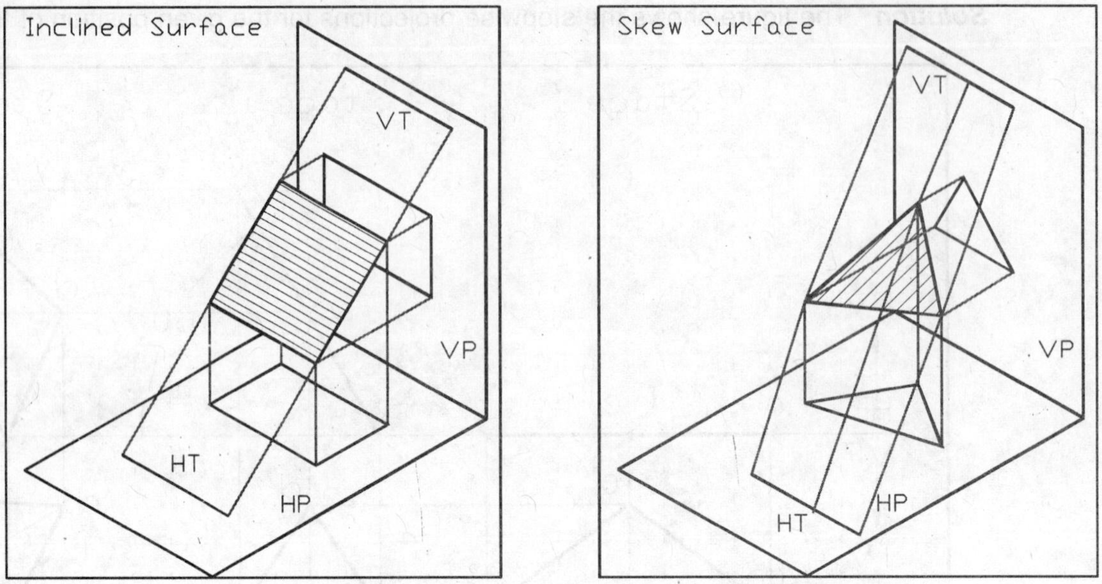


**Fig. 10.7** Traces of inclined and skew surfaces.

# Exercise

**Assumptions:**

1. All the laminas are in I-quadrant unless it is specified.
2. Plate has a negligible thickness.

**10.1** A rectangular plate of 200 mm × 100 mm is resting on one of its smaller sides on the VP. It is tilted about that side so that, it appears as a square in the elevation (FV). Find the inclination of the plate with the VP.

**10.2** A circular lamina is resting on the HP on its circumferential point. It is inclined with the HP such that its TV appears as an ellipse with minor axis 100 mm and major axis 150 mm. Find the surface inclination with the HP as well as the diameter of the circular lamina.

**10.3** A semi-circular lamina is resting on the HP on its straight edge having dimension 150 mm. It is inclined with the VP such that in the FV straight edge is reduced to half of it's original dimension. Draw projections of the lamina and find the surface inclination with the VP.

**10.4** A plate of equilateral triangular shape (side = 100 mm) is resting on one of its sides on the VP. It is tilted about that edge so that the FV appears as an isosceles triangle with 75 mm of height. Draw projections of the plate and find the surface inclination with the VP.

**10.5** A plate of rhombus shape (major diagonal = 250 mm and minor diagonal = 150 mm) is resting on one of its corners on the VP. It is tilted about that corner in such a way that the FV appears as another rhombus with major diagonal equal to the minor diagonal of the plate and minor diagonal equal to half of the major diagonal of the plate. Draw projections of the plate and find the surface inclination with the VP. Mention dimensions of the major and minor diagonals of the new rhombus.

**10.6** A plate of square shape (side = 150 mm) is resting on one of its sides on the HP. It is tilted about that side in such a way that it appears as a rectangle with smaller side equal to half of the side of the square plate in TV. Draw projections of the plate and find the surface inclination with the HP. Mention dimensions of the plate view, which appears in a rectangular form.

**10.7** A plate with an elliptical shape (major axis = 200 mm and minor axis = 150 mm) is resting on its peripheral point on HP in such a way that it appears as a circle of maximum dimension in the TV. Draw projections of the plate. Find the surface inclination with the HP and the diameter of the circle.

**10.8** Draw the projections of a regular hexagonal plate of 75 mm side, having one of its edges in the HP, inclined 60° to the VP and its surface makes an angle of 40° with the HP.

**10.9** A plate of rhombus shape ABCD has its diagonal AC = 150 mm and BD = 250 mm. The side AB of the plate is in the HP. The side BC is in the VP and the plate makes an angle of 30° with the HP. Obtain the projections of the plate and find the inclination with the VP.

**10.10** An isosceles triangular plate of 100 mm base and 150 mm altitude appears as an equilateral triangle. Draw the projections of the plate if its 100 mm long edge is on the HP and inclined at 45° to the VP. What are the inclinations of the plate with the HP and the VP?

**10.11** A circular plate of 150 mm diameter has its imaginary diameter AB perpendicular to HP and diameter CD parallel to and 100 mm above the HP. The surface is inclined 45° to the VP. Draw the projections of the plate if the diameter CD is inclined at 50° to the HP.

**10.12** A square lamina ABCD (diagonal = 200 mm) is resting on its corner A on the VP. The diagonal BD is perpendicular to the HP and in front of the VP such that the FV appears as a rhombus having one diagonal 150 mm. Draw the projections of the square lamina. Find the surface inclination with the VP.

**10.13** A hexagonal plate (side = 100 mm) with an elliptical hole at the centre (major axis = 100 mm and minor axis = 50 mm) is resting on one of its corner on VP with the minor axis of the hole perpendicular to the HP. In the rest condition, the elliptical hole appears as a circular hole of maximum possible diameter. Draw the projections of this lamina in given rest condition. Find the surface inclination with the VP.

**10.14** A hexagonal plate (side = 100 mm) with a circular hole at the centre (diameter = 100 mm) is resting on one of its corner on the HP. In the rest condition, the circular hole appears as an ellipse (major axis = 100 mm and minor axis = 75 mm). Draw the projections of this lamina in given rest condition. Find the surface inclination with the HP.

**10.15** A 30°–60° set square having smallest side 100 mm long is resting on its smallest side on the HP and perpendicular to the VP. The surface is inclined to the HP such that:

(a) in the rest condition, it appears as another 30°–60° set square in TV.

(b) in the rest condition, it appears as a triangle with the smallest side equal to half the smallest side of the actual lamina in TV.

(c) in the rest condition, it appears as a 45°–90°–45° set square in TV.

In all the cases, draw the projections of this lamina in given rest condition. Find the surface inclination with the HP.

**10.16** A rhombus has smaller diagonal of 100 mm and longer diagonal of 150 mm. Draw projections of the rhombus, if the longer diagonal is perpendicular to the profile plane and the smaller diagonal makes 35° to the VP.

**10.17** Find out the surface inclination that a square plate of 150 mm diagonal makes with the HP if it is resting on one of its corner on the HP and appears as a rhombus with smaller diagonal 100 mm and longer diagonal 150 mm.

**10.18** Obtain a true shape of a plate, the surface of which is perpendicular to the HP, makes inclination of 60° to the VP and appears as a circle with diameter 100 mm in FV. Find the key dimensions of the plate.

**10.19** A pentagonal plate of side 100 mm has its surface inclined 30° to the VP with one side parallel to the VP and makes an inclination 30° to the HP. Draw the projections of the plate.

**10.20** A regular hexagonal plate of side 75 mm has its corner on VP and surface inclined 40° to VP. The FV of the diagonal passing through that corner makes inclination of 30° to HP. Draw the projections of the plate.

**10.21** A rectangular plate of 150 mm × 100 mm is resting on one of its sides on the HP. It is inclined about that side so that it appears as a square of maximum dimension in the TV. Draw projections of the plate, when one of its diagonals is parallel to the VP.

**10.22** A pentagonal lamina of side 100 mm is resting on one of its corner on the HP and the VP. Draw its projections if an edge opposite to that corner is 100 mm above the HP and parallel to the VP. Find the inclinations of the lamina with the HP and VP.

**10.23** A hexagonal plate ABCDEF of side 100 mm is resting on its corner A on the HP and VP. Draw its projections when a corner B is 25 mm above the HP and on the VP. Find inclination of the plate with the HP and VP.

**10.24** PQRS and ABCD are the two square plates with their diagonals 75 mm and 100 mm, respectively. Both the plates are touching the HP at their corners P and A, respectively and touching each other at their corners R and C, respectively. The plates are perpendicular to each other and are also perpendicular to the VP. Draw the projections. Find the inclinations of these plates with the HP.

**10.25** A triangular lamina ABC (side 100 mm) having its corner C 50 mm above the HP and 50 mm in front of the VP. Edge AB is 25 mm above HP and parallel to the VP. Draw the projections of the lamina and find its inclinations with the HP and the VP.

**10.26.** Draw the projections of a rhombus having diagonals 125 mm and 50 mm long. The smaller diagonal of which is parallel to the HP and the VP, while the other is inclined at 30° to the HP. Find its inclinations with the HP and VP.

# 11

# Various Measurements

## Objectives

- Learn the various measurements required in engineering graphics.
- Know the graphical methods to obtain the solution for various measurements.

## Introduction

An understanding of the space relationships of points, lines and planes are necessary for the application of graphical methods for many purposes in engineering and science. In this chapter, graphical methods are explained to obtain various measurements between lines, lines and points and points and planes.

## 11.1 Various Measurements

Various measurements required are listed below:

1. Distance between point and line
2. Shortest distance between two parallel lines
3. Inclination between two lines
4. Distance between points and planes
5. Angle between two skew planes
6. Angle between lines and planes

In order to find out the required measurements we need to know the graphical methods to convert a line into a point view and to convert a plane into a line view. In addition to this, we are required to know the precedence of intersecting or non-intersecting lines in order to know the visibility in FV as well as in TV.

## 11.2 End View of a Line

If the direction of the visual ray is along the direction of the line then we get an "end view" or a "point view" of the line. In graphics, the line is made perpendicular to HP or VP so that we get a point view in TV or FV, respectively. The following examples show how to obtain a point view for a principal line, inclined line and a skew line.

**EXAMPLE 11.1** Line ab is in III-quadrant and parallel to HP and VP (Fig. 11.1). Obtain the end view of this line.

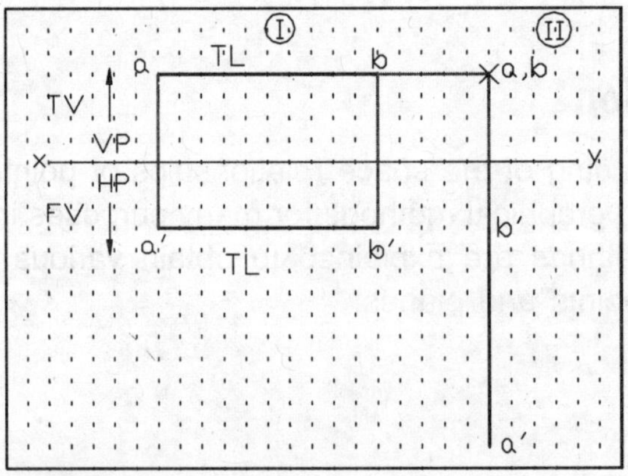


**Fig. 11.1** End view of a principal line.

***Solution***

1. TV and FV show a true length. COPY the FV (a′b′) and place it in Stage II.
2. ROTATE the copied FV with either a′ or b′ as base point and make it perpendicular to the xy line.
3. Project the rotated FV and Stage I TV. The intersection of these two projection lines is the end view of the line ab (x a, b).

**EXAMPLE 11.2** Line AB is in III-quadrant. The line is parallel to VP and inclined to HP (Fig. 11.2). Obtain the end view of this line.

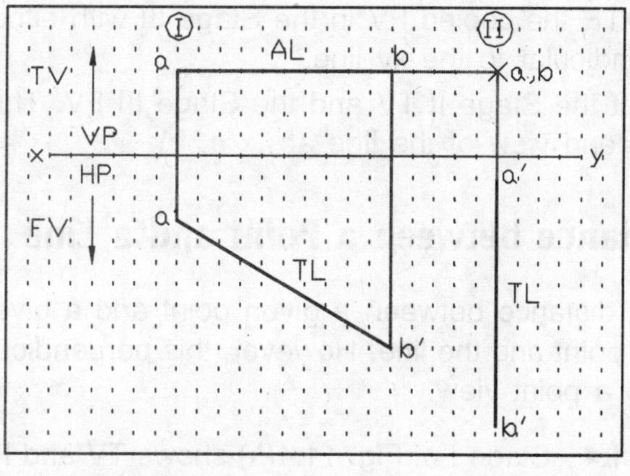


**Fig. 11.2** End view of an inclined line.

***Solution***

1. FV shows a true length. COPY the FV (a′b′) and place it in the Stage II FV.
2. ROTATE the copied FV with either a′ or b′ as base point and make it perpendicular to the xy line.
3. Project the rotated FV and the Stage I TV. The intersection of these two projection lines is the end view of the line ab (x a, b).

**EXAMPLE 11.3** Line AB is in III quadrant and inclined to HP and VP. Obtain the end view of this line.

***Solution*** Line ab is parallel to the VP and inclined to the HP (Fig. 11.3). We can now obtain the end view of this line.

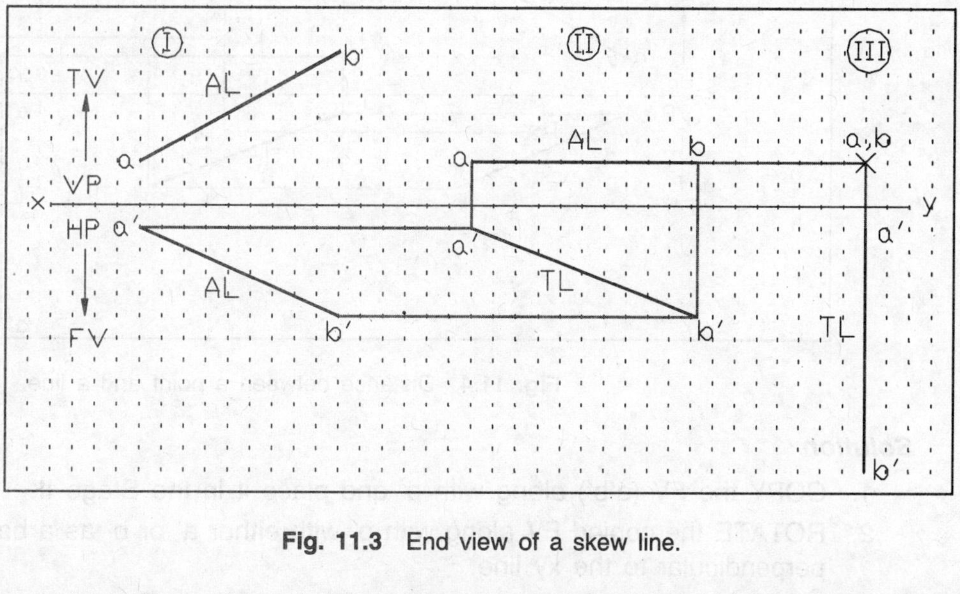


**Fig. 11.3** End view of a skew line.

1. Both the views in the Stage I do not show a true length. Therefore, in order to obtain an end view a true length is required in either of the views.
2. COPY the TV (ab) and place it in the Stage II TV.
3. ROTATE the copied TV with a as a base point till the TV (ab) becomes parallel to the xy line.
4. Project the Stage II TV and the Stage I FV to obtain a Stage II FV. The Stage II FV shows a true length. Now, this problem is similar to EXAMPLE 11.2.
5. COPY the Stage II FV (a′b′) which shows a true length and place it in the Stage III FV.

6. ROTATE the copied FV in the Stage III with either a′ or b′ as base point and make the FV perpendicular to the xy line.
7. Project the Stage II TV and the Stage III FV. The intersection of these two projection lines is the end view of the line ab (x a, b).

## 11.3 Distance between a Point and a Line

The shortest distance between a given point and a given straight line is a perpendicular distance between the point and the line. However, this perpendicular must be drawn in the view in which the line falls into a point view.

**EXAMPLE 11.4** Stage I of Fig. 11.4(a) shows TV and FV view of a line ab, which is parallel to HP and VP and a point P. Find out a true distance between the point P and the line ab.

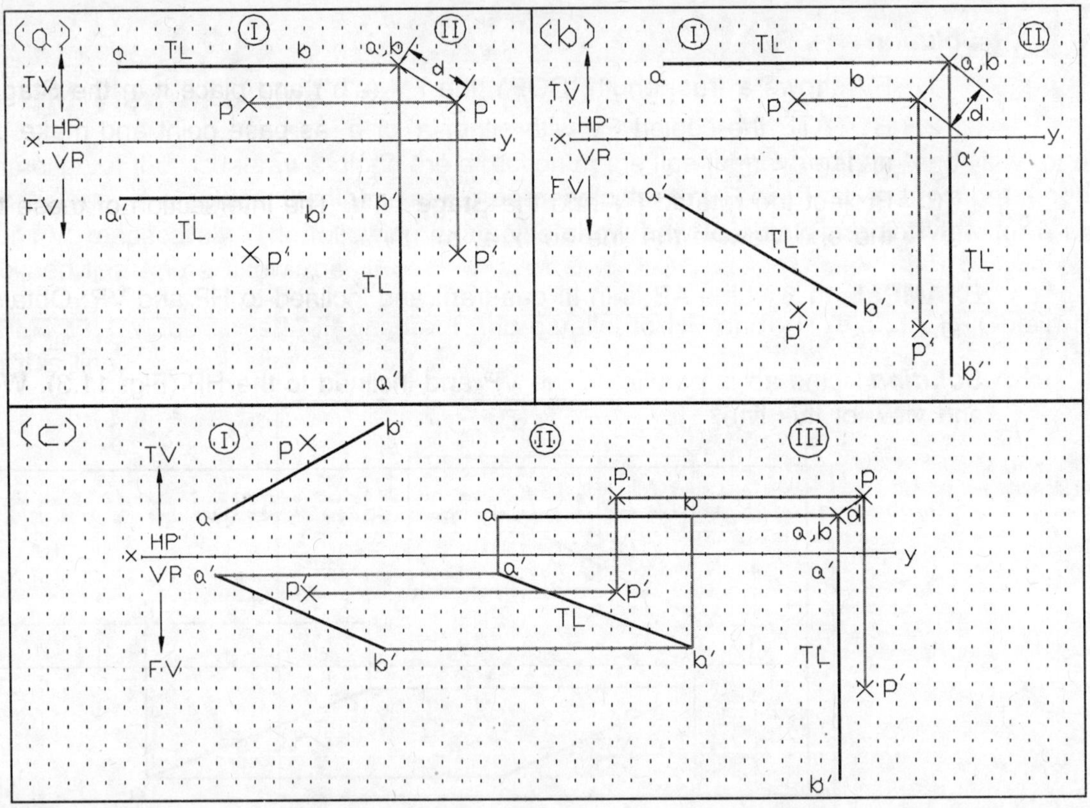


**Fig. 11.4** Distance between a point and a line.

***Solution***

1. COPY the FV (a′b′) along with p′ and place it in the Stage II.
2. ROTATE the copied FV along with p′ with either a′ or b′ as a base point and make the FV perpendicular to the xy line.
3. Project the rotated FV and the Stage I TV of line ab. The intersection of these two projection lines is the end view of the line ab. Also, project the Stage II FV and Stage I TV of the point P.
4. Draw a line from end view of line ab and point P in the Stage II TV.
5. Use the DIM command with ALIGNED option to measure the distance d between the end view of line ab and the projected point P.

***Note:*** Figure 11.4(b) shows how to obtain a true distance between a point and an inclined line. Figure 11.4(c) shows how to obtain a true distance between a point and a skew line.

## 11.4 Shortest Distance between Two Parallel Lines

In order to find out the shortest distance between two (inclined/skew) parallel lines, the lines should be converted into their point view. Then these point views are joined by a line, which can be measured for the true and shortest distance.

**EXAMPLE 11.5** Stage I of Fig. 11.5 shows the principal views of two parallel lines ab and pq, both of which are parallel to HP and VP. Find out a true distance between these two lines.

***Solution***

1. COPY the FV a′b′ and p′q′ and place it in the Stage II FV. (This is not shown here).
2. ROTATE the copied FV with either of the endpoints of line ab as a base point and make the FV perpendicular to the xy line.
3. Project the rotated FV and Stage I TV in the Stage II. The intersections of these projection lines are the end views of the line ab and the line pq.
4. Use the DIM command with ALIGNED option to measure the distance d between the two end views.

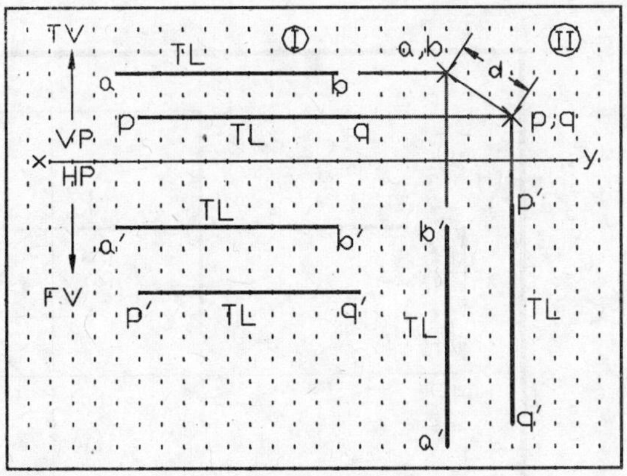


Fig. 11.5

**EXAMPLE 11.6** Stage I of Fig. 11.6 shows principal views of two skew lines ab and pq. Find out a true distance in between these two lines.

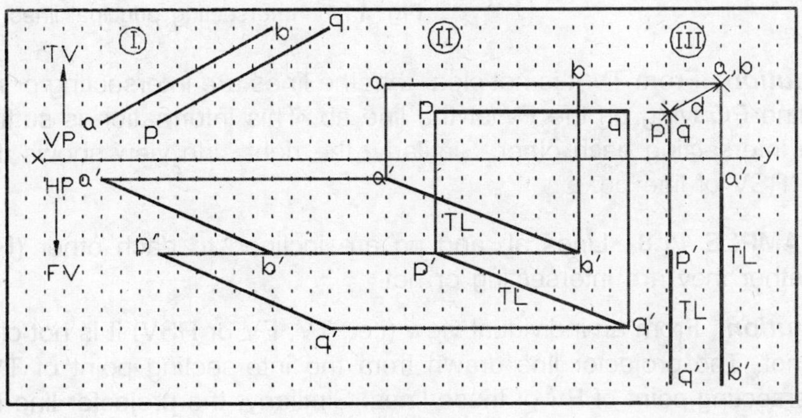


Fig. 11.6

***Solution***

1. COPY the TV ab and pq and place it in the Stage II TV.
2. ROTATE the copied TV with either of the endpoints of line ab as a base point and make the TV parallel to the xy line, in Stage II.

3. Project the rotated TV of line ab and pq and the Stage I FV in the Stage II. The intersectior of these projection lines are the true lengths of the line ab and line pq.
4. COPY the true lengths and place it in the Stage III FV.
5. ROTATE the copied true lengths with either of the endpoints of line ab as a base point ar make them perpendicular to the xy line.
6. Project the true lengths of Stage III and the Stage II TV. The intersections of these projectic lines are the end views of the line ab and line pq.
7. Use the DIM command with ALIGNED option to measure the distance d between the two en views.

## 11.5 Intersecting and Non-intersecting Lines

It is difficult to find out from the TV and FV whether lines are intersecting or not. In order to chec this we take help of the side view. This is illustrated in examples given below.

**EXAMPLE 11.7** Lines ab and pq are perpendicular to each other (Fig. 11.7). It is required to kno whether they are intersecting or not.

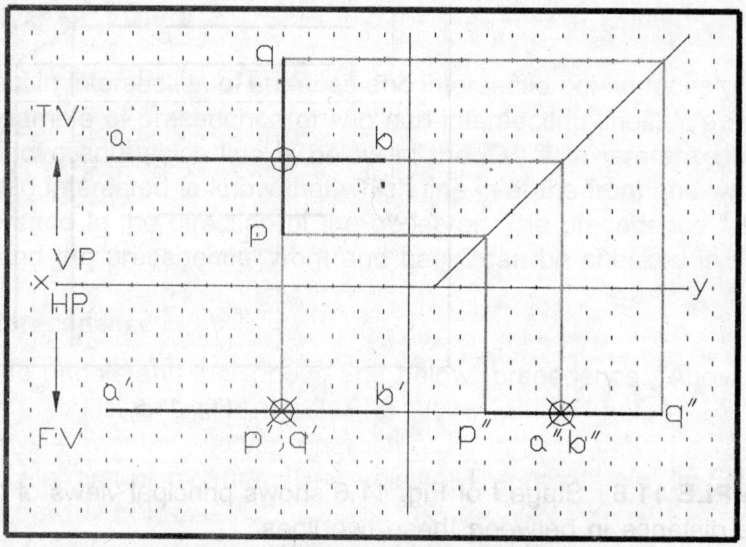


**Fig. 11.7** Intersecting principal lines.

***Solution*** From TV it is not clear that the lines are intersecting or not. The FV shows the point view of line PQ lying on the FV of the line ab. This information is sufficient to know that both the line are intersecting each other. Similarly, the right side view shows the point view of line ab lying o the RSV of line pq.

**EXAMPLE 11.8** Lines ab and pq are inclined to each other (Fig. 11.8). It is required to know whether they are intersecting or not.

***Solution*** From an individual view (i.e., TV, FV or RSV) it is not clear that the lines are intersectin or not. The projector line drawn from the intersecting point of TV of lines ab and pq meets th intersecting point of FV of these lines. Similarly, the projector line drawn from the intersecting poir of TV of lines ab and pq meets the intersecting point of RSV of these lines. From all the views, is clear that there is a single projector line between TV and FV, FV and RSV and TV and RSV. Thi information is sufficient to know that both the lines are intersecting each other.

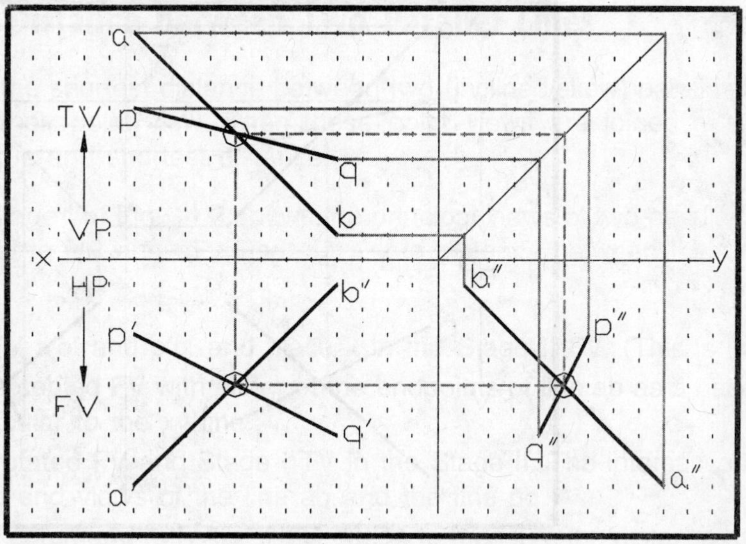


**Fig. 11.8** Intersecting skew lines.

Figure 11.9 shows example of non-intersecting lines. In this example, the projector line drawn from the intersecting point of TV of lines ab and pq meet the intersecting point of FV of these lines. However, from right side view it is clear that the lines are not intersecting each other.

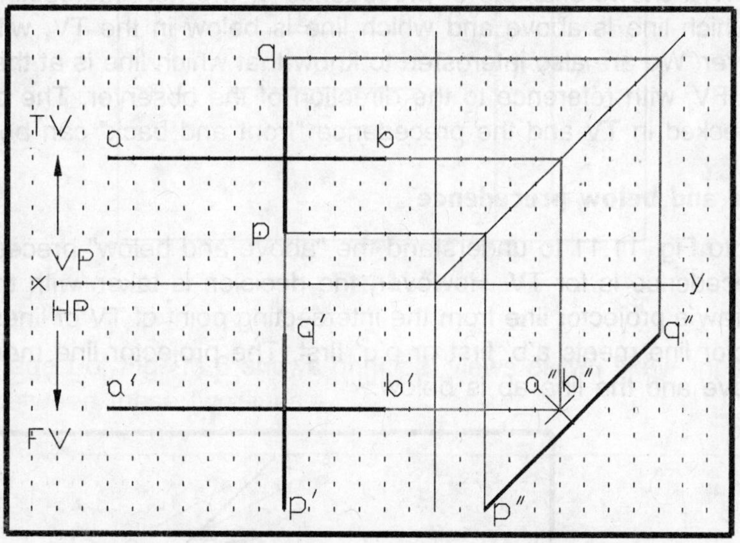


**Fig. 11.9** Non-intersecting lines.

Figure 11.10 shows another example of non-intersecting lines. In this example, the projector line drawn from the intersecting point of TV of lines ab and pq does not meet the intersecting point of FV of these lines. Similarly, the projector line drawn from the intersecting point of TV of lines ab and pq does not meet the intersecting point of RSV of these lines.

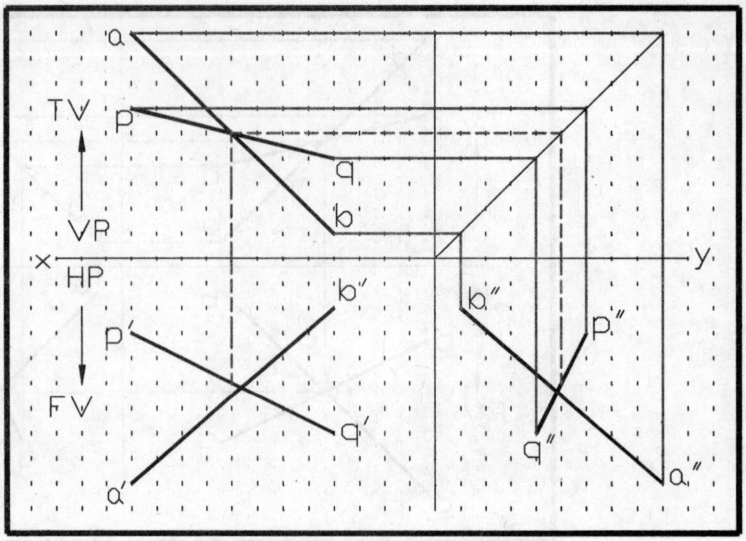


**Fig. 11.10** Non-intersecting lines.

## 11.6 Precedence of Lines

This concept is useful in intersection of surfaces and intersection of surfaces of solids. Figures 11.1 and 11.12 shows example of precedence of two non-intersecting lines. We are interested to kno that which line is above and which line is below in the TV, with reference to the direction of th observer. We are also interested to know that which line is at the front and which line is at the bac in the FV, with reference to the direction of the observer. The precedence "above and below" ca be checked in TV and the precedence "front and back" can be checked in FV.

### Above and below precedence

Refer to Fig. 11.11 to understand the "above and below" precedence. "Above and below" decisio of precedence is for TV. However, the decision is taken with the help of FV.

Draw a projector line from the intersecting point of TV of lines ab and pq and check whether th projector line meets a′b′ first or p′q′ first. The projector line meets p′q′ first. Therefore, the line p is above and the line ab is below.

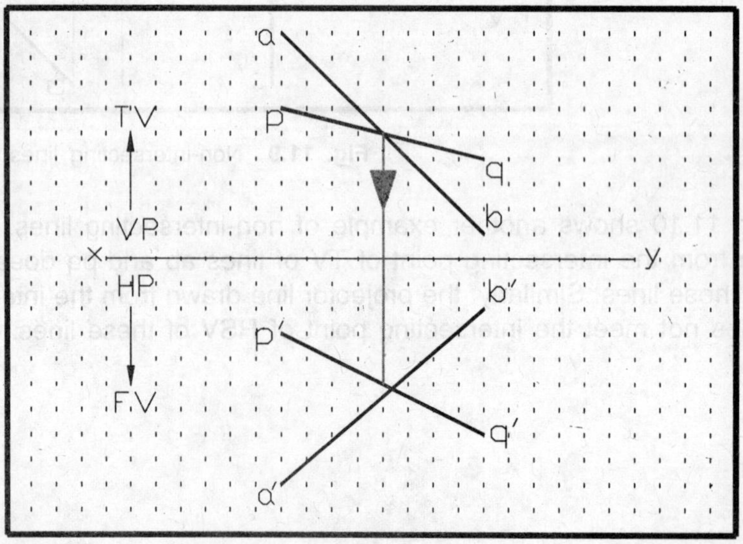


**Fig. 11.11** Above and below precedence.

### Front and back precedence

Refer to Fig. 11.12 to understand "front and back" precedence. "Front and back" decision of precedence is for FV. However, the decision is taken with the help of TV.

Draw a projector line from the intersecting point of FV of lines ab and pq and check whether the projector line meets ab first or pq. The projector line meets ab first. Therefore, the line ab is at the front and pq is at the back.

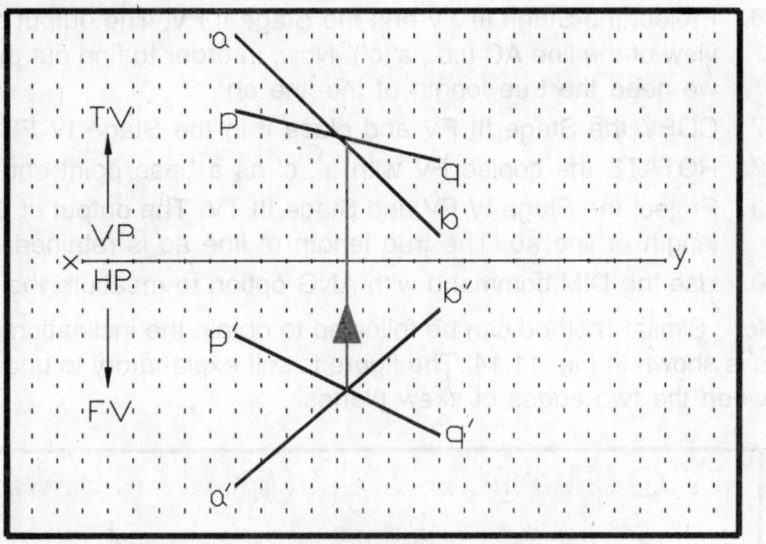


**Fig. 11.12** Front and back precedence.

## 11.7 Inclination between Two Lines

Inclination between two lines can be obtained if true lengths are available for both the lines. Apparent length gives an apparent inclination. True inclination can be obtained from true length.

**EXAMPLE 11.9** Stage I of Fig. 11.13 shows TV and FV of two inclined and intersecting lines ab and ac. Find out an angle in between these two lines.

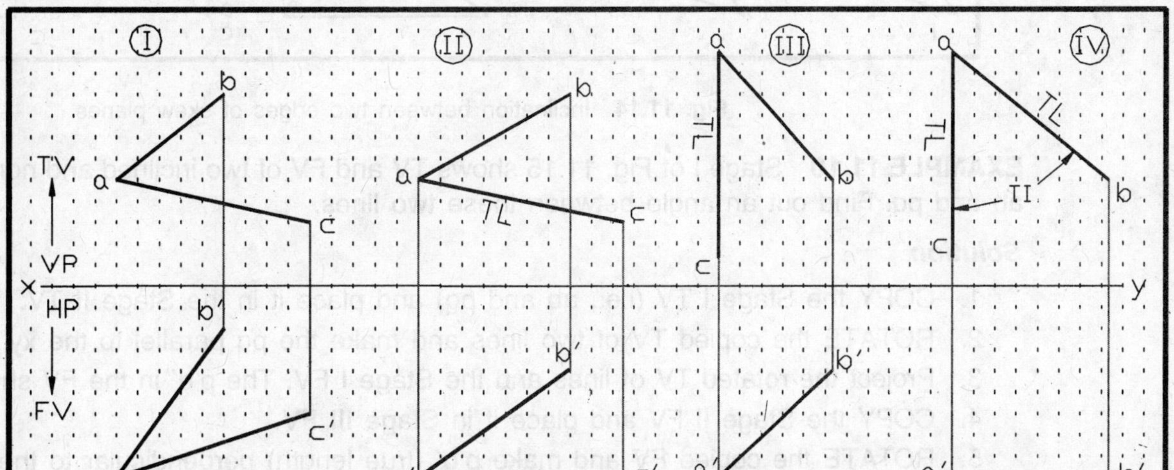


**Fig. 11.13** Inclination between two intersecting lines.

***Solution***

1. COPY the Stage I FV (i.e., a′b′ and a′c′) and place it in the Stage II FV.
2. ROTATE the copied FV of two lines and make the a′c′ parallel to the xy line.
3. Project the rotated FV of lines and the Stage I TV. The ac in the TV shows a true lengt
4. COPY the Stage II TV and place it in Stage III TV region.
5. ROTATE the copied TV and make ac (true length) perpendicular to the xy line.
6. Project the Stage III TV and the Stage II FV. The output of these projection lines shows poi view of the line AC (i.e., a′,c′). Now, in order to find out the true inclination between two line we need the true length of the line ab.
7. COPY the Stage III FV and place it in the Stage IV FV.
8. ROTATE the copied FV with a′, c′ as a base point and make it parallel to the xy line.
9. Project the Stage IV FV and Stage III TV. The output of these projection lines shows the tru length of line ab. The true length of line ac is retained.
10. Use the DIM command with ANG option to measure the angle (TI) between these two line

***Note:*** Similar method can be followed to obtain the inclination between two edges of skew plane This is shown in Fig. 11.14. The figure is self explanatory to understand how to obtain the inclinatic between the two edges of skew planes.

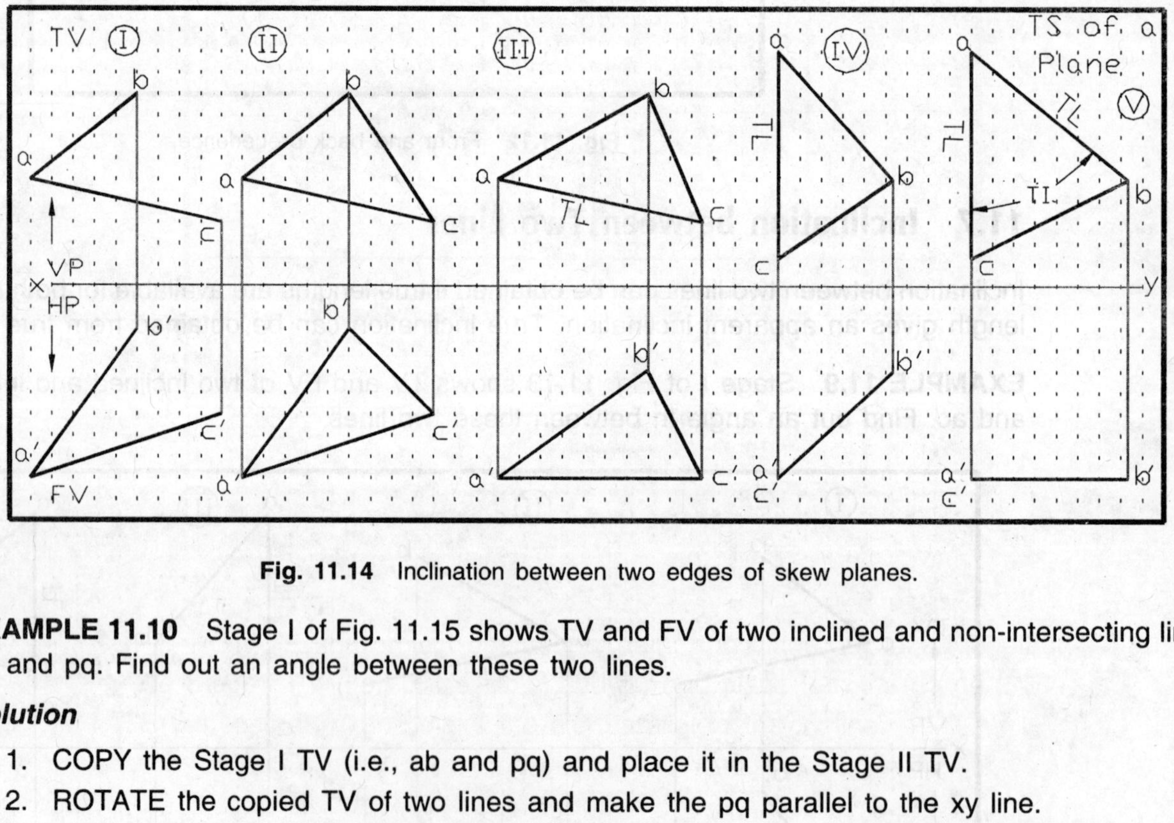


**Fig. 11.14** Inclination between two edges of skew planes.

**EXAMPLE 11.10** Stage I of Fig. 11.15 shows TV and FV of two inclined and non-intersecting line ab and pq. Find out an angle between these two lines.

***Solution***

1. COPY the Stage I TV (i.e., ab and pq) and place it in the Stage II TV.
2. ROTATE the copied TV of two lines and make the pq parallel to the xy line.
3. Project the rotated TV of lines and the Stage I FV. The p′q′ in the FV shows a true lengtl
4. COPY the Stage II FV and place it in Stage III FV.
5. ROTATE the copied FV and make p′q′ (true length) perpendicular to the xy line.
6. Project the Stage III FV and the Stage II TV. The output of these projection lines shows poi view of the line pq. Now, in order to find out the true inclination between the two lines, w need the true length of line ab.
7. COPY the Stage III TV and place it in the Stage IV TV.
8. ROTATE the copied TV with b as a base point and make it parallel to the xy line.

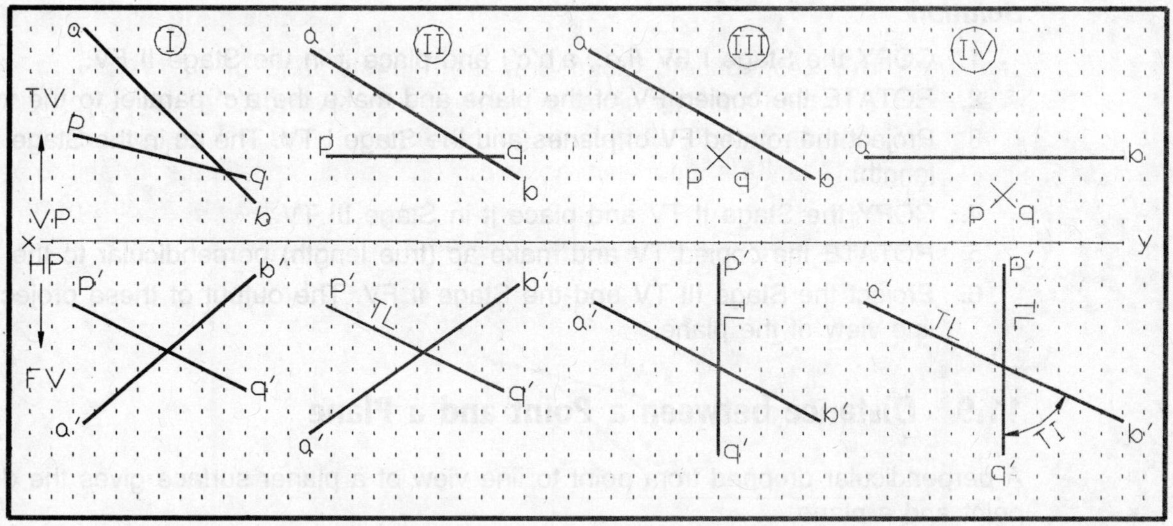

Fig. 11.15 TV and FV of two inclined and non-intersecting lines.

9. Project the Stage IV TV and Stage III FV. The output of these projection lines shows true length of line ab. The true length of line pq is retained.
10. Use the DIM command with ANG option to measure the angle (TI) between these two lines.

## 11.8 Line View of a Plane

If direction of the observer is parallel to the planer surface and lie in the same plane then the planer surface appears as a line view. It is also called an edge view of a plane.

**EXAMPLE 11.11** Stage I of Fig. 11.16 shows TV and FV of a skew plane ABC. Obtain a line view of the plane.

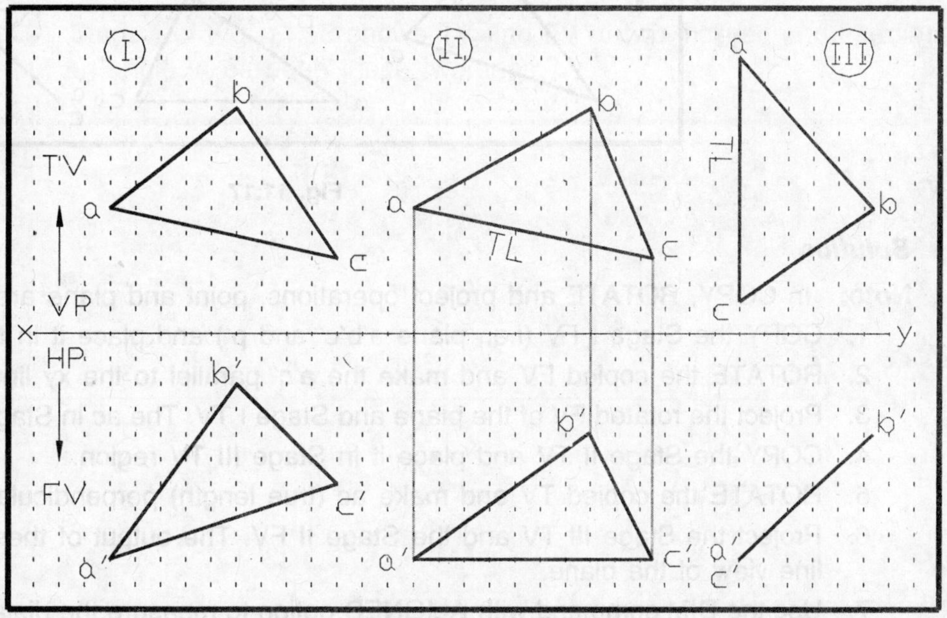

Fig. 11.16

***Solution***

1. COPY the Stage I FV (i.e., a′b′c′) and place it in the Stage II FV.
2. ROTATE the copied FV of the plane and make the a′c′ parallel to the xy line.
3. Project the rotated FV of planes and the Stage I TV. The ac in the Stage II TV shows a tru length.
4. COPY the Stage II TV and place it in Stage III TV.
5. ROTATE the copied TV and make ac (true length) perpendicular to the xy line.
6. Project the Stage III TV and the Stage II FV. The output of these projection lines shows line view of the plane.

## 11.9 Distance between a Point and a Plane

A perpendicular dropped from point to line view of a planer surface gives the distance betwean point and a plane.

**EXAMPLE 11.12** Stage I of Fig. 11.17 shows TV and FV of a skew plane abc and point p. Fin out the distance between a point and a plane.

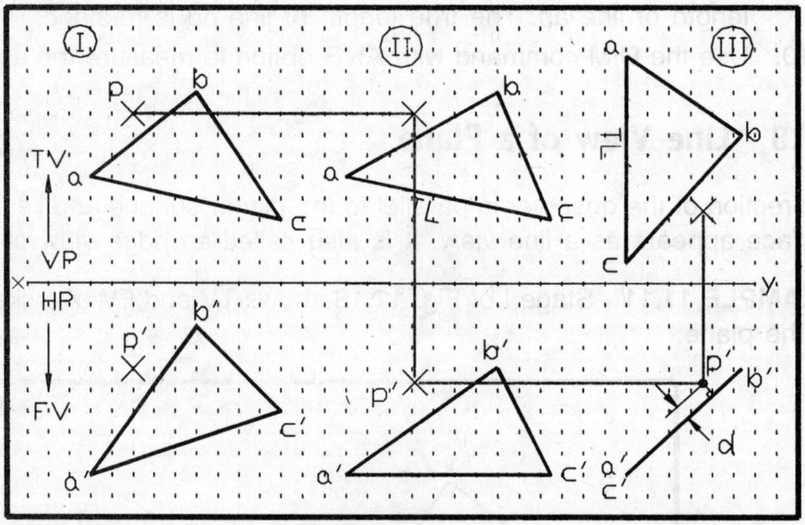


**Fig. 11.17**

***Solution***

**Note:** In COPY, ROTATE and project operations, point and plane are taken together.

1. COPY the Stage I FV (i.e., plane a′b′c′ and p′) and place it in the Stage II FV.
2. ROTATE the copied FV and make the a′c′ parallel to the xy line.
3. Project the rotated FV of the plane and Stage I TV. The ac in Stage I TV shows a true lengt
4. COPY the Stage II TV and place it in Stage III TV region.
5. ROTATE the copied TV and make ac (true length) perpendicular to the xy line.
6. Project the Stage III TV and the Stage II FV. The output of these projection lines shows line view of the plane.
7. Use the DIM command with ALIGNED option to measure the distance between the point ar the plane.

## 11.10 Angle between Two Skew Planes

The angle between two skew planes can be obtained if both the planes are converted into line views.

**EXAMPLE 11.13** Stage I of Fig. 11.18 shows TV and FV of two skew planes abc and pqr. Find out an angle between these planes.

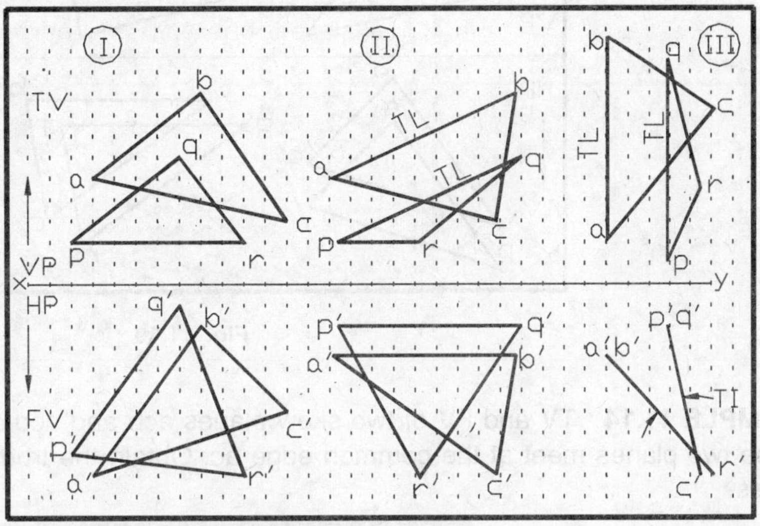


**Fig. 11.18**

***Solution***

**Note:** In COPY, ROTATE and project operations both the planes are taken together.

From the Stage I views, it is observed that edges ab and pq are parallel to each other.

1. COPY the Stage I FV (i.e., a′b′c′ and p′q′r′) and place it in the Stage II FV location.
2. ROTATE the copied FV of two planes and make the a′b′ and p′q′ parallel to the xy line.
3. Project the rotated FV of planes and the Stage I TV. The ab and pq in the TV shows a true length.
4. COPY the Stage II TV and place it in Stage III TV.
5. ROTATE the copied TV and make ac and pq (true length) perpendicular to the xy line.
6. Project the Stage III TV and the Stage II FV. The output of these projection lines shows line views of the planes.
7. Use the DIM command with ALIGNED option to measure the angle (TI) between the line and the plane.

***Note:*** With this technique, it can be checked whether the two planes are parallel or not. This is shown in Fig. 11.19. In this case, the planer surfaces are not only parallel but they are lying in the same plane.

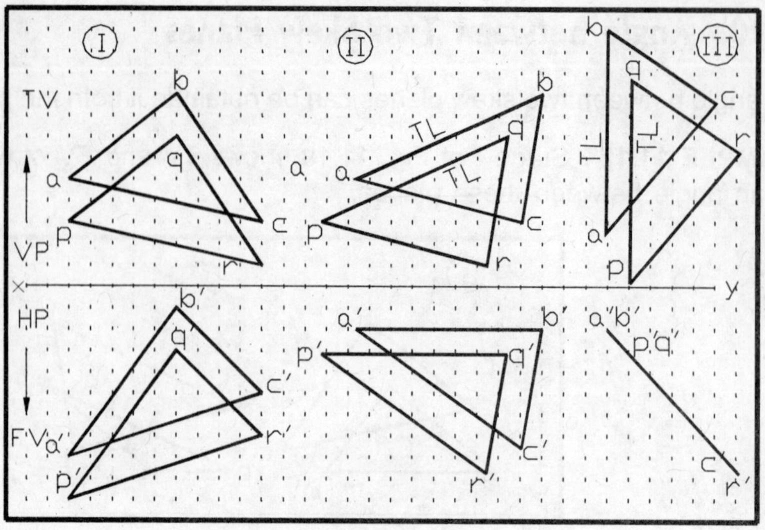

Fig. 11.19

**EXAMPLE 11.14** TV and FV of two skew planes acb and acd are shown in Stage I of Fig. 11.20. These two planes meet at the common edge ac. Obtain the true inclination (TI) between these two planes.

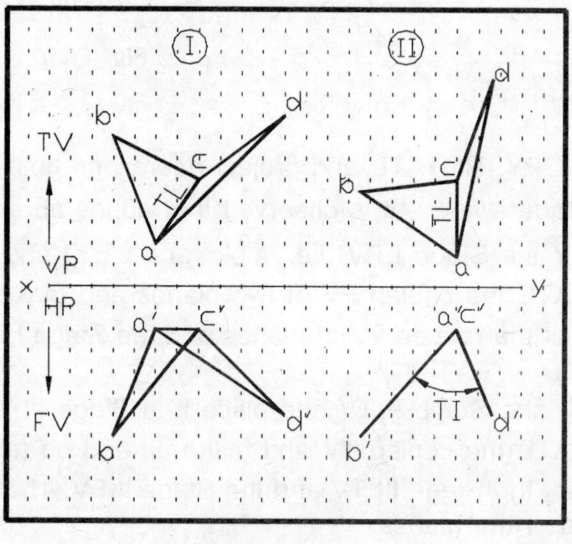

Fig. 11.20

***Solution***

1. FV of the edge ac is parallel to the xy line. Therefore, the TV of the line ac, which is inclined to the xy line, shows a true length.
2. COPY the entire TV of these two planes and place it in the TV region of the Stage II.
3. ROTATE the copied TV with a as a base point and make the ac perpendicular to the xy line.
4. Project the Stage II TV and the Stage I FV. The output of these projection lines gives the line views of these two planes.
5. With DIM command and ANG option obtain the true inclination in between these two planes as TI.

## 11.11 Angle between a Line and a Plane

The angle between a line and a plane can be obtained if the plane is converted into a line view and a true length is obtained for a line. The angle between a line view of plane and true length of the line is the angle between a line and a plane.

**EXAMPLE 11.15** Stage I of Fig. 11.21 shows TV and FV of a skew line pq and a skew plane abc. Find out an angle between the line and the plane.

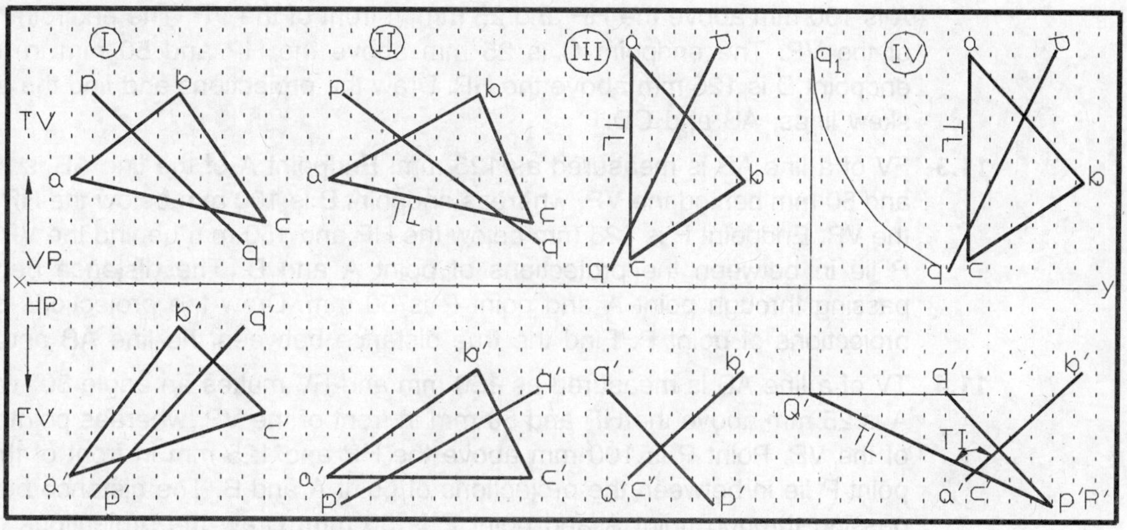


**Fig. 11.21**

***Solution***

**Note:** In COPY, ROTATE and project operations the line and the plane are taken together.

1. COPY the Stage I FV (i.e., a′b′c′ and p′q′) and place it in the Stage II FV location.
2. ROTATE the copied FV of the plane and the line with c as a base point and make the a′c′ parallel to the xy line.
3. Project the rotated FV of the plane and the line and the Stage I TV. The ac in the TV shows a true length.
4. COPY the Stage II TV and place it in Stage III TV.
5. ROTATE the copied TV and make ac (true length) perpendicular to the xy line.
6. Project the Stage III TV and the Stage II FV. The intersections of these projection lines show a line view of the plane abc along with the FV of the line. In order to obtain the angle between the plane and the line we need true length of line and a line view of the plane.
7. Stage IV shows the procedure to obtain the true length of the line. The Stage IV FV shows the true length of the line pq (p′q′) and a line view of the plane abc.
8. Use the DIM command with ALIGNED option to measure the angle (TI) between the line and the plane (i.e., p′q′ and a′,c′-b′) as shown.

***Note:*** With this technique, it can be checked whether a line is normal to the plane or not.

## Exercise

**11.1** Find the shortest distance between a line AB and a point P. The distance between end projectors of this line is 125 mm. The endpoint A is in the VP and 125 mm below the HP. Endpoint B is 175 mm behind the VP and 25 mm below the HP. The point P is 75 mm behind the VP, 150 mm below the HP and 75 mm from the end projector for the point A.

**11.2** Two lines AB and CD have common end projectors. The line AB is parallel to the HP and line CD is parallel to the VP. The distance between the end projectors is 150 mm. The end A is 100 mm above the HP and 25 mm in front of the VP. The endpoint B is 100 mm in front of the VP. The endpoint C is 25 mm above the HP and 50 mm in front of the VP. The endpoint D is 125 mm above the HP. Draw the projections and find the distance between the skew lines: AB and CD.

**11.3** FV of a line AB is measured as 125 mm. Endpoint A of the line AB is 25 mm below the HP and 50 mm behind the VP, whereas endpoint B is 100 mm below the HP and 125 mm behind the VP. Endpoint P is 125 mm below the HP and 150 mm behind the VP. Projections of point P lie in between the projections of point A and B. The distance between the projectors passing through point A and point P is 50 mm. Draw the projections of line AB. Draw the projections of point P. Find the true distance between the line AB and point P.

**11.4** TV of a line AB is measured as 150 mm and FV makes an angle 30° with the HP. Endpoint A is 25 mm above the HP and 50 mm in front of the VP, whereas point B is 100 mm in front of the VP. Point P is 100 mm above the HP and 125 mm in front of the VP. Projections of point P lie in between the projections of point A and B. The distance between the projectors passing through point A and point P is 50 mm. Draw the projections of line AB. Draw the projections of point P. Find the true distance between the line AB and point P.

**11.5** Line AB is parallel to the HP and its FV is 100 mm long. Point A is 75 mm above the HP and 25 mm in front of the VP, whereas point B is 125 mm in front of the VP. Line PQ is parallel to the PP. Point P is 25 mm above the HP and 125 mm in front of the VP, whereas point Q is 125 mm above the HP and 25 mm in front of the VP. Projections of point P lie in between the projections of point A and B. The distance between the projectors passing through point A and point P is 50 mm. Draw the projections of line AB and PQ. Find the true angle between the line AB and the line PQ.

# 12

# Projections of Solids

## Objectives

- To learn projections of engineering solids in a given position.
- To know how to transfer a point from one view to another view.

## Introduction

In this chapter, we are going to study the projections of engineering solids in a given position. The fundamentals of projection of engineering solids help all science and engineering students to represent any engineering solid in a given position. A solid has three dimensions: length, depth and height. In order to represent a solid on a plane paper, at least two orthographic views are necessary. However, if required, additional views are projected on auxiliary planes to complete the description of a solid. We have studied the fundamentals of projections, i.e., orthographic projections of engineering components.

## 12.1 Division of Engineering Solids

Solids may be divided into two main groups:

(i) Polyhedra
(ii) Solids of revolution

## 12.2 Polyhedra

A polyhedra is defined as a solid bounded by planes called faces. When all the faces are equal and regular, the polyhedron is said to be regular, otherwise it is an irregular polyhedra.

### 12.2.1 Regular Polyhedra

There are five regular polyhedra which may be defined as follows:

1. ***Tetrahedron:*** It has four equal faces, each an equilateral triangle [Fig. 12.1(a)].
2. ***Cube or hexahedron:*** It has six equal faces, all equal squares [Fig. 12.1(b)].
3. ***Octahedron:*** It has eight equal equilateral triangles as faces [Fig. 12.1(c)].
4. ***Dodecahedron:*** It has twelve equal and regular pentagons as faces [Fig. 12.1(d)].
5. ***Icosahedron:*** It has twenty faces, all equal equilateral triangles.

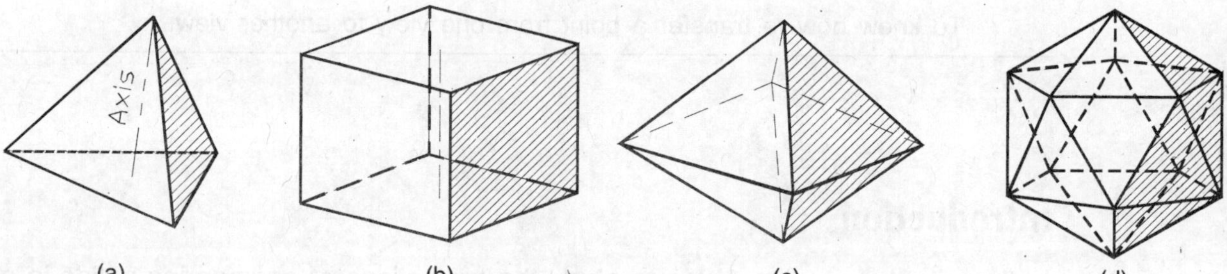


**Fig. 12.1** Regular polyhedra.

### 12.2.2 Irregular Polyhedra

Prism is a polyhedron having two equal and similar faces called its ends or bases, parallel to each other and joined by vertical faces, which are parallelograms. The imaginary line joining the centres of the bases is called axis.

#### Prism

A polyhedron having two bases and axis perpendicular to bases is called a right and regular prism (Fig. 12.2). The right and regular prism has equal rectangular faces. The corners and edges of bases are called base-corners and base-edges, respectively. The edges of faces are called vertical edges. Based on the number of base-edges, the prisms are called triangular, square, pentagonal and hexagonal prisms.

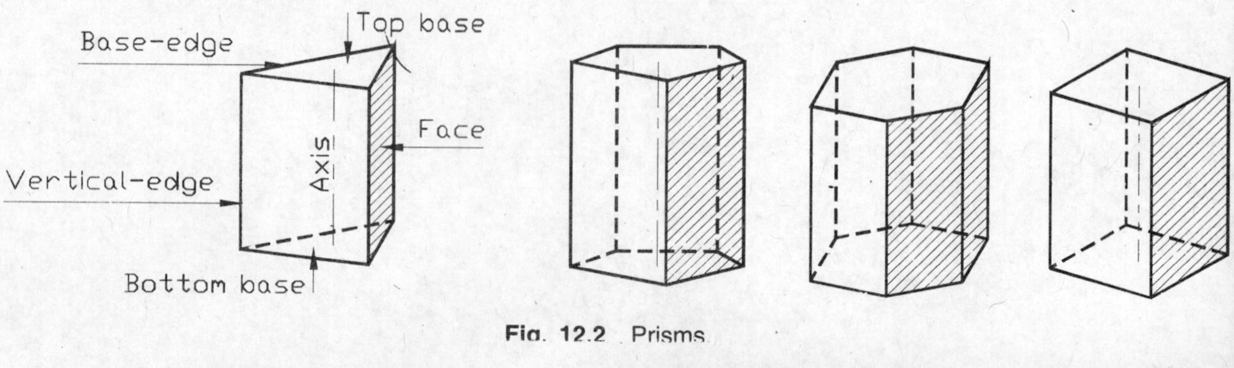


**Fig. 12.2** Prisms

### Pyramid

A polyhedron having one base and many slant faces is called a pyramid. The number of slant faces meeting at a point is called apex. The imaginary line joining apex and centre of base is called axis. This is shown in Fig. 12.3. The right and regular pyramid has equal triangular (isosceles triangle) faces. The corners and edges of bases are called base-corners and base-edges, respectively. The edges joining apex and base-corners are called slant edges. Based on the number of base-edges, the pyramids are called triangular, square, pentagonal and hexagonal pyramids.

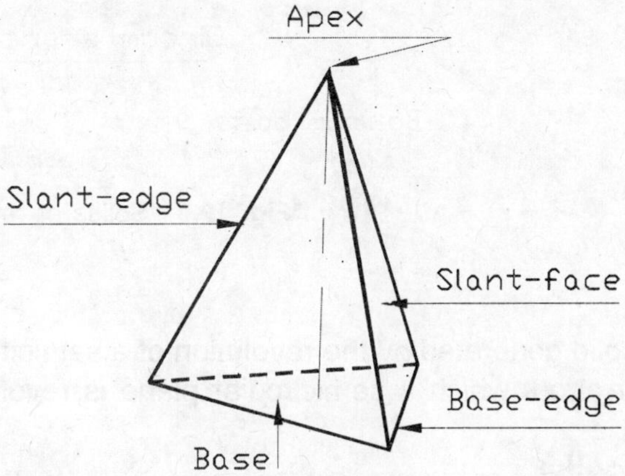


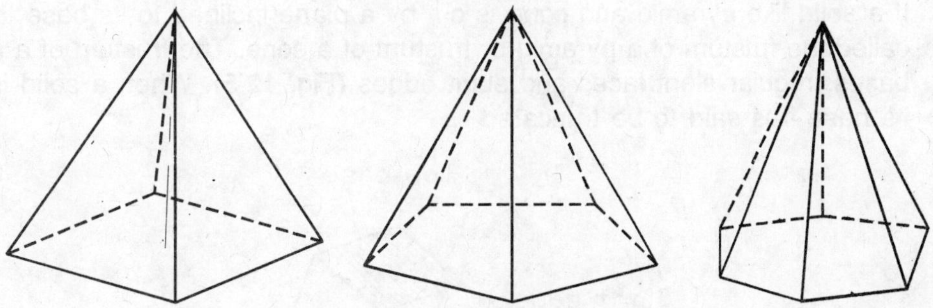

Fig. 12.3 Pyramids.

Oblique prisms and pyramids have their axis inclined to their bases. Prisms and pyramids are named according to the shape of their bases, as triangular, square, pentagonal, hexagonal etc.

## 12.3 Solids of Revolution

Revolution of a straight or curved edge about an axis generates a hollow solid. Revolution of a plane figure about its own edge generates a solid.

### Cylinder

Cylinder is a solid generated by the revolution of a rectangle about one of its sides [Fig. 12.4(a)]. It has two bases namely, top and bottom. The fixed side about which a rectangle is revolved is called an axis. The side which is parallel to the axis is called a generator as it generates a curved surface. The base-edge is called a circumference.

### Cone

Cone is a solid generated by the revolution of a right angled triangle about one of its perpendicular sides [Fig. 12.4(b)]. It has one base. The fixed side about which a rectangle is revolved is called an axis. The side which is inclined to the axis is called a slant generator as it generates a slant curved surface. The base-edge is called a circumference.

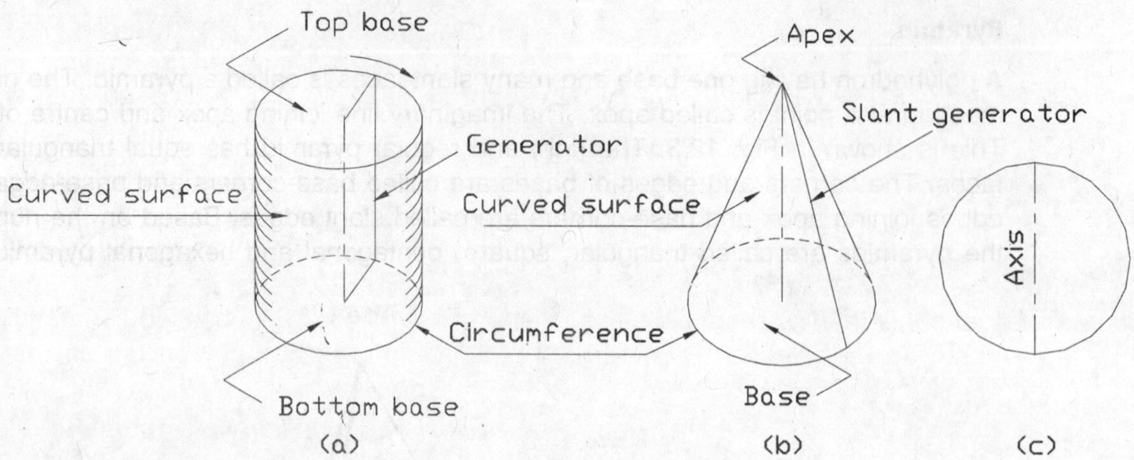


**Fig. 12.4** Solids of revolution.

### Sphere

Sphere is a solid generated by the revolution of a semicircular plane about its diameter [Fig. 12.4(c)]. The diameter about which a semicircular plane is revolved is called an axis.

### Frustums of solids

If a solid like pyramid and cone is cut by a plane inclined to its base then the remaining portion is called the frustum of a pyramid or frustum of a cone. The frustum of a pyramid has top and bottom bases, regular slant faces and slant edges (Fig. 12.5). When a solid is cut by a plane inclined to its base it is said to be truncated.

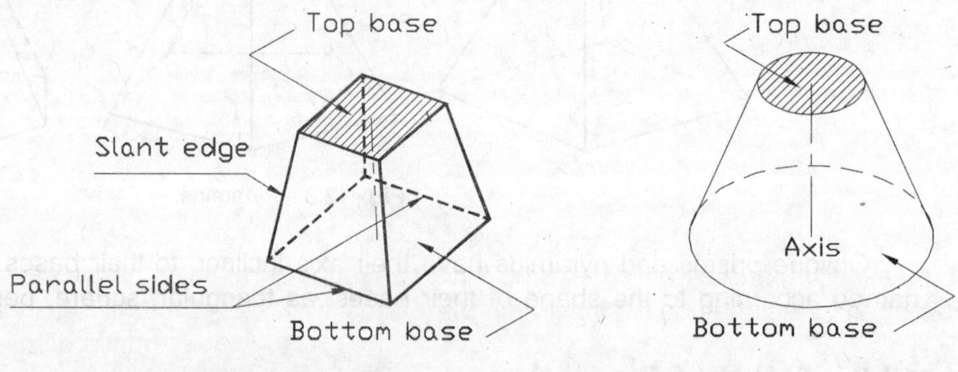


**Fig. 12.5** Frustum of square pyramid and cone.

## 12.4 Projections of Solids

A solid in simple position may have its axis perpendicular to one of the reference planes (HP, VP or PP). When the axis is perpendicular to one reference plane, it is parallel to the other. Also, when the axis of a solid is perpendicular to a plane, its base will be parallel to that plane. We have seen in Chapter 9 that if a plane is parallel to a reference plane, its projection on that plane shows its true shape and size. Therefore, the projection of a solid on the plane, to which its axis is perpendicular, will show the true shape and size.

### If axis is perpendicular to HP or ground

The TV will show the true shape and size of the base. The TV is drawn first and then the FV is projected from it. This is shown in Fig. 12.6.

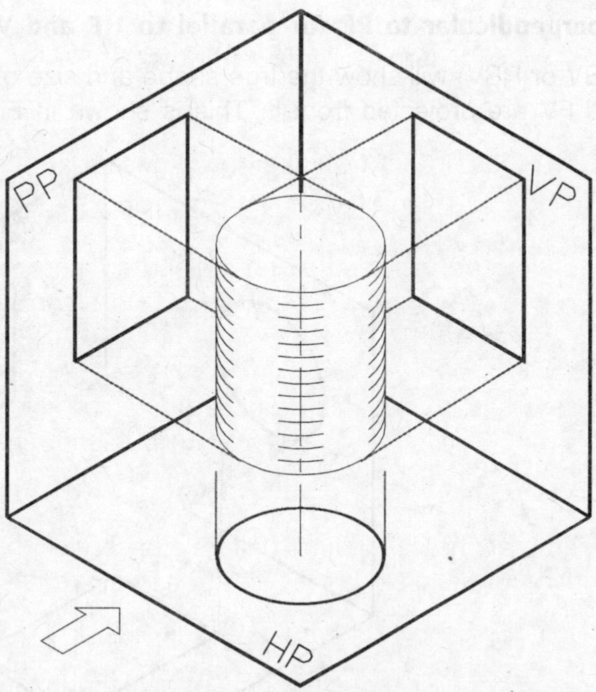


Fig. 12.6 Projections of a cylinder with axis perpendicular to HP.

**If axis is perpendicular to VP**

The FV will show the true shape and size of the base. The FV is drawn first and then the TV is projected from it. This is shown in Fig. 12.7.

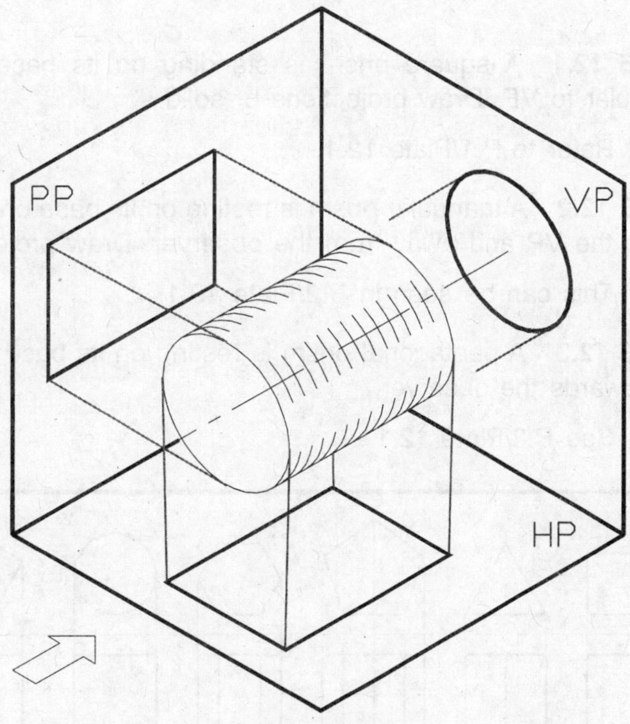


Fig. 12.7 Projections of a cylinder with axis perpendicular to VP.

### If axis is perpendicular to PP (or parallel to HP and VP)

The SV (LSV or RSV) will show the true shape and size of the base. The SV is drawn first and then the TV and FV are projected from it. This is shown in Fig. 12.8.

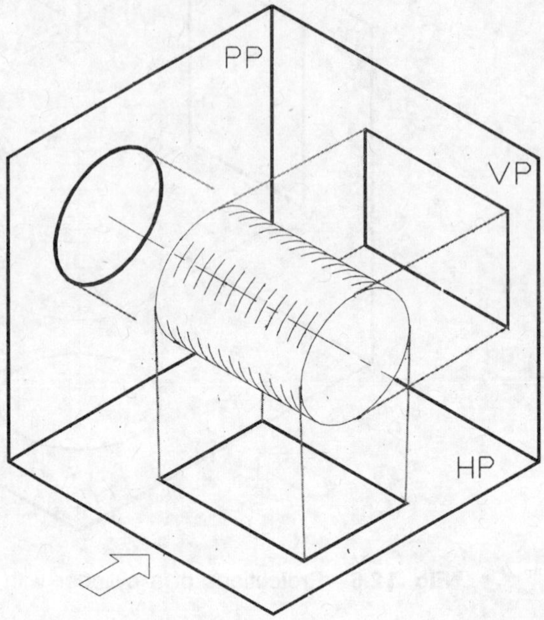


**Fig. 12.8** Projections of a cylinder with axis perpendicular to PP.

### Examples based on projections of prism and cylinder (with axis perpendicular to ground or HP)

**EXAMPLE 12.1** A square prism is standing on its base on the ground with two of its faces perpendicular to VP. Draw projections of solid.

***Solution*** Refer to P.1/Plate 12.1.

**EXAMPLE 12.2** A triangular prism is resting on its base on the ground with two of its faces equally inclined to the VP and away from the observer. Draw projections of solid.

***Solution*** This can be seen in P.2/Plate 12.1.

**EXAMPLE 12.3** A pentagonal prism is resting on its base on the ground with one face parallel to VP and towards the observer.

***Solution*** See P.3/Plate 12.1.

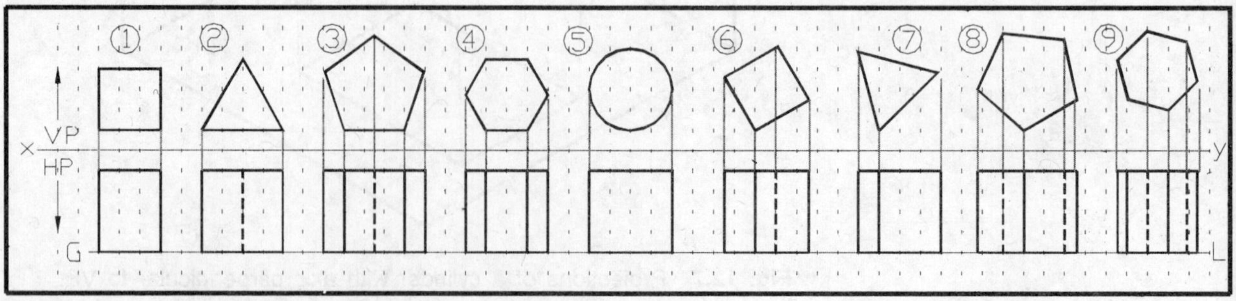


**PLATE 12.1** Projections of prism and cylinder with axis perpendicular to HP.

**EXAMPLE 12.4** A hexagonal prism is resting on its base on the ground with two of its alternate faces equally inclined to the VP. Draw projections of solid.

***Solution*** This is evident in P.4/Plate 12.1.

**EXAMPLE 12.5** A cylinder is resting on its base on the ground. Draw projections of solid.

***Solution*** Observe P.5/Plate 12.1.

**EXAMPLE 12.6** A square prism is resting on its base on the ground. Draw projections of solid if one of its faces makes an angle of 30° to VP.

***Solution*** Refer to P.6/Plate 12.1.

**EXAMPLE 12.7** A triangular prism is resting on its base on the ground. Draw projections of solid if one of its faces makes an angle of 45° to VP.

***Solution*** See P.7/Plate 12.1.

**EXAMPLE 12.8** A pentagonal prism is resting on its base on the ground. Draw projections of solid if one of its faces makes an angle of 30° to VP and exposed to observer.

***Solution*** We can see this in P.8/Plate 12.1.

**EXAMPLE 12.9** A hexagonal prism is resting on its base on the ground. Draw projections of solid if one of its faces makes an angle of 45° to VP and exposed to observer.

***Solution*** Refer to P.9/Plate 12.1.

**Examples based on projections of prism and cylinder (with axis perpendicular to VP)**

**EXAMPLE 12.10** A square prism is resting on one of its faces on the ground with axis perpendicular to VP. Draw projections of solid.

***Solution*** This is seen in P.10/Plate 12.2.

**EXAMPLE 12.11** A triangular prism is resting on one of its faces on the ground with axis perpendicular to VP. Draw projections of solid.

***Solution*** Refer to P.11/Plate 12.2.

**EXAMPLE 12.12** A pentagonal prism is resting on one of its faces on the ground with axis perpendicular to VP. Draw projections of solid.

***Solution*** This is shown in P.12/Plate 12.2.

**EXAMPLE 12.13** A hexagonal prism is resting on one of its vertical edges on the ground with axis perpendicular to VP and faces associated with vertical edge equally inclined to the ground. Draw projections of solid.

***Solution*** See P.13/Plate 12.2.

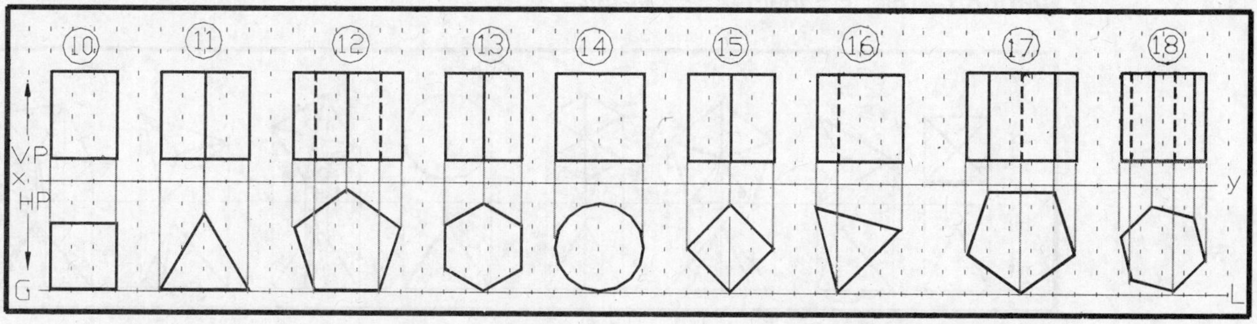


**PLATE 12.2** Projections of prism and cylinder with axis perpendicular to VP.

**EXAMPLE 12.14** A cylinder is resting on one of its generator on the ground with axis perpendicular to VP. Draw projections of solid.

***Solution*** Refer to P.14/Plate 12.2.

**EXAMPLE 12.15** A square prism is resting on one of its vertical edges on the ground with axis perpendicular to VP and all faces equally inclined to the ground. Draw projections of solid.

***Solution*** This is exhibited in P.15/Plate 12.2.

**EXAMPLE 12.16** A triangular prism is resting on one of its vertical edges on the ground with axis perpendicular to VP and one face is inclined at 45° to the ground. Draw projections of solid.

***Solution*** See P.16/Plate 12.2.

**EXAMPLE 12.17** A pentagonal prism is resting on one of its vertical edges on the ground with axis perpendicular to VP and opposite face is parallel to the ground. Draw projections of solid.

***Solution*** Observe P.17/Plate 12.2.

**EXAMPLE 12.18** A hexagonal prism is resting on one of its vertical edges on the ground with axis perpendicular to VP and one of its faces inclined at 45° to the ground. Draw projections of solid.

***Solution*** We can see this in P.18/Plate 12.2.

**Examples based on projection of pyramids and cones (with axis perpendicular to ground or HP)**

**EXAMPLE 12.19** A square pyramid is resting on its base on the ground. Draw projections of solid if two of its base edges are perpendicular to VP.

***Solution*** We can see this in P.19/Plate 12.3.

**EXAMPLE 12.20** A triangular pyramid is resting on its base on the ground. Draw projections of solid if one of its base edges is perpendicular to VP.

***Solution*** This is evident in P.20/Plate 12.3.

**EXAMPLE 12.21** A pentagonal pyramid is resting on its base on the ground. Draw projections of solid if one of its base edges is parallel to VP and towards (↑) the observer.

***Solution*** Refer to P.21/Plate 12.3.

**EXAMPLE 12.22** A hexagonal pyramid is resting on its base on the ground. Draw projections of solid if two of its adjacent base edges are equally inclined to VP.

***Solution*** See P.22/Plate 12.3.

**EXAMPLE 12.23** A cone is resting on its base on the ground. Draw projections of solid.

***Solution*** This is shown in P.23/Plate 12.3.

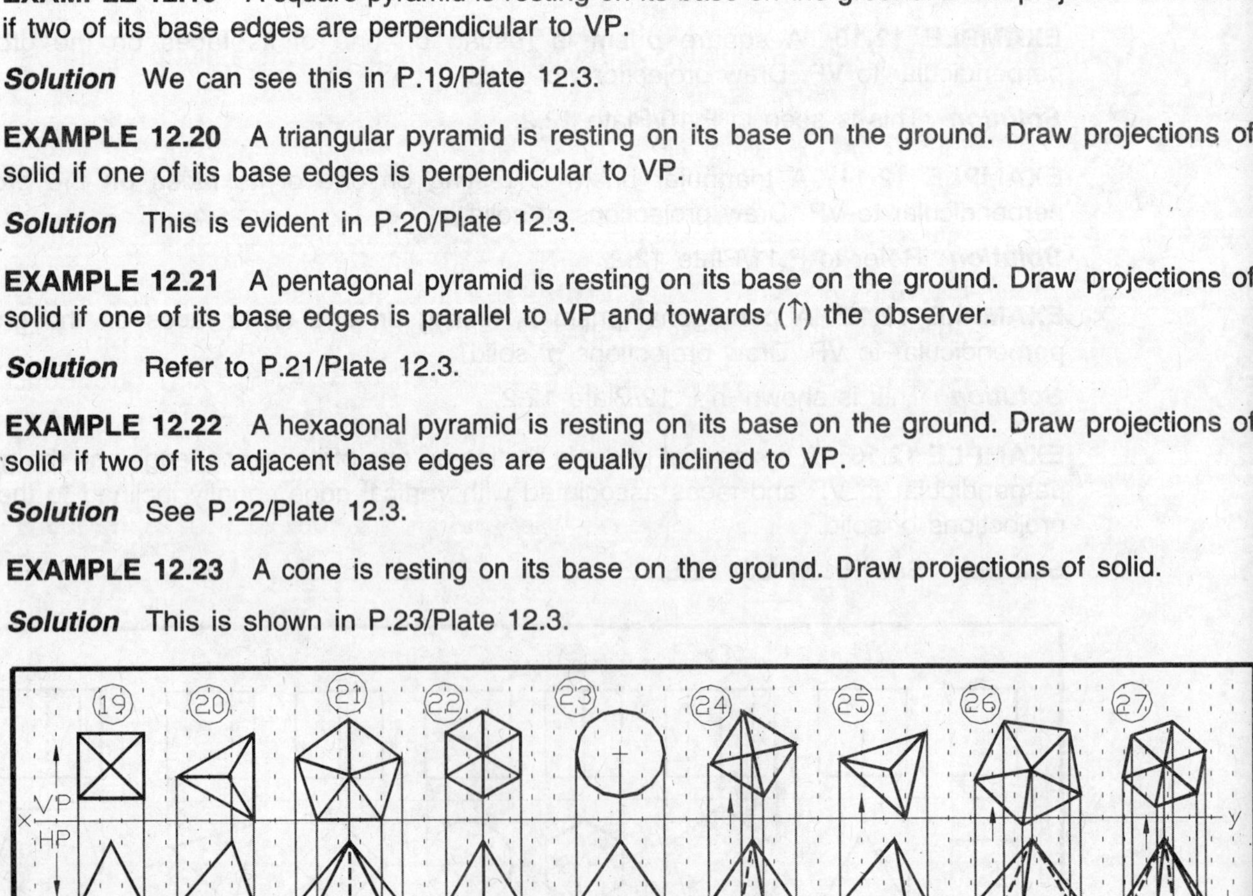


**PLATE 12.3** Projections of pyramids and cones with axis perpendicular to HP.

**EXAMPLE 12.24** A square pyramid is resting on its base on the ground. Draw projections of solid if one of its base edges is inclined 18° to VP and exposed to the observer.

***Solution*** See P.24/Plate 12.3.

**EXAMPLE 12.25** A triangular pyramid is resting on its base on the ground. Draw projections of solid if one of its base edges is inclined at 45° to VP and exposed (↑) to the observer.

***Solution*** Observe P.25/Plate 12.3.

**EXAMPLE 12.26** A pentagonal pyramid is resting on its base on the ground. Draw projections of solid if one of its base edges is inclined at 45° to VP and exposed (↑) to the observer.

***Solution*** This is shown in P.26/Plate 12.3.

**EXAMPLE 12.27** A hexagonal pyramid is resting on its base on the ground. Draw projections of solid if one of its base edges is inclined 45° to VP and exposed (↑) to the observer.

***Solution*** See P.27/Plate 12.3.

**Examples based on projection of pyramids and cones (with axis perpendicular to VP)**

**EXAMPLE 12.28** A square pyramid is resting on one of its base edges on the ground. Draw projections of solid if the base is parallel to VP and away from the observer.

***Solution*** Refer to P.28/Plate 12.4.

**EXAMPLE 12.29** A triangular pyramid is resting on one of its base corners on the ground and base-edge associated with that corner is inclined 14° to the ground. Draw projections of solids if the base is parallel to VP and away from the observer.

***Solution*** See P.29/Plate 12.4.

**EXAMPLE 12.30** A pentagonal pyramid is resting on one of its base corners on the ground and one of the base edges opposite to that corner is parallel to the ground. Draw projections of solid if the base is parallel to VP and away from the observer.

***Solution*** This is exhibited in P.30/Plate 12.4.

**EXAMPLE 12.31** A hexagonal pyramid is resting on one of its base-corner on the ground. Draw projections of solid if the base is parallel to VP and away from the observer.

***Solution*** This is shown in P.31/Plate 12.4.

**EXAMPLE 12.32** A cone is resting on its circumference of base on the ground. Draw projections of solid if the base is parallel to VP and away from the observer.

***Solution*** See P.32/Plate 12.4.

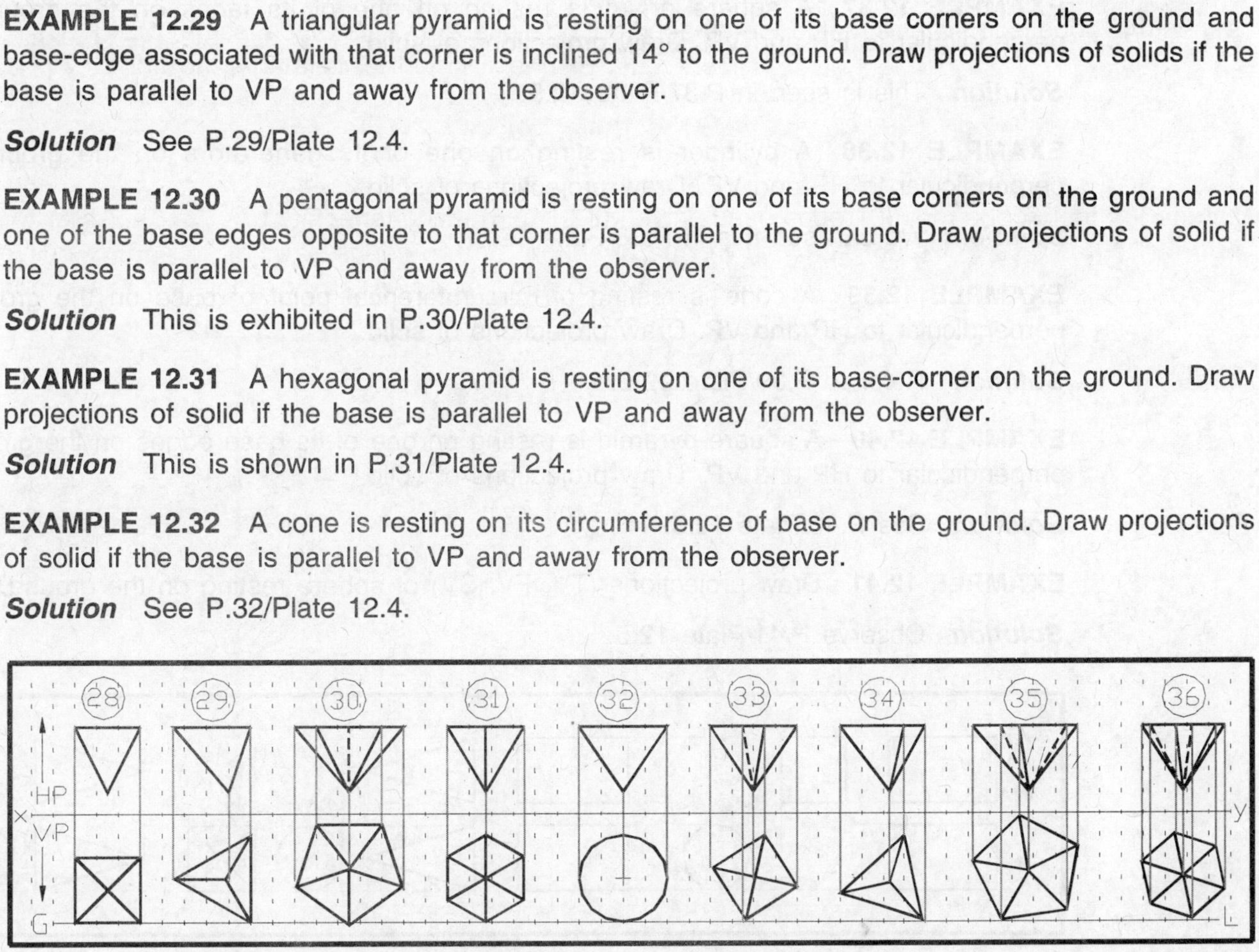


**PLATE 12.4** Projections of pyramids and cones with axis perpendicular to VP.

**EXAMPLE 12.33** A square pyramid is resting on one of its base corners on the ground and base-edge associated with that corner is inclined 30° to the ground. Draw projections of solid if the base is parallel to VP and away from the observer.

***Solution*** Observe P.33/Plate 12.4.

**EXAMPLE 12.34** A triangular pyramid is resting on one of its base corners on the ground and base-edge associated with that corner is inclined 18° to the ground. Draw projections of solid if the base is parallel to VP and away from the observer.

***Solution*** See P.34/Plate 12.4.

**EXAMPLE 12.35** A pentagonal pyramid is resting on one of its base corners on the ground and base-edge associated with that corner is inclined 45° to the ground. Draw projections of solid if the base is parallel to VP and away from the observer.

***Solution*** Observe P.35/Plate 12.4.

**EXAMPLE 12.36** A hexagonal pyramid is resting on one of its base corners on the ground and base edge associated with that corner is inclined 45° to the ground. Draw projections of solid if the base is parallel to VP and away from the observer.

***Solution*** This is shown in P.36/Plate 12.4.

**Examples based on projection of solids (with axis parallel to HP and VP)**

**EXAMPLE 12.37** A square prism is resting on one of its faces on the ground with bases perpendicular to HP and VP. Draw projections of solid.

***Solution*** This is seen in P.37/Plate 12.5.

**EXAMPLE 12.38** A cylinder is resting on one of its generators on the ground with bases perpendicular to HP and VP. Draw projections of solid.

***Solution*** We can see this in P.38/Plate 12.5.

**EXAMPLE 12.39** A cone is resting on circumferential point of base on the ground with base perpendicular to HP and VP. Draw projections of solid.

***Solution*** Refer to P.39/Plate 12.5.

**EXAMPLE 12.40** A square pyramid is resting on one of its base edges on the ground with base perpendicular to HP and VP. Draw projections of solid.

***Solution*** See P.40/Plate 12.5.

**EXAMPLE 12.41** Draw projections (TV, FV, SV) of sphere resting on the ground.

***Solution*** Observe P.41/Plate 12.5.

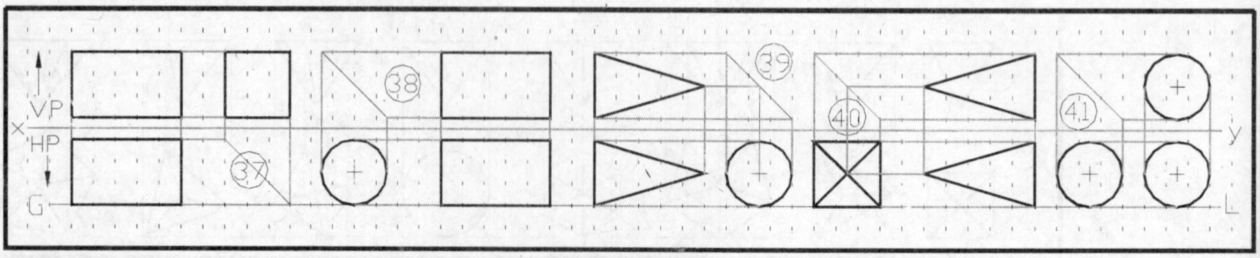


**PLATE 12.5** Projections of solids with axis parallel to HP and VP.

## 12.5 Axis Inclined to One Reference Plane and Parallel to the Other

If a solid has its axis inclined to one reference plane and parallel to the other, its projections are drawn in two stages. In the initial stage, the solid is assumed to be in a simple position, i.e., its axis is perpendicular to one of the reference planes (HP or VP).

If the axis is inclined to the ground, i.e., the HP, it is assumed to be perpendicular to the HP. In the initial stage, accordingly, the true shape of the base of the solid is drawn in the TV. Similarly, if the axis is to be inclined to the VP, it is assumed to be perpendicular to the VP. In the initial stage, accordingly, the true shape of the base of the solid is drawn in the FV. This is discussed from **EXAMPLE 12.42** to **EXAMPLE 12.51**.

As the solid is tilted, it may rest on one of its: base-corner, base-edge, vertical faces and vertical edges (in case of prism); base-corners, apex, base-edges, slant faces and slant edges (in case of pyramid); circumferential point of base and generator (in case of cylinder); circumferential point of base, slant generator and apex (in case of cone).

### Rest condition

While drawing the true shape (TS) of the base, the following conventions are considered. If it is resting on:

(i) Base-edge: Draw that base-edge perpendicular to the xy line.
(ii) Base-corner: Draw a line joining that base-corner and the centre of the base parallel to the xy line.
(iii) Slant face: Draw a base-edge of that slant face perpendicular to the xy line.
(iv) Slant/vertical edge: Draw a line joining the base-corner of that slant/vertical edge and centre of the base parallel to the xy line.

### Example based on axis inclined to one reference plane and parallel to the other reference plane

**EXAMPLE 12.42** A hexagonal prism (Fig. 12.9) having base edge 100 mm and height 150 mm is resting on one of its base edges on the ground in such a way that the axis is inclined 70° to the ground and parallel to VP. Draw projections of solid.

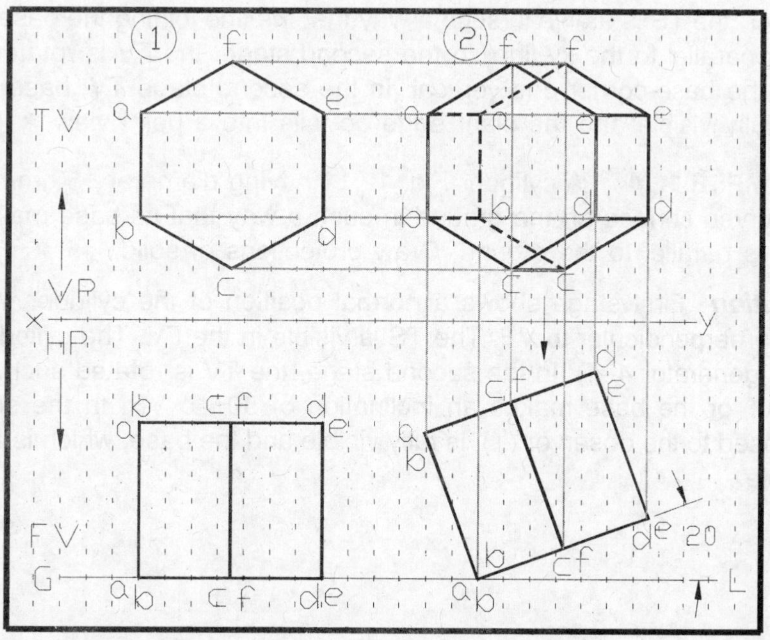


Fig. 12.9

***Solution*** First stage shows a normal position of the prism. As axis is inclined to the ground, it is made perpendicular to the ground. The TS is visible in the TV. As the prism is resting on its base-edge the TS is drawn in such a way that the edge on which it is resting, is made perpendicular to the xy line. In the second stage, the FV is rotated such that the axis makes an inclination of 70° or the base makes an inclination of 20° to the G-L line. In the second stage TV, the top base is fully visible and the bottom base is partially visible. The vertical edges aa and bb are not visible.

**EXAMPLE 12.43** A square pyramid (Fig. 12.10) having base diagonal 200 mm and height 175 mm is resting on one of its base corner on the ground in such a way that, the slant edge associated with that corner is vertical. Draw projections of solid if the axis is parallel to the VP.

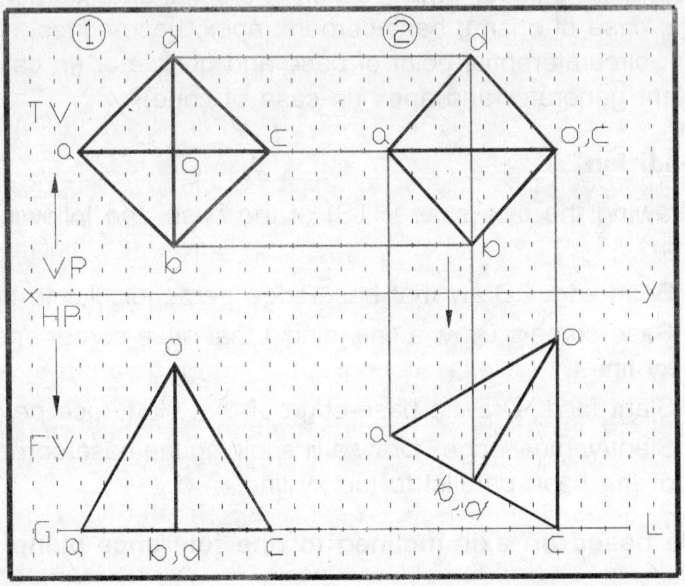


**Fig. 12.10**

***Solution*** First stage shows a normal position of the pyramid. The pyramid is resting on the ground on one of its corner. Therefore, the TS is visible in the TV. As the pyramid is resting on its base-corner, the TS is drawn in such a way that the line joining the base-corner and the centre of the base co is parallel to the xy line. In the second stage, the FV is rotated such that the line oc associated with the base-corner c is vertical. In the second stage TV, base edges as well as the slant edges are fully visible and the slant edge oc falls into a point view.

**EXAMPLE 12.44** A cylinder Fig. 12.11 having diameter 150 mm and height 175 mm is resting on its curved surface on the ground in such a way that its base makes an angle of 50° to VP and the axis is parallel to the ground. Draw projections of solid.

***Solution*** First stage shows a normal position of the cylinder. As the axis is inclined to VP, it is made perpendicular to VP. The TS is visible in the FV. The cylinder is resting on its curved surface (i.e., generator 4–4). In the second stage, the TV is rotated such that the axis makes an inclination of 40° or the base makes an inclination of 50° to VP. In the second stage FV, a base which is exposed to the observer (↑), is fully visible and the base, which is away from the observer, is partially visible.

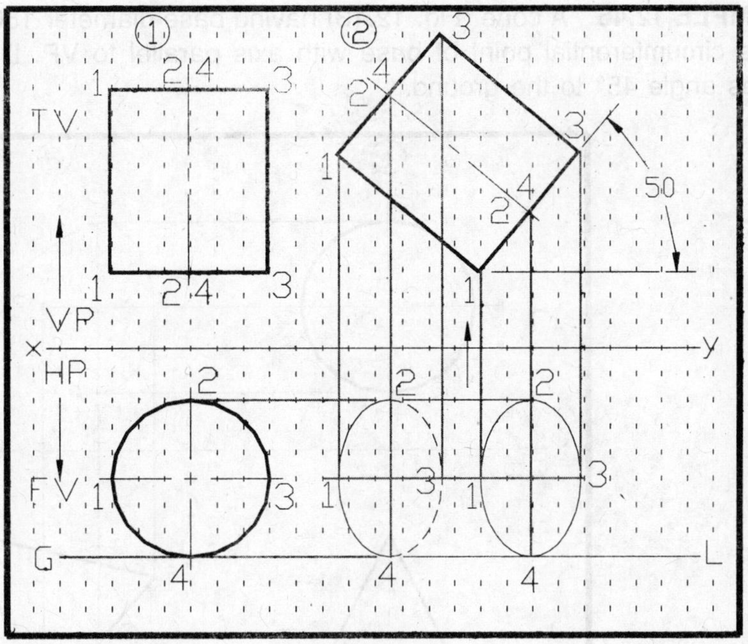


Fig. 12.11

**EXAMPLE 12.45** A cone (Fig. 12.12) having base diameter 200 mm and height 200 mm is resting on circumferential point of base on the ground with axis inclined 50° to VP and parallel to the ground. Draw projections of solid if the apex is towards the observer.

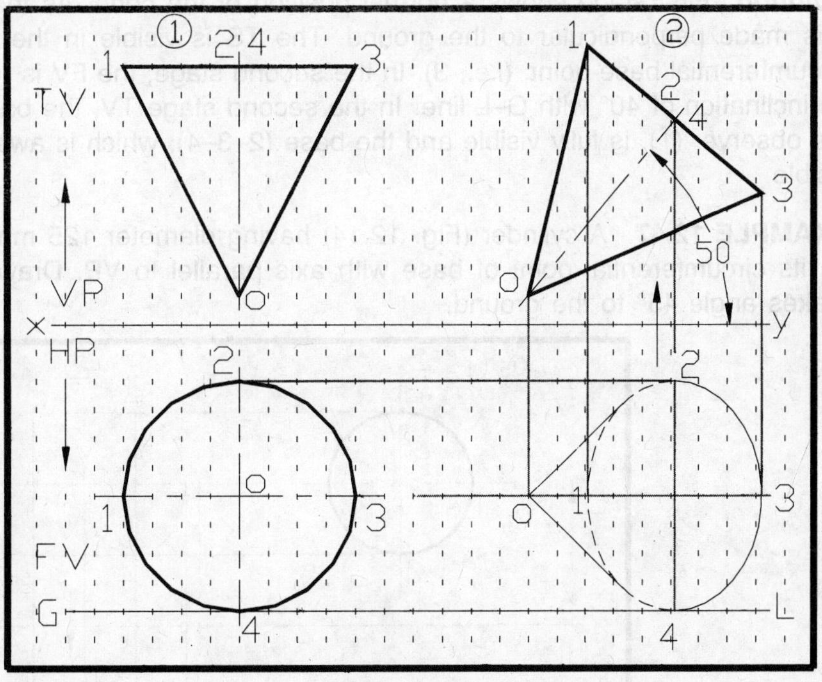


Fig. 12.12

***Solution*** First stage shows a normal position of the cone. As axis is inclined to VP, it is made perpendicular to VP. The TS is visible in the FV. The cone is resting on its circumferential base-point, (i.e., 4). In the second stage, the TV is rotated such that the axis makes an inclination of 50° with the xy line. In the second stage FV, the base (2–3–4), which is nearer to the observer (↑) is fully visible and the base (4–1–2), which is away from the observer is partially visible.

**EXAMPLE 12.46** A cone (Fig. 12.13) having base diameter 150 mm and height 150 mm is resting on its circumferential point of base with axis parallel to VP. Draw projections of solid if the axis makes angle 45° to the ground.

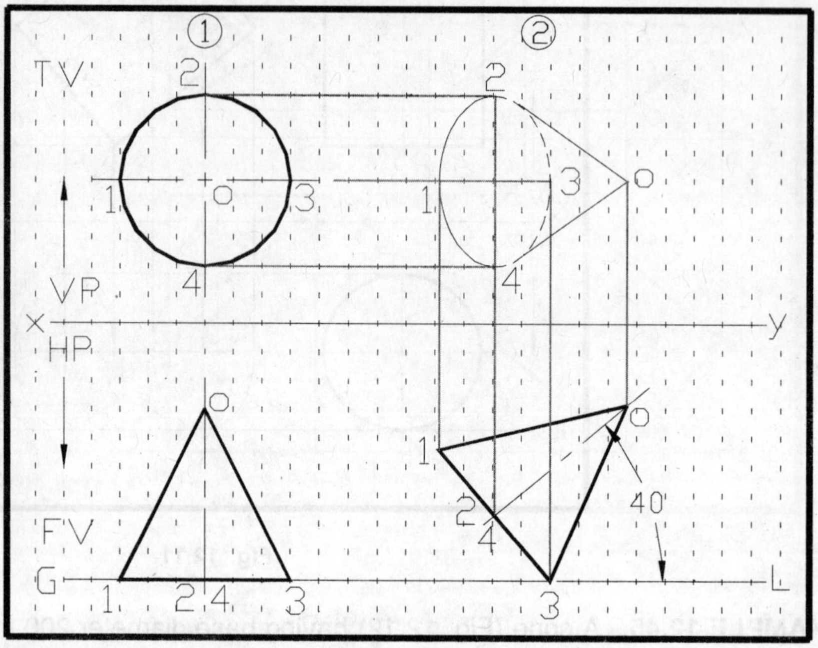


**Fig. 12.13**

***Solution*** First stage shows a normal position of the cone. As the axis is inclined to the ground, it is made perpendicular to the ground. The TS is visible in the TV. The cone is resting on its circumferential base-point, (i.e., 3). In the second stage, the FV is rotated such that the axis makes an inclination of 40° with G–L line. In the second stage TV, the base (2–1–4) which is exposed to the observer (↑), is fully visible and the base (2–3–4), which is away from the observer, is partially visible.

**EXAMPLE 12.47** A cylinder (Fig. 12.14) having diameter 125 mm and height 125 mm is resting on its circumferential point of base with axis parallel to VP. Draw projections of solid if the axis makes angle 45° to the ground.

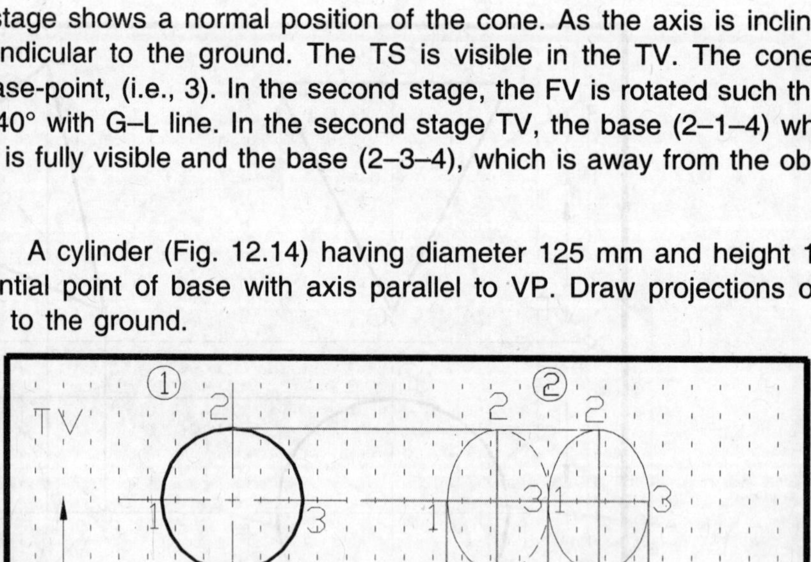


**Fig. 12.14**

***Solution*** First stage shows a normal position of the cylinder. As axis is inclined to the ground, it is made perpendicular to the ground. The TS is visible in the TV. The cylinder is resting on its circumferential base point, (i.e., 3). In the second stage, the FV is rotated such that the axis makes an inclination of 45° or the base makes an inclination of 45° to line GL. In the second stage TV, the top base is fully visible and the bottom base is partially visible.

**EXAMPLE 12.48** A square prism (Fig. 12.15) having base diagonal 100 mm and height 150 mm is resting on one of its vertical edges on the ground with axis parallel to the ground. Draw projections of solid if axis makes angle 30° to VP.

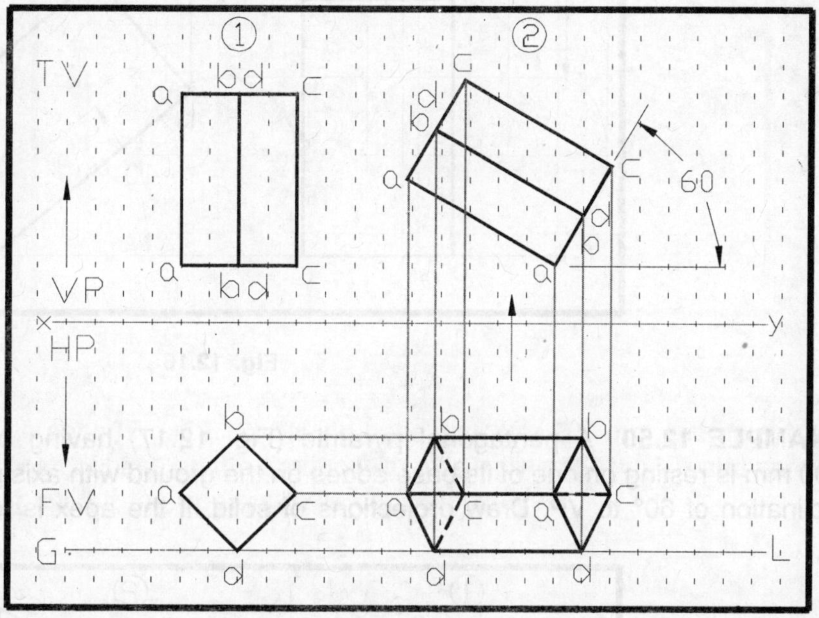


**Fig. 12.15**

***Solution*** First stage shows a normal position of the prism. As axis is inclined to VP, it is made perpendicular to VP. The TS is visible in the FV. The prism is resting on its vertical edge dd. In the second stage, the TV is rotated such that the axis makes an inclination of 30° or the base makes an inclination of 60° to xy line. In the second stage FV, the base, which is exposed to observer (↑), is fully visible and the base, which is away from the observer, is partially visible. Vertical edge cc is not visible.

**EXAMPLE 12.49** A triangular prism Fig. (12.16) having base edge 150 mm and height 225 mm is resting on one of its base edges on the ground with axis parallel to VP. Draw projections of solid if the axis makes angle of 40° to the ground.

***Solution*** First stage shows a normal position of the prism. As the axis is inclined to the ground, it is made perpendicular to the ground. The TS is visible in the TV. As the prism is resting on its base-edge, the TS is drawn in such a way that the edge on which it is resting, is made perpendicular to the xy line. In the second stage, the FV is rotated such that the axis makes an inclination of 40° or the base makes an inclination of 50° to GL line. In the second stage TV, the top base is fully visible and the bottom base is partially visible. All vertical edges are visible.

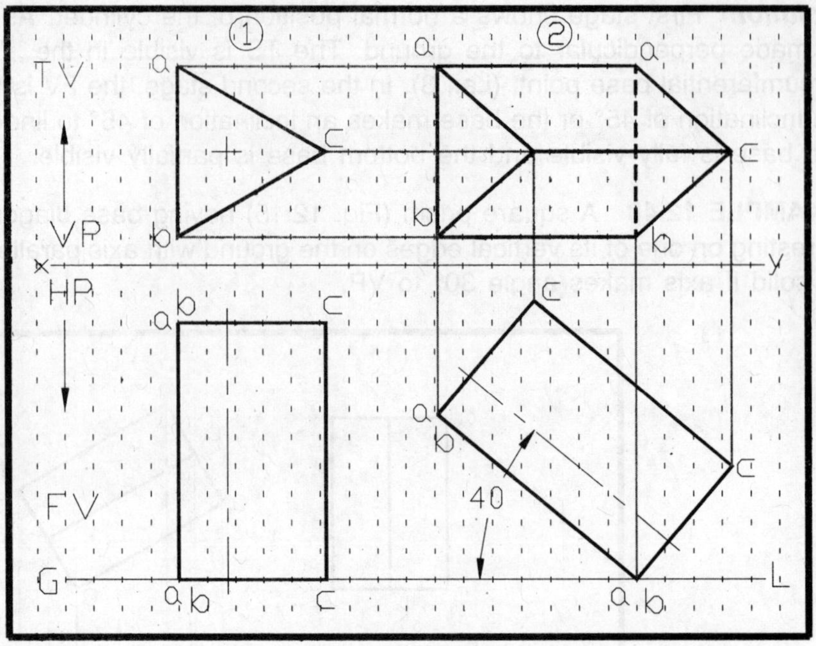


Fig. 12.16

**EXAMPLE 12.50** A pentagonal pyramid (Fig. 12.17) having base-edge 100 mm and height 200 mm is resting on one of its base edges on the ground with axis parallel to the ground and makes inclination of 60° to VP. Draw projections of solid, if the apex is towards the observer.

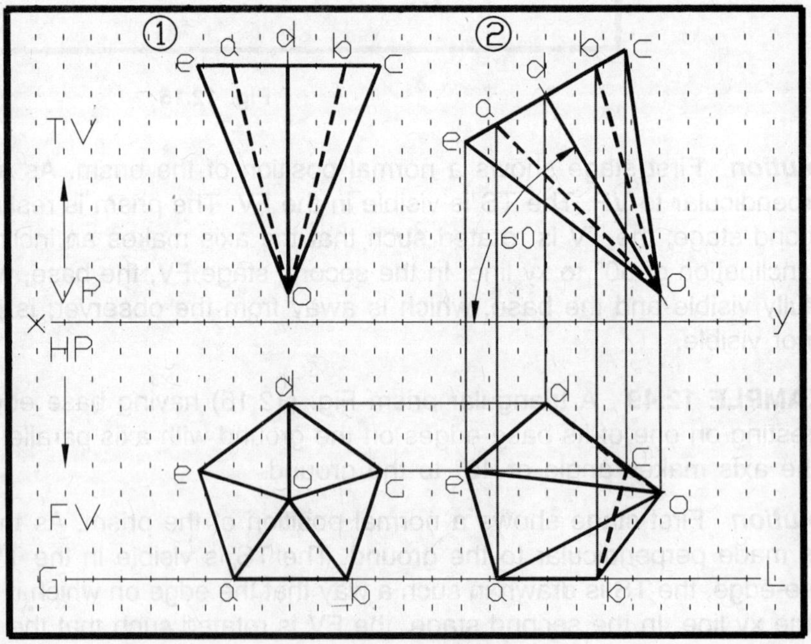


Fig. 12.17

***Solution*** First stage shows a normal position of the pyramid. As axis is inclined to VP, it is made perpendicular to VP. The TS is visible in the FV. The pyramid is resting on its base-edge a-b. In the second stage, the TV is rotated such that the axis makes an inclination of 60° to VP. In the second stage FV, base-edges bc and cd are not visible.

**EXAMPLE 12.51** A hexagonal prism (Fig. 12.18) having base edge 50 mm and height 175 mm is resting on one of its base corner on the ground with axis parallel to VP. Draw projections of solid if the axis makes angle 50° to the ground.

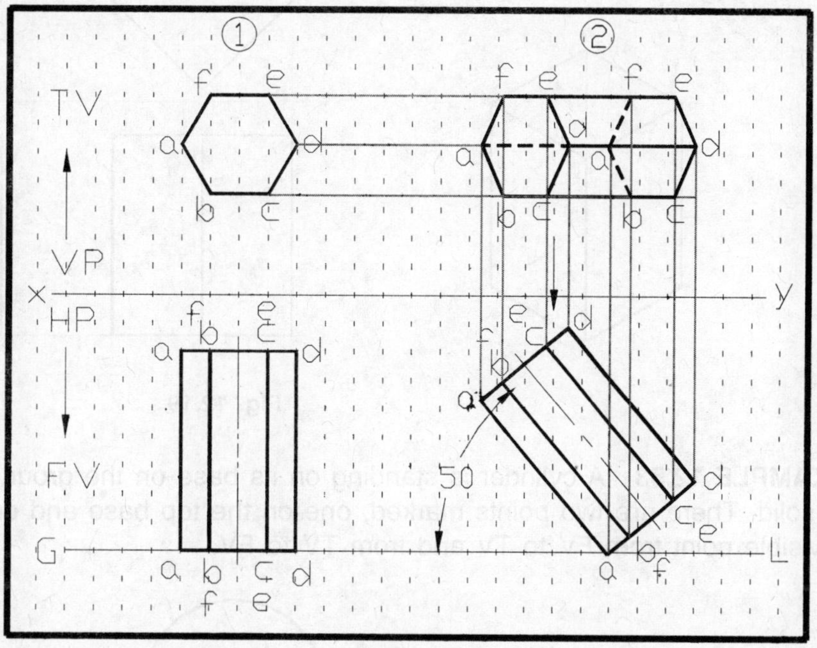


**Fig. 12.18**

***Solution*** First stage shows a normal position of the prism. As axis is inclined to the ground, it is made perpendicular to the ground. The TS is visible in the TV. As the prism is resting on its base-corner the TS is drawn in such a way that the line joining the corner a and the centre of the base is parallel to xy line. In the second stage, the FV is rotated such that the axis makes an inclination of 50° or the base makes an inclination of 40° to GL line. In the second stage TV, the top base is fully visible and the bottom base is partially visible. The vertical edge aa associated with the corner a is not visible.

## 12.6 Transfer of Point(s) from One View to Another

The transfer of point(s) from one view to another is frequently required. In case of prisms and cylinders, the points can be directly transferred. This concept is useful in section of solids, development of surfaces of solids and intersection of surfaces of solids for the point transfer from one view to another.

The following examples discuss on the transfer of points in case of prisms, cylinders, pyramids and cones.

**Example based on the transfer of point from one view to another**

**EXAMPLE 12.52** A square prism is standing on its base on the ground with all four faces equally inclined to VP (Fig. 12.19). Draw projections of solid. There are two points marked, one on the top base and one on the face AADD. Transfer a face point from FV to TV. Transfer a base point from TV to FV.

***Solution*** First stage shows the projections of solids with four faces AABB, BBCC, CCDD and DDAA along with the location of point views. The visible point in the FV is on the face AADD. In the second stage, this is directly transferred by drawing a projector from the visible point to the base edge AD, which represents the face AADD in the TV. The visible point in the TV is on the top base ABCD. In the second stage, this is directly transferred by drawing a projector from the visible point to the top base in the FV.

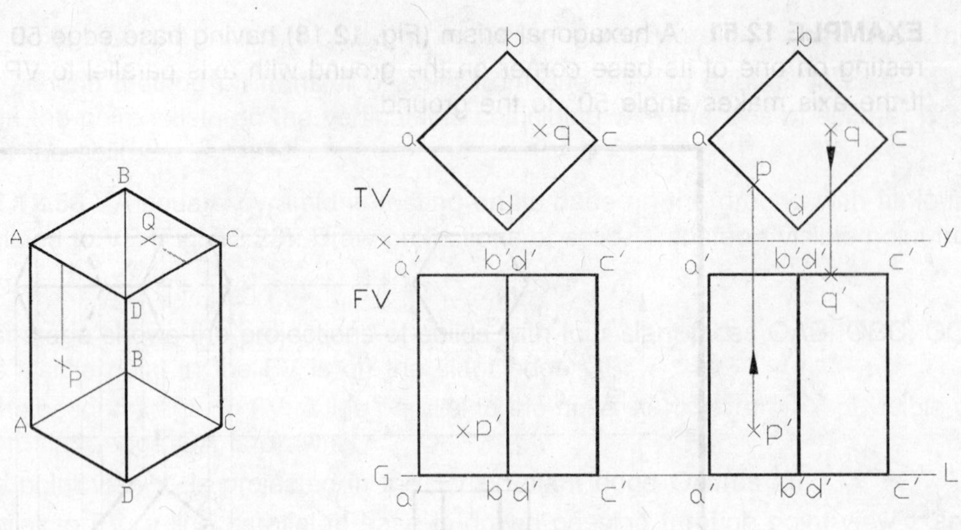


Fig. 12.19

**EXAMPLE 12.53** A cylinder is standing on its base on the ground (Fig. 12.20). Draw projections of solid. There are two points marked, one on the top base and one on the curved face. Transfer a visible point from FV to TV and from TV to FV.

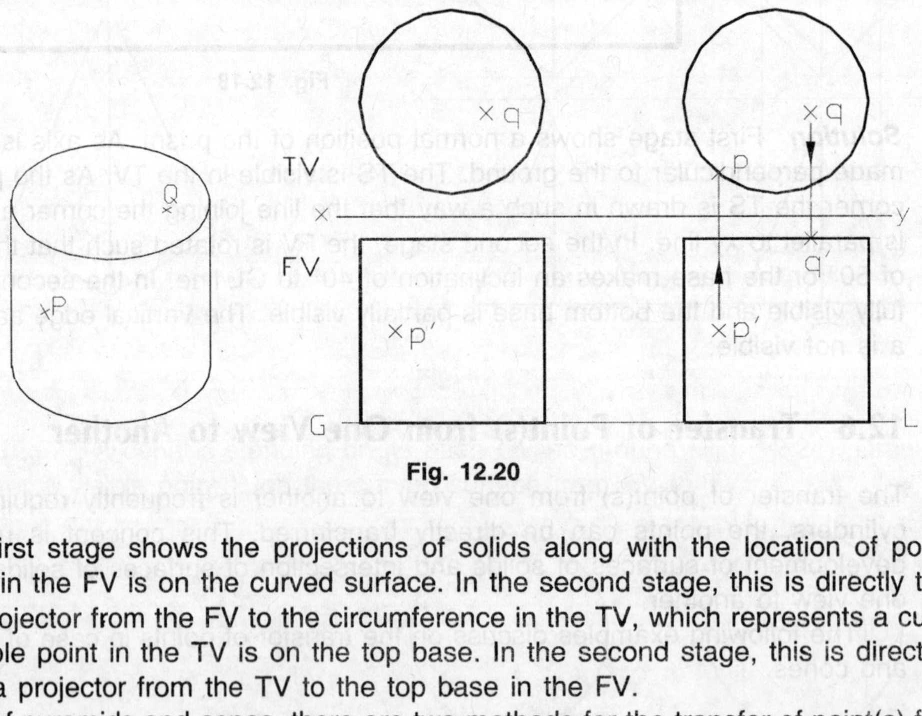


Fig. 12.20

***Solution*** First stage shows the projections of solids along with the location of point view. The visible point in the FV is on the curved surface. In the second stage, this is directly transferred by drawing a projector from the FV to the circumference in the TV, which represents a curved surface.

The visible point in the TV is on the top base. In the second stage, this is directly transferred by drawing a projector from the TV to the top base in the FV.

In case of pyramids and cones, there are two methods for the transfer of point(s) from one view to another:

(i) Drawing a slant line starting from apex to the base or from base to the apex

(ii) Drawing a line parallel to the base. (This method is also called cutting plane method).

**Examples based on the transfer of point by using a slant line method**

**EXAMPLE 12.54** A square pyramid is resting on its base on the ground with all four base edges equally inclined to VP (Fig. 12.21). There is a visible point p′ marked on a slant surface OAB. Draw projections of solid. Transfer a visible point from FV to TV.

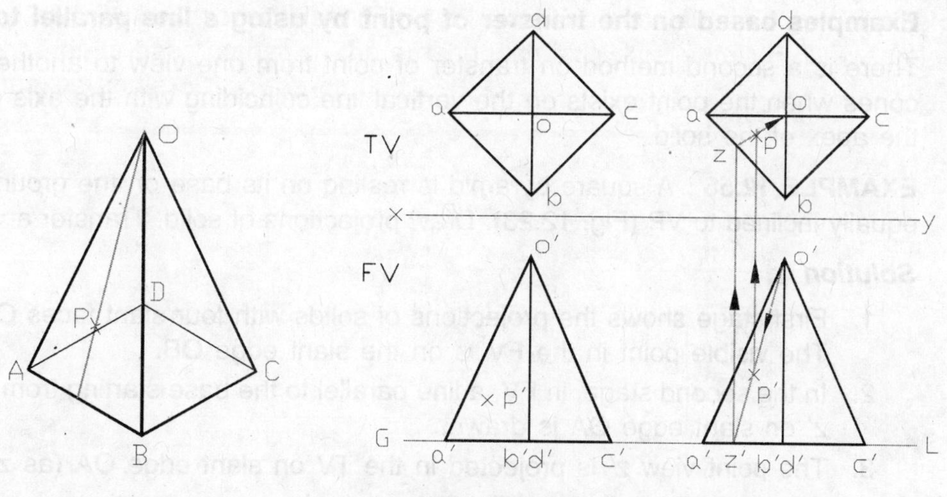


Fig. 12.21

***Solution***

1. First stage shows the projections of solids with four slant faces OAB, OBC, OCD and ODA. The visible point view p′ in the FV is on the face o′a′b′.
2. In the second stage, in FV, an imaginary line is drawn starting from apex o′, passing through the FV of visible point and meeting the base at z′.
3. The z′ is projected in the TV on front base-edge ab and then same line is continued to o′. o′z′ and oz are FV and TV of slant line oz.
4. Draw a projector line from the FV of visible point to intersect the TV of slant line i.e., oz.

**EXAMPLE 12.55** A cone is standing on its base on the ground (Fig. 12.22). Draw projections of solid. Transfer a visible point from FV to TV.

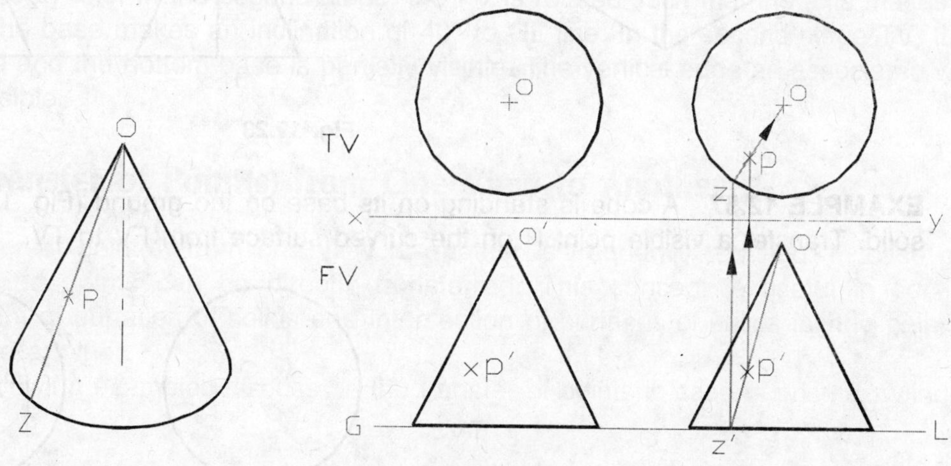


Fig. 12.22

***Solution***

1. First stage shows the projections of solid. The visible point in the FV is on the curved surface.
2. In the second stage, in FV, an imaginary line is drawn starting from apex o′, passing through the FV of visible point and meeting the base at z′.
3. The z′ is projected in the TV on the circular base z and then the same line is continued to o. o′z′ and oz are FV and TV of slant line oz.
4. Draw a projector line from the FV of visible point to intersect the TV of slant line i.e., oz.

***Note:*** With the help of single LINE command this transfer can be achieved. Line from point o′, to point z′, to point z, to point o.

**Examples based on the transfer of point by using a line parallel to the base method**

There is a second method on transfer of point from one view to another in case of pyramids and cones when the point exists on the vertical line coinciding with the axis of solid or passing through the apex of the solid.

**EXAMPLE 12.56** A square pyramid is resting on its base on the ground with all four base edges equally inclined to VP (Fig. 12.23). Draw projections of solid. Transfer a visible point from FV to TV.

***Solution***

1. First stage shows the projections of solids with four slant faces OAB, OBC, OCD and ODA. The visible point in the FV is on the slant edge OB.
2. In the second stage, in FV, a line parallel to the base starting from FV of visible point, to point z′ on slant edge OA is drawn.
3. The point view z′ is projected in the TV on slant edge OA (as z).
4. Similar to FV, a line parallel to base is drawn passing through point view z and cutting the slant edge OB in the TV to obtain the required transfer. This transfer can also be obtained by drawing an arc from z on oc to ob. This is shown in the same figure.

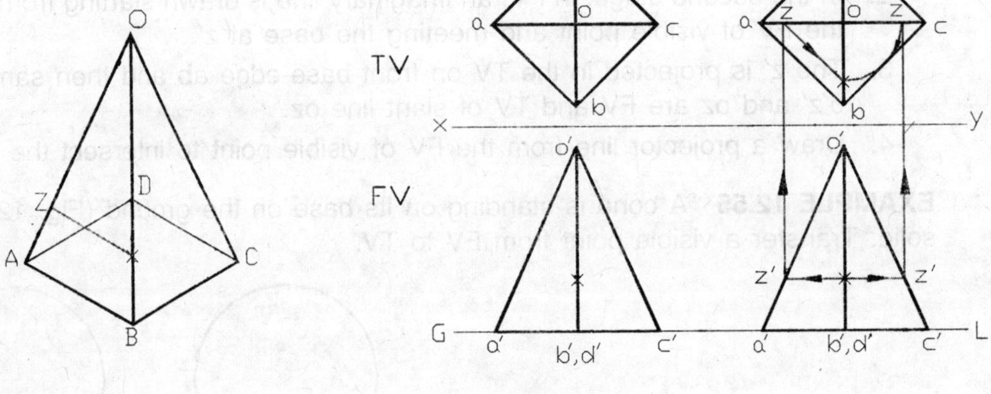


**Fig. 12.23**

**EXAMPLE 12.57** A cone is standing on its base on the ground (Fig. 12.24). Draw projections of solid. Transfer a visible point P on the curved surface from FV to TV.

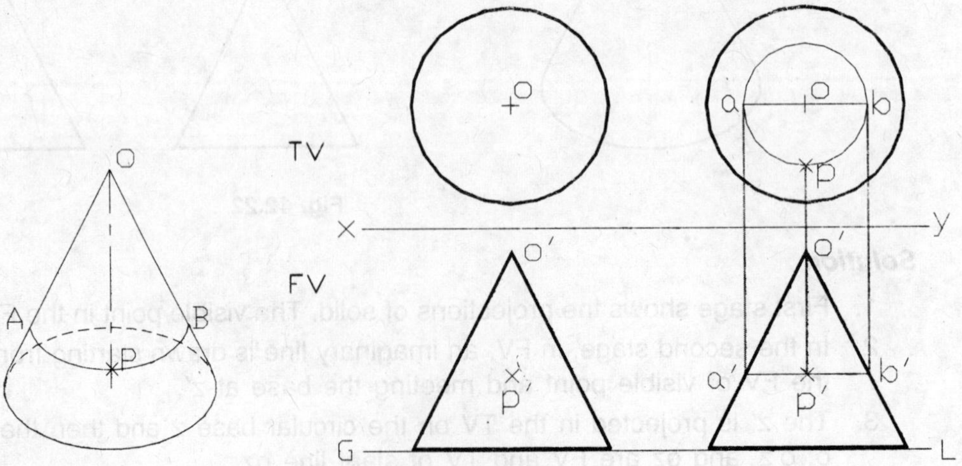


**Fig. 12.24**

***Solution***

1. First stage shows the projections of solid. The visible point in the FV is on the curved surface lying on the axis of the cone.
2. In the second stage, in FV, a line a′b′ parallel to the base of the cone is drawn passing through the FV of visible point. This line is considered as a cutting plane.
3. If the cone is cut with the help of this cutting plane, the section in the TV will appear as a circular plane with diameter ab in the TV.
4. The FV of the visible point is now directly projected on the circumference of the cut section in the TV.

This method is called "cutting plane method" or "line (curve) parallel to base method".

**EXAMPLE 12.58** A square prism is resting on its vertical edge on the ground with its axis parallel to HP and VP (Fig. 12.25). There are two points, one V on vertical edge CC and another U on base edge BC. Transfer these points on the respective edges in the TV.

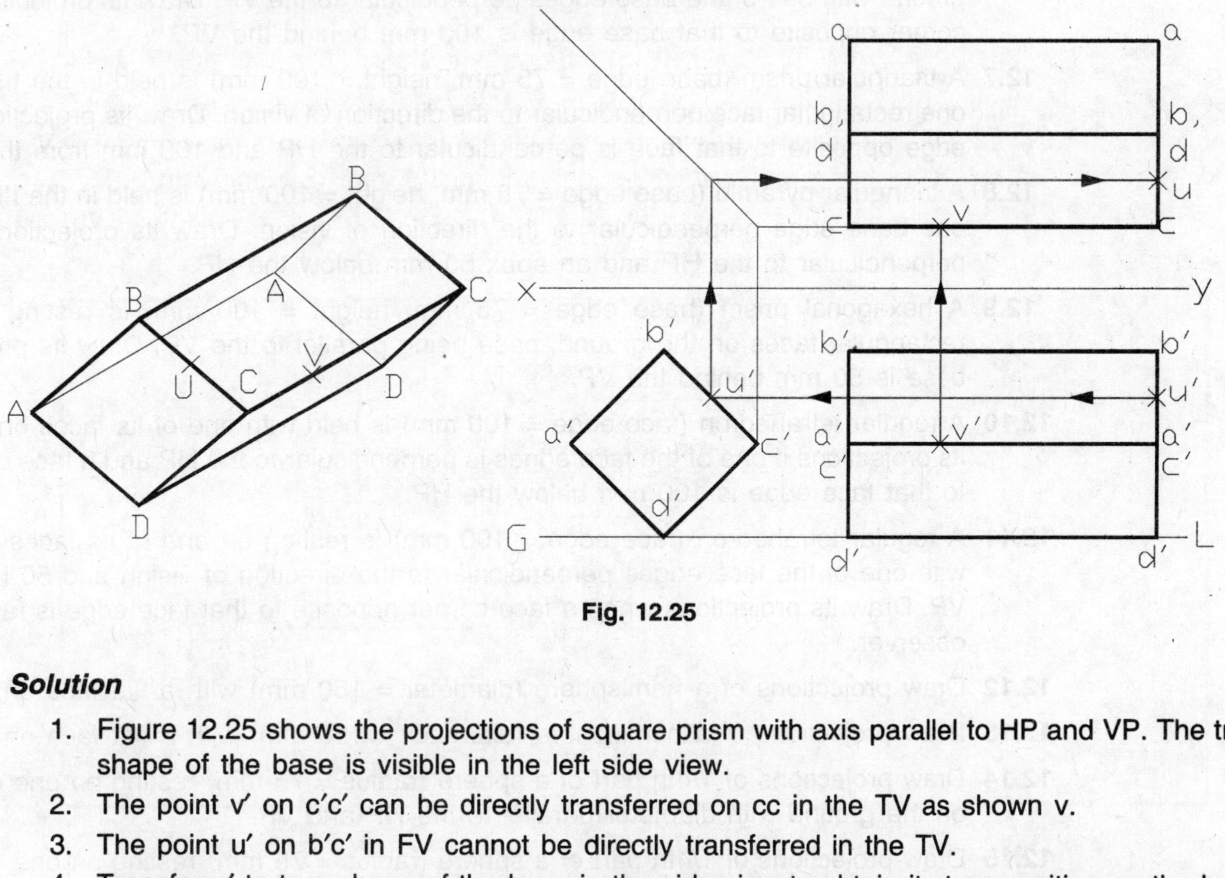


**Fig. 12.25**

***Solution***

1. Figure 12.25 shows the projections of square prism with axis parallel to HP and VP. The true shape of the base is visible in the left side view.
2. The point v′ on c′c′ can be directly transferred on cc in the TV as shown v.
3. The point u′ on b′c′ in FV cannot be directly transferred in the TV.
4. Transfer u′ to true shape of the base in the side view to obtain its true position on the base edge b′c′.
5. Transfer the u′ by using a 45° line method to the TV and obtain u as shown.

# Exercise A (Projection of Solids in Simple Position)

**Instructions:**

1. 1 unit = 4 grid spacing = 100mm.
2. Use point for locating the centre with PDMODE: 2 and PDSIZE: 0.15.
3. Draw a polygon by edge method and locate the centre of the polygon/circle.
4. Show the axis by a centre line (–·–·–·–·–·–·–·–·–·–·–·–).

**Solve for the following positions of solids in III-quadrant**

**12.1** A square prism (base edge = 75 mm, height = 100 mm) is resting on its base on the ground. Draw its projections if two of its adjacent faces are equally inclined to the VP.

**12.2** A square prism (base edge = 75 mm, height = 100 mm) is resting on its base on the ground. Draw its projections if one of its rectangular faces makes an angle 30° to the VP.

**12.3** A square pyramid (base edge = 75 mm, height = 100 mm) is resting on its base on the ground. Draw its projections if two of its adjacent base edges are equally inclined to the VP.

**12.4** A square pyramid (base edge = 75 mm, height = 100 mm) is resting on its base on the ground. Draw its projections if one of its base edges makes an angle 30° to the VP.

**12.5** A pentagonal prism (base edge = 75 mm, height = 100 mm) is resting on its base on the ground with one rectangular face perpendicular to the VP. Draw its projections if a vertical edge opposite to that face is 100 mm behind the VP.

**12.6** A pentagonal pyramid (base edge = 75 mm, height = 100 mm) is resting on its base on the ground with one of the base edges perpendicular to the VP. Draw its projections if the base corner opposite to that base edge is 100 mm behind the VP.

**12.7** A triangular prism (base edge = 75 mm, height = 100 mm) is held in the III-quadrant with one rectangular face perpendicular to the direction of vision. Draw its projections if a vertical edge opposite to that face is perpendicular to the HP and 100 mm from the VP.

**12.8** A triangular pyramid (base edge = 75 mm, height = 100 mm) is held in the III-quadrant with one base edge perpendicular to the direction of vision. Draw its projections if an axis is perpendicular to the HP and an apex 50 mm below the HP.

**12.9** A hexagonal prism (base edge = 75 mm, height = 100 mm) is resting on one of its rectangular faces on the ground, base being parallel to the VP. Draw its projections if the base is 50 mm behind the VP.

**12.10** A regular tetrahedron (face edge = 100 mm) is held with one of its faces on the VP. Draw its projections if one of the face edges is perpendicular to the HP and a face corner opposite to that face edge is 100 mm below the HP.

**12.11** A regular tetrahedron (face edge = 100 mm) is resting on one of its faces on the ground with one of the face edges perpendicular to the direction of vision and 50 mm behind the VP. Draw its projections when a face corner opposite to that face edge is farthest from the observer.

**12.12** Draw projections of a hemisphere (diameter = 150 mm) with a flat face on the VP.

**12.13** Draw projections of a hemisphere (diameter = 150 mm) with a flat face on the ground.

**12.14** Draw projections of 1/4th part of a sphere (radius = 75 mm) resting on one of its flat faces on the ground with diameter parallel to the HP and VP.

**12.15** Draw projections of 1/4th part of a sphere (radius = 75 mm) resting on one of its flat face on the ground with diameter perpendicular to the VP.

**12.16** A right circular cone (diameter = 100 mm, height = 125 mm) is held with its base on the VP. Draw its projections when the apex is 50 mm below the HP.

**12.17** A right circular cylinder (diameter = 100 mm, height = 125 mm) is resting on its base on the ground. Draw its projections.

**12.18** A hexagonal pyramid (base edge = 75 mm and height = 125 mm) is held in the III-quadrant with an axis perpendicular to the VP and an apex farthest from the observer. Draw its projections if one of the base edges is parallel to the HP.

**12.19** A square prism (base edge = 75 mm, height = 125 mm) is held in the III-quadrant. Draw its projections if an axis is parallel to both the HP and VP and rectangular faces equally inclined to the ground.

**12.20** A square pyramid (base edge = 75 mm, height = 125 mm) is held in the III-quadrant. Draw its projections if an axis is parallel to both the HP and VP and base edges are equally inclined to the ground.

**12.21** A square pyramid (base edge = 100 mm, height = 125 mm) is resting on its base on the ground with a base edge parallel to the VP. It is surrounded by four spheres, resting on the ground, each having a radius 68.65 mm. In this arrangement each sphere touches the adjacent two spheres and a triangular face of a square pyramid. Draw projections of square pyramid and spheres. Find the distance between the centres of two opposite spheres.

**12.22** Three equal spheres each having 50 mm radius are resting on the ground so that each touches the other two with the line joining the centres of any two spheres, is parallel to the VP and one sphere is towards the observer. A fourth sphere of 75 mm radius is placed centrally on the three spheres. Draw the TV, FV and RSV of the arrangement. Find the distance of the centre of the fourth sphere above the ground.

**12.23** Three cones similar in dimensions having base radius 50 mm and height 100 mm are resting on the ground on their bases. Each touches the other two with a line joining the centres of the two cones and is parallel to the VP. The third cone is towards the observer. A sphere of 37.5 mm radius is placed centrally among them. Draw the FV, TV and RSV of the arrangement. Find the height of the centre of the sphere above the ground.

**12.24** Draw projections of a frustum of a square pyramid (with the top base side 75 mm, bottom base side 125 mm and height 200 mm) with all base edges equally inclined to the VP and the axis perpendicular to the ground.

## Exercise B (Projection of Solids with Axis Inclined to One Reference Plane and Parallel to the other)

**Solve for the following positions of the engineering solids in III-quadrant.**

**12.1** A right circular cone of diameter 150 mm and height 200 mm is resting on its circumferential base point on the ground with an axis inclined at 45° to the HP and parallel to the VP. Draw its projections.

**12.2** A right circular cylinder of diameter 150 mm and height 200 mm is resting on its circumferential base point on the ground with an axis inclined at 45° to the HP and parallel to the VP. Draw its projections.

**12.3** A right circular cone of diameter 150 mm and height 150 mm is resting on its circumferential base point on the ground with an apex 125 mm above the ground and axis parallel to the VP. Draw its projections.

**12.4** A right circular cylinder of diameter 150 mm and height 200 mm is resting on its curved surface on the ground with the axis inclined 45° to the VP and parallel to the HP. Draw its projections.

**12.5** A triangular prism of base edge 150 mm and height 200 mm is resting on one of its rectangular faces on the ground with axis inclined 30° to the VP and parallel to the HP. Draw its projections.

**12.6** A square pyramid, base edge 100 mm and height 150 mm is resting on one of its base corners on the ground with a slant edge passing through the resting corner, in vertical position and axis parallel to the VP. Draw its projections.

**12.7** A hexagonal prism, base edge 100 mm and height 150 mm is resting on one of its base edges on the ground with base inclined at 20° to the ground and axis parallel to the VP. Draw its projections.

**12.8** A triangular prism, base edge 150 mm and height 225 mm is resting on one of its base edges on the ground with an axis inclined 40° with the ground and parallel to the VP. Draw its projections.

**12.9** A pentagonal pyramid, base edge 100 mm and height 200 mm is resting on one of its base edges on VP. Draw its projections if an axis makes 35° to the VP and parallel to the HP.

**12.10** A right circular cone of diameter 150 mm and height 150 mm is held in the III-quadrant with an axis 45° to the VP and parallel to the HP. Draw its projections if an apex is towards the observer.

**12.11** A regular tetrahedron, face edge 100 mm is resting on one of its face edges with a face associated with resting edge in a vertical position and perpendicular to the VP. Draw its projections.

## Exercise C (Transfer of Points)

**12.1** Transfer the points from one view to another in problems shown in Plate 12.6.

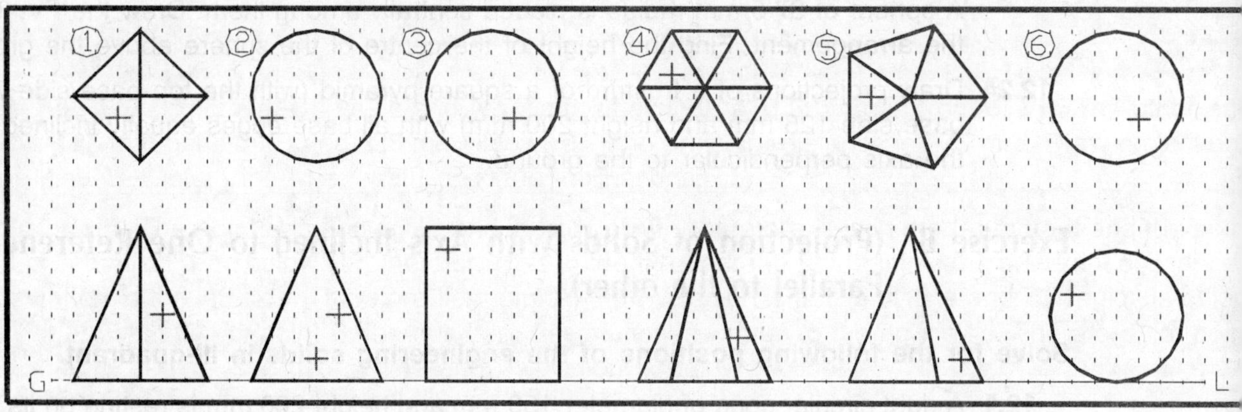


**PLATE 12.6**

# 13

# Sections of Solids

## Objectives

- To learn the concepts of section plane and sectional views.
- To know how to obtain the typical shape of a cut section from an engineering solid.

## Introduction

Invisible features of an object are shown by dotted lines in their projected views. However, when such features are too many, then the number of hidden lines makes the views more complicated and difficult to interpret. In such cases, it is customary to imagine the object as being cut by a cutting plane or section plane. The part of the object between the cutting plane and the observer is assumed to be removed and the view is then shown as a sectional view. This will help us in better understanding of an object.

## 13.1 Section and Section Planes

The surface produced by cutting the object by the section plane is called the section. The section is represented by thin section lines, uniformly spaced and inclined at 45°, to the cutting plane. The imaginary plane is called a section plane or a cutting plane.

The projection of the section along with the remaining portion of the object is called a sectional view. Sometimes the word section is also used to denote a sectional view. If the plan represents the section along with the remaining object, then this plan is called a sectional plan or sectional top view. And if the elevation represents the section along with the remaining object, then this elevation is called a sectional elevation or sectional front view.

The section plane is represented by a cutting plane line. The arrows represent the direction of an observer.

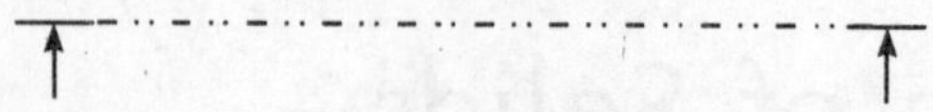

Figure 13.1(a) shows an isometric view of a solid cylinder with its orthographic projections, before the cutting plane (A-A-A-A) cuts the same passing through its vertical diameter. Figure 13.1(b) shows an isometric view of a cut portion of the cylinder with hatched lines and orthographic views with cutting plane and sectional FV.

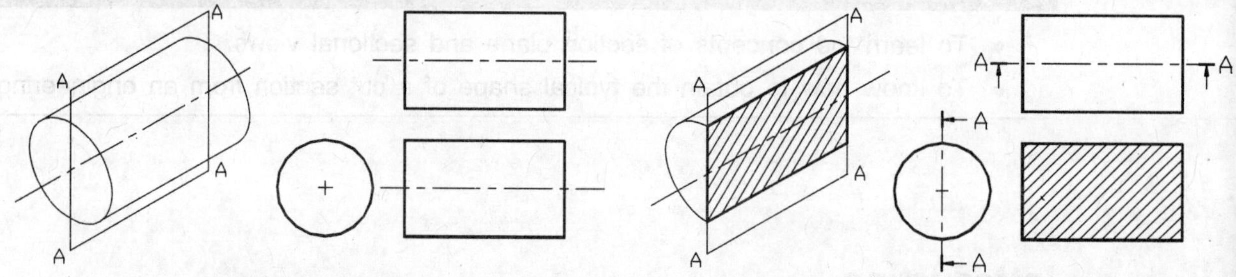


(a) Before section (b) After section

**Fig. 13.1** Section of a solid cylinder.

Figure 13.2(a) shows an isometric view of a hollow cylinder with its orthographic projections before the cutting plane (A-A-A-A) cuts the same passing through its vertical diameter. Figure 13.2(b) shows an isometric view of a cut portion of the cylinder with hatched lines and orthographic views with cutting plane and sectional FV.

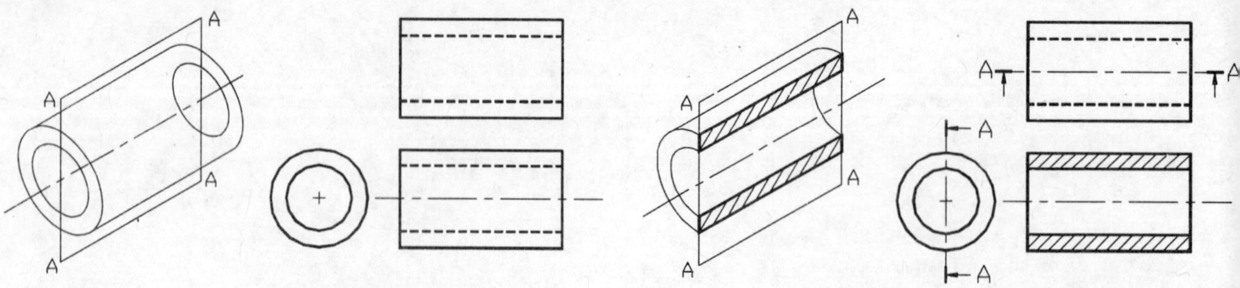


(a) Before section (b) After section

**Fig. 13.2** Section of a hollow cylinder.

Figure 13.3 shows isometric and orthographic views of a complicated cylindrical machine element. The figure is self explanatory for the interpretation.

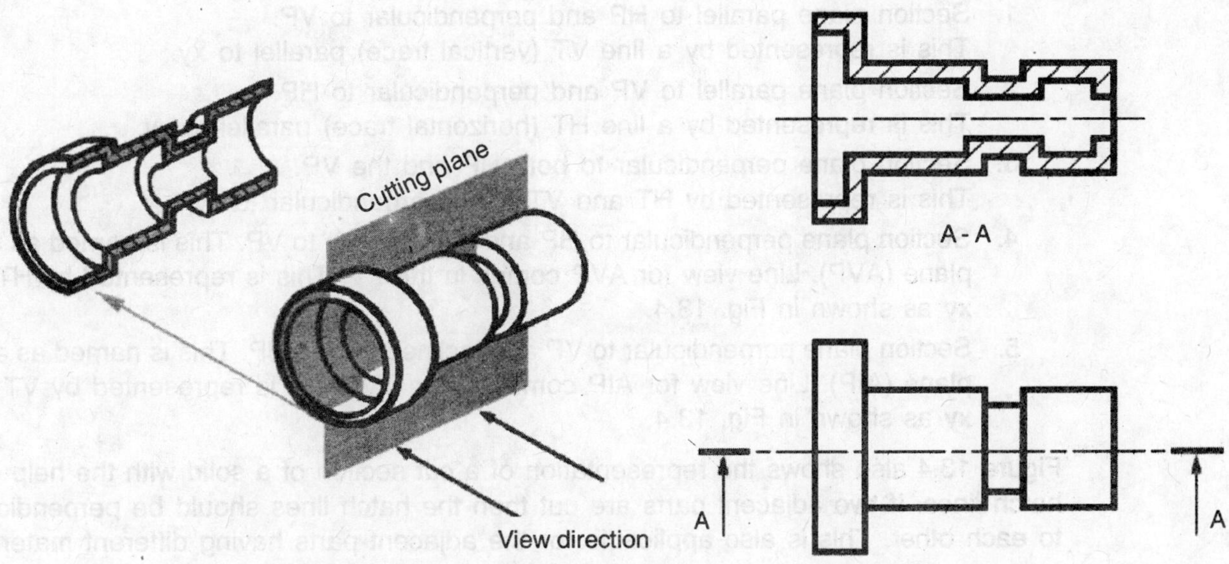


**Fig. 13.3** Cut section of a cylindrical machine element.

## 13.2 Types of Section Planes

Section planes are generally perpendicular planes. They may be perpendicular to one of the reference planes and either perpendicular, parallel or inclined to the other planes. They are usually described by their traces. It is important to remember that the projection of a section plane, on the plane to which it is perpendicular, is a straight line called a trace line. The trace line can be parallel, perpendicular or inclined to xy, depending upon the orientation of section plane with respect to reference planes. Figure 13.4 shows the various types of section planes.

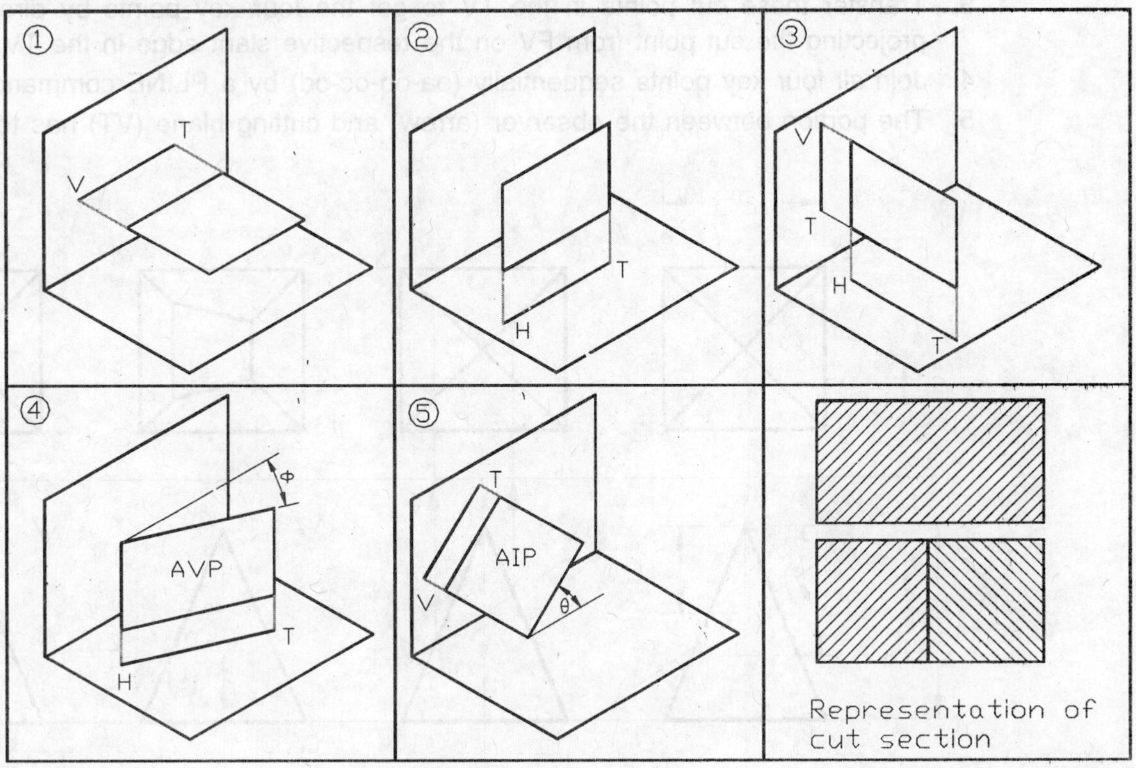


**Fig. 13.4** Types of section planes.

1. Section plane parallel to HP and perpendicular to VP.
   This is represented by a line VT (vertical trace) parallel to xy.
2. Section plane parallel to VP and perpendicular to HP.
   This is represented by a line HT (horizontal trace) parallel to xy.
3. Section plane perpendicular to both HP and the VP.
   This is represented by HT and VT, both perpendicular to xy.
4. Section plane perpendicular to HP and inclined ($\phi°$) to VP. This is named as auxiliary vertica plane (AVP). Line view for AVP comes in the FV. This is represented by HT inclined ($\phi°$) tc xy as shown in Fig. 13.4.
5. Section plane perpendicular to VP and inclined ($\theta°$) to HP. This is named as auxiliary inclinec plane (AIP). Line view for AIP comes in the TV. This is represented by VT inclined ($\theta°$) tc xy as shown in Fig. 13.4.

Figure 13.4 also shows the representation of a cut section of a solid with the help of 45° universa hatch lines. If two adjacent parts are cut then the hatch lines should be perpendicular (45°–135° to each other. This is also applicable for the adjacent parts having different materials.

**EXAMPLE 13.1** A square pyramid, base edge 125 mm and height 150 mm is standing on its base on the ground with two base edges parallel to VP. It is cut by an AIP, which makes an inclinatior of 35° to the ground and passing through a point on the axis 60 mm down the apex. Draw a sectiona view.

***Solution***

1. Draw projections (TV and FV) of solid in given position. Name the base as abcd and apex as o.
2. Locate a cutting plane as a line view VT, inclined 35° to the ground and passing through a point on the axis 60 mm down the apex. This VT cuts the slant edges o′a′, o′b′, o′c′ and o′d in the FV. Therefore, in total there are four cut points.
3. Transfer these cut points in the TV to get the four key points by direct transfer, i.e., by projecting the cut point from FV on the respective slant edge in the TV.
4. Join all four key points sequentially (oa-ob-oc-od) by a PLINE command.
5. The portion between the observer (arrow) and cutting plane (VT) has to be removed.

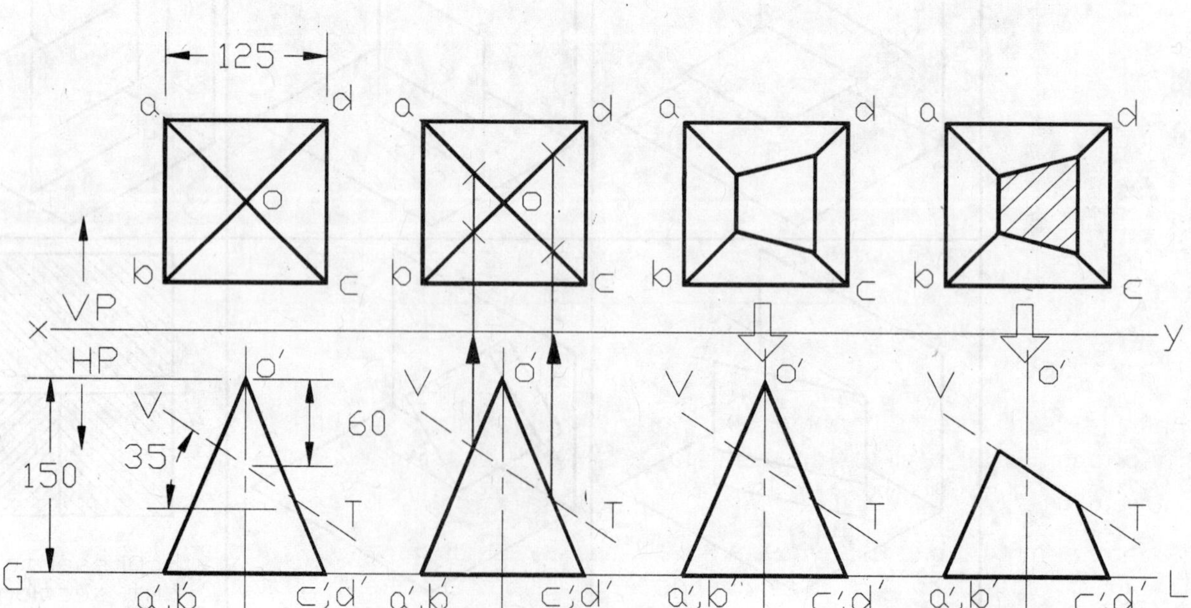

6. As we observe the apex is being cut, part of all four slant edges has to be removed inside the polyline boundary. Trim the inside portion of the slant edges by clicking the boundary defined by PLINE command.
7. HATCH the portion with 45°, universal pattern and 0.125 spacing.

**EXAMPLE 13.2** A square pyramid, base diagonal 150 mm and height 175 mm is standing on its base on the ground with all four base edges equally inclined to VP. It is cut by an AVP, which makes an inclination of 35° to the VP and passing through a point on the closest slant edge, at a top view distance of 25 mm from the apex. Draw a sectional view.

***Solution***

1. Draw projections (TV and FV) of solid in given position. Name the base as abcd and apex as o.
2. Locate a cutting plane as a line view HT, inclined 35° to VP and passing through a point on the slant edge ob 25 mm down the apex in the TV. This HT cuts the base edges: ad, bc and slant edges oa, ob in the TV. Therefore, in total there are four cut points.

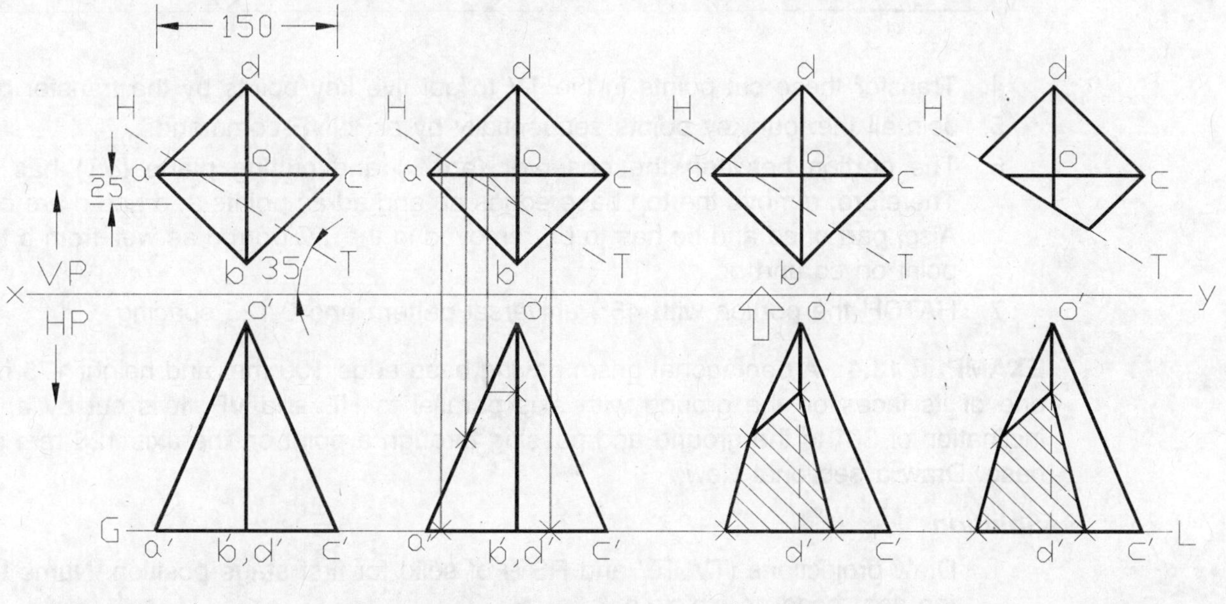


3. Transfer these cut points in the FV to get four key points by transfer procedure. The cut points on ad, oa and bc can be directly transferred. However, line parallel to base method is used for the transfer of cut point on ob.
4. Join all the four key points sequentially by a PLINE command.
5. The portion between the observer (arrow) and cutting plane (HT) has to be removed. As we observe, the base corners a and b are being cut, they have lost their identity. Trim the lines starting from a and b till the key points.
6. Now draw a separate line for segment of line od, which was trimmed in step (5). CHANGE into hidden line, as it is part of od, which is not visible.
7. HATCH the cut portion with 45°, universal pattern and 0.125 spacing.

**EXAMPLE 13.3** A square prism having base diagonal 125 mm and height 150 mm is standing on its base corner on the ground with axis inclined at 45° to the ground. It is cut by an AIP parallel to and 125 mm away from the ground. Draw a sectional view.

***Solution***

1. Draw projections (TV and FV) of solid with axis perpendicular to the ground and line joining the corner c to centre of the base parallel to xy. Name the top base and the bottom base as abcd.

2. COPY the FV and bring it in FV region of Stage II. ROTATE the view about corner c, suc that the axis makes 45° to the ground.
3. Locate a cutting plane as a line view VT, parallel to and 125 mm away from the ground. Thi VT cuts vertical edges: a′a′, b′b′, d′d′ and base edges: b′c′, c′d′ in the FV. In the F therefore, in total there are three cut points.

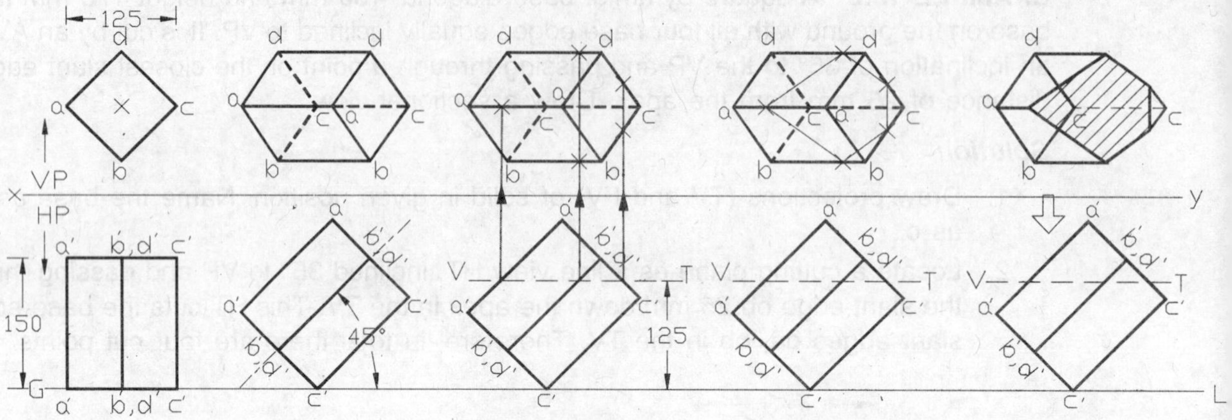


4. Transfer these cut points in the TV to get five key points by the transfer procedure.
5. Join all the four key points sequentially by a PLINE command.
6. The portion between the observer (arrow) and cutting plane (VT) has to be remove Therefore, remove the top base edges ab and ad as points a, b and d are cut in TV and F Also, part of cd and bc has to be removed in the TV from d as well from b till respective ke point on cd and bc.
7. HATCH the portion with 45°, universal pattern and 0.125 spacing.

**EXAMPLE 13.4** A pentagonal prism having base edge 100 mm and height 175 mm, is resting o one of its faces on the ground with axis parallel to HP and VP. It is cut by an AIP making a inclination of 30° to the ground and passing through a point on the axis 125 mm away from eithe base. Draw a sectional view.

***Solution***

1. Draw projections (TV, FV and RSV) of solid for first stage position. Name the left base an the right base as a′b′c′d′e′.
2. Locate a cutting plane as a line view VT, 30° to the ground passing through a point on th axis 125 mm away from the left base on vertical edge, which is 100 mm above the groun This VT cuts the base edges: ae, de, ab, cd and vertical edges: aa, dd in the FV. In the F in total there are three cut points.
3. Transfer the cut points on the vertical edges a′a′ and d′d′ from FV to the respective edge in the TV to get two key points by direct transfer procedure.

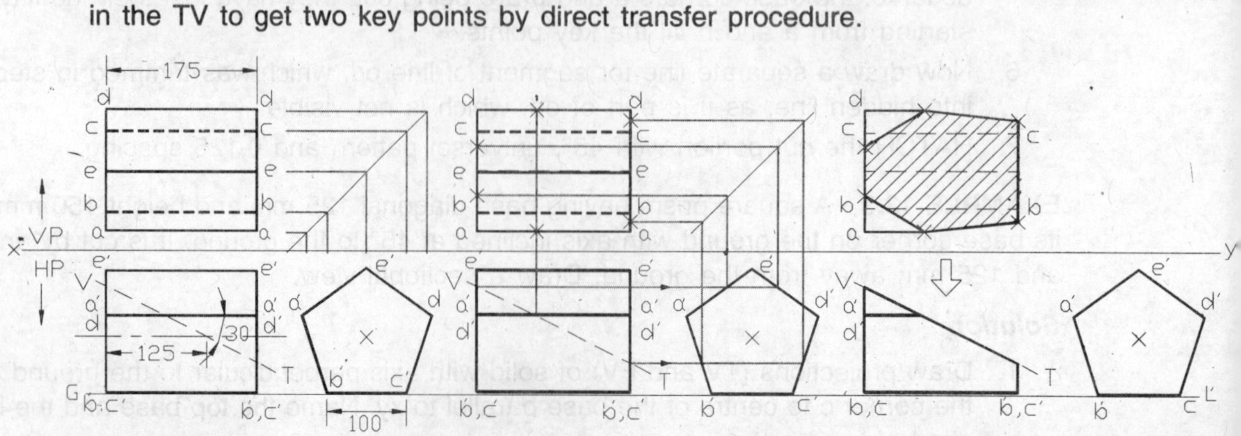

4. Transfer the cut points on the left and right base to the true shape of the base in the RSV to get their true location. Then by a 45° line procedure, these cut points are transferred on the respective base edges to get four key points in the TV.
5. Join all six key points sequentially by a PLINE command.
6. The portion between the observer (arrow) and the cutting plane (VT) has to be removed. We observe that the vertical edge ee is completely removed. Also, the points a and d of the right base have lost their identity. TRIM the lines starting from a and d till the key points in the sectional view.
7. HATCH the portion with 45°, universal pattern and 0.125 spacing.

**EXAMPLE 13.5** A hexagonal pyramid having base edge 100 mm and height 255 mm is standing on its base on the ground with its alternate base edges equally inclined to VP. It is cut by two AIPs. One AIP is parallel to the ground and another is inclined 60° to the ground. These two cutting planes meet each other as shown in the figure. Draw a sectional view.

***Solution***

1. Draw projections (TV and FV) of solid in a given position. Name the base as abcdef and apex as o.
2. There are two cutting planes VT and V′T′. The VT cuts the slant edges oa, ob, of and V′T′ cuts the slant edges oc, od, oe. The line views of both the cutting planes VT and V′T′ meet at a point on the slant surface obc in front and slant surface oef at the back. Therefore, in total there are five cut points.
3. Transfer the cut points on the slant edges directly from FV to the respective slant edges in the TV to get six key points. The cut point on slant surface is transferred by a slant line method to get two key points in the TV, one on slant surface obc and another on slant surface oef.
4. Join all the eight key points sequentially (oa-ob-surface point-oc-od-oe- surface point-of) by a PLINE command. The key points on the front slant surface (OBC) and on the back slant surface (OBF) in the TV are to be joined as it represents an intersection of two cutting planes: VT and V′T′.
5. The portion between the observer (arrow) and the cutting planes (VT and V′T′) has to be removed.
6. As we observe the apex is being cut, part of all six slant edges has to be removed inside the polyline boundary. TRIM the inside portion of the slant edges by clicking the boundary defined by PLINE.
7. HATCH the portion with 45°, universal pattern and 0.125 spacing.

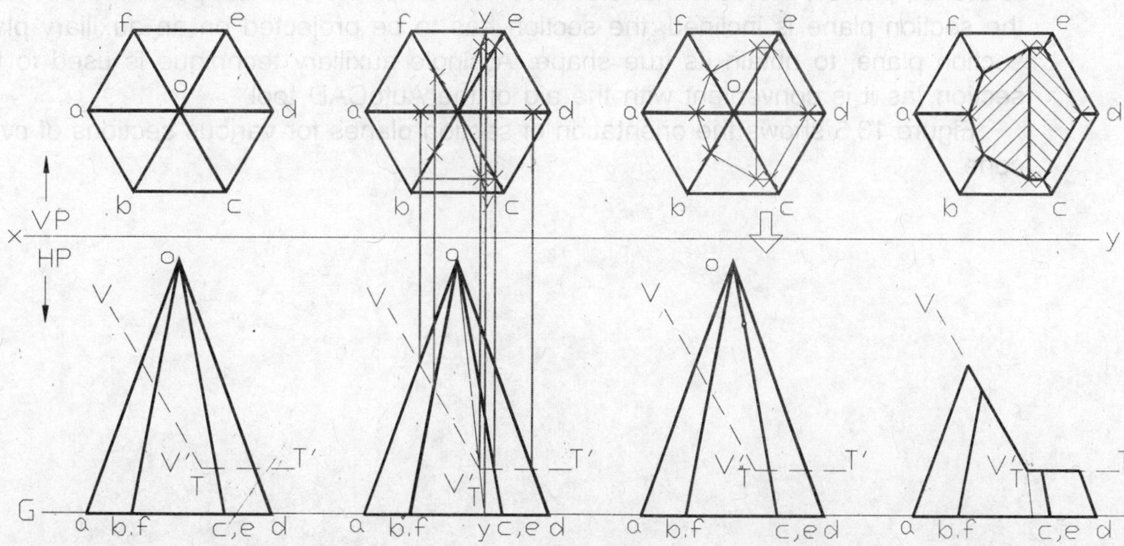

Plate 13.1 shows few more examples on sections of solids. The stepwise procedure is self explanatory.

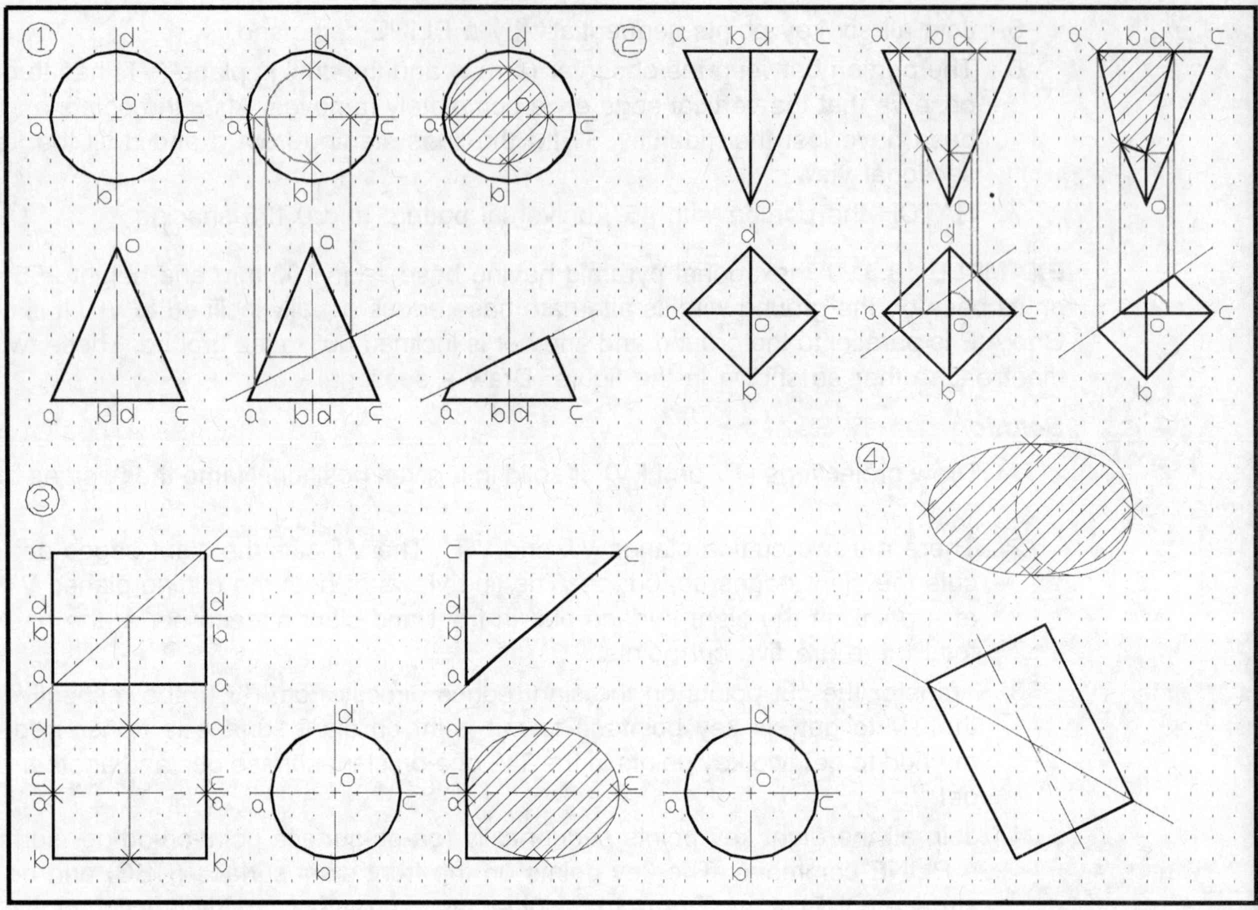

PLATE 13.1

## 13.3 True Shape of a Section

The projection of a cut section on a plane parallel to the section plane will show a true shape (TS) of the section. Therefore, when the section plane is parallel to the HP (or the ground), true shape of the section will be visible in the sectional TV. When it is parallel to the VP, the true shape of the section will be visible in the sectional FV. When the section plane is perpendicular to both the reference planes, the sectional side view will show the true shape of the section. However, when the section plane is inclined, the section has to be projected on an auxiliary plane parallel to the section plane, to obtain its true shape. A single auxiliary technique is used to find the TS of the section, as it is convenient with the aid of the AutoCAD tool.

Figure 13.5 shows the orientation of section planes for various sections of cylinder, sphere and cone

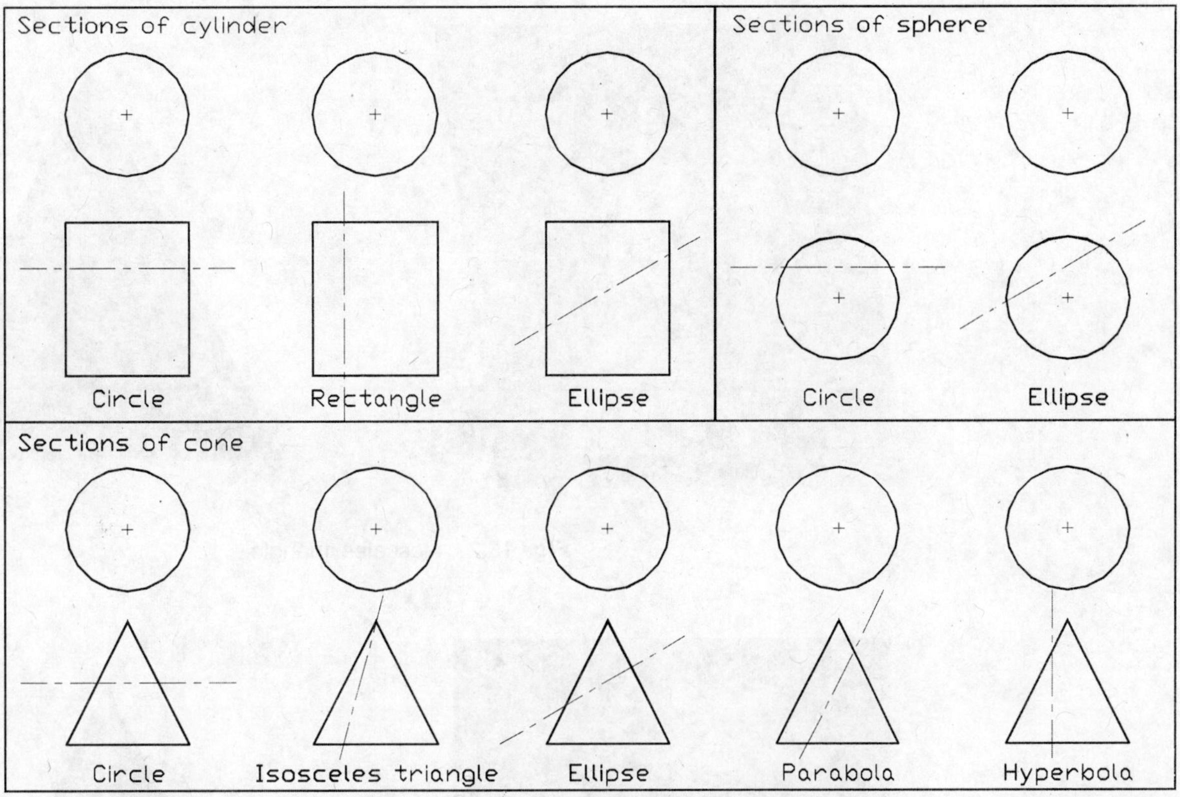


**Fig. 13.5** Section plane orientation of true shape.

The following figures show various sections of a cone (Figs. 13.6–13.9) and cylinder (Fig. 13.10) obtained by using a Pro-Engineer software.

**Fig. 13.6** Parabola.

Fig. 13.7 Isosceles triangle.

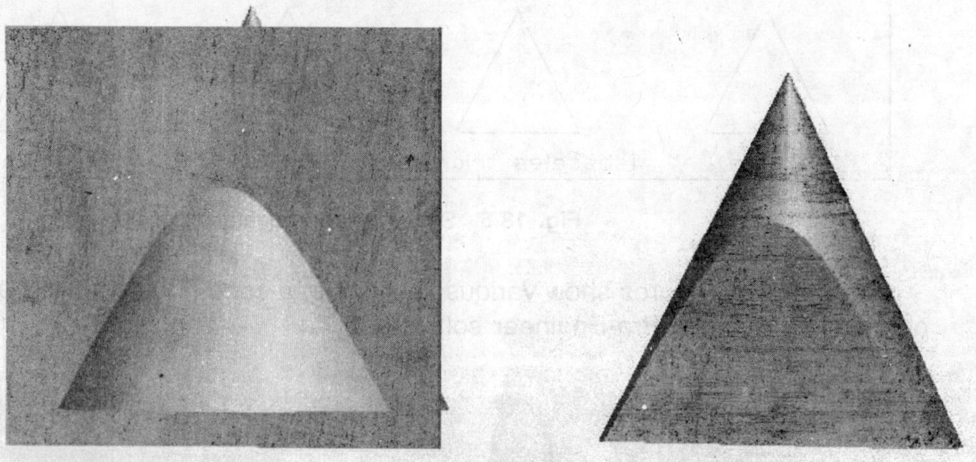

Fig. 13.8 Hyperbola.

Fig. 13.9 Ellipse from cone.

Fig. 13.10 Ellipse from cylinder.

## Exercise A (Section of Solids)

**13.1** A cube of 150 mm long edges is resting on one of its faces on the ground with vertical faces equally inclined to the VP. It is cut by an AIP 45° to the HP and passing through the mid point of axis. Draw a sectional view.

**13.2** A hexagonal pyramid, base edge 100 mm and height 150 mm is lying on the ground on one of its triangular faces with the axis parallel to the VP. It is cut by a vertical section plane (AVP) inclined 30° to the VP and passing through the centre of the base. Draw a sectional view.

**13.3** A square pyramid, base edge 100 mm and height 200 mm is resting on one of its triangular faces on the ground with axis parallel to the VP. It is cut by a horizontal section plane, which intersects the axis 75 mm from the ground plane. Draw a sectional view.

**13.4** A square pyramid base edge 100 mm and height 200 mm is resting on one of its triangular faces on the ground. It is cut by a section plane passing through the centre of the gravity (CG) of the solid and inclined 30° to the VP. Draw a sectional view if an apex is being retained.

**13.5** A cone of 150 mm base diameter and height 200 mm rests on the ground on its circumferential base point with axis 60° to the HP and parallel to the VP. The cone is cut by a horizontal section plane passing through a mid-point of the axis. Draw a sectional view.

**13.6** A sphere of 200 mm diameter is resting on the ground. It is cut by a section plane perpendicular to the VP and inclined 45° to the HP and passing through the top most point of the sphere. Draw a sectional view and true shape (TS) of the section.

**13.7** A hexagonal pyramid, base edge 100 mm and height 175 mm long is resting on its base on the ground with one base edge parallel to the VP. Draw its projections. It is cut by two section planes: $V_1T_1$ and $V_2T_2$ meeting together on an axis 50 mm from the base. The section plane $V_1T_1$ is 45° to the base and $V_2T_2$ is parallel to the base. Draw a sectional top view if a common smaller portion of the solid, contained by two section planes, has to be removed.

**13.8** A pentagonal pyramid, base edge 125 mm and height 175 mm long is resting on its base on the ground with one base edge perpendicular to the VP. Draw its projections. It is cut by a section plane parallel to the VP and 37.5 mm from an apex towards the observer. Draw a sectional elevation.

**13.9** A square prism, base diagonal 125 mm long and height 175 mm is resting on one of its base corners on the ground with an axis inclined 45° to the ground. Draw its projections. It is cut by a section plane making an angle 30° to the VP and passing through a centre of one of the base. Draw a sectional view.

**13.10** A pentagonal prism base edge 100 mm and height 200 mm long is resting on one of its faces on the ground with axis parallel to the HP and VP. An AIP inclined at 30° to the ground cuts the solid at a point on the axis 50 mm from either of the bases. Draw a sectional plan if the smaller portion is to be removed.

**13.11** The following figure shows the FV of a cylinder of 200 mm diameter with a hemisphere of 75 mm radius centrally placed on a top base of the cylinder. The formed composite solid is cut by two cutting planes: AB and PQ inclined 45° to the ground and passing through a top most point of the sphere. Draw a sectional TV.

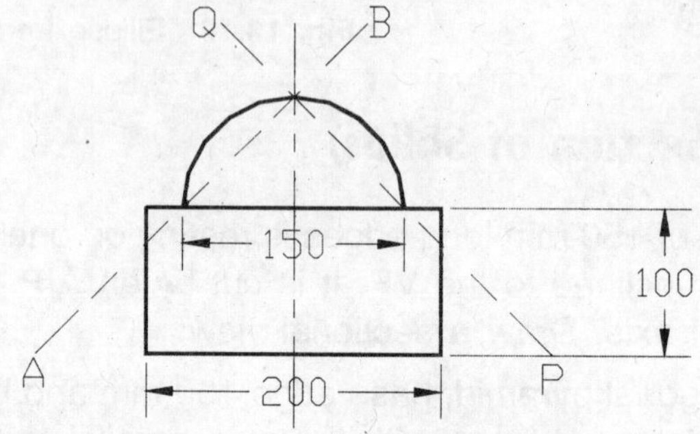


**13.12** A square prism of base diagonal 150 mm and height 200 mm is standing on the ground with its adjacent faces equally inclined to the VP. A sphere of diameter 150 mm is placed centrally on the top base of the prism. Both the solids are cut by an AIP at 45° and passing through a common touching point of these two solids. Draw the projections of the combined solids. Locate the position of the section plane. Draw its sectional view and true shape of the cut section.

## Exercise B (True Shape of Section)

In this exercise, students are asked to orient a section plane in order to obtain a typical true shape of the cut section as discussed in section 13.3. In order to orient a section plane, a general rule is given below.

***Note:*** The base of the TS of the section is to be adjusted in the TS of the base of the solid and the height of the TS of the section is to be adjusted in the other view of the solid.

**13.1** A square pyramid, base edge 150 mm and height 200 mm stands vertically on the ground on its base with base edges equally inclined to the VP. The solid is cut by a section plane in such a way that a sectional TV appears as an isosceles triangle with base equal to 100 mm. Locate the section plane and draw a sectional TV.

**13.2** A cube of 150 mm long edges is resting on one of its faces on the ground with four vertical faces equally inclined to the VP. It is cut by an AIP passing through the body diagonal. Draw a TS of the section using the single auxiliary method.

**13.3** A cube of 150 mm sides is resting on one of its faces with four vertical faces equally inclined to the VP. It is cut by an AIP so that a true shape of the section is a regular hexagon, having distance between two opposite corners equal to the length of the base diagonal of the solid. Find the inclination of the AIP with the ground. Draw a sectional TV as well as a TS of the section.

**13.4** A solid crucible, height 200 mm has a top square base of 200 mm side and bottom square base 100 mm side. Draw its projections when it is resting on the ground on its bottom base with one of its base edges parallel to the VP. It is cut by a vertical section plane, parallel to the profile plane such that the TS of the section is a trapezium with parallel sides 200 mm and 150 mm. Locate the section plane and draw a TS of the section. Find the distance between the parallel sides of the trapezium.

**13.5** A cylinder of 150 mm diameter and 200 mm height rests on the ground on one of its bases. It is cut by an AIP passing through the mid-point of the axis and 45° to the HP. Draw a sectional TV and TS of the section.

**13.6** A cylinder of 150 mm diameter and 200 mm height rests on the ground on one of its bases. It is cut by an AIP such that the TS of the section is a rectangle of 200 mm × 100 mm in the side view. Locate the section plane and find out the inclination of the AIP with the HP. Draw TS of the section.

**13.7** A triangular pyramid base edge 150 mm and height 200 mm stands on its base on the ground with one of the base edge perpendicular to the the VP. It is cut by an AIP such that a sectional TV appears as a trapezium with parallel sides as 100 mm and 50 mm. Locate the section plane and find the inclination of the AIP with the HP. Draw a sectional TV. (Note: The original TV and FV can be moved to adjust the parallel sides of the trapezium).

**13.8** A triangular prism base edge 150 mm and height 200 mm stands on its base on the ground with one of the faces perpendicular to the VP. It is cut by an AIP such that a TS of the section is an isosceles triangle of base 100 mm and maximum height. Locate the section plane and find the inclination of AIP with HP. Draw a sectional TV. (Note: The original TV and FV can be moved to adjust the parallel sides of the trapezium).

**13.9** A cone 150 mm diameter and 200 mm height stands on its base on the ground. It is cut by an AIP such that a TS of the section is an isosceles triangle with base 100 mm. Locate the AIP. Draw TS of the section and find the height of the isosceles triangle.

**13.10** A semi cylinder of 150 mm diameter and 200 mm height rests on its flat face on the ground with its base parallel to the VP. It is cut by a vertical section plane parallel to the PP such that a TS of the section is:

(a) A rectangle with maximum dimensions (length and height)

(b) A rectangle with maximum length but height equal to half of the maximum height. Draw a sectional view.

**13.11** A sphere with 200 mm diameter is resting on the ground. It is cut by an AIP 45° to the HP and passing through the top most point of the sphere. Draw a sectional TV and TS of a section.

**13.12** A cone of base diameter 200 mm is resting on its base on the ground. It is cut by an AIP such that a true shape of the cut section is an equilateral triangle with 175 mm side. Draw a sectional view. Draw a TS of the section by using an appropriate technique. Specify the height of the cone by using a DIM mode.

**13.13** A cone of base diameter 200 mm and height 150 mm is resting on its base on the ground. It is cut by an AIP such that a true shape of the cut section is an isosceles triangle with maximum base and minimum height. Draw a sectional view. Draw the TS of the section by using an appropriate technique. Specify the height of the isosceles triangle by using a DIM mode.

**13.14** A pentagonal pyramid (base edge 125 mm and height 250 mm) is resting on its base on the ground with one of the base edge perpendicular to the VP. It is cut by an AIP such that a true shape of the cut section is an isosceles triangle with maximum base and minimum height. Draw a sectional view, if the larger portion of the solid is removed. Draw the TS of the section by using an appropriate technique. Specify the base and height dimension of the isosceles triangle (TS) by using a DIM mode.

**13.15** A pentagonal pyramid (base edge 125 mm and height 225 mm) is resting on its base on the ground with one of the base edges perpendicular to the VP. It is cut by an AIP such that a true shape of a cut section is an equilateral triangle with maximum side. Draw a sectional view by removing the smaller portion of the solid. Draw the TS of the section by using an appropriate technique. Specify the base and height dimension of the equilateral triangle (TS) by using a DIM mode.

**13.16** A glass tumbler of height 225 mm is in the form of a frustum of a cone with top and bottom base diameters 200 mm and 100 mm, respectively. It is completely filled with water when it is resting on the ground on its bottom base. It is then tilted on the rim of the bottom base in such a way that the axis remains at 40° to the ground. Draw the projections of a glass tumbler when it is resting on the ground on its bottom base and when the axis remains at 40° to the ground. Show a water level in the elevation and the plane.

**13.17** A cylinder (diameter 140 mm and height 75 mm is resting on its base on the ground. A square prism (base edge 100 mm and height 75 mm) is centrally placed vertically on the top base of the cylinder with one rectangular face parallel to the VP. Draw the projections of the combined solids. Locate a cutting plane such that the sectional FV will be a rectangle of maximum dimensions. Draw the sectional FV. Mention the dimensions of the rectangle in mm.

**13.18** A hexagonal pyramid (base edge 75 mm and height 200 mm) is cut with a plane passing through its apex and any two opposite base corners. This half hexagonal pyramid is resting on its cut surface on the ground with apex farthest from the observer and base parallel to VP. Draw the projections of the half hexagonal pyramid.

It is further cut by an AIP such that the sectional TV appears as a right angle triangle with height equal to height of pyramid. Locate the cutting plane. Draw the sectional TV. Mention the base length of the right angle triangle in mm.

**13.19** A square pyramid (base edge = 150 mm and height = 250 mm) is standing on its base on the ground such that the two of its base edges are perpendicular to the VP. It is cut by two cutting planes: one parallel to extreme left slant face and the other parallel to extreme right slant face. Both these cutting planes intersect each other on an axis of the pyramid 100 mm below the apex. Draw projections of the solid. Locate these cutting planes. Draw sectional top view. Mention the dimensions of the cut section.

**13.20** Two frustums of a cone having same dimensions (smaller base diameter 75 mm, larger base diameter 150 mm and height 150 mm) are joined at their smaller bases such that their axes coincide. Draw the projections of a composite solid when it is standing on larger base on the ground with combined axis perpendicular to the ground. It is cut by an AIP such that the sectional view appears as a rhombus of maximum dimensions. Locate the section plane. Draw the sectional view. Specify the dimensions of the major and minor diagonals of the rhombus.

**13.21** A frustum of a cone has its bottom base diameter 200 mm, top base diameter 100 mm and height 200 mm is resting on its bottom base on the ground. It is cut by an AIP in such a way that the true shape of a section appears as an ellipse withmaximum dimension of the major axis. Draw the projections of solids. Locate the section plane. Draw the sectional view. Specify the major and minor axes dimension using DTEXT command.

**13.22** A frustum of a cone (bottom base-diameter 200 mm, top base-diameter 100 mm and height 200 mm) is cut by a vertical plane passing through top base and bottom base diameters. Draw the projections of this half frustum with cut surface exposed to the direction of observer and standing on its bigger base on ground.

It is further cut by an AVP such that true shape of a cut section is a figure which if mirrored one can obtain the original FV. Locate the section plane. Draw the sectional view.

**13.23** A combination of a hexagonal pyramid (base edge = 100 mm, height = 150 mm) and a cone (height = 150 mm) is made in such a way that their bases meet each other with the base corners of the pyramid exactly touching a circumference of the cone. The combined solid is resting on the apex of cone on the ground with combined axis perpendicular to the ground. Draw the projections (TV and FV) of the combined solid. It is cut by an AIP such that a true shape of the section is a rhombus with maximum dimensions of major and minor diagonals. Locate the cutting plane. Draw a sectional view. Specify the dimensions of major and minor diagonals using DTEXT command.

**13.24** Frustums of two cones, similar in dimensions (height = 100 mm, bigger base diameter = 200 mm, smaller base diameter = 100 mm) are attached and aligned along their bigger bases. The combination of solid is kept on the ground on either of the smaller bases. Draw the projections (TV and FV) of the combined solid.

It is cut by an AIP such that the true shape of the section is two trapeziums having similar dimensions, with minimum distance between parallel sides and meeting at their bigger sides. Locate the cutting plane. Draw a sectional view. Find the dimensions of the parallel sides and height of the trapeziums.

**13.25** A combination of solid consists of the hexagonal pyramid (base edge = 100 mm, height = 150 mm) and a rectangular pyramid (height = 150 mm). The combination is made in such a way that their bases meet each other. The base of the hexagonal pyramid accommodates the complete base of the rectangular pyramid with two opposite base edges of rectangular pyramids aligned and matched with the base edges of the hexagonal pyramid. The combined solid is resting on the apex of the rectangular pyramid with combined axis perpendicular to the ground and adjacent faces of the hexagonal pyramid equally inclined to VP. Draw the projections (TV and FV) of the combined solid. It is cut by an AIP passing through the nearest and farthest base corner of hexagonal pyramid and 45° to the ground. Locate the cutting plane. Draw the sectional top view.

**13.26** A frustum of a cone of height 100 mm has 150 mm bottom base diameter of 150 mm and top base diameter of 100 mm. The frustum is standing on its bigger base on the ground. A cylinder of base diameter 100 mm and height 125 mm is placed centrally on the top base of the frustum. This combination of solids is cut by two section planes which meet at a common base to both the solids such that a cut section appears as a trapezium with one of the parallel sides 125 mm and a rectangle of B × H dimensions in side view.

Draw projections of both the solids. Locate the position of the section planes and find their inclination with the reference plane. Draw sectional top view and sectional side view. Find B and H dimensions of the rectangle, which appears in sectional side view.

**13.27** A frustum of a square pyramid of height 100 mm has bottom base diagonal of 150 mm and top base diagonal of 100 mm. The frustum is standing on its bigger base on the ground with base edges equally inclined to the VP. A sphere of diameter 100 mm is placed centrally on the top base of the frustum. Both the solids are cut by an AIP such that the true shape of a cut section appears as a triangle with 50 mm base and 75 mm height and a circle of diameter D. Draw projections of both the solids. Locate the position of the section plane and find its inclination with the reference plane. Draw a sectional top view and true shape of the section. Find the diameter D of the circle in the true shape of the section.

**13.28** Two identical cones of base diameter 150 mm are attached base to base. This combination is resting on the ground such that the common axis is perpendicular to the ground. These solids are cut by two cutting planes (meeting at their common base) such that the cut section appears to be an isosceles triangle of base 100 mm in the top view. Draw the projection of the combined solids. Locate the position of the cutting planes. Find out their inclination with the reference plane. Draw a sectional plane and sectional side view.

# 14

# Development of Surfaces of Solids

## Objectives

- To learn the theory and principles of development.
- To study the applications of development.
- To know the applications of reverse development.

## Introduction

A layout of the complete surface of an object is called a pattern. When the surfaces of a solic are laid out on a plane, the figure obtained is called development. The development of a solid thus, represents the actual shape of its lateral surfaces and base surfaces which, when ben or folded at the edges would form a hollow solid. Therefore, it is very important to note that every line on the development must be the true length of the corresponding edge on the surface. The surfaces are also true shapes and points at their true positions.

Many manufactured items can be made by rolling, folding or pressing. These operations require the use of accurately developed views. The knowledge of development of surfaces is essential particularly in sheet metal work. In construction of boilers, funnels, chimneys, packing boxes, buckets, air-conditioning ducts, space capsules, cake pans, furnace pipes etc. the plates are marked and cut according to the developments which, when folded, form the desired object

## 14.1 Theory of Development

Any solid can be enclosed in a wrapper of thin material, such as aluminium sheet metal. If this wrapper is opened at any folding edge and laid on a flat plane, the flattened out sheet metal is the development of complete surfaces of the solid. The development includes the lateral surfaces as well as base surfaces. Thus, when surfaces of a solid are laid out on a plane, the successive surfaces of the object are called its development.

Figure 14.1 shows a square prism enclosed with sheet metal. The development consists of four equal rectangular surfaces for the faces and two similar squares for its bases. Each figure shows the true size and shape of the corresponding surface of the prism. The development of a solid thus, represents the successive face surfaces and base surfaces with true area and shape. This development, when bent or folded at the vertical edges and base edges, would form a hollow square prism. It is very important to note that every line on the development must be the true length of the corresponding edge on the surface.

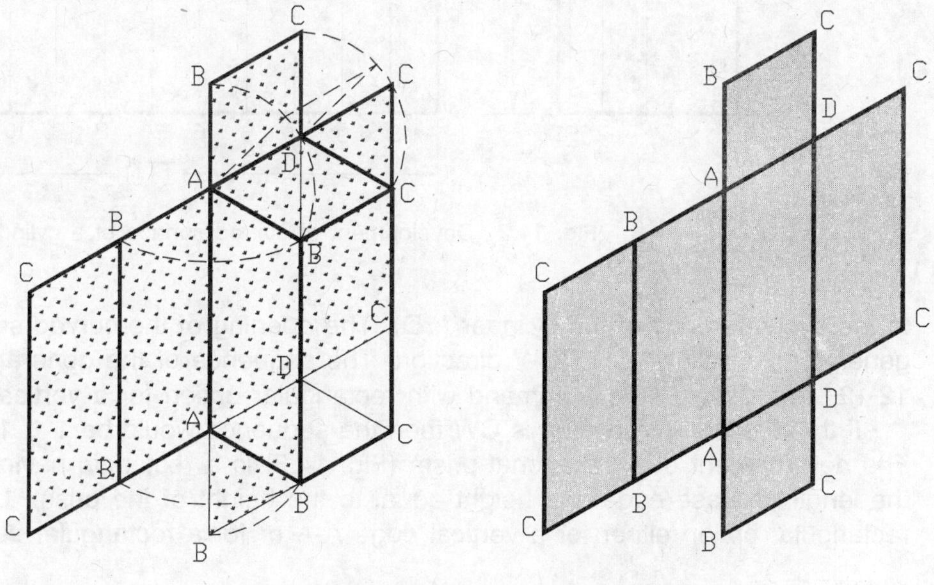

Fig. 14.1 Development of a square prism.

The flat surface of prisms and pyramids, and single curved surfaces like cones and cylinders can be accurately developed. However, the surface of the sphere cannot be accurately developed. The surface of the sphere is a double-curved surface which is undevelopable. The sphere can be approximately developed by dividing it into a number of parts.

Usually, only the lateral surfaces of the solids are developed and the ends or bases are omitted from the developments. The bases can be easily incorporated whenever required.

## 14.2 Methods of Development

The wrapper can be opened at any folding edge and in either clockwise (CW) or counter clockwise (CCW) direction. There are two principal methods to draw developments of surfaces of various solids:

### 14.2.1 Parallel-line Development

This method is used in case of cylinders and prisms, in which, edges of the top base and bottom base are considered parallel to each other. Rectangular ARRAY can be used to develop the surfaces. The development of a right circular cylinder (Fig. 14.2) is a rectangle having a width equal

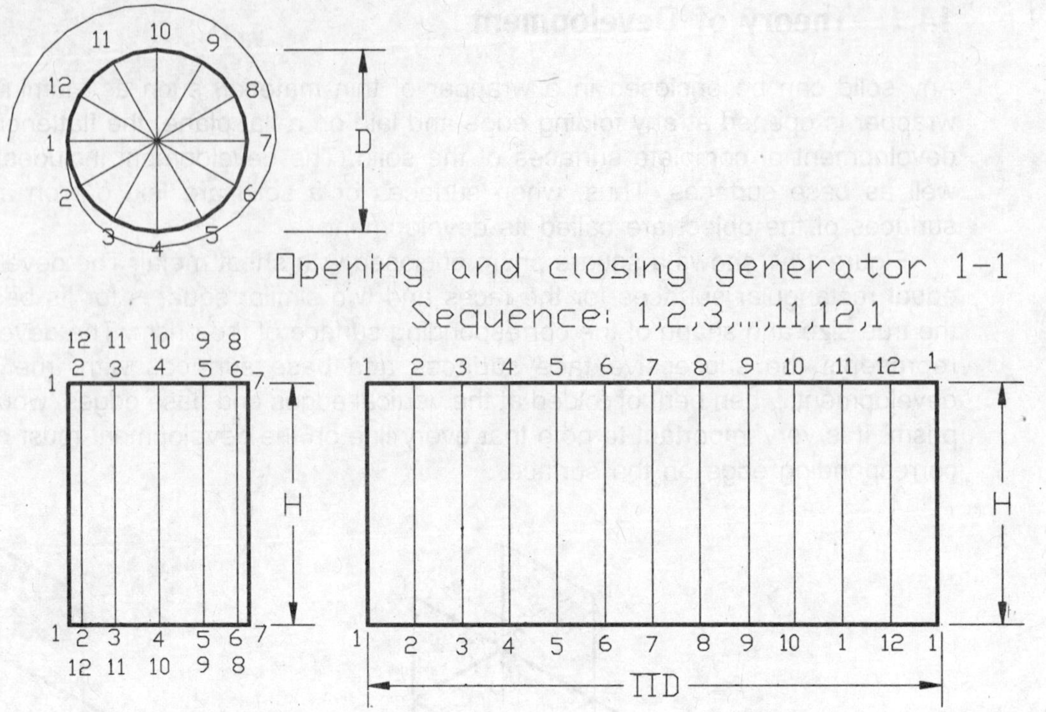


**Fig. 14.2** Development of curved surface of a cylinder.

to the circumference of the cylinder ($\pi$D). The opening of the curved surface is at the extreme left generator 1-1 following a CCW direction. The sequence of the generators is 1-1, 2-2, 3-3, 11-11, 12-12, 1-1. Use ARRAY command with rectangular option for a vertical generator 1-1.

If the direction of opening is CW then the sequence would be 1-1, 12-12, 11-11... 3-3, 2-2, 1-1. The development of a hexagonal prism (Fig. 14.3) is a rectangle having length equal to six times the length of base edge and height equal to the height of the prism. Use ARRAY command with rectangular option either for a vertical edge A-A or for a rectangular surface AABB.

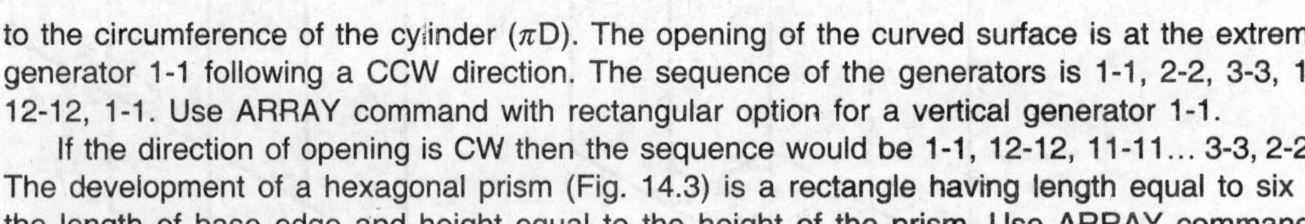


**Fig. 14.3** Development of lateral surfaces of a hexagonal prism.

### 14.2.2 Radial-line Development

This method is used in case of cones and pyramids, in which the true length of the generator or slant edge is used as the radius. Polar ARRAY can be used to develop the surfaces.

The development of a right circular cone (Fig. 14.4) is a sector of circle having a radius equal to the slant height S and arc length equal to the circumference of the base. The angle of the sector of circle is to be found by using equation $\theta = r/S \times 360°$. Draw an ARC with apex as a centre and radius equal to slant height S. Use MEASURE command to divide the length of arc into 12 equal parts. The segment distance to be given in MEASURE command is $S\theta/12$. The opening generator is 0-1 and the sequence is 0-1, 2, 3, ..., 11, 12, 1 following a CCW direction.

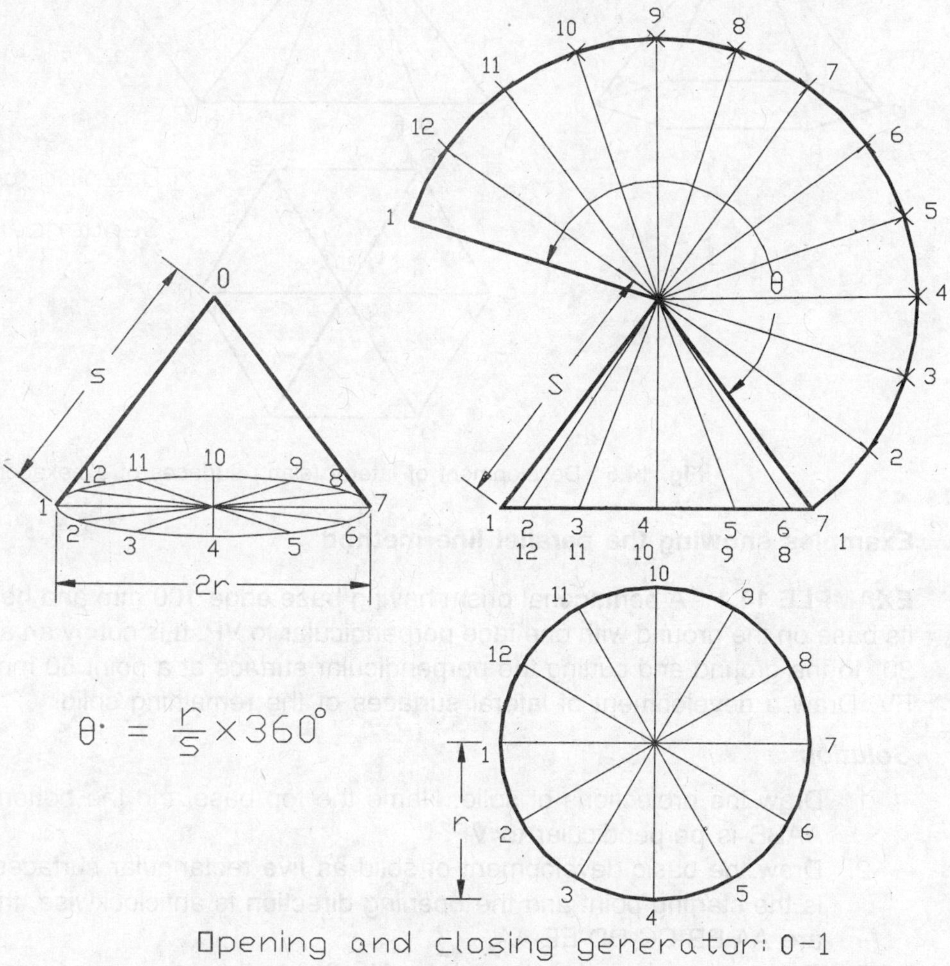


**Fig. 14.4** Development of curved surface of a cone.

The development of a hexagonal pyramid (Fig. 14.5) is six similar isosceles triangles having base equal to the base edge and two similar sides equal to the length of the slant height of the pyramid.

Draw an ARC with centre as an apex and radius equal to the true length of slant edge (od or oa). Other slant edges represent apparent lengths. Use MEASURE command to set up the segment distance of base edge drawn on the arc. Join apex with the division points in order to get isosceles triangular surfaces. Set the point with PDMODE 3 and PDSIZE as 0.10 before the use of MEASURE command. The opening edge is O-A with the sequence O-A, B, C, D, E, F, A following a CCW direction.

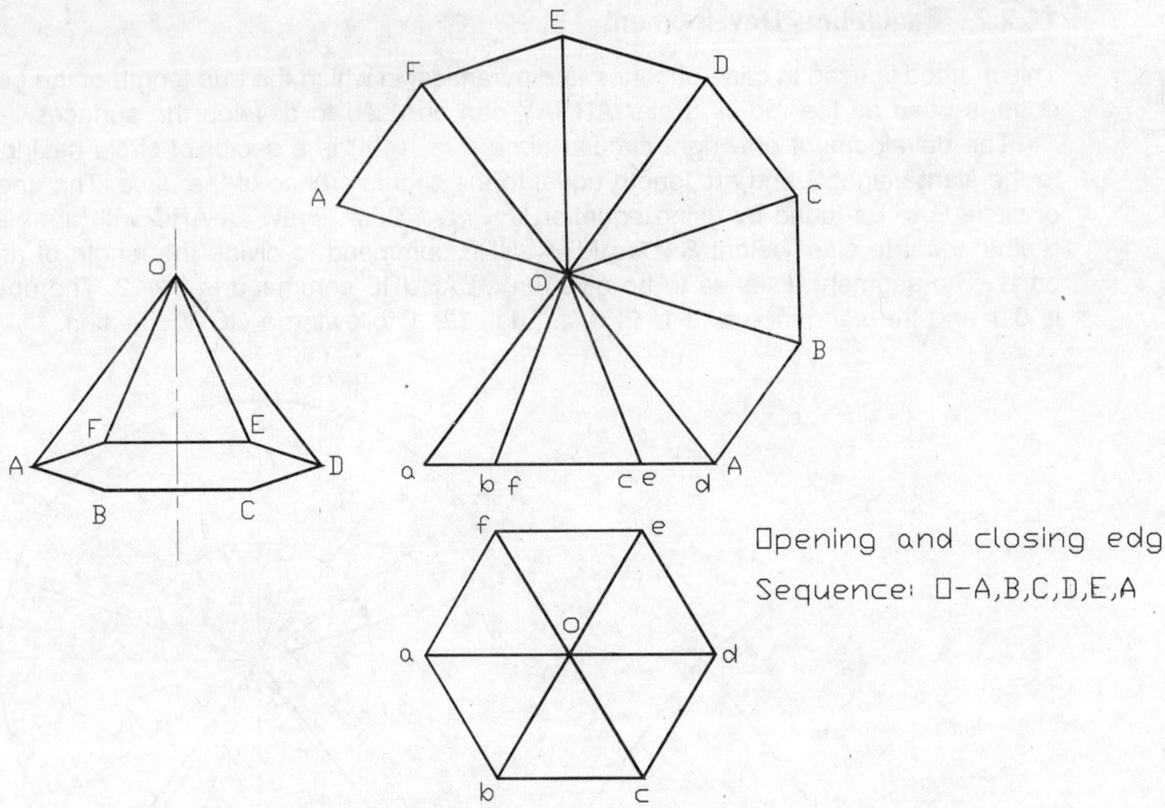


**Fig. 14.5** Development of lateral (slant) surfaces of a hexagonal pyramid.

### Examples showing the parallel line method

**EXAMPLE 14.1** A pentagonal prism having base edge 100 mm and height 175 mm is standing on its base on the ground with one face perpendicular to VP. It is cut by an auxiliary inclined plane (AIP) 30° to the ground and cutting the perpendicular surface at a point 50 mm below the top base in the FV. Draw a development of lateral surfaces of the remaining solid.

***Solution***

1. Draw the projections of solid. Name the top base and the bottom base as abcde. The face AABB is perpendicular to VP.
2. Draw the basic development of solid as five rectangular surfaces (175 mm × 100 mm). If a is the starting point and the opening direction is anticlockwise, then the development edges are AA-BB-CC-DD-EE-AA.
3. Draw a line view of cutting plane AIP 30° to the ground and passing through a point on the perpendicular surface AABB 50 mm below the top base in the FV. There are three cut points, which will give us six key points in the development. Five key points for five vertical edges and key point on edge AA is repeated twice as AA in closing as well as opening edge.

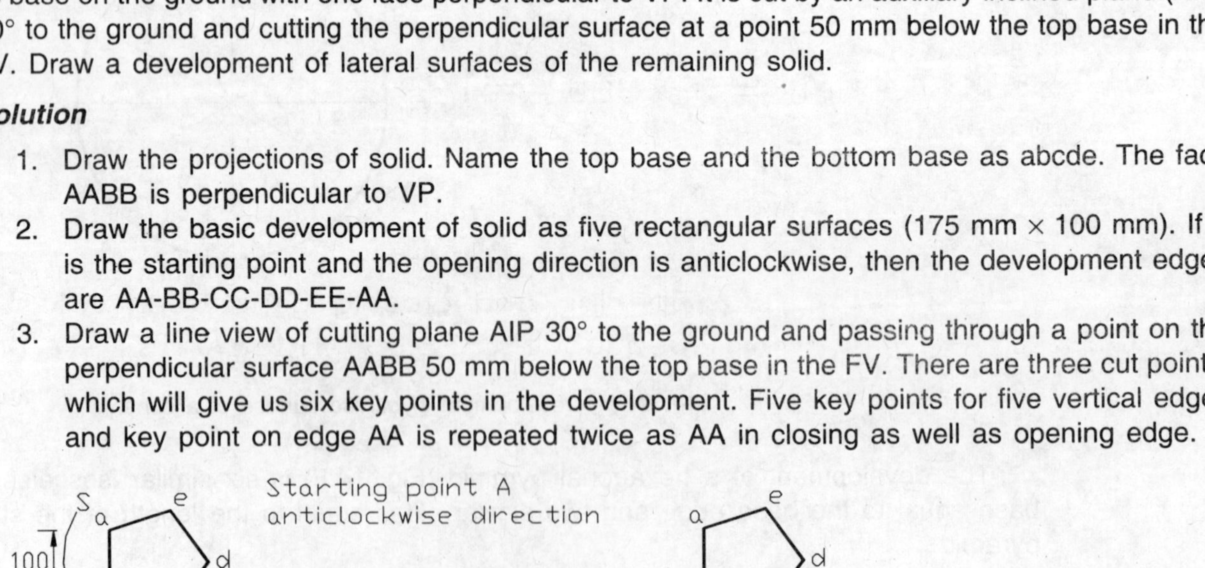

4. Transfer all the cut points from the FV to the development by a parallel line method.
5. Mark the points by using a POINT command and get the six key points.
6. Join all the key points by using a PLINE command.
7. Trim the portion of solid in the development.

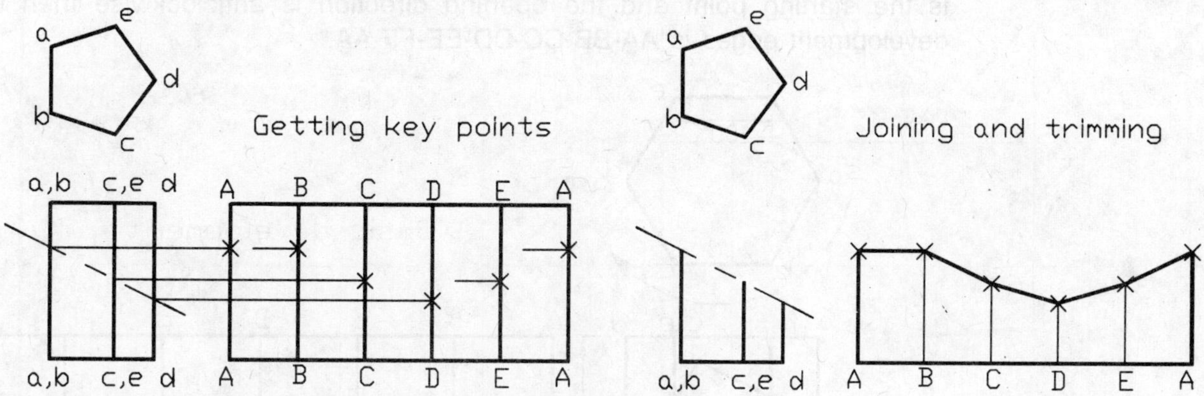


Figure 14.6 shows four cases of the same example.

Case-1 Starting point A, CW opening direction
Case-2 Starting point B, CCW opening direction
Case-3 Starting point C, CCW opening direction
Case-4 Starting point D, CCW opening direction

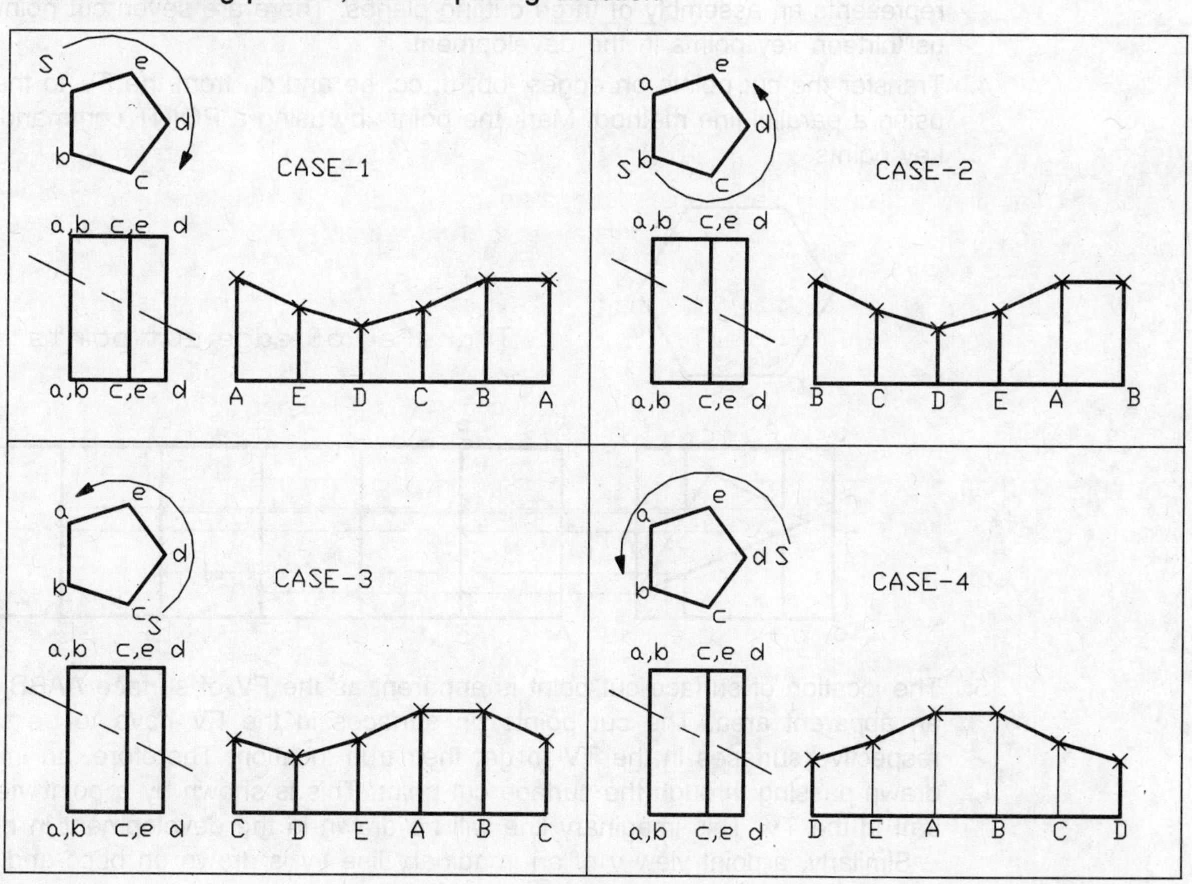


**Fig. 14.6**

**EXAMPLE 14.2** A hexagonal prism having base edge 100 mm and height 175 mm is standing on its base on the ground with alternate faces equally inclined to VP. A triangular hole is cut through the solid as shown in the figure. Draw the development of lateral surface of the remaining portion of the solid.

***Solution***

1. Draw the projections of hexagonal prism with axis perpendicular to the ground and alternate faces equally inclined to VP. Name the top base and the bottom base as abcdef.
2. Draw the basic development of solid as six rectangular surfaces (175 mm × 100 mm). If a is the starting point and the opening direction is anticlockwise then the sequence of development edges is AA-BB-CC-DD-EE-FF-AA.

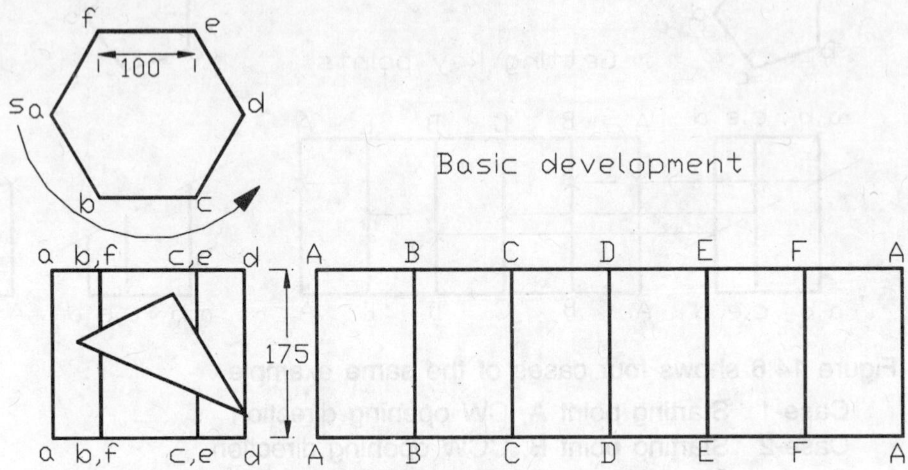


3. Draw a triangle by locating the corner points on grids in the FV. This triangular hole represents an assembly of three cutting planes. There are seven cut points, which will give us thirteen key points in the development.
4. Transfer the cut points on edges: bb, ff, cc, ee and dd from the FV to the development by using a parallel line method. Mark the points by using a POINT command and get the nine key points.

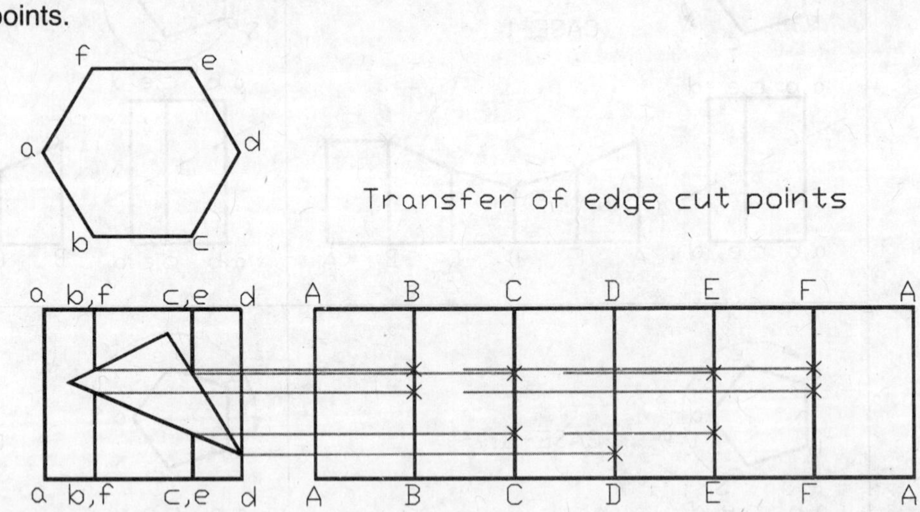


5. The location of surface cut point is apparent as the FV of surface AABB and AAFF shows an apparent area. The cut points on surfaces in the FV have to be transferred on the respective surfaces in the TV to get their true location. Therefore, an imaginary line xx is drawn passing through the surface cut point. This is shown by a point view x on aabb and aaff in the TV. This imaginary line will be drawn in the development in next step.
   Similarly, a point view y of an imaginary line yy is drawn on bbcc and eeff in the TV.
6. Use MEASURE command to locate the distance of imaginary line XX on surface AABB and AAFF, YY on BBCC and EEFF in the development. The imaginary lines are XX on surface AABB, AAFF and YY on surface BBCC and EEFF.
7. Now similar to Step (4), the cut points on surface are to be transferred from FV on the respective imaginary lines by a parallel line method. Mark the points by using a POINT command and obtain four key points.

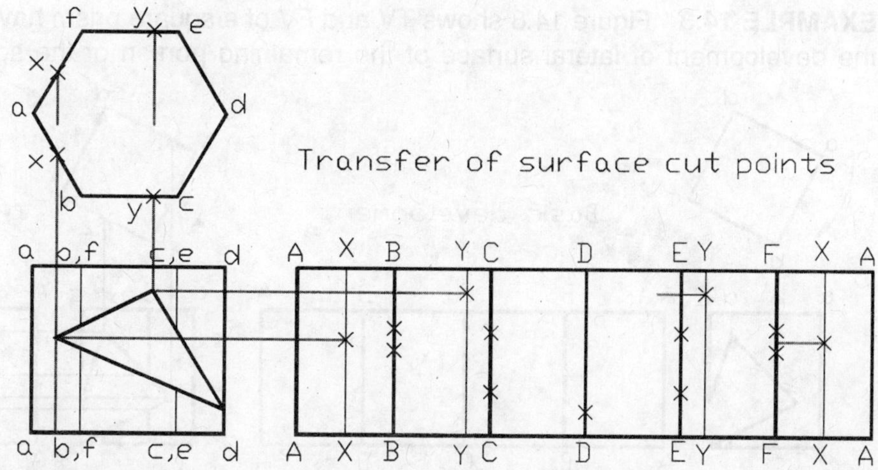


8. While joining the key points, a triangular path has to be followed for the front visible surfaces: AABB, BBCC, CCDD and back invisible surfaces: DDEE, EEFF, FFAA. Join all the key points by using a PLINE command.

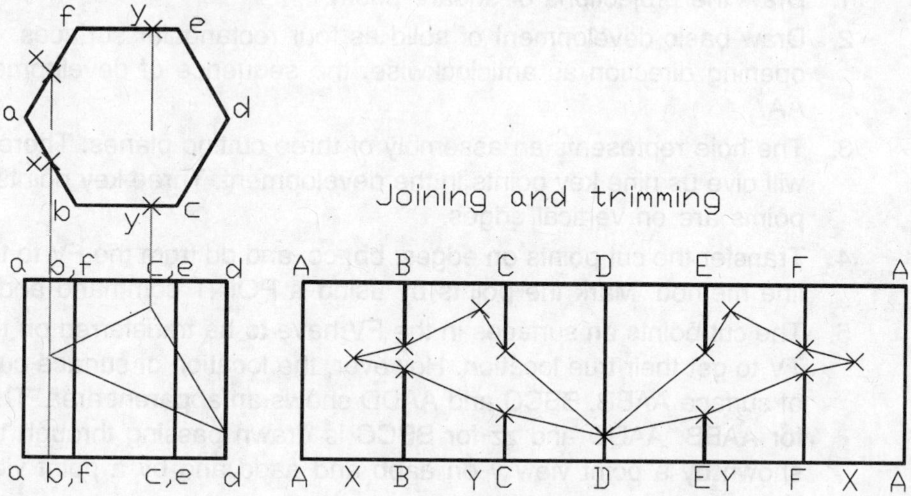


9. The edges: BB, FF, CC and EE are cut in the FV. Therefore, trim the edges: BB, FF, CC and EE in the development.
10. Remove the imaginary lines XX and YY from the development.

Figure 14.7 shows the same solution with D as a starting point and anticlockwise opening direction.

**Fig. 14.7**

**EXAMPLE 14.3** Figure 14.8 shows TV and FV of a square prism having a hole cut through it. Draw the development of lateral surface of the remaining portion of the solid.

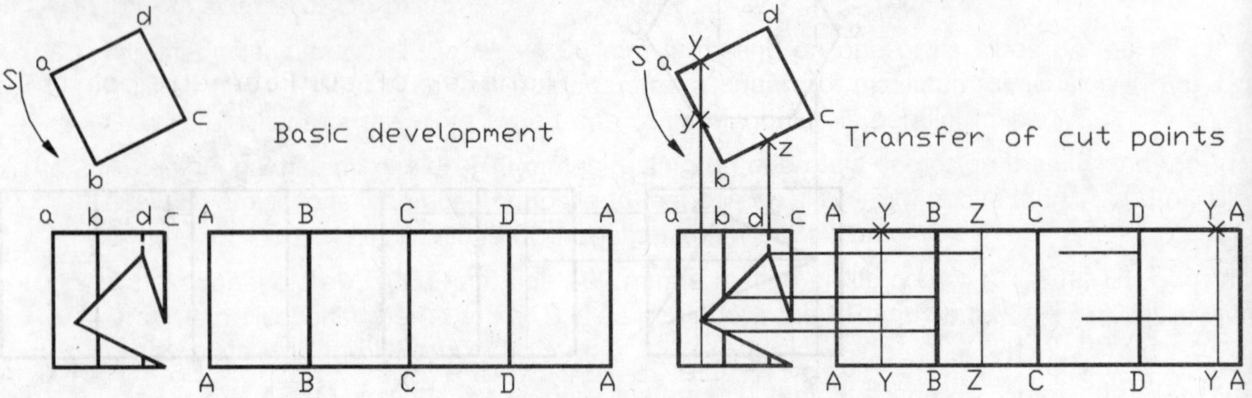


**Fig. 14.8**

***Solution***

1. Draw the projections of square prism.
2. Draw basic development of solid as four rectangular surfaces. With A as starting point and opening direction as anticlockwise, the sequence of development edges is AA-BB-CC-DD-AA.
3. The hole represents an assembly of three cutting planes. There are seven cut points, which will give us nine key points in the development. Three key points are on surfaces and six key points are on vertical edges.
4. Transfer the cut points on edges: bb, cc, and dd from the FV to the development by a parallel line method. Mark the points by using a POINT command and obtain the six key points.
5. The cut points on surfaces in the FV have to be transferred on the respective surfaces in the TV to get their true location. However, the location of surface cut point is apparent as the FV of surface AABB. BBCC and AADD shows an apparent area. Therefore, an imaginary line yy for AABB, AADD and zz for BBCC is drawn passing through the surface cut point. This is shown by a point view y on aabb and aadd and by a point view z on BBCC in the TV.
6. Use MEASURE command to locate the distance of imaginary line YY on surface AABB and AADD, ZZ on BBCC in the development.
7. Now similar to step (4), the cut points from FV are to be transferred on respective imaginary lines by a parallel line method. Mark the points by using a POINT command and get three key points.

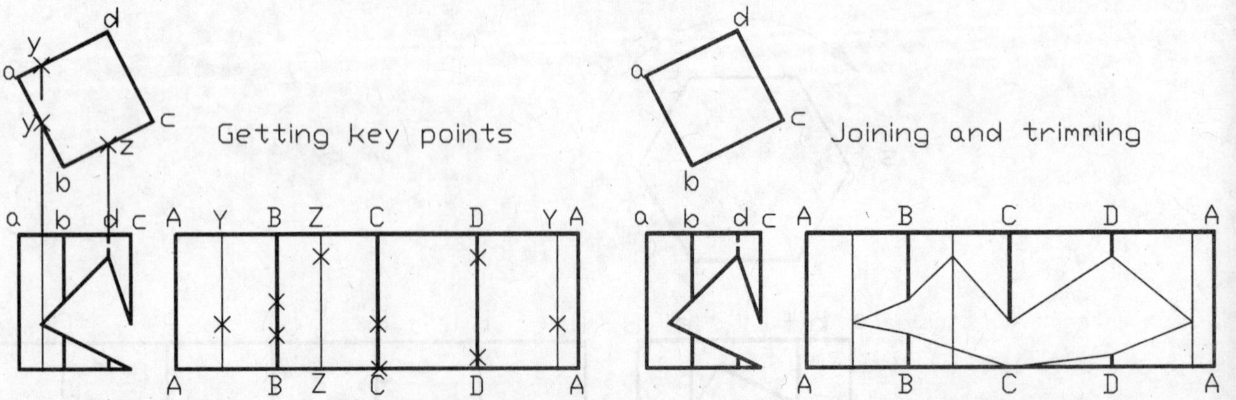


8. While joining the key points, a path has to be followed for the front visible surfaces: AABB BBCC and back invisible surfaces: CCDD and DDAA. Join all the key points by using a PLINE command.

   The edges: BB, DD and CC are cut in the FV. Therefore, trim the edges: BB, DD and CC in the development.

**EXAMPLE 14.4** Draw the development of lateral surfaces of the cut portion of an irregular pentagonal prism with respect to A as a starting point and following a CCW direction.

***Solution*** Self explanatory.

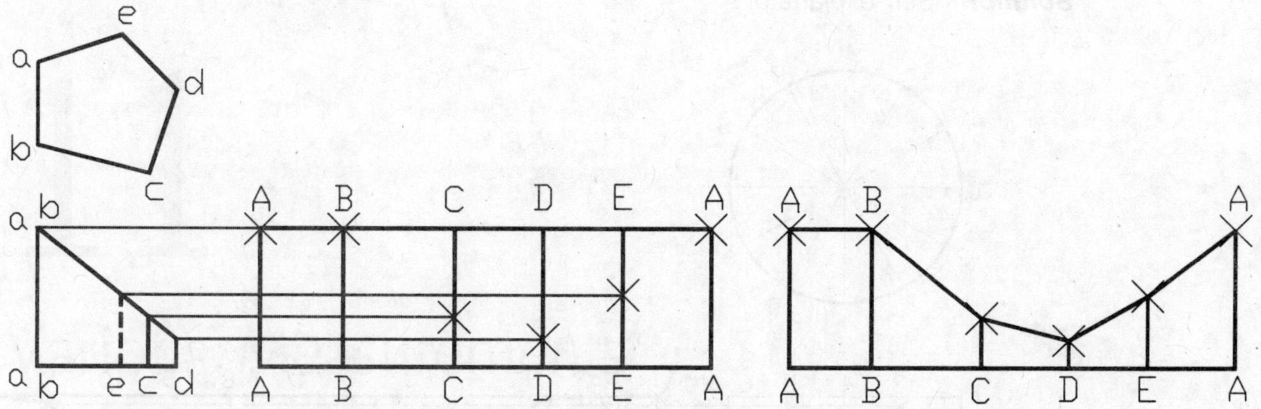


**EXAMPLE 14.5** A square prism is standing on its base on the ground with two rectangular faces equally inclined to VP. The prism is cut as shown in the FV of its projections. Draw the development of lateral surfaces of the prism with respect to A as a starting point and following a CCW direction.

***Solution*** Self explanatory.

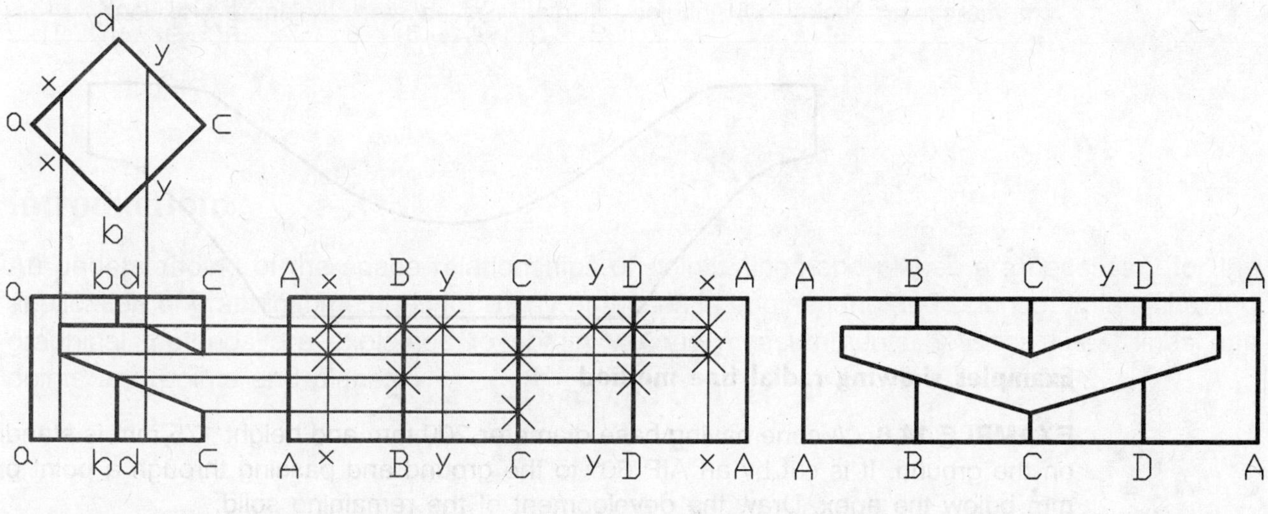


**EXAMPLE 14.6** Irregular pentagonal prism is standing on its base on the ground There is a triangular hole cut in it as shown in the FV of its projections. Draw the development of the lateral surfaces of the prism with respect to A as a starting point and following a CCW direction.

***Solution*** Self explanatory.

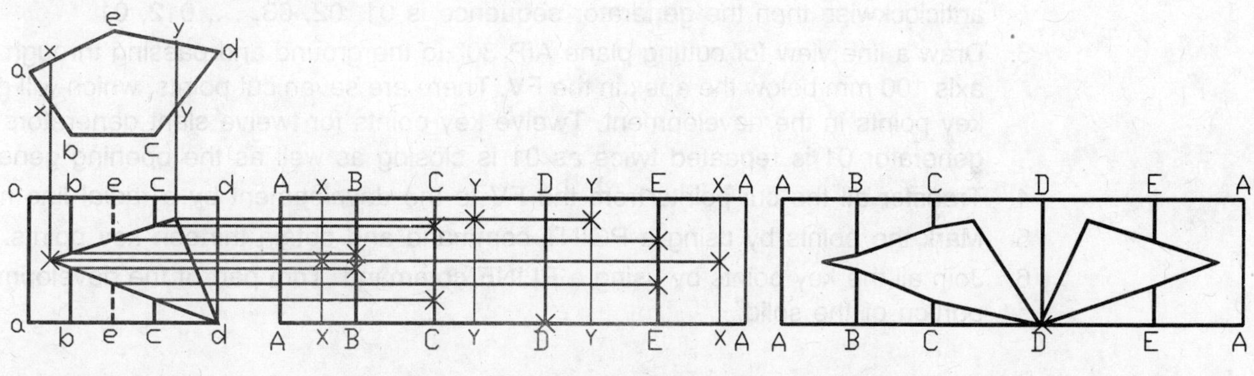

**EXAMPLE 14.7** A cylinder is standing on its base on the ground. It is cut as shown in the FV of its projections. Draw the development of the curved surfaces of the cylinder starting from the extreme left generator in CCW direction.

***Solution*** Self explanatory.

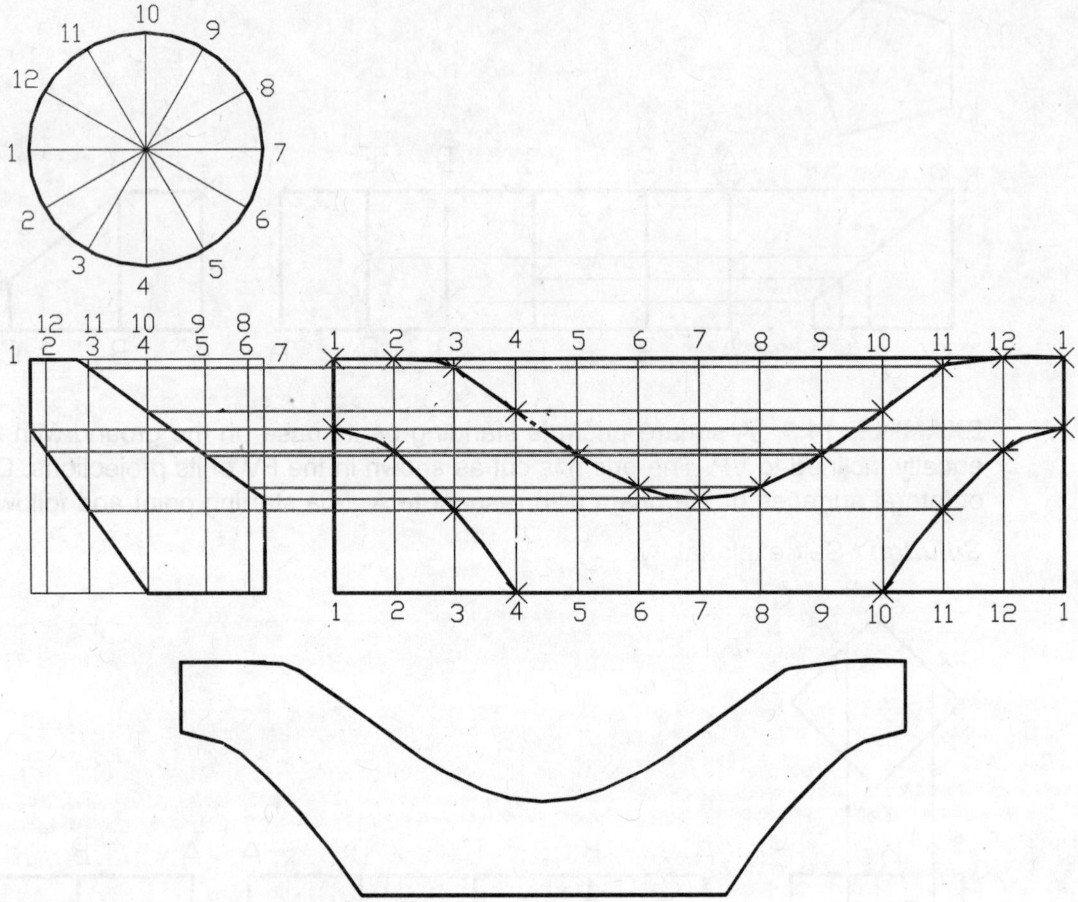


**Examples showing radial line method**

**EXAMPLE 14.8** A cone having base diameter 200 mm and height 175 mm is standing on its base on the ground. It is cut by an AIP 30° to the ground and passing through a point on the axis 100 mm below the apex. Draw the development of the remaining solid.

***Solution***

1. Draw the projections of solid. Divide the circular base into 12 equal number of parts using 12 slant generators. Name the circumferential points 1, 2, 3, ..., 11, 12.
2. Draw the basic development of solid. If 1 is the starting point and opening direction is anticlockwise then the generator sequence is 01, 02, 03, ..., 012, 01.
3. Draw a line view for cutting plane AIP 30° to the ground and passing through a point on the axis 100 mm below the apex in the FV. There are seven cut points, which will give us thirteen key points in the development. Twelve key points for twelve slant generators and key point generator 01 is repeated twice as 01 is closing as well as the opening generator.
4. Transfer all the cut points from the FV to the development by a radial line method.
5. Mark the points by using a POINT command and obtain thirteen key points.
6. Join all the key points by using a PLINE command. Trim part of the development for the cut portion of the solid.

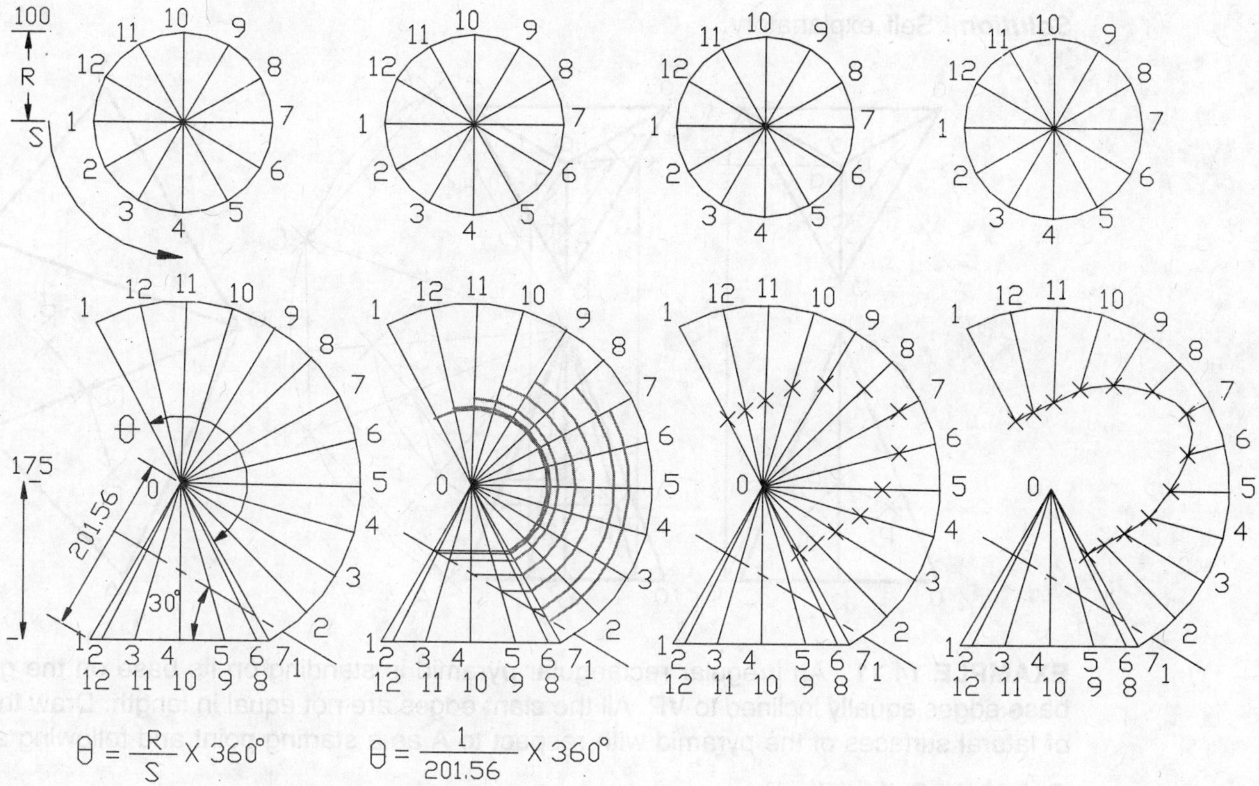


**EXAMPLE 14.9** Square pyramid is standing on its base on the ground with two base edges parallel to VP. It is cut by an AIP as shown in the figure. Draw the development of lateral surfaces of the cut portion of the pyramid with respect to A as a starting point and following a CCW direction.

***Solution*** Self explanatory.

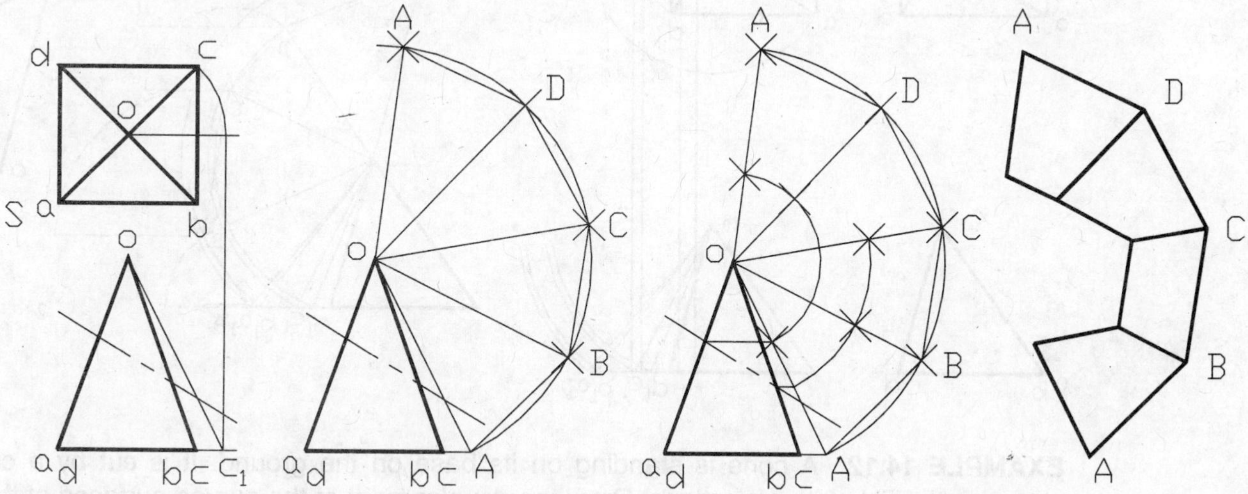


**EXAMPLE 14.10** A triangular pyramid is standing on its base on the ground with two base edges equally inclined to VP. There are three points P, Q on slant surfaces and R on the slant edges. Draw the development of lateral surfaces of the pyramid with respect to A as a starting point and following a CCW direction. Show lines connecting these three points on the development.

***Solution*** Self explanatory.

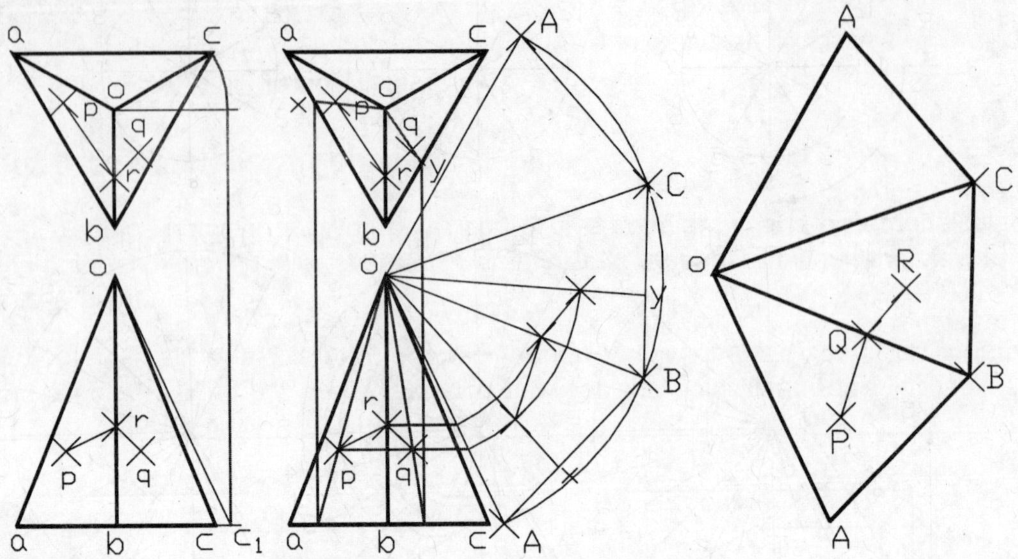


**EXAMPLE 14.11** An irregular rectangular pyramid is standing on its base on the ground with two base edges equally inclined to VP. All the slant edges are not equal in length. Draw the development of lateral surfaces of the pyramid with respect to A as a starting point and following a CW direction.

***Solution*** Self explanatory.

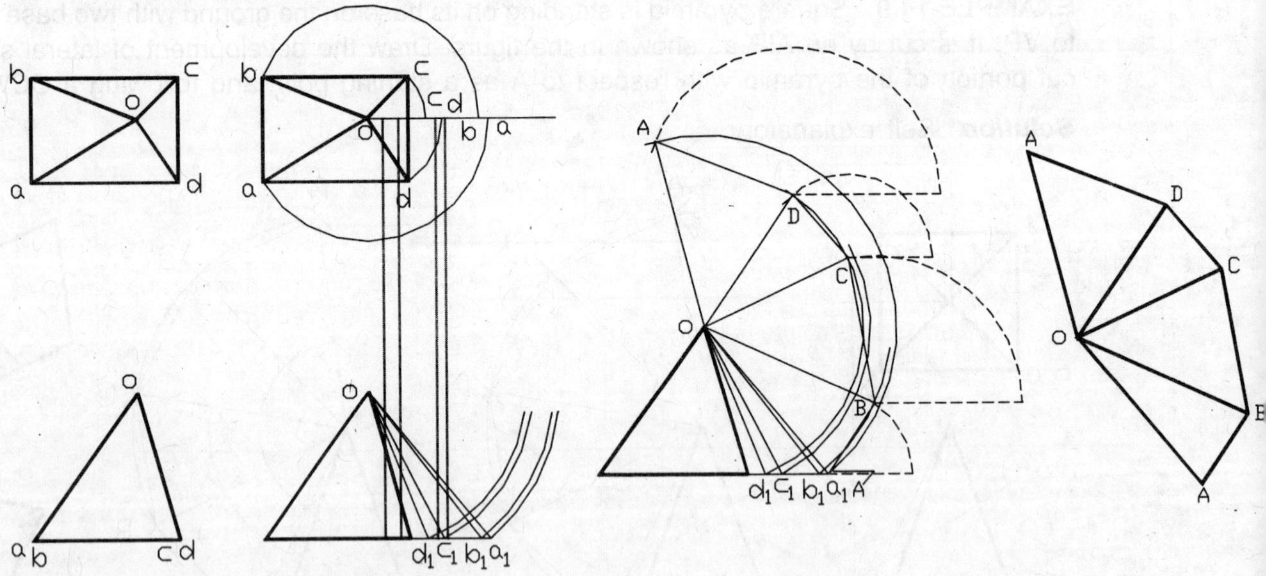


**EXAMPLE 14.12** A cone is standing on its base on the ground. It is cut by a cutting plane as shown in the FV of its projections. Draw the development of the curved surfaces of the cone starting from the extreme right generator and in CW direction.

***Solution*** Self explanatory.

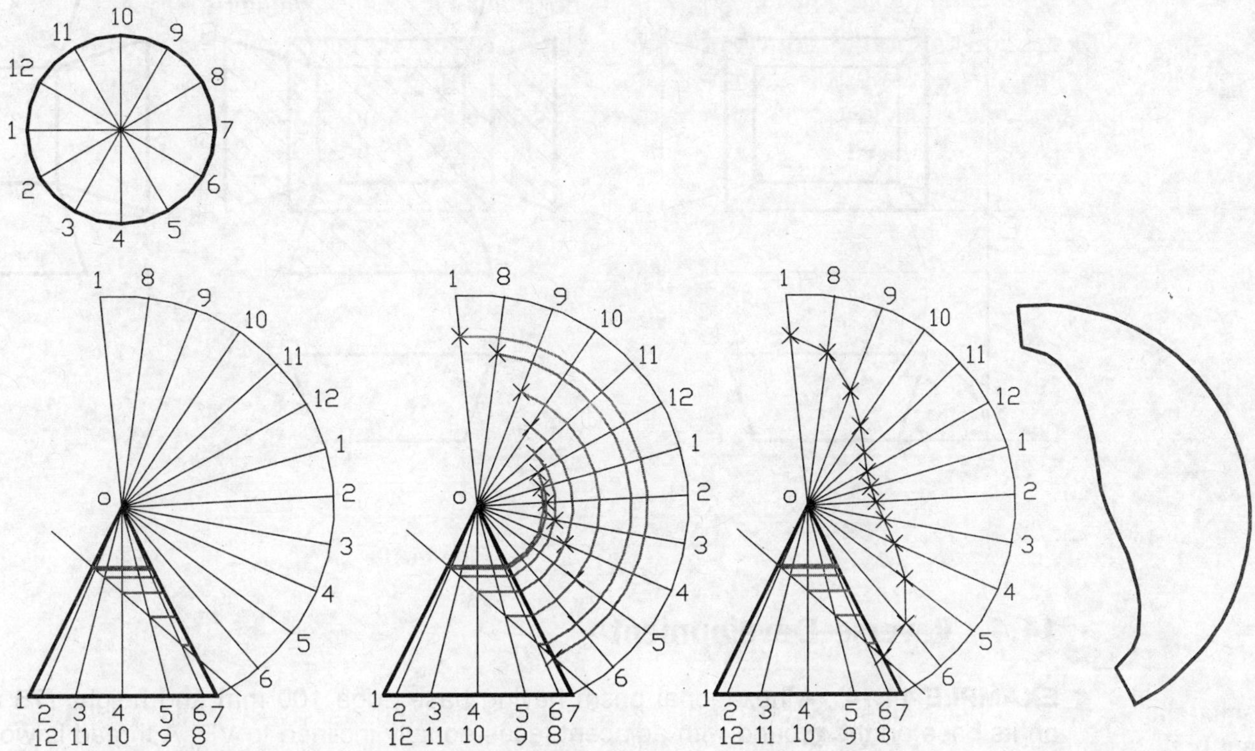

## 14.3 Development of Lateral Surfaces Along with Base

**EXAMPLE 14.13** Figure 14.9 shows a hollow crucible as well as its orthographic views (TV, FV and RSV). Draw a development of lateral as well as base surfaces.

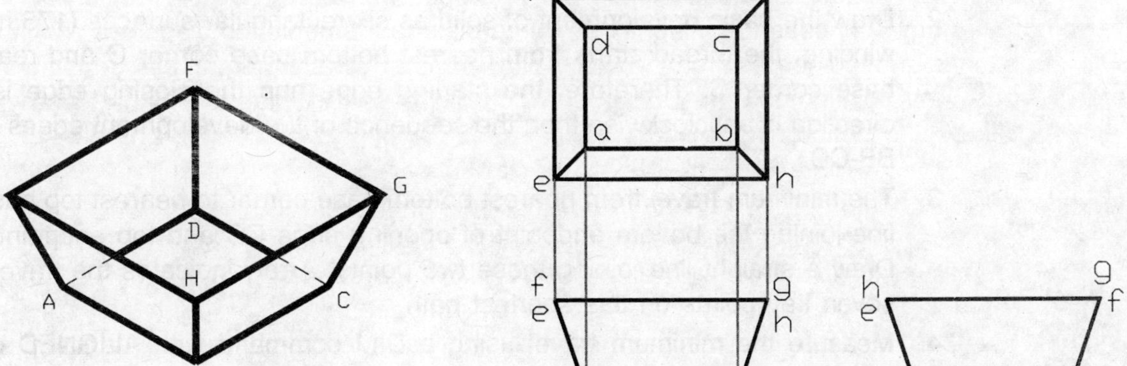


**Fig. 14.9**

***Solution*** The TV shows the true shape as well as the true dimensions of the base surface ABCD. However, all orthographic views show an apparent shape and dimensions of lateral surfaces ABHE, BCGH, CDFG, DAEF. In order to find a true shape of the lateral surfaces we need to find the true lengths of all slant edges. The crucible is regular and symmetric. Therefore, finding one true length of slant edge is sufficient. Figure 14.10(a) shows finding of the true length of slant edge ae i.e AE. We can follow MIRROR command to get mirror image for other slant edges: DF, CG and BH. Figure 14.10(b) shows the shaded region as the development of base and lateral surfaces along with TV and FV of the crucible. Figure 14.10(c) shows only development.

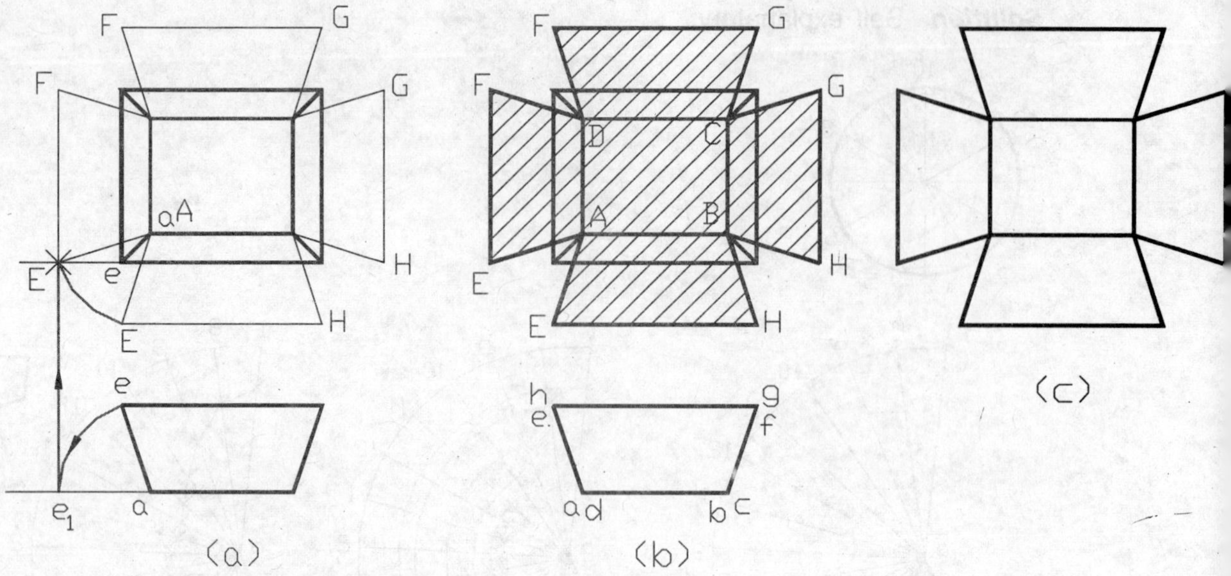


Fig. 14.10

## 14.4 Reverse Development

**EXAMPLE 14.14** A hexagonal prism having base edge 100 mm and height 175 mm is standin on its base on the ground with adjacent faces equally inclined to VP. A thread is wound around th prism, starting from the nearest corner of the bottom base and is brought back to the nearest corne of the top base in CCW direction, completing one round. Find the minimum length of the thread an show it on the elevation.

***Solution***

1. Draw the projections of solid with axis perpendicular to the ground and adjacent faces equall inclined to VP. Name the top base and the bottom base as abcdef.
2. Draw the basic development of solid as six rectangular surfaces (175 mm × 100 mm). Whi winding, the thread starts from nearest bottom base corner C and reaches the nearest to base corner C. Therefore, the starting edge and the closing edge is CC. If the openin direction is anticlockwise then the sequence of the development edges is CC-DD-EE-FF-A BB-CC.
3. The minimum travel from nearest bottom base corner to nearest top base corner is a straig line joining the bottom endpoint of opening edge CC and top endpoint of closing edge C Draw a straight line joining these two points. Arrow indicates the travel direction. There a seven key points on this shortest path.
4. Measure the minimum travel using a DIM command with ALIGNED option. The minimu travel is found to be 625 mm.

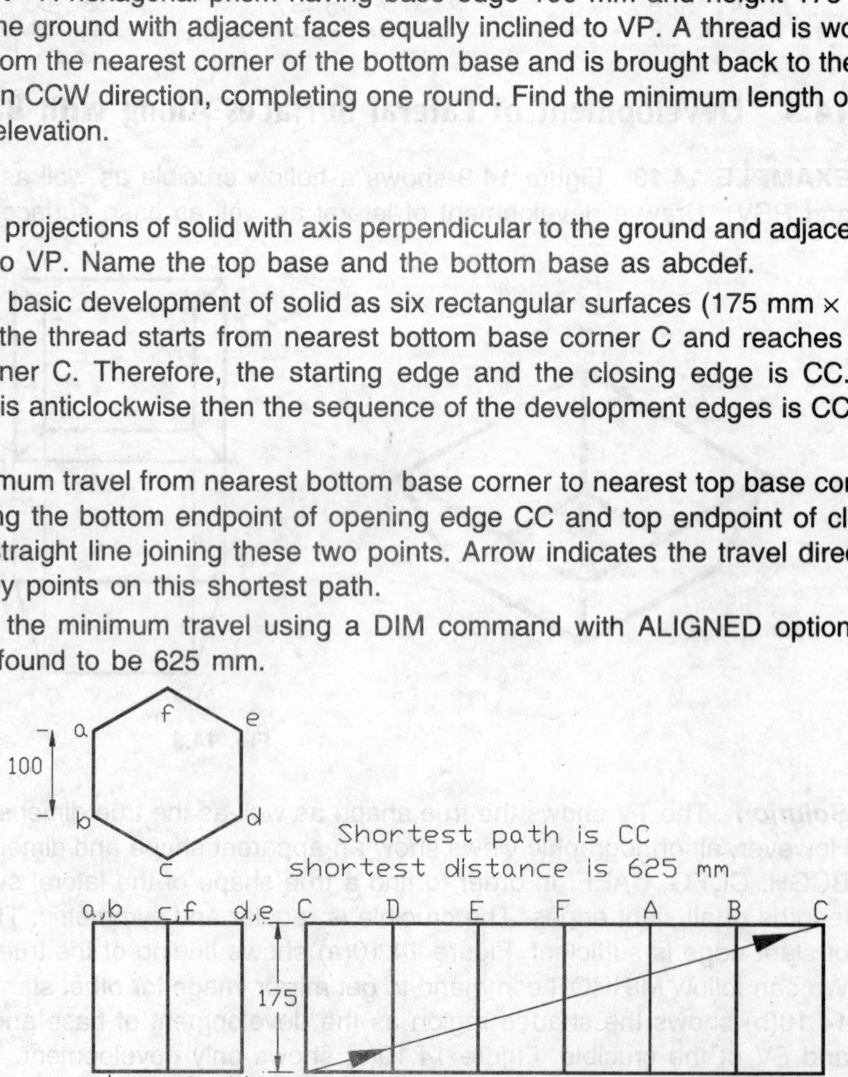

5. Transfer all the seven key points from the development to the FV by a parallel line method.
6. Mark the points by using a POINT command and get the seven key points.
7. Join all the key points by using a LINE command. The travel on the surface CCDD, DDEE, AABB and BBCC is visible to the observer. Therefore, this travel is shown as a solid line. However, the travel on the surfaces EEFF and FFAA is not visible to the observer. Therefore, this travel is shown as a hidden line.

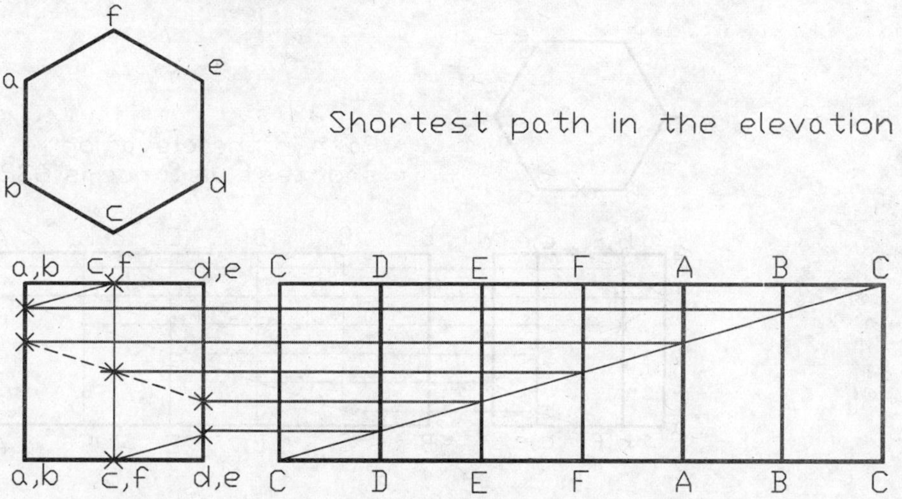


**EXAMPLE 14.15** A hexagonal prism having base edge 100 mm and height 175 mm is standing on its base on the ground with alternate faces equally inclined to VP. A thread is wound around the prism, starting from the mid-point of nearest base edge of the bottom base and is brought back to the mid-point of the nearest base edge of the top base in CCW direction completing one round. Find the minimum length of the thread and show it on the elevation.

***Solution***

1. Draw the projections of solid with axis perpendicular to ground and alternate faces equally inclined to VP. Name the top base and the bottom base as abcdef.
2. Draw the basic development of solid as six rectangular surfaces (175 mm × 100 mm). While winding, the thread starts from the mid-point of the nearest bottom base edge BC and reaches the mid-point of nearest top base edge BC after completing one round. Therefore, the starting edge is BB and closing edge is CC. If the opening direction is anticlockwise then the sequence of development edges is BB-CC-DD-EE-FF-AA-BB-CC.
3. The minimum travel from the mid-point of the nearest bottom base edge to the mid-point of nearest top base edge is a straight line joining mid-point of bottom base edge BC in the left to the mid-point of top base edge BC in the right. Draw a straight line joining these two points. Arrows indicate the travel direction. There are eight key points on this shortest path.
4. Measure the minimum travel using a DIM command with ALIGNED option. The minimum travel is found to be 632.5 mm.

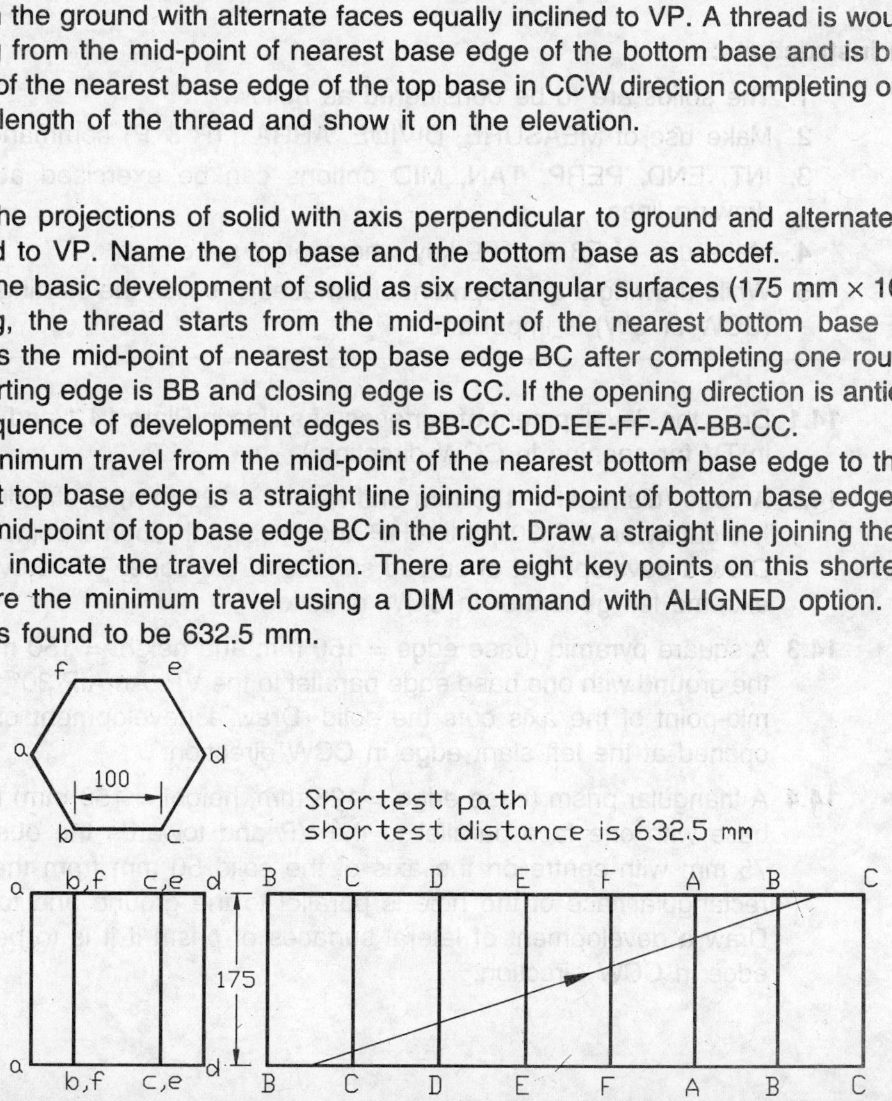

5. Transfer all the eight key points from the development to the FV by a parallel line methoc
6. Mark the points by using a POINT command and obtain the eight key points.
7. Join all the eight key points by using a LINE command. The travel on the surface AABE BBCC and CCDD is visible to the observer. Therefore, this travel is shown as a solid line However, the travel on the surfaces DDEE, EEFF and FFAA is not visible to the observe Therefore, this travel is shown as a hidden line.

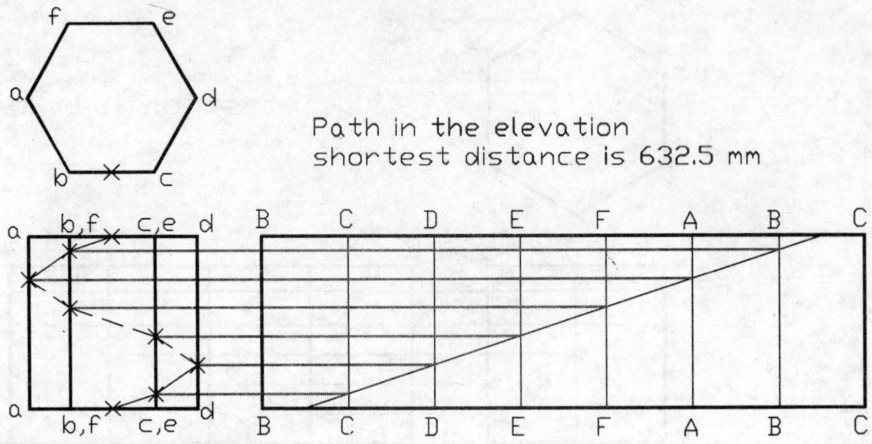


## Exercise A (Development of Surfaces of Solids)

**Instructions:**

1. The solids are to be considered as hollow.
2. Make use of MEASURE, DIVIDE, ARRAY (R & P) commands wherever possible.
3. INT, END, PERP, TAN, MID options can be exercised at the command prompt whi drawing lines.
4. Make use of F8 and F9 key wherever, if necessary.
5. While drawing a development of surfaces of solids the starting point S and opening directic (CCW or CW) is important.

---

**14.1** Draw the development of surfaces of solids in Plate 14.1 and 14.2. (Take the left-most poi in TV for opening in CCW direction.)

**14.2** A cone (diameter = 150 mm and height = 125 mm) is standing on its base on the groun It is cut by an AIP 30° to the HP and passing through a point on axis 75 mm from the bas Draw a development of curved surface of the cone. The curved surface is to be opened extreme left generator in CCW direction.

**14.3** A square pyramid (base edge = 150 mm and height = 150 mm) is standing on its base c the ground with one base edge parallel to the VP. An AIP 30° to the HP and passing throug mid-point of the axis cuts the solid. Draw a development of lateral surfaces if it is to b opened at the left slant edge in CCW direction.

**14.4** A triangular prism (base edge = 125 mm, height = 150 mm) is resting on the ground on i base with one face parallel to the VP and towards the observer. A triangular hole, sic 75 mm with centre on the axis of the solid 50 mm from the base is made through. Or rectangular face of the hole is parallel to the ground and towards the base of the prisr Draw a development of lateral surfaces of prism if it is to be opened from the left vertic edge in CCW direction.

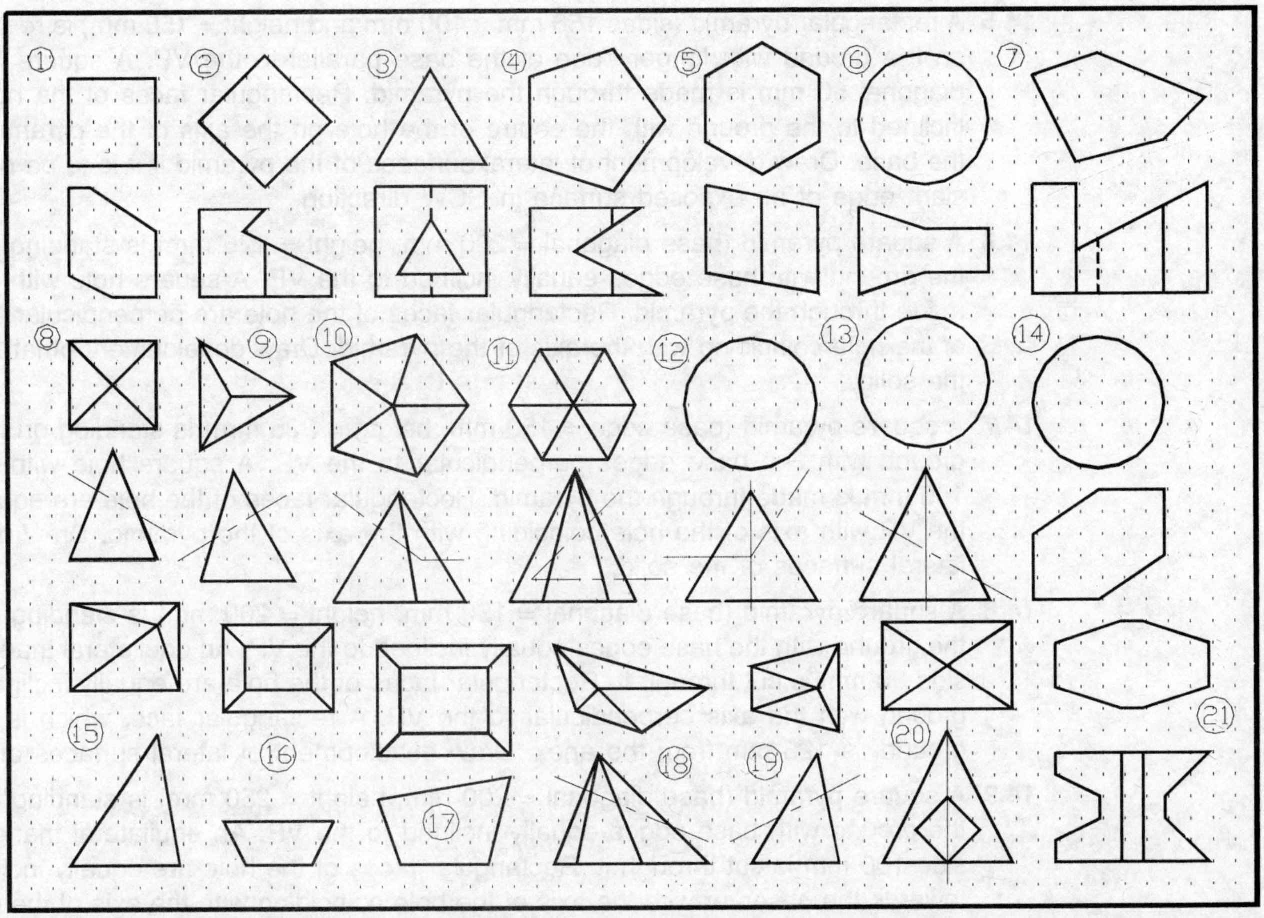


**PLATE 14.1**

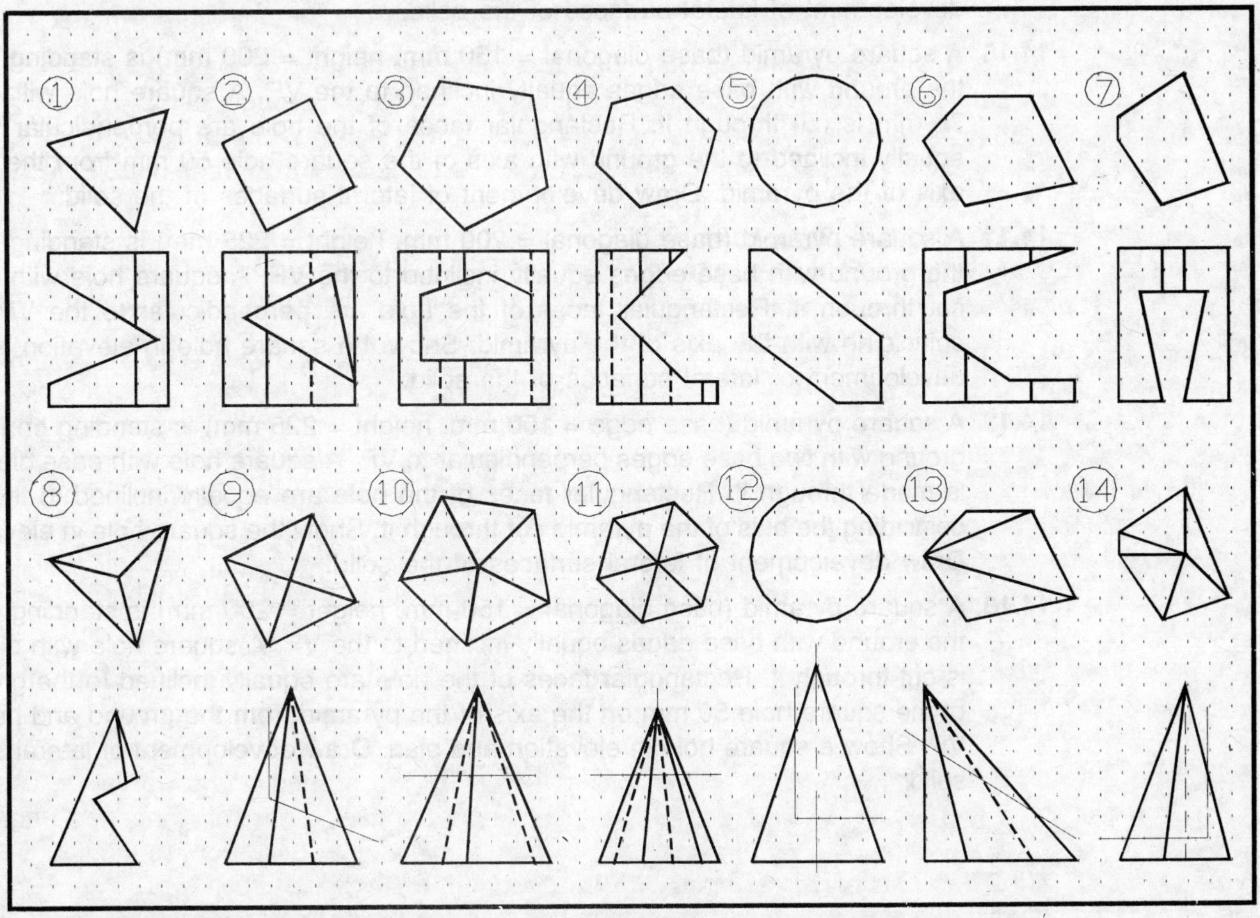


**PLATE 14.2**

14.5 A rectangular pyramid (sides 150 mm × 100 mm and height = 150 mm) is resting on its base on the ground with longer edge of the base parallel to the VP. A square hole with base diagonal 50 mm is made through the pyramid. Rectangular faces of the hole are equally inclined to the ground with the centre of the hole on the axis of the pyramid 25 mm from the base. Draw development of lateral surfaces of the pyramid if it is to be opened at a left slant edge of an exposed surface in CCW direction.

14.6 A square pyramid (base diagonal = 200 mm, height = 225 mm) is standing on its base on the ground with base edges equally inclined to the VP. A square hole with side 75 mm is made through the pyramid. Rectangular faces of the hole are perpendicular to VP with axis of the hole coinciding with the axis of the pyramid. Draw development of lateral surfaces of the solid.

14.7 A square pyramid (base edge = 150 mm, height = 225 mm) is standing on its base on the ground with two base edges perpendicular to the VP. A square hole with base diagonal 100 mm is made through the pyramid. Rectangular faces of the hole are equally inclined to the VP with axis of the hole coinciding with the axis of the pyramid. Draw development of lateral surfaces of the solid.

14.8 A square pyramid (base diagonal = 150 mm, height = 200 mm) is standing on its base on the ground with the base edges equally inclined to the VP. An equilateral triangular hole with side 50 mm is cut through it. Rectangular faces of the hole are equally inclined towards the ground with the axis perpendicular to the VP. A rectangular face, which is parallel to the ground, is 125 mm from the apex. Draw development of lateral surfaces of the solid.

14.9 A square pyramid (base diagonal = 200 mm, height = 250 mm) is standing on its base on the ground with base edges equally inclined to the VP. An equilateral triangular hole with side 100 mm is cut through it. Rectangular faces of the hole are equally inclined to the VP towards the observer with the axis of the hole coinciding with the axis of the pyramid. Draw development of lateral surfaces of the solid.

14.10 A square pyramid (base diagonal = 150 mm, height = 200 mm) is standing on its base on the ground with base edges equally inclined to the VP. A square hole with base diagonal 75 mm is cut through it. Rectangular faces of the hole are perpendicular to the VP and equally inclined to the ground with axis of the square hole 50 mm from the ground on the axis of the pyramid. Draw development of lateral surfaces of the solid.

14.11 A square pyramid (base diagonal = 200 mm, height = 225 mm) is standing on its base on the ground with base edges equally inclined to the VP. A square hole with side 75 mm is cut through it. Rectangular faces of the hole are perpendicular to the VP with the axis coinciding with the axis of the pyramid. Show the square hole in elevation and plan. Draw development of lateral surfaces of the solid.

14.12 A square pyramid (base edge = 150 mm, height = 225 mm) is standing on its base on the ground with two base edges perpendicular to VP. A square hole with base diagonal 100 mm is made through it. Rectangular faces of the hole are equally inclined to the VP with axis coinciding the axis of the pyramid cut through it. Show the square hole in elevation and plan. Draw development of lateral surfaces of the solid.

14.13 A square pyramid (base diagonal = 150 mm, height = 200 mm) is standing on its base on the ground with base edges equally inclined to the VP. A square hole with diagonal 75 mm is cut through it. Rectangular faces of the hole are equally inclined to the ground with axis of the square hole 50 mm on the axis of the pyramid from the ground and perpendicular to VP. Show a square hole in elevation and plan. Draw development of lateral surfaces of the solid.

**14.14** A square pyramid (base diagonal = 200 mm, height = 250 mm) is standing on its base on the ground with base edges equally inclined to the VP. An equilateral triangular hole with side 100 mm is made through it with its axis coinciding with the axis of the pyramid. Rectangular faces of the hole are equally inclined to the VP and towards the observer with axis coinciding with the axis of the pyramid. Show the triangular hole in elevation and plan. Draw development of lateral surfaces of the solid.

**14.15** A square pyramid (base diagonal = 200 mm, height = 250 mm) is standing on its base on the ground with the base edges equally inclined to the VP. A triangular hole with side 100 mm is made through it. Rectangular faces of the hole are equally inclined to the VP and away from the observer with the axis coinciding with the axis of the pyramid. Show the triangular hole in elevation and plan. Draw development of lateral surfaces of the solid.

**14.16** Determine graphically, the length of a body diagonal of a cube having edge length 100 mm.

**14.17** Determine graphically, the height of a triangular pyramid having length of base edges equal to the length of slant edges which is equal to 150 mm.

**14.18** Two square pyramids having equal dimensions (base edge = slant edge = 150 mm, height = 100 mm) are meeting at their bases with base edges aligned. The combination of these two solids is standing on one of the apexes on the ground with common axis perpendicular to the ground and two aligned base edges perpendicular to the VP. Determine a distance between the two apexes.

**14.19** A hexagonal pyramid of base edge 100 mm and height 200 mm stands on the ground with two of its base edges parallel to the VP. A triangular hole of side 75 mm is made through the hexagonal pyramid with axis of the hole coinciding with axis of the pyramid. One side of the hole is parallel to the VP and a corner opposite to this side is towards the observer.

Draw the projections of the solid. Show the hole in the FV. Develop the lateral surfaces of the hexagonal pyramid from the left most base-corner in CCW direction. Show the profile of intersection on the development.

**14.20** A pentagonal pyramid of base edge 100 mm and height 150 mm has apex on the ground with axis perpendicular to the ground and one of its base edges perpendicular to the VP. A square hole of side 50 mm is made through the pentagonal pyramid, the axis of which coincides with the axis of the pyramid. Sides of the square hole are equally inclined to the VP.

Draw the projections of the solid. Show the profile of the square hole in the FV. Develop lateral surfaces of the pentagonal pyramid from the left base corner nearer to the observer in CCW direction. Show the profile of the square hole on the development.

**14.21** A cylindrical water-tank of base diameter 1.5 m and height 3 m is resting on the ground with axis perpendicular to the ground. A squirrel starts to climb the tank from a circumferential bottom base point nearest to the observer and makes two rounds along the curved surface. Finally, it reaches to a circumferential point on the top base nearest to the observer. While climbing, the vertical distance covered in the first round is 1 m and in the second round it is 2 m. In each round, it covers shortest distances along the curved surface.

Draw projection of the water tank. Develop its lateral surface. Show the path followed by the squirrel in development. Show the path followed by squirrel in the FV (Scale: 0.25 AutoCAD unit = 0.5m.)

**14.22** A frustum of square pyramid has its bottom base edge 100 mm, top base edge 50 mm and height 125 mm. It is resting on the ground with two of its base edges perpendicular to the VP. An ant starts from a mid-point of the top base edge, nearest to the observer and reaches a mid-point of bottom base edge nearest to the observer after completing one round in CCW direction. While travelling on every new surface, the vertical height is reduced by 25 mm. Each path follows a shortest distance.

Draw the projections of the frustum. Develop lateral surfaces of the solid. Show the path followed by the ant in the FV, TV and on the developed surfaces.

**14.23** Draw a development of a sheet metal tray with top square 150 mm side, bottom square 100 mm side and height 75 mm along the base.

## Exercise B (Reverse Development)

**14.1** A pentagonal prism, side 100 mm, height 150 mm is resting on the ground on its base with one vertical face away from the observer and parallel to the VP. A thread is wound around the prism, starting from the nearest bottom base corner and is brought back to the nearest top base corner in CCW direction following a shortest path. Find the length of the thread and show it on the elevation.

**14.2** A cone (base diameter = 150 mm and height = 200 mm) is kept on the ground on its base with an imaginary diameter AB parallel to VP. A point P starting from A moves along the curved surface of the cone and comes back to A moving around an axis of the cone following the shortest path. Show this shortest path in the elevation and plan of the cone.

**14.3** The development of a cylinder is a rectangle of 472 mm length. A semicircle of 472 mm diameter is perfectly inscribed in the rectangle. Draw the projections of the cylinder showing the semicircle on elevation. (Take the diameter of the cylinder as a nearest rounded value).

**14.4** A square pyramid with 150 mm base diagonal and 300 mm height is resting on the ground with base edges equally inclined to the VP. A point P, starts from the base corner nearest to the observer, moves on the slant surfaces of the pyramid in CCW direction and reaches to its original position after making two rounds around the axis by the shortest route. Draw development of the slant surfaces of the square pyramid. Show the shortest path on the development. Represent this shortest path in two views of the pyramid.

**14.5** A frustum of a square pyramid has its bottom base edge 150 mm, top base edge 50 mm and height 300 mm is resting on its bottom base on the ground with two base edges parallel to the VP. A point P, starts from a mid-point of the bottom base edge nearest to the observer, moves on the slant surfaces of the frustum in CCW direction and reaches to the mid-point of the top base edge farthest from the observer by the shortest route. How many rounds the point P makes around the axis?

Draw a development of the slant surfaces of the frustum of a square pyramid. Show the shortest path on the development. Represent this shortest path in two views of the frustum of a pyramid.

**14.6** A composite solid consists of a half-square pyramid (half-base diagonal 100 mm, height 200 mm) and half-hexagonal pyramid (base edge = 100 mm, height = 200 mm). The combination is joined at the interface containing both the apex and aligning the bottom base. It is resting on the common formed base on the ground with the half-hexagonal pyramid on the right side and the interface perpendicular to the VP.

A point P, starts from a base corner at the interface nearest to the observer, moves on slant surfaces of the half-hexagonal pyramid in CCW direction and reaches to a base corner at the interface farthest from the observer by following a shortest route.

Draw development of slant surfaces of this composite solid. Show the shortest path on the development. Represent this shortest path in two views of the composite solid.

**14.7** A square pyramid (base diagonal = 200 mm and height = 250 mm) is standing on its base on the ground with two base edges equally inclined to the VP. Nails are put on the centre of every slant face and all base corners. A string is tied at the nearest base corner and passed over all the nails sequentially in CCW direction reaching to the nearest base corner. Draw a path followed by the string on a lateral surfaces and find a shortest distance for the string in mm. Show the path of the string on the projections of solid (TV and FV).

**14.8** A hexagonal pyramid (base edge = 100 mm and height = 200 mm) is standing on its base on the ground with two of its base edges perpendicular to the VP. Draw a development of lateral surfaces with farthest edge as an opening edge, following CCW direction.

A square of maximum dimension has to be inscribed in the development of lateral surfaces of this solid. One corner of the square is at apex and an opposite corner is on one of the base corners of the pyramid such that the development picture looks symmetric. Find dimension of side of the square. Draw the square on projections of solid (TV and FV).

**14.9** A square pyramid (base diagonal = 200 mm and height = 250 mm) is standing on its base on the ground with two base edges equally inclined to the VP. A point P starts its travel from a nearest base corner, meets mid-point of three slant edges, in between, with a shortest path and then reaches to the same base corner following a shortest distance. This way it completes the first round.

In the second round, it starts from the nearest base corner, meets a farthest base corner and then returns back to the nearest base corner following a shortest path in CCW direction. Draw the shortest path showing the full travel for the two rounds and mention the distance of the total path in mm. Show the path of point P on projections of solid (TV and FV).

**14.10** A frustum of a cone has smaller base diameter 75 mm, larger base diameter 150 mm and height 150 mm. It is cut by a vertical plane passing through the diameters of both the bases. Draw the projection of this half frustum with flat cut surface exposed to the direction of the observer and standing on its bigger base on the ground.

A point P, starts from a mid-point of top base edge, reaches to a right bottom base corner. Further, completing half-round around an axis of the solid in CCW direction, it reaches to the left bottom base corner and finally reaches the starting point. In all this travel, it follows a shortest path. Show the shortest path on the development. Represent this shortest path in elevation. Specify the shortest distance by using TEXT command.

**14.11** Figure 14.11 shows a FV of a sheet metal tray standing on its base on the ground. Draw a development of lateral surfaces with left most slant edge as an opening edge, following CCW direction.

Point P starts from a bottom left corner of the front face, reaches to a mid-point of right slant edge of the same face. Further, it travels at the same elevation on right face, back face and finally reaches the starting point by following a shortest path. Show the path of the point P in the development. Represent this shortest path in elevation. Specify the shortest distance by using TEXT command.

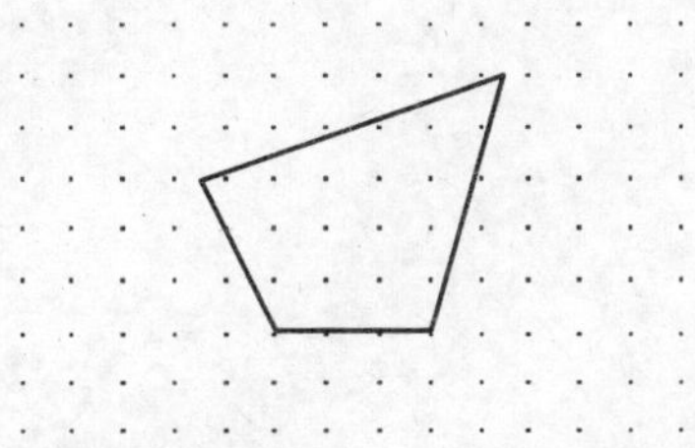

**Fig. 14.11**

**14.12** A square pyramid is standing on its base on the ground with base edges equally inclined to the VP. Draw development of lateral surfaces with left most slant edge as an opening edge, following CCW direction.

A development of this solid is inscribed in a semicircle of diameter 375 mm. A point P starts from a nearest base corner, completes one round around the axis of the solid following a shortest path on every slant surface and then reaches the starting point. Show the shortest path on the development. Represent this shortest path in elevation of the pyramid. Specify the shortest distance by using TEXT command.

# 15

# Intersection of Surfaces

## Objectives

- To learn the concepts of intersection and interpenetration of two solids.
- To know how to derive the lines of intersection and interpenetration of surfaces of two solids.

## Introduction

In practice, science and engineering student is required to know the line of intersections resulting from intersection or interpenetration of two solids. In many applications, these solids are hollow one. The line of intersection is basically an output of intersection of surfaces of solids. There can be intersection of flat surface versus flat surface, flat surface versus curved surface and curved surface versus curved surface.

## 15.1 Line of Intersection

In engineering practice, objects constructed may have constituent parts, the surfaces of which intersect one another in lines, which are called lines of intersection. A dome fitted on a boiler as shown in Fig. 15.1 is one such example. The surface of the dome extends upto the line of intersection only. For accurate development of the surface of the dome, this line of intersection must be accurately located. The shape of the hole to be cut in the boiler-shell is also determined from the shape of the same line of intersection.

**Fig. 15.1** Dome fitted on a cylindrical boiler.

Thus, the line of intersection of the two surfaces is a line common to both. It is composed of points at which the lines of one surface intersect with the other surface. The line of intersection may be straight or curved, depending upon the nature of intersecting surfaces. Two plane surfaces (e.g. faces of prisms and pyramids) intersect in a straight line. The line of intersection between two curved surfaces (e.g. faces of cylinders and cones) or between a plane surface and a curved surface (e.g. faces of cylinder and prism) is a curve.

### Intersection of Line versus Line

**EXAMPLE 15.1** Figure 15.2 shows TV and FV of two skew lines: AB and PQ. In the TV, find out which line is above and which is below.

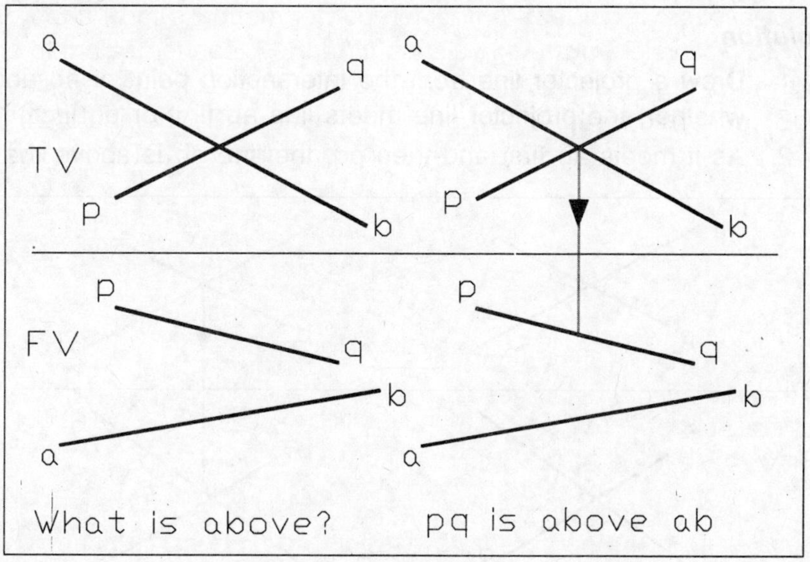


**Fig. 15.2**

***Solution***

1. Draw a projector line from the intersection point of ab and pq in the TV to the FV. Check whether, the projector line meets ab first or pq first.
2. As it meets pq first and then ab, the line PQ is above the line AB.

**EXAMPLE 15.2** Figure 15.3 shows TV and FV of two skew lines: mn and rs. In the FV, find ou which line is at front and which line is at back.

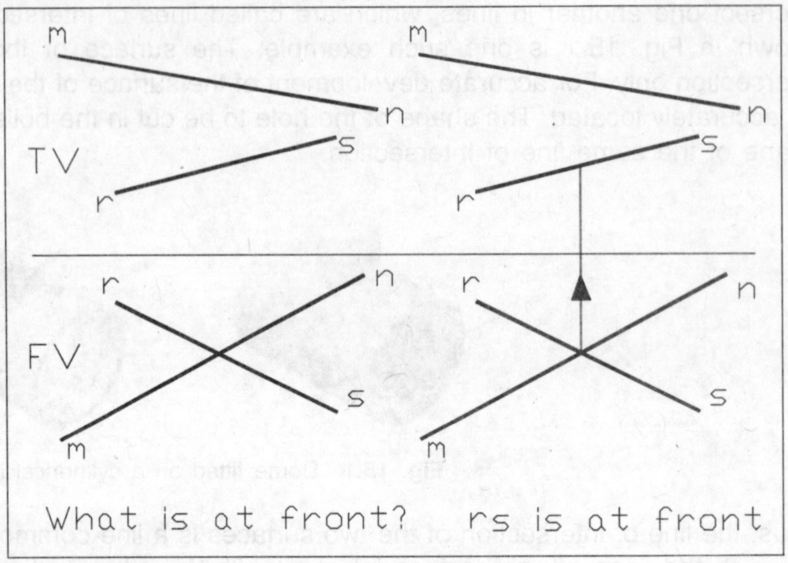


**Fig. 15.3**

***Solution***

1. Draw a projector line from the intersection point of mn and rs in the FV to the TV. Checl whether, the projector line meets mn first or rs first.
2. As it meets rs first and then mn, the line RS is at the front and the line MN is at the back

**EXAMPLE 15.3** Figure 15.4 shows TV and FV of two skew lines ab and pq. In the TV, find ou which line is above and which is below and in the FV, find out which line is at front and which line is at back.

***Solution***

1. Draw a projector line from the intersection point of ab and pq in the TV to the FV. Checl whether, the projector line meets line ab first or pq first.
2. As it meets ab first and then pq, the line ab is above the line pq.

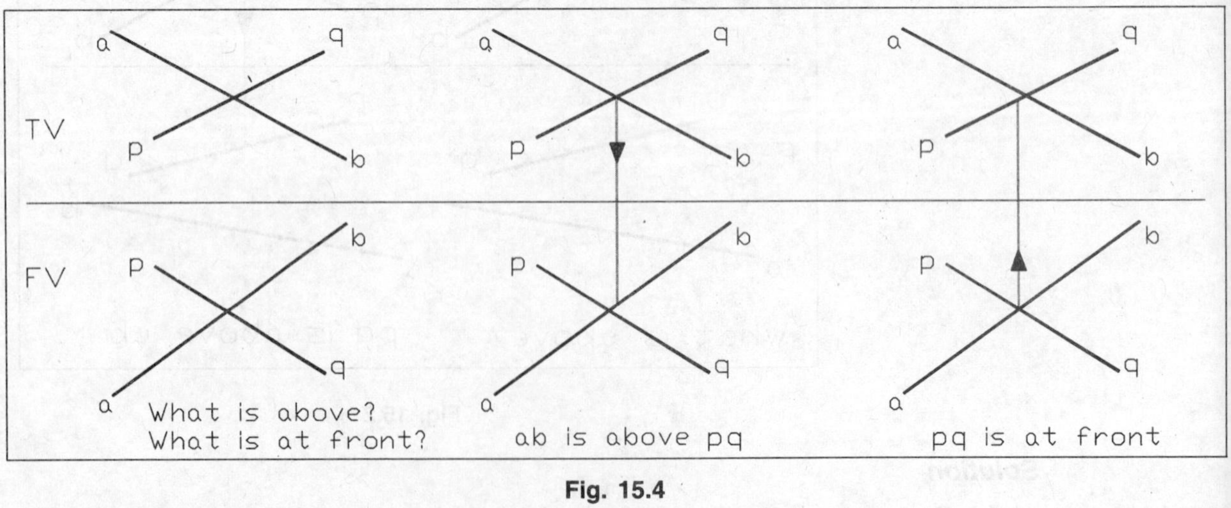


**Fig. 15.4**

3. Draw a projector line from the intersection of lines ab and pq in the FV to the TV. Checl whether, the projector line meets line ab first or pq first.

As it meets pq first and then ab, the line pq is at the front and the line ab is at the back.

## 15.2 Rule for Visibility

Based upon the discussions in Examples 15.1, 15.2 and 15.3, a rule of visibility can be constructed. In the TV, to decide which line is above and which line is below, project the intersection point on respective lines in the FV. The line to which, projector meets first is above and the other one is below.

In the FV, to decide which line is at front and which line is at back, project the intersection point on respective lines in the TV. The line to which, the projector meets first is at front and the other one is at back.

### Intersection of Line and Plane

**EXAMPLE 15.4** Figure 15.5 shows intersection of line PQ with skew plane ABC. Find out the location of intersection point.

***Solution***

1. Consider line PQ as a cutting plane either in the TV or FV. In this exercise, it is considered in the TV. Line cuts the edges ac and bc in the TV.

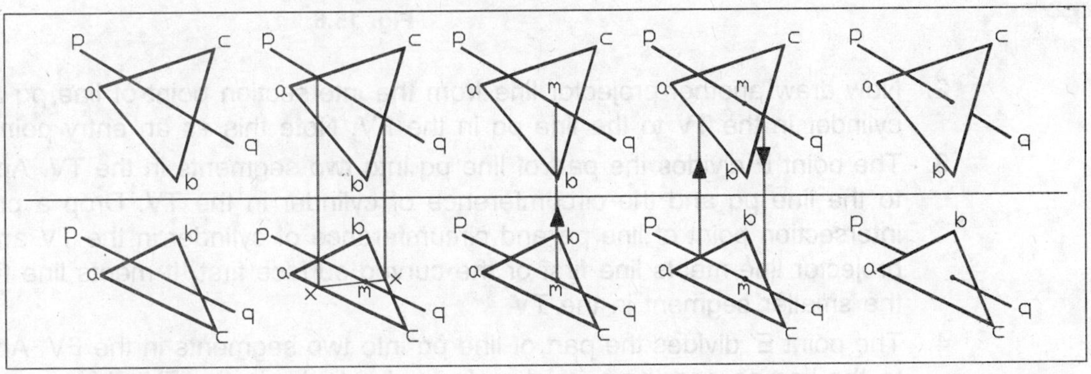


**Fig. 15.5**

2. Transfer the cut points from TV to FV on respective edges of plane ABC and obtain two key points. Name both the key points as x on ac as well as x on bc.
3. Join both the key points as a line xx in the FV. The intersection of line xx and the line pq in the FV gives intersection point m.
4. Draw a projector line from m in the FV to the line pq in the TV. Point m divides the part of line pq into two segments in both the views.
5. Apply a rule for visibility to these segments in the FV as well as TV. The last stage shows the final solution of visibility.

## 15.3 Intersection of Lines and Solids

### Intersection of Line and Cylinder

**EXAMPLE 15.4** Orthographic view of Fig. 15.6 shows intersection of line with cylinder. Find out the location of entry point and exit point of line on cylinder from the orthographic views.

***Solution***

1. Draw a projector line from the intersection point of line pq and top base in the FV to the line pq in the TV. Note this as an entry point E.

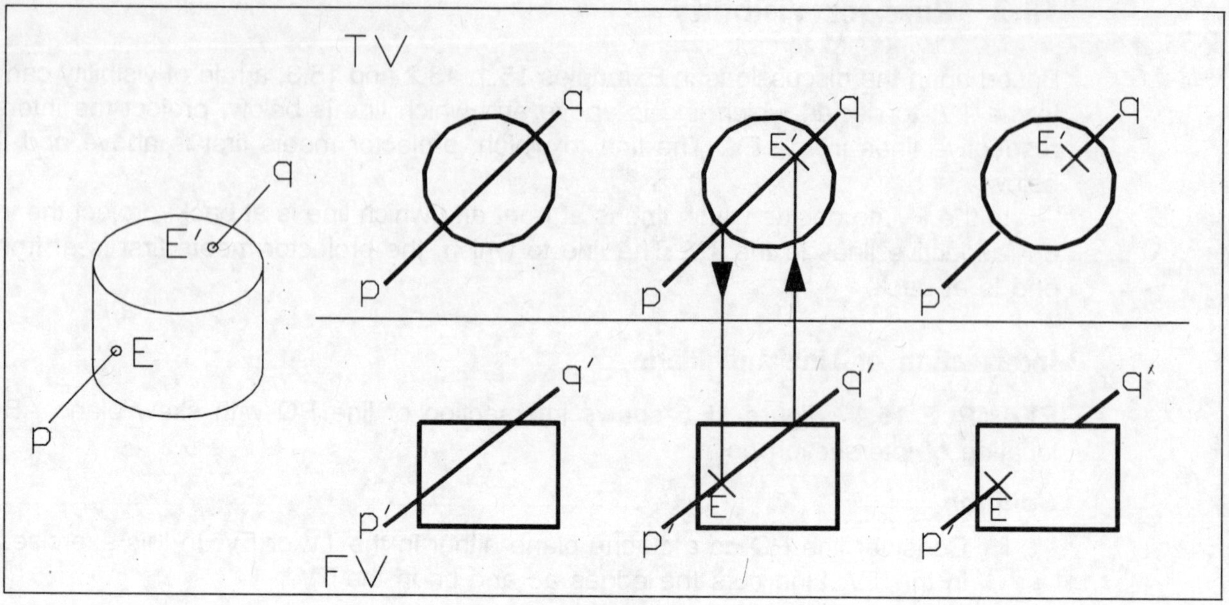


Fig. 15.6

2. Now draw another projector line from the intersection point of line pq and circumference of cylinder in the TV to the line pq in the FV. Note this as an entry point E′.
3. The point E divides the part of line pq into two segments in the TV. Apply a rule of visibility to the line pq and the circumference of cylinder in the TV. Drop a projector line from the intersection point of line pq and circumference of cylinder in the TV and check whether the projector line meets line first or the curved surface first. It meets line first. Therefore, retain the smaller segment in the TV.
4. The point E′ divides the part of line pq into two segments in the FV. Apply a rule of visibility to the line pq and the curved surface of cylinder in the FV. Draw a projector line from the intersection point of line pq and curved surface of cylinder in the FV and check whether the projector line meets the line first or the circumference (curved surface) first. It meets the line first. Therefore, retain the smaller segment in the FV.

## Intersection of Line and Pyramid

**EXAMPLE 15.5** Orthographic view of Fig. 15.7 shows intersection of a line with pyramid. Find out the location of point of entry and exit.

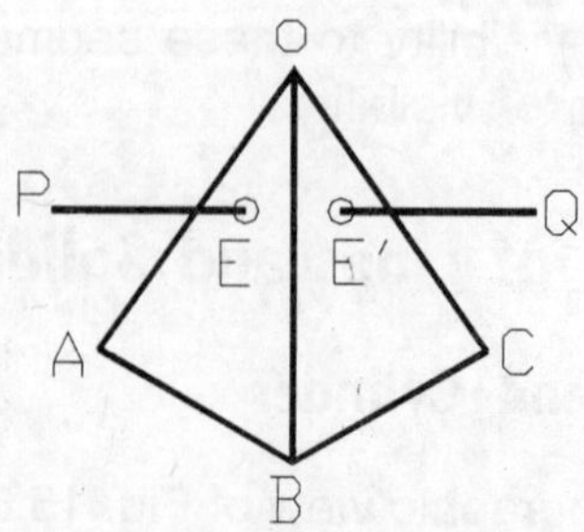


***Solution***

1. Consider line pq as a cutting plane either in the TV or FV. In this exercise, it is considered in the FV.
2. There are three cut points in the FV, which give four key points in the TV. Transfer the cut points from FV to TV and obtain four key points on four slant edges: oa, ob, oc and od. Join four key points and show the cut section profile in the TV.

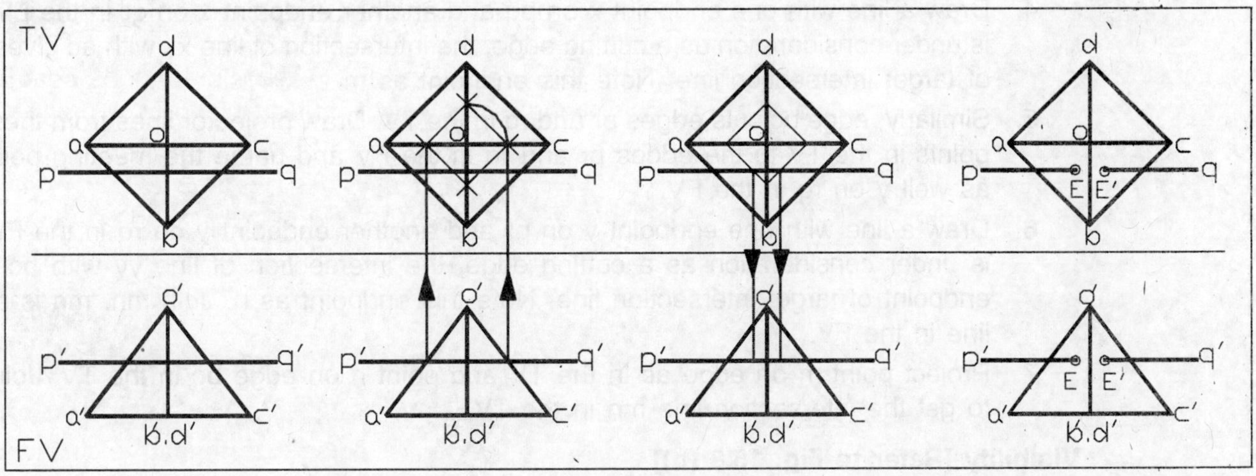


Fig. 15.7

Now consider intersection of the line pq and the cut section profile. The line pq cuts the section profile at two points. One point is considered as an entry point and another point is considered as exit point. The segment EE′ of the line pq is not visible in TV and FV. Trim the same in both the views.

## 15.4 Intersection of Plane versus Plane

### One Plane Intersects Another

**EXAMPLE 15.6** Figure 15.8(a)(1) shows a plane abc intersecting plane pqr. Determine the line of intersection and show the visible portion of these planes.

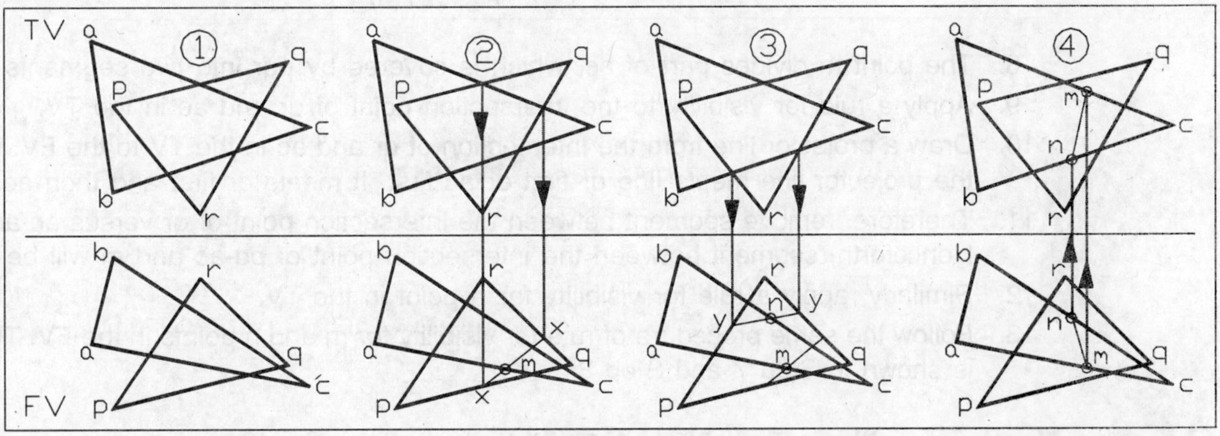


Fig. 15.8(a)

***Solution***

1. As plane abc intersects plane pqr consider the edges of plane abc as cutting edges. Edge ab cannot be considered as a cutting edge because it is completely outside and does not cut edges of plane pqr in both the views. The cutting edges are ac and bc.
2. Consider the cutting edges in either of the views. In this solution, TV is considered for the cutting edges of plane abc.
3. Cutting edge ac cuts edges pq and qr in the TV. Draw projector lines from the intersection points ac-pq and ac-qr in the TV to the edges pq and qr in the FV. Name the meeting points as x on pq as well x on qr in the FV.

4. Draw a line with one endpoint x on pq and another endpoint x on qr in the FV. As edge AC is under consideration as a cutting edge, the intersection of line xx with ac gives one endpoint of target intersection line. Note this endpoint as m.
5. Similarly, edge bc cuts edges pr and rq in the TV. Draw projector lines from these intersection points in the TV to the edges pr and rq in the FV and name the meeting points as y on pr as well y on rq in the FV.
6. Draw a line with one endpoint y on pr and another endpoint y on rq in the FV. As edge ac is under consideration as a cutting edge, the intersection of line yy with bc gives another endpoint of target intersection line. Note this endpoint as n. Join mn. mn is an intersection line in the FV.
7. Project point m on edge ac in the TV and point n on edge bc in the TV. Join these points to get the intersection line mn in the TV.

**Visibility [Refer to Fig. 15.8 (b)]**

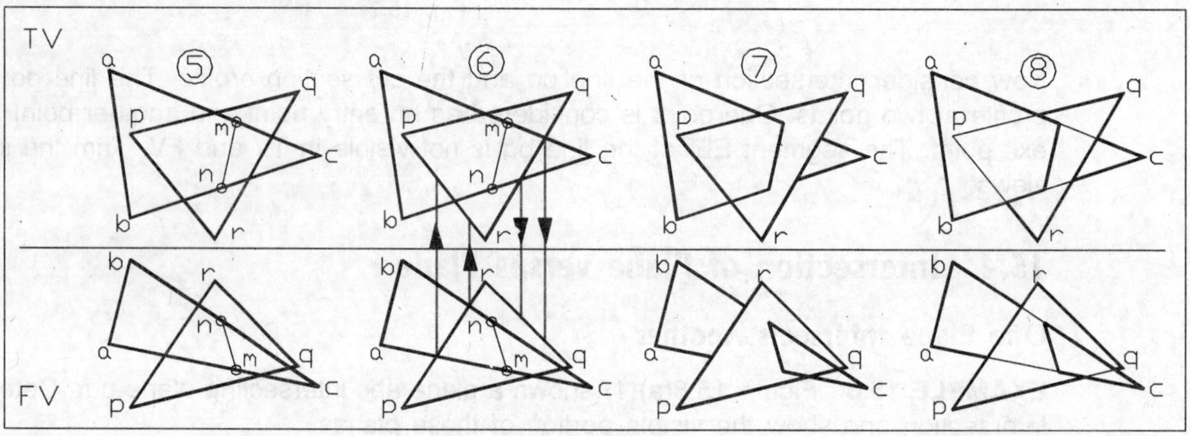


**Fig. 15.8(b)**

8. The point m divides part of ac, which is covered by pqr into two segments in the TV.
9. Apply a rule for visibility to the intersection point of qr and ac in the TV.
10. Draw a projector line from the intersection of qr and ac in the TV to the FV. Check whether, the projector line meets line qr first or ac first. It meets qr first and then ac.
11. Therefore, remove segment between the intersection point of qr versus ac and m in the TV. Henceforth, segment between the intersection point of pq-ac and m will be retained.
12. Similarly, apply a rule for visibility for n point in the TV.
13. Follow the same procedure of rule for visibility for m and n points in the FV. The final solution is shown in Step 7 and Step 8.

## Plane versus Plane (Two Plane Intersects Each Other)

**EXAMPLE 15.7** Figure 15.9(a) show planes abc and pqr intersecting each other. Determine the line of intersection and show the visibility.

***Solution*** Trial and error procedure is adapted to solve the problem as both the planes intersect each other.

Firstly, consider plane abc intersecting plane pqr. This will give one end point of intersection line.

1. As plane abc intersects plane pqr, consider edges of plane abc as cutting edges. The cutting edges could be ac, ab, and bc.
2. Consider cutting edges in either of the views. In this solution, TV is considered for the cutting edges of plane abc.
3. Trial could be done using either of the cutting edges: ab, ac, and bc.

4. Cutting edge ab cuts edges pq and qr in the TV. Draw projector lines from the intersection points of ab-pq and ab-qr in the TV to the edges pq and qr in the FV. Name the meeting points as x on pq as well x on qr in the FV.
5. Draw a line with point x on pq and point x on qr in the FV. As cutting edge ab is under consideration, the intersection of line xx with ab gives one endpoint of intersection line. Note this endpoint as m.

   The trial with cutting edges ac and bc can be tried. However, it will not give us endpoint of intersection line.

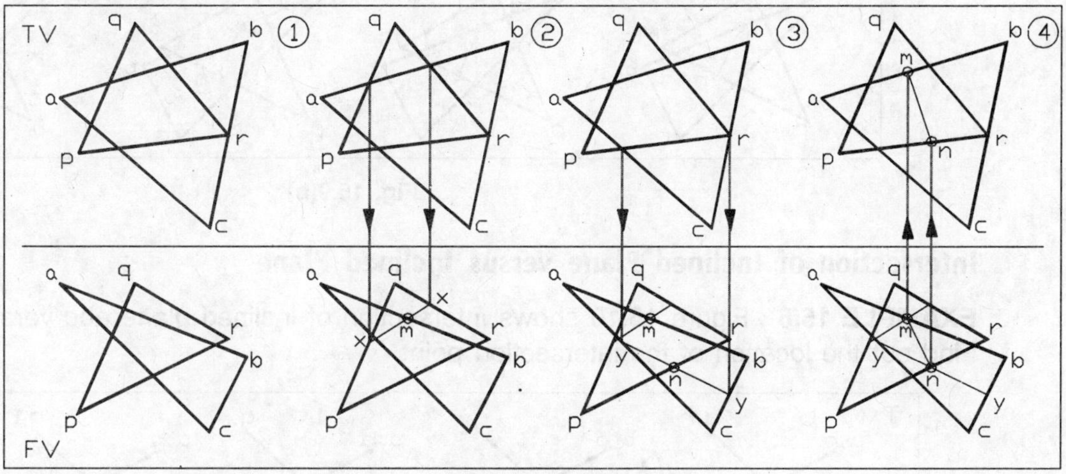


**Fig. 15.9(a)**

Secondly, consider plane pqr intersecting plane abc. This will give another endpoint of intersection line.

6. Trial could be done using either of the cutting edges: pq, pr, and qr. Edge pq, and qr will not give the desired result (another endpoint of intersection n). Therefore, the edge pr will be considered as a cutting edge in the TV.
7. Cutting edge pr cuts edges ac and bc in the TV. Draw projector lines from these intersection points of pr-ac and in the TV to the edges ac and bc in the FV and name the meeting points as y on ac as well y on bc in the FV.
8. Draw a line with one endpoint y on ac and another endpoint y on bc in the FV. The intersection of line yy with pr gives one endpoint of intersection line. Note it as n.
9. mn is an intersection line in the FV.
10. Project point m on edge ab in the TV and point n on edge pr in the TV. Join these points to get intersection line mn in the TV.

**Visibility [Refer to Fig. 15.9 (b)]**

11. The point m divides part of ab, which is covered by pqr into two segments in the TV.
12. Apply a rule for visibility to the intersection point of qr versus ab or pq versus ab in the TV.
13. Draw a projector line from the intersection of qr versus ab in the TV to the FV. Check whether, the projector line meets line qr first or ab first. It meets qr first and then ab.
14. Therefore, remove segment between the intersection point of pq versus ab and m in the TV. Henceforth, segment between the intersection point of qr-ab and m will be retained.
15. Similarly, apply a rule for visibility for n point in the TV.

Follow the procedure of rule for visibility for m and n points in the FV. The final solution is shown in Step 7 and Step 8.

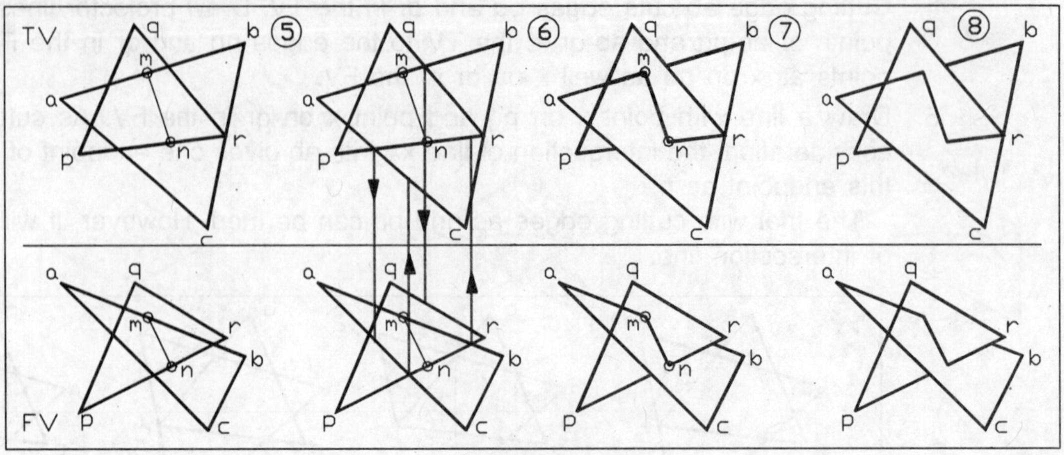

Fig. 15.9(b)

## Intersection of Inclined Plane versus Inclined Plane

**EXAMPLE 15.8** Figure 15.10 shows intersection of inclined plane abc versus inclined plane pqr. Find out the location of the intersection point.

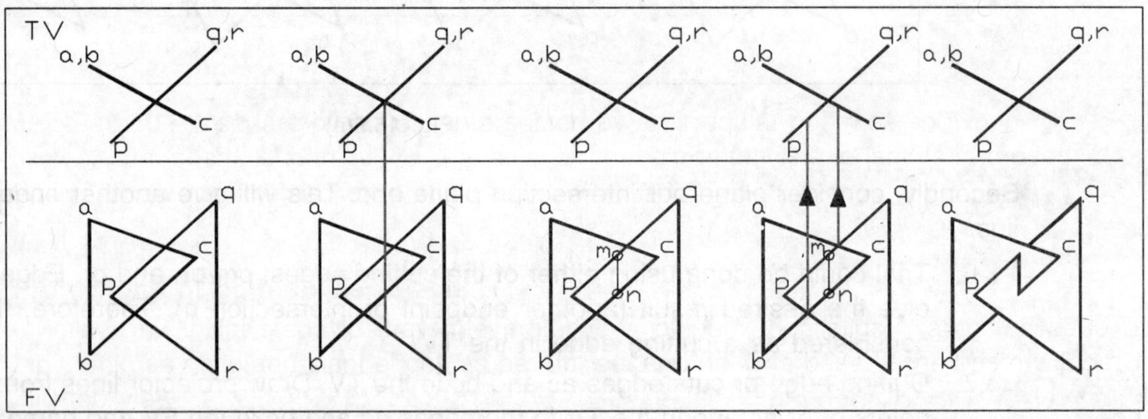

Fig. 15.10

***Solution***

1. The TV shows the line views of both plane abc and pqr. The intersection point of these two line views is a point view of line of intersection.
2. Draw a projector line from intersection point of the two line views to the FV.
3. In the FV, the line common to both the planes is a line of intersection mn.
4. Apply a rule of visibility to obtain the final solution.

## 15.5 Intersection of Surfaces of Two Solids

In case of intersection of two solids, one solid is taken out after the intersection. Student is required to draw the identities that are produced by the other solid during intersection.

### 15.5.1 Initial Steps

1. Draw the projections (TV and FV) of main solid.
2. Locate the TS of the base of the smaller solid. TS of the base of the smaller solid may be:
   (i) In the TV, if it is a case of vertical interpenetration.
   (ii) In the FV, if it is a case of horizontal interpenetration.

Accordingly, draw the other view.

## Horizontal Intersection

**EXAMPLE 15.9** A hexagonal pyramid is intersected by a square prism. The axis of the prism is parallel to HP, perpendicular to VP and intersects the axis of the pyramid at a point 75 mm above the ground. The prism is removed after the intersection. Draw the intersection profile if the faces of the prism are equally inclined to the ground.

***Solution***

(A) **Projections of both solids (Stage 1)**

1. Follow the initial steps to draw the projections of both the solids. Name the base of the pyramid as abcdef and the slant edges as oa, ob, oc, od, oe and of. The base of the square prism will be visible in the front view. The entry of the square prism is on the slant faces obc, ocd and the exit of square prism is on slant faces oaf and ofe.

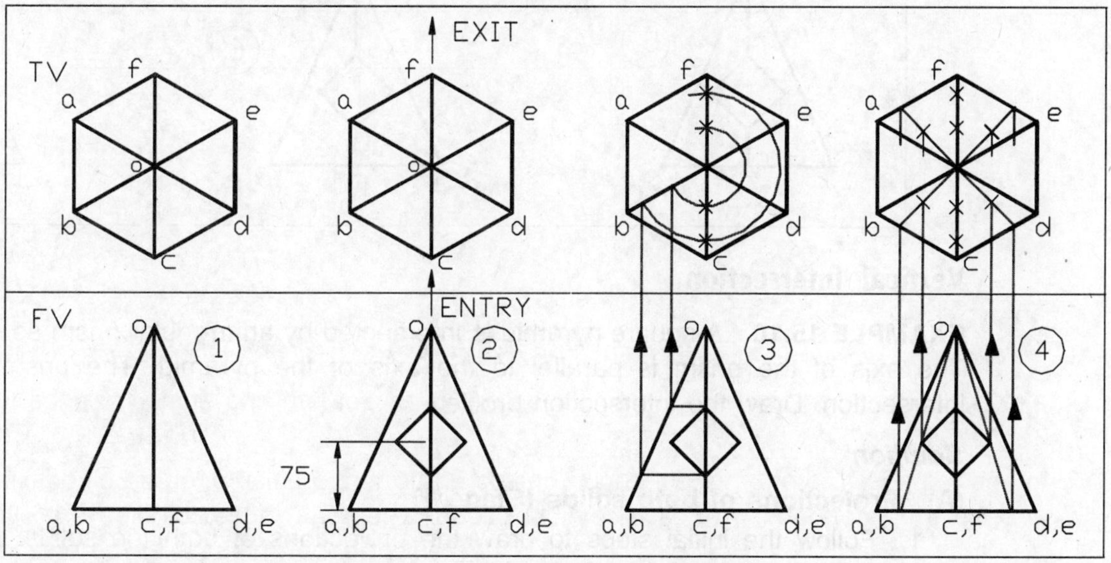


(B) **Determination of key points from the cut points (Stage 2)**

2. With reference to the base of the square prism, there are in total four cut points, two are on surface and two are on the edges: oc, of. Two surface cut points gives four key points, two in the entry zone and two in the exit zone, of. The cut points on edges: oc and of gives four key points, two in entry zone and two in the exit zone. Therefore, four cut points gives us eight key pòints, four key points are in the entry zone and four key points are in the exit zone.

(C) **Transfer of cut points to obtain the key points (Stages 3 and 4)**

3. The cut points on the edges: oc and of are transferred by using a line parallel to the base method. Mark these points by using a POINT command (Stage 3).
4. The two cut points, which are on the surfaces, are transferred in the TV by using a slant line method. Mark these points by using a POINT command (Stage 4).

(D) **Joining the key points for the entry zone and exit zone (Stage 5)**

5. Refer to the smaller base of the square prism for sequential joining of key points in the entry as well as exit zone.

   **Entry zone**

6. The entry is on surfaces obc and ocd. Follow the path of square in the FV to join the key points for the entry profile in the TV.
7. **Exit zone**
8. The exit is on surfaces oef and ofa. Follow the path of square in the FV to join the key points for the exit profile in the TV.

(E) **Trimming and visibility (Stages 6 and 7)**

9. Trim the part of the edges oc and of between the respective two key points in the TV (Stage 6).
10. The final solution is shown in Stage 7.

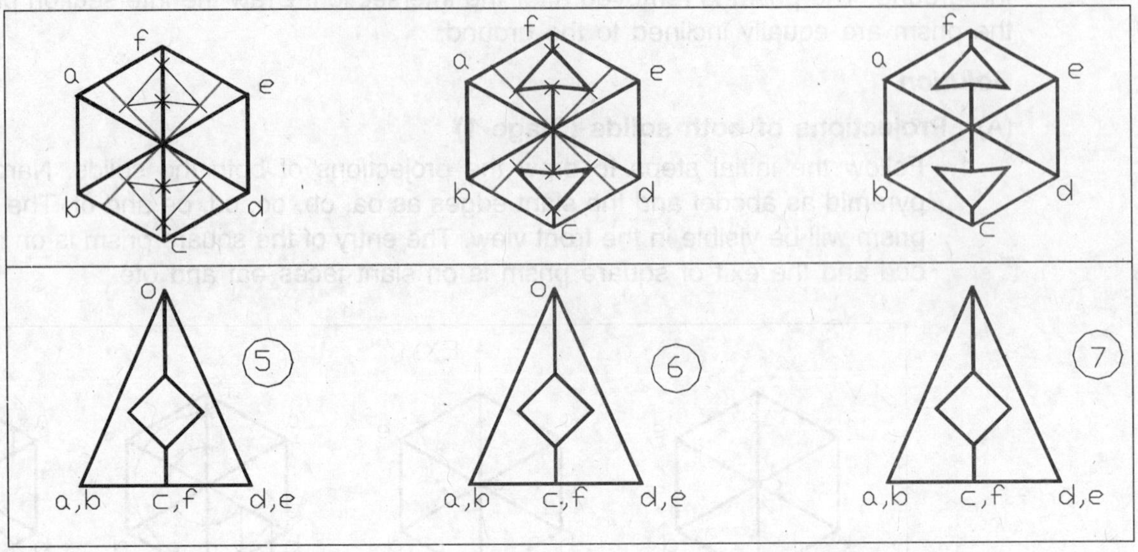


## Vertical Intersection

**EXAMPLE 15.10** A square pyramid is intersected by an irregular prism as shown in Fig. 15.11(2). The axis of the prism is parallel to the axis of the pyramid. The prism is removed after the intersection. Draw the intersection profile.

***Solution***

(A) **Projections of both solids (Stage 1)**

1. Follow the initial steps to draw the projections of both the solids. Name the base of the pyramid as abcd and slant edges as oa, ob, oc, and od. The base of the prism will be visible in top view.

(B) **Determination of key points from cut points (Stage 2)**

2. With reference to the base of the square prism, there are eight cut points, four are on four surfaces and four are on the edges: oa, ob, oc, and od. Eight cut points gives eight key points in FV, four key points on four edges and four key points on slant surfaces.

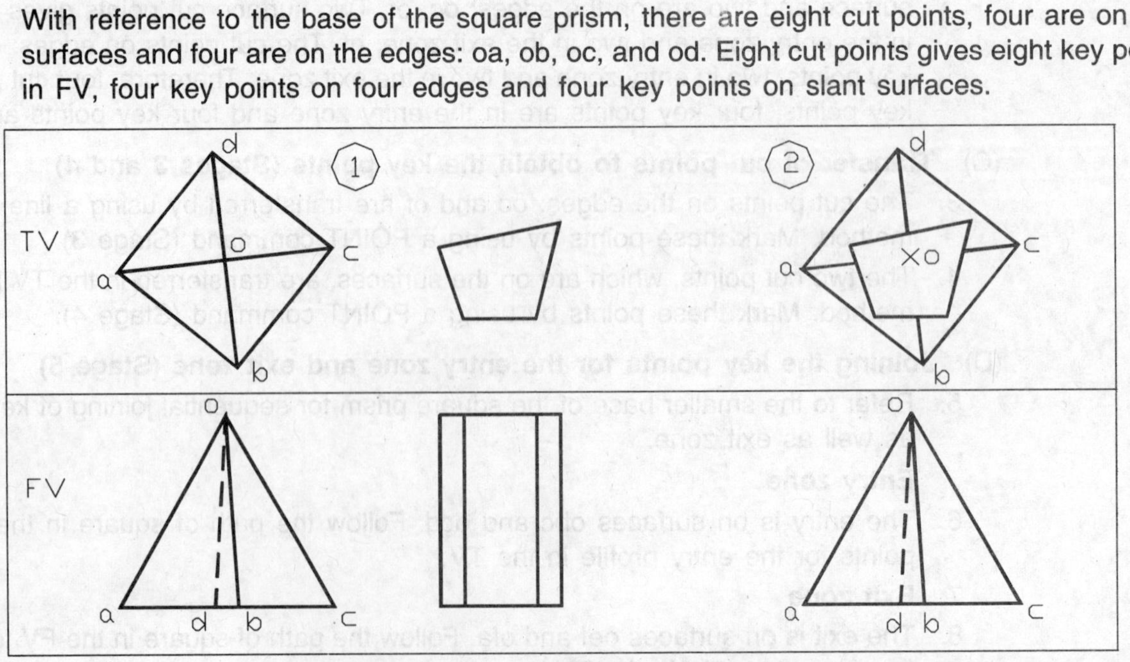


Fig. 15.11

(C) **Transfer of cut points to obtain key points (Stages 3 and 4)**

3. The cut points on the edges are directly transferred by drawing the projection lines from the cut point on respective slant edges. Mark these key points by using a POINT command (Stage 3).

   The four cut points, which are on the surfaces, are transferred in the TV by using a slant line method. Mark these points by using a POINT command (Stage 4).

(D) **Joining the key points (Stage 5)**

4. Refer to the true shape of the base of the pyramid. The visible slant surfaces are oab and obc in the TV. The invisible slant surfaces are ocd and oda in the TV.
5. For sequential joining of key points refer to the base of the prism.
6. Join the key points which are on visible surfaces by a solid line. Join the key points which are on invisible surfaces by a hidden line.

(E) **Trimming and visibility (Stages 5 and 6)**

7. Trim the part of slant edges from apex o to key points on the respective slant edges (Stage 5). Following this procedure, it is observed that the boundary lines of the views are hidden. However, no boundary line appears as a hidden line. Therefore, change only the boundary part of the hidden line as a solid line.
   The final solution is shown in Stage 6.

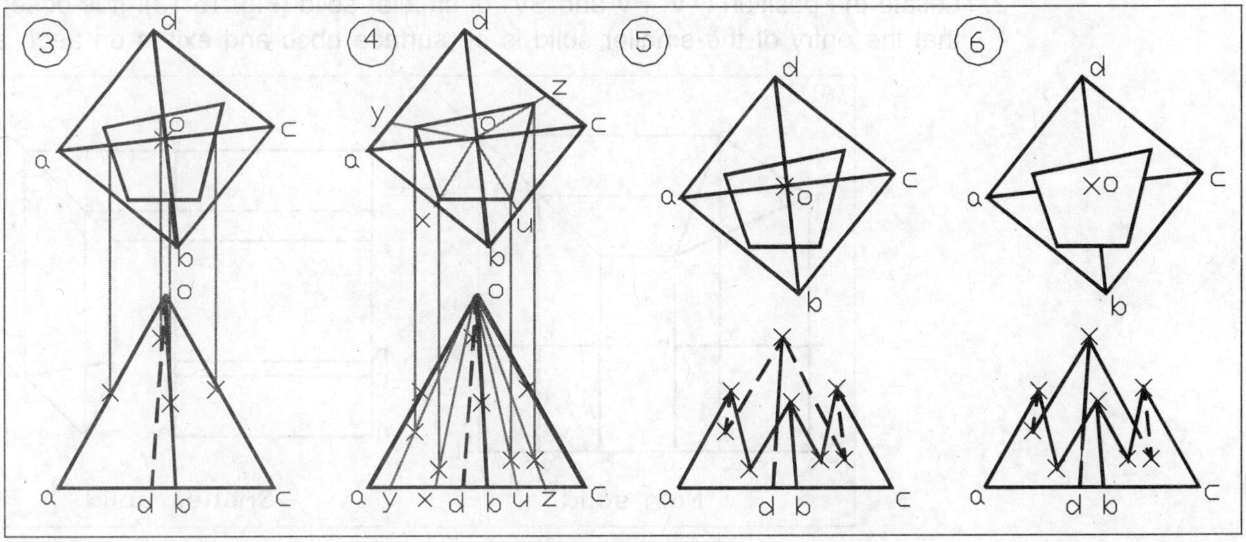


**Fig. 15.11**

## 15.6 Interpenetration of Two Solids

In case of interpenetration of two solids, both the solids are present. Student is required to draw the identities that are produced out of intersection of surfaces of both the solids. Figure 15.12 shows few applications of interpenetration of two solids.

**Fig. 15.12**

### 15.6.1 Initial Steps to Solve the Interpenetration Problem

1. Draw the projections (TV, FV) of main solid. Decide about the requirement of the side view. The side view is required only if axis of the smaller solid is parallel to HP and VP.
2. Locate TS of the base of the smaller solid. TS of the base of the smaller solid may be
   (a) In the TV, if it is a case of vertical intersection.
   (b) In the FV/SV, if it is a horizontal intersection.

Accordingly draw the other views.

### Horizontal Interpenetration (Axis of Smaller Solid Perpendicular to PP)

**EXAMPLE 15.11** A bigger triangular prism having axis perpendicular to the ground is interpenetrated by a smaller triangular prism. The axis of the smaller triangular prism is parallel to HP and VP. Draw an interpenetration profile.

***Solution***

(A) **Projections of both solids**

1. Follow the initial steps to draw the projections of both the prisms. The base of the smaller triangular prism will be visible in the side view as the axis is parallel to HP and VP. Name the base of the bigger prism as abc and vertical edges as aa, bb and cc. Name the base of the smaller prism as pqr and vertical edges as pp, qq and rr.
2. Locate the position (TV, FV and SV) of smaller solid (Fig. 15.13). It is observed from the TV that the entry of the smaller solid is on surface bbcc and exit is on aabb and aacc.

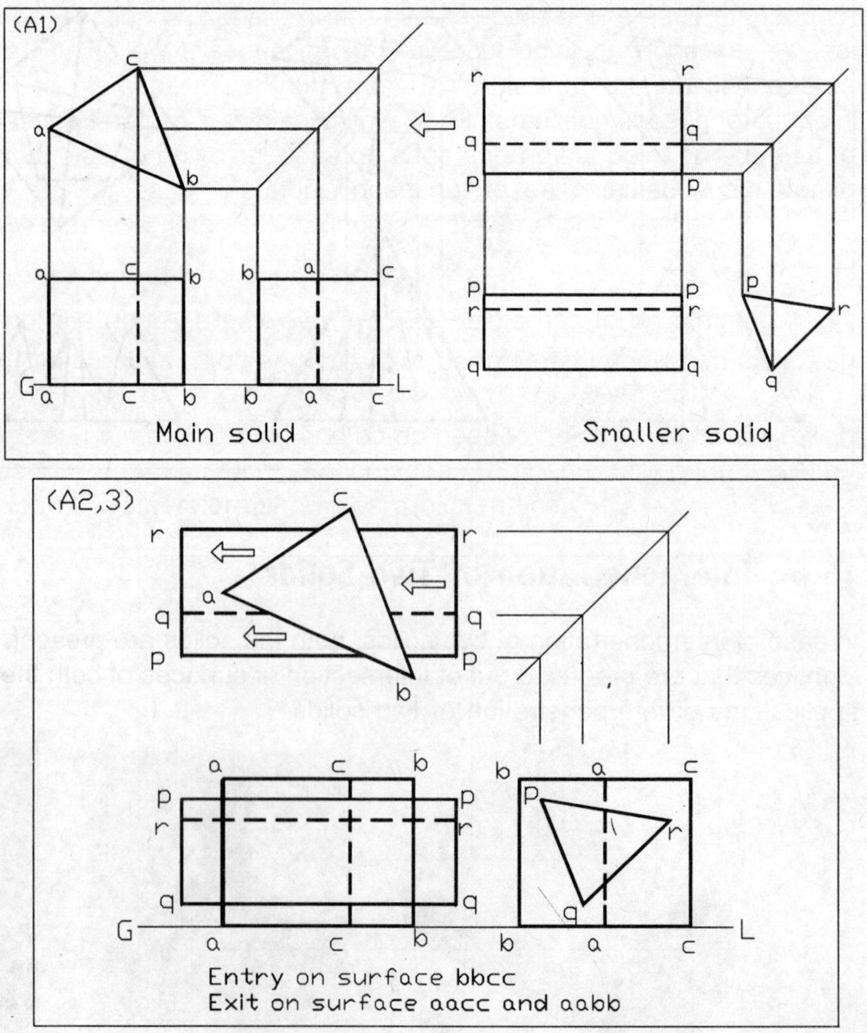


Fig. 15.13

(B) **Determination of key points from cut points**

3. With reference to the smaller base pqr in the side view, there are in total five cut points, three are on surface (bbcc) of bigger prism and two are on the edge: aa. Three surface cut points gives six key points, three in the entry zone and three in the exit zone. The cut points on edge aa give two key points, which come in exit zone. Therefore, five cut points gives us eight key points, three key points are in the entry zone and five key points are in the exit zone.

(C) **Transfer of cut points to obtain key points**

4. The cut points on the edge aa are transferred directly in the FV by drawing a projector line.
5. The three cut points on the surfaces have to be located for their true position on the true shape of the base of the main solid in the TV by a 45° line method. These locations are the intersection points, where, the edges rr, qq and pp cut the surface bbcc in the entry zone and surface aabb, aacc in the exit zone in the TV.
6. Transfer these true locations in the FV on respective edges rr, qq and pp and locate the key points using POINT command.

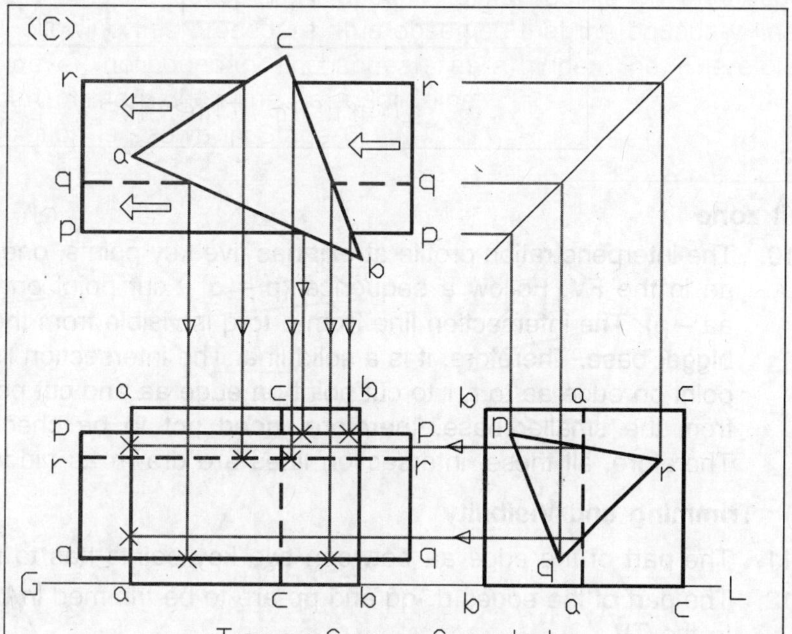

Transfer of points

(D) **Joining the key points for the entry zone and exit zone**

7. Refer to the base of the smaller solid in RSV for sequential joining of key points in the entry as well as in the exit zone in FV.
8. While joining the key points in the FV, both the true shape of the bases are observed (↑→) to know the common visibility of the individual key points. Referring to the smaller base pqr in the SV, surface ppqq is visible and surfaces qqrr, rrpp are not visible to the observer (→). Referring to the bigger base, surface aabb is visible and surfaces bbcc, ccaa are not visible to the observer (↑).

**Entry zone**

9. The interpenetration profile at entry has three key points, one each on pp, qq and rr in the FV. Follow a sequence {p – q – r – p}. The intersection line joining key point on p to key point on q is visible from the smaller base, however, not visible from the bigger base. Therefore, it is a hidden line. The intersection lines joining key point on q to key point on r and from key point on r to joining key point on p are not visible from the smaller base, therefore, need not to be checked again from the bigger base. Therefore, intersection lines: q to r and r to p are hidden.

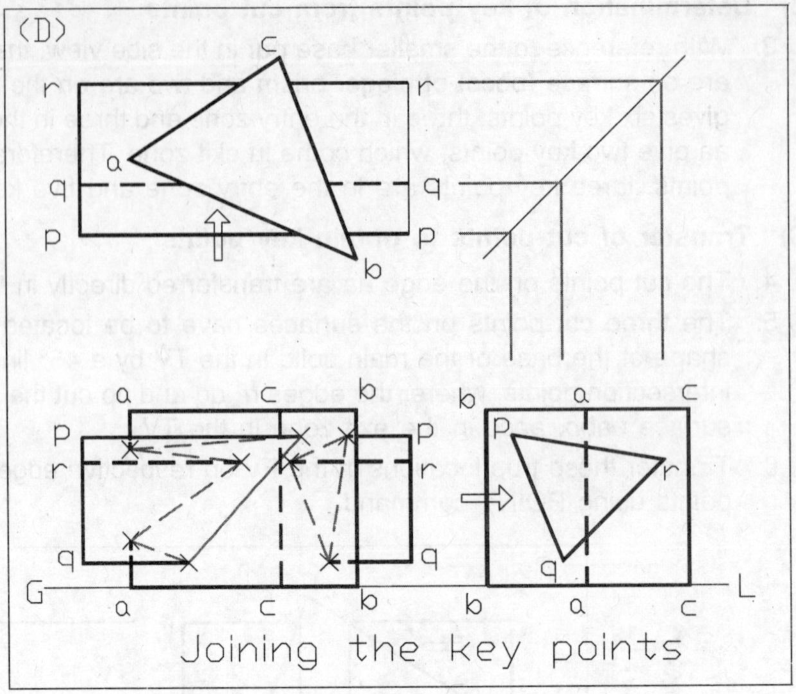


**Exit zone**

10. The interpenetration profile at exit has five key points, one each on pp, qq and rr and two o aa in the FV. Follow a sequence {p – q – cut point on edge aa – r – cut point on edg aa – p}. The intersection line from p to q is visible from the smaller base as well as from th bigger base. Therefore, it is a solid line. The intersection lines: q to cut point on edge aa, cu point on edge aa to r, r to cut point on edge aa and cut point on edge aa to p are not visibl from the smaller base, therefore, need not to be checked again from the bigger base Therefore, all these, intersection lines are drawn as hidden lines.

(E) **Trimming and Visibility**

11. The part of the edge aa between two key points has to be trimmed in the FV.
12. The part of the edges rr, qq and pp are to be trimmed in between their respective key point in the TV.
13. The edges aa-pp, aa-qq, bb-qq and bb-pp need to be checked for their visibility (by using th rule of visibility) in the interpenetration zone. The part of aa, qq and pp are not visible in th intersection zone, therefore shown as a hidden line.

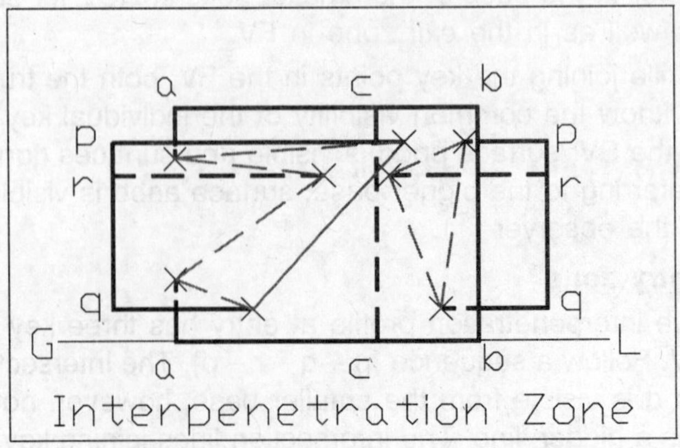

Figure 15.14 shows graphical display of interpenetration of prism versus prism derived from the Pro-Engineer software.

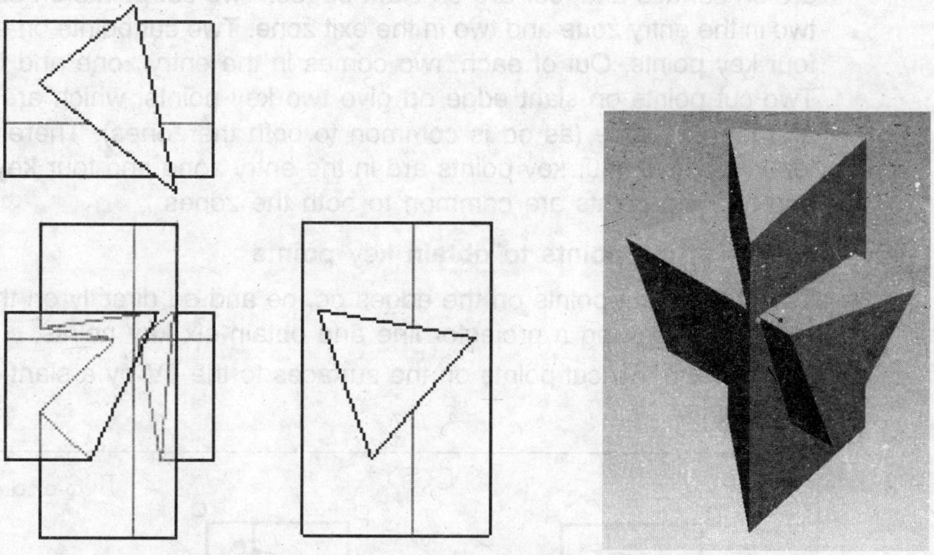

**Fig. 15.14** Interpenetration of prism versus prism.

## Horizontal Interpenetration (Axis of Smaller Solid Perpendicular to VP)

**EXAMPLE 15.12** A pentagonal pyramid is interpenetrated by a triangular prism. While interpenetrating, axis of the prism is perpendicular to VP. Draw a profile of interpenetration.

***Solution***

(A) **Projections of solids**

1. Follow the initial steps to draw the projections of both the solids. The base of the triangular prism will be visible in the front view as axis of the prism is perpendicular to VP. Name the base of the pyramid as abcde and apex as o. Name the base of the prism as pqr and vertical edges as pp, qq and rr.
2. It is observed from the TV that the entry of the smaller solid is on surfaces obc and ocd and exit is on oae and ode.

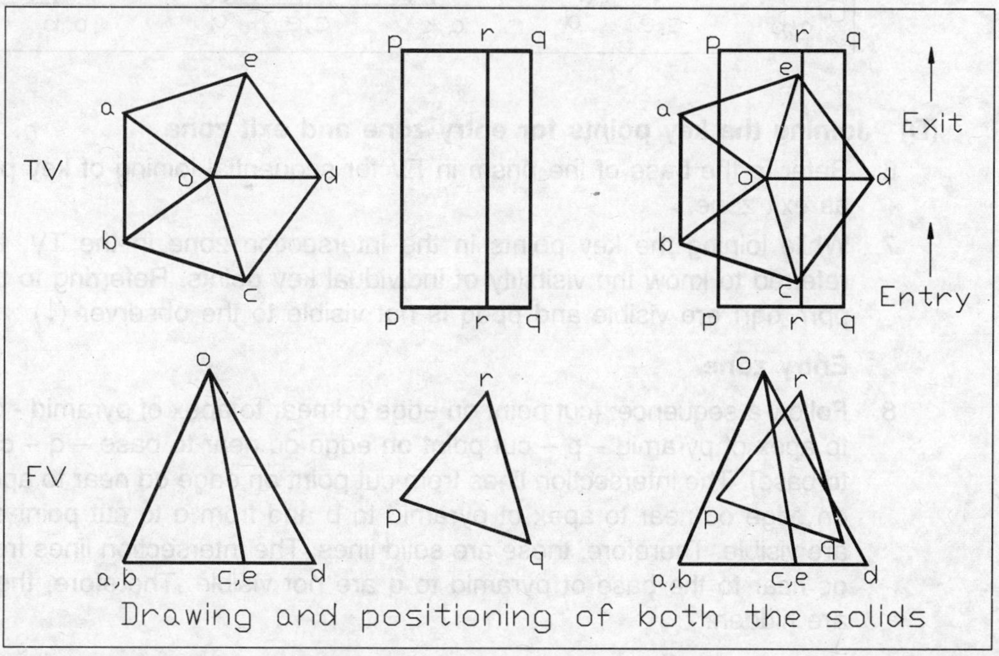


Drawing and positioning of both the solids

(B) **Determination of key points from cut points**

3. With reference to the base of the prism pqr in the FV, there are in total six cut points, two are on surface and four are on slant edges. Two cut points on surface gives four key points, two in the entry zone and two in the exit zone. Two cut points on slant edges: oc and oe gives four key points. Out of each, two comes in the entry zone and two comes in the exit zone. Two cut points on slant edge od give two key points, which are common to the entry zone and the exit zone (as od is common to both the zones). Therefore, six cut points gives us ten key points, four key points are in the entry zone and four key points are in the exit zone and two key points are common to both the zones.

(C) **Transfer of cut points to obtain key points**

4. Transfer the cut points on the edges oc, oe and od directly on the respective slant edges in the TV by drawing a projector line and obtain six key points.
5. Transfer the two cut points on the surfaces to the TV by a slant line method and obtain four key points.

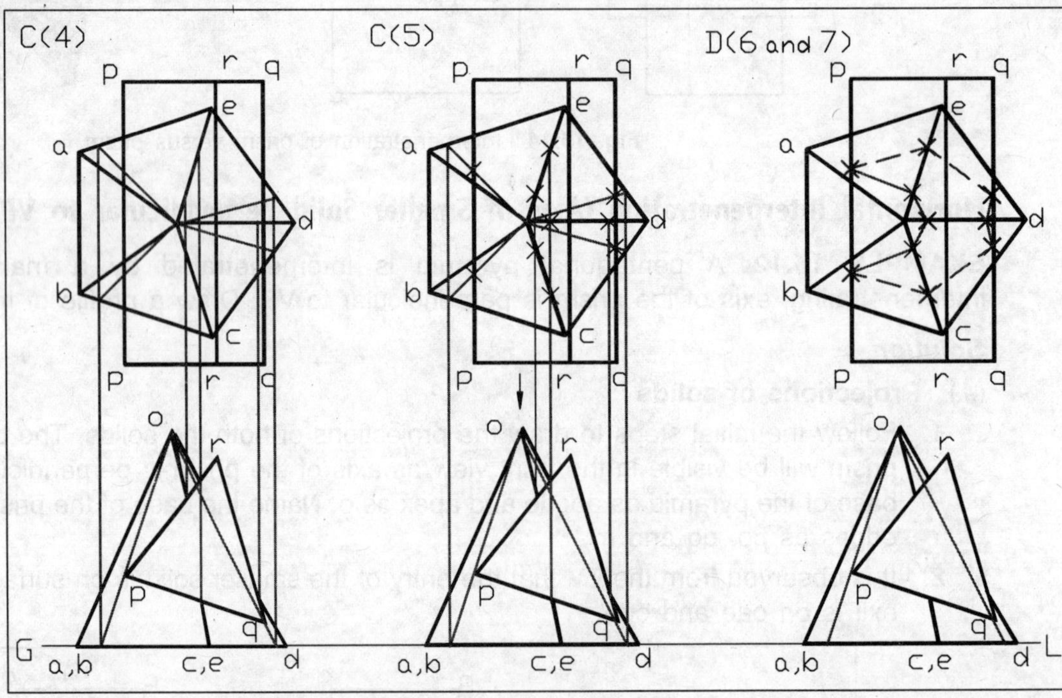


(D) **Joining the key points for entry zone and exit zone**

6. Refer to the base of the prism in FV for sequential joining of key points in the entry as well as exit zone.
7. While joining the key points in the intersection zone in the TV, the base of the prism is referred to know the visibility of individual key points. Referring to the smaller base, surfaces pprr, qqrr are visible and ppqq is not visible to the observer (↓).

**Entry zone**

8. Follow a sequence: {cut point on edge od near to apex of pyramid - cut point on edge oc near to apex of pyramid – p – cut point on edge oc near to base – q – cut point on edge od near to base}. The intersection lines from cut point on edge od near to apex of pyramid to cut point on edge oc near to apex of pyramid to p and from q to cut point on edge od near to base are visible. Therefore, these are solid lines. The intersection lines from p to cut point on edge oc near to the base of pyramid to q are not visible. Therefore, these two intersection lines are hidden.

**Exit zone**

9. Follow a sequence: {cut point on edge od near to apex of pyramid - cut point on edge oe near to apex of pyramid – p – cut point on edge oe near to base – q – cut point on edge od near to base}. The intersection lines from cut point on edge od near to apex to cut point on edge oe near to apex to p and from q to cut point on edge od near to base are visible. Therefore, these are solid lines. The intersection lines from p to cut point on edge oe near to the base to q are not visible. Therefore, these two intersection lines are hidden.

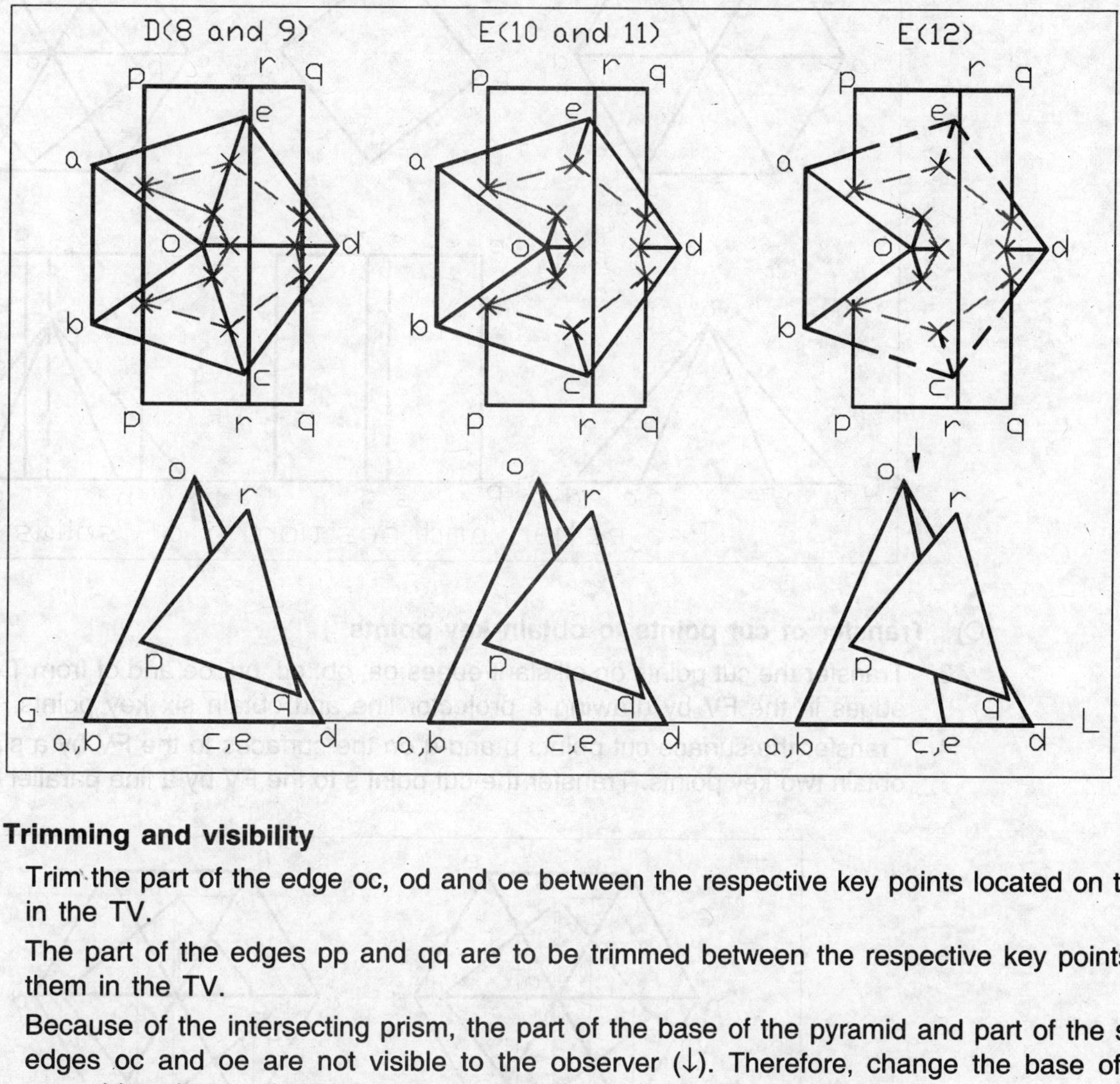


(E) **Trimming and visibility**

10. Trim the part of the edge oc, od and oe between the respective key points located on them in the TV.
11. The part of the edges pp and qq are to be trimmed between the respective key points on them in the TV.
12. Because of the intersecting prism, the part of the base of the pyramid and part of the slant edges oc and oe are not visible to the observer (↓). Therefore, change the base of the pyramid and part of slant edges oc and oe into hidden lines in the TV.

## Vertical Interpenetration (Axis of Smaller Solid Perpendicular to HP)

**EXAMPLE 15.13** A hexagonal pyramid is interpenetrated by a square prism. While interpenetrating, axis of the prism is perpendicular to the ground. Draw a profile of interpenetration.

***Solution***

(A) **Projections of solids**

1. Follow the initial steps to draw the projections of both the solids. The base of the square prism will be visible in the top view as axis of the prism is perpendicular to the ground. Name the base of the pyramid as abcdef and apex as o. Name the base of the prism as pqrs and vertical edges as pp, qq, rr and ss.

(B) **Determination of key points from cut points**

2. With reference to the base of the prism pqrs in the TV, in total there are nine cut points. Three cut points (p, q and s) are on surface. Remaining six cut points are on the slant edges of the pyramid. These nine cut points in the TV give nine key points in the FV.

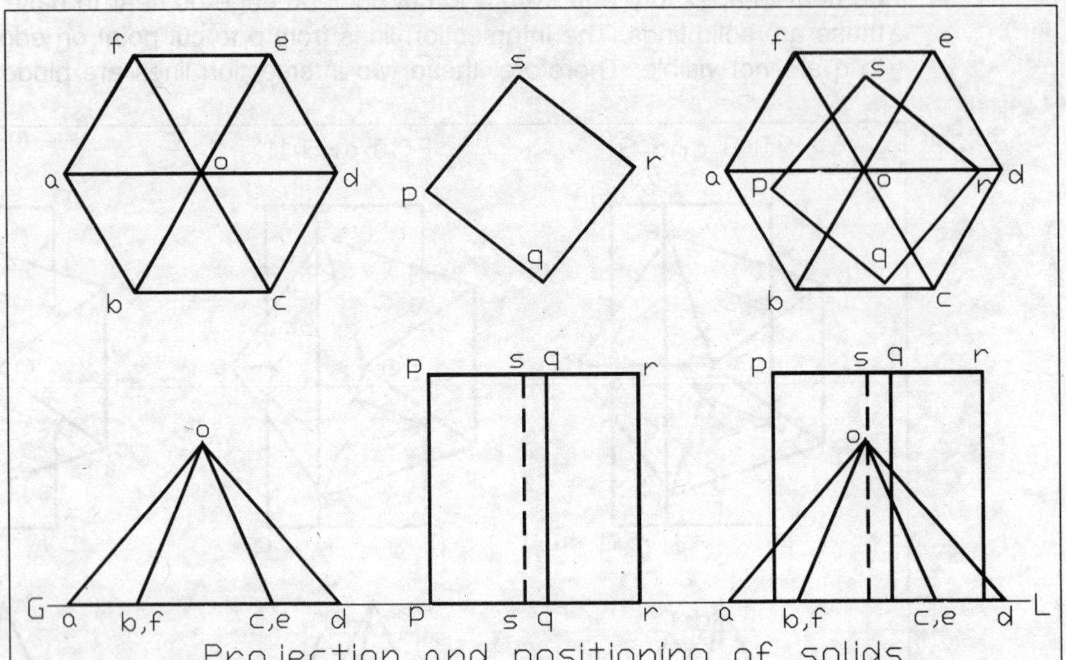

Projection and positioning of solids

(C) **Transfer of cut points to obtain key points**

3. Transfer the cut points on all slant edges oa, ob, oc, od, oe and of from TV to respective slant edges in the FV by drawing a projector line and obtain six key points.
4. Transfer the surface cut points p and q on the surfaces to the FV by a slant line method and obtain two key points. Transfer the cut point s to the FV by a line parallel to the base method.

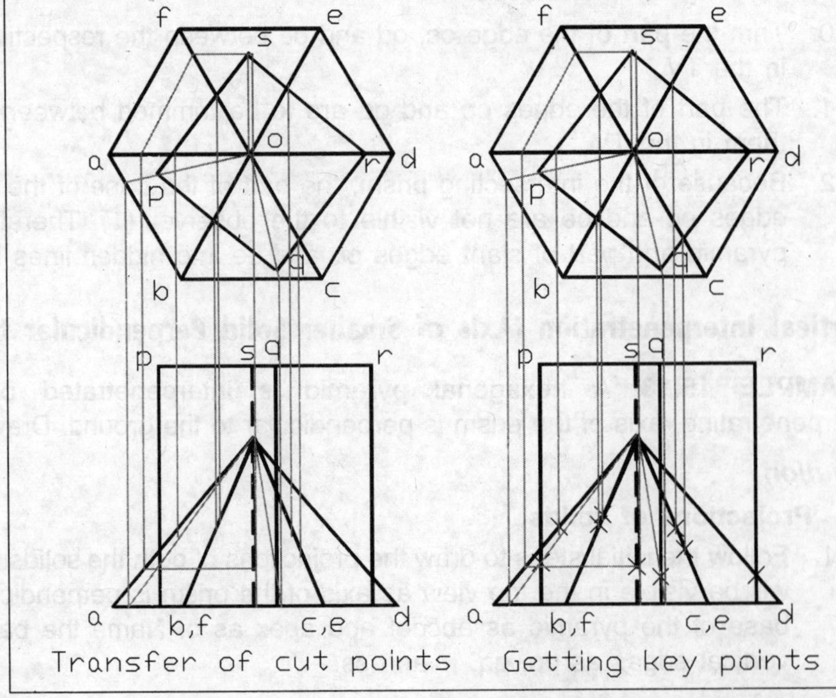

Transfer of cut points Getting key points

(D) **Joining the key points**

5. For sequential joining of key points in the FV, refer to the base of the prism in the TV.
6. For the visibility of individual key points, in the interpenetration zone, the true shape of the base of the prism as well as true shape of the base of the pyramid is referred. Referring to the true shape of the base of the prism in the TV, surfaces ppqq, qqrr are visible and rrss, sspp are not visible to the observer (↑) i.e., from p to q to r. Referring to true shape of the base of the pyramid, slant surfaces oab, obc, ocd are visible and slant surfaces ode, oef, ofa are not visible to the observer (↑). The common visibility has to be considered here. According to the common visibility, visible cut points are: cut point on slant surface oab, cut point on slant edge ob, cut point on slant surface obc and cut point on slant edge oc and cut point on slant edge od. Remaining four cut points are not visible to the observer (↑).
7. Follow a sequence {p – cut point on edge ob – q – cut point on edge oc – r. The intersection lines for this path are visible according to common visibility criterion. Therefore, these intersection lines are solid lines.
8. Follow a sequence: r-cut point on edge oe – s – cut point on edge of – cut point on edge oa – p. The intersection lines for this path are not visible according to common visibility criterion. Therefore, these intersection lines are hidden lines.

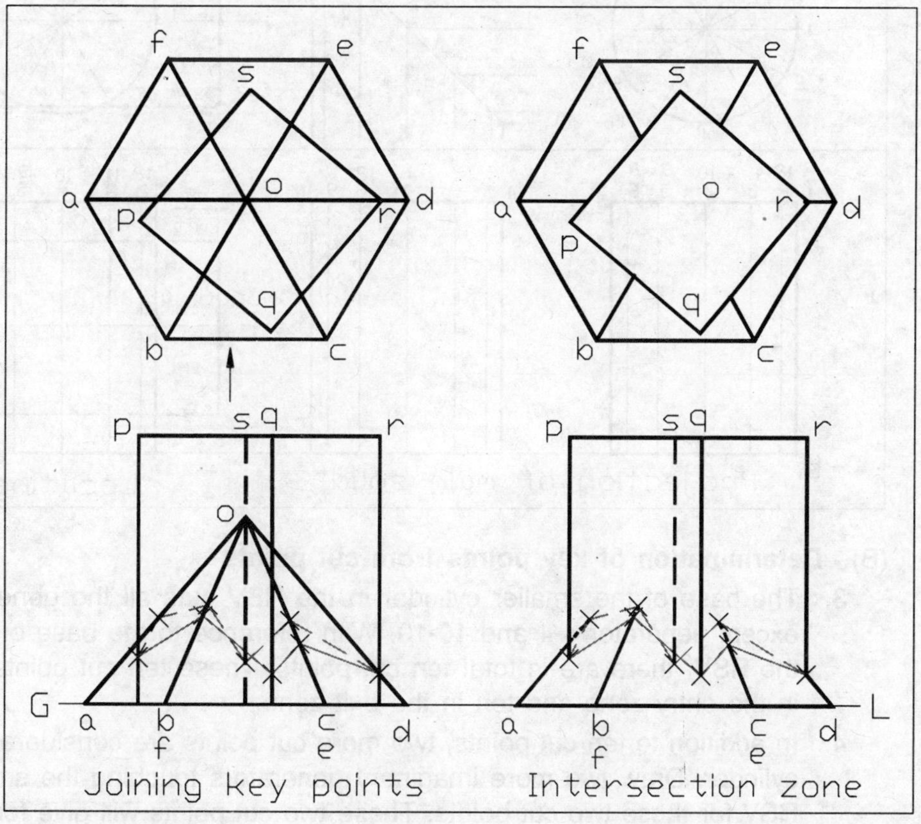


(E) **Trimming and visibility**

9. Trim part of all six slant edges of pyramid: oa, ob, oc, od, oe and of from apex to the respective key points on each of them in the FV. The part of vertical edges: pp, qq, rr and ss are trimmed beyond key points.
10. The slant edges oe and of are basically hidden. However, as they are coinciding with oc and ob, the FV shows solid lines. In the interpenetration zone, key points on oc and ob are nearer to base. If the position of key point on the visible slant edge is nearer to the base and key point position of non-visible slant edge is nearer to the apex then the part between these two key points will be changed into hidden lines. This is done here for the non-visible slant edges oe and of.

11. Visibility rule is applied for slant edge oa versus vertical edge pp for their visibility in the interpenetration zone. The part of oa is not visible as edge pp is in front of slant edge oa.

## Cylinder versus Cylinder (Horizontal Interpenetration)

**EXAMPLE 15.14** A bigger cylinder having axis perpendicular to HP is interpenetrated by a smaller cylinder. Axis of the smaller cylinder is parallel to HP and VP. Draw the interpenetration profile.

***Solution***

(A) **Projections of solids**

1. Follow initial steps to draw the projections of both the cylinders. Divide the base of the bigger cylinder into 12 equal numbers of parts. Draw twelve generators 1-1 to 12-12 in the FV as well as in the RSV. The base of the smaller cylinder will be visible in the side view as axis is parallel to HP and VP.
2. Draw TV, FV and SV of smaller solid.

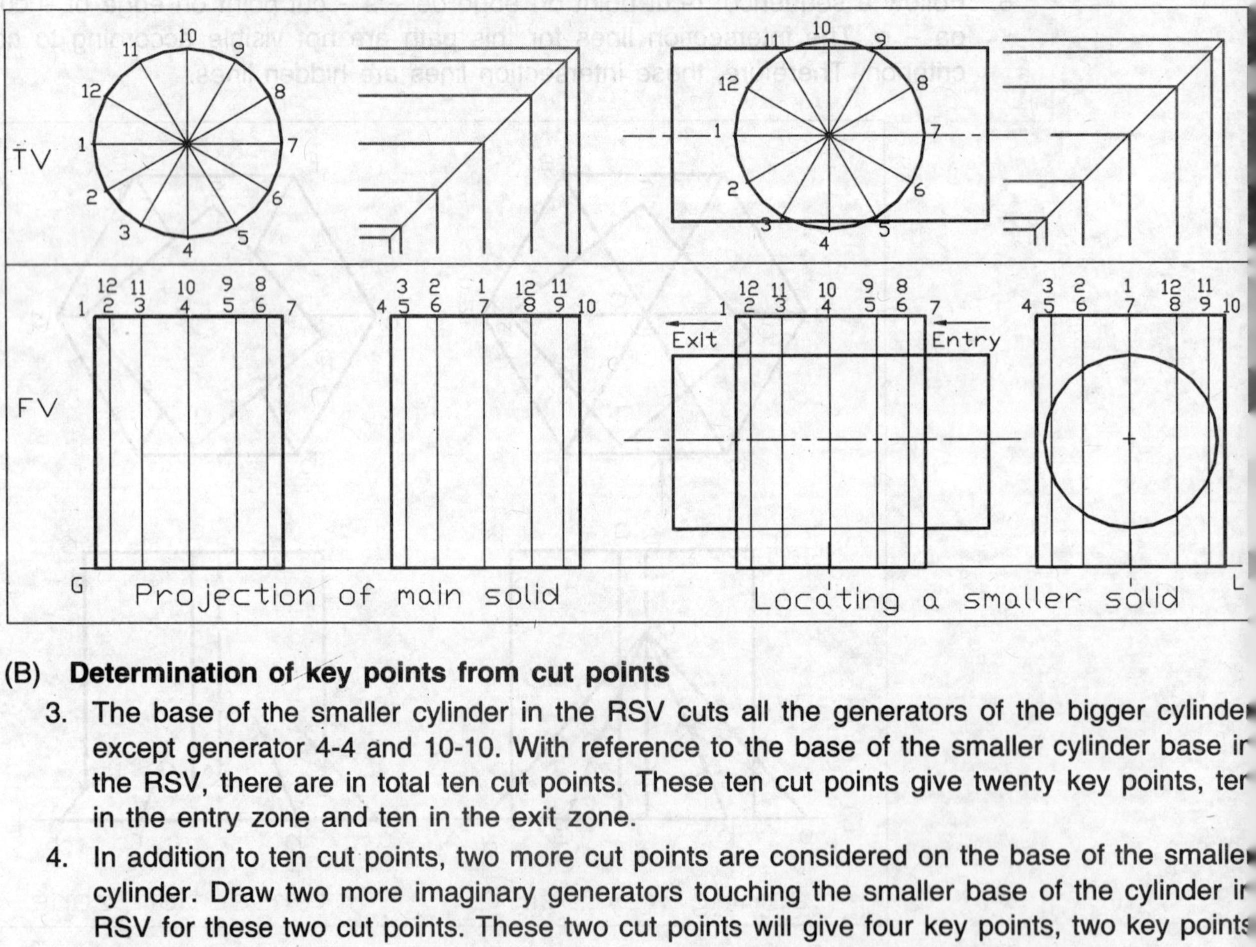


(B) **Determination of key points from cut points**

3. The base of the smaller cylinder in the RSV cuts all the generators of the bigger cylinder except generator 4-4 and 10-10. With reference to the base of the smaller cylinder base in the RSV, there are in total ten cut points. These ten cut points give twenty key points, ten in the entry zone and ten in the exit zone.
4. In addition to ten cut points, two more cut points are considered on the base of the smaller cylinder. Draw two more imaginary generators touching the smaller base of the cylinder in RSV for these two cut points. These two cut points will give four key points, two key points in entry zone and two key points in exit zone. Therefore, in total there will be twenty-four key points.

(C) **Transfer of cut points to obtain key points**

5. The original ten cut points on all generators in the RSV are directly transferred in the FV by drawing the projector lines.
6. For additional two cut points in the RSV, the imaginary generators have to be drawn in the FV to obtain respective key points. The two cut points on the additional generators have to be located for their true position on the true shape of the base of the bigger cylinder in TV by a 45° line method. These locations are the intersection points, where, the extreme generators of smaller cylinder intersect the base of the bigger cylinder in the TV.

7. Transfer the true location of two generators in the FV by drawing projector lines from TV to FV.

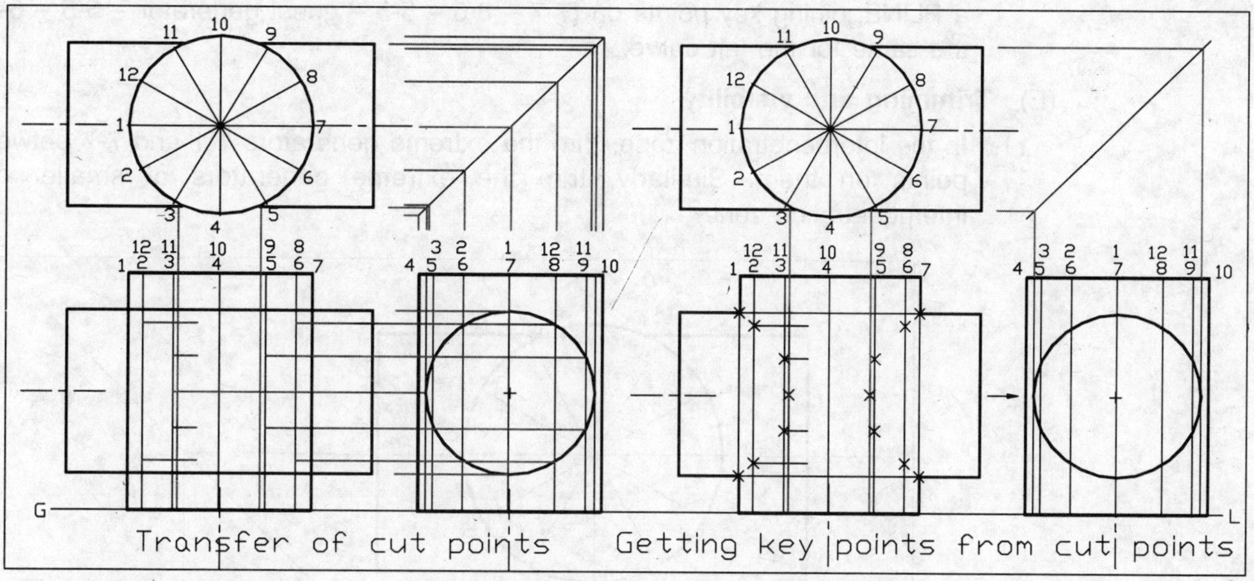


**(D) Joining the key points for the entry zone and exit zone**

8. Refer to the base of the smaller solid for sequential joining of key points in the entry as well as exit zone.
9. With reference to the base of the smaller cylinder in RSV, the cut points on the left-half of the cylinder are visible to the observer (→). Similarly, with reference to the base of the bigger cylinder in the TV, the cut points on the front-half of the cylinder are visible to the observer. However, as the interpenetration is very symmetric, key points at front side will coincide with the key points on the back side.

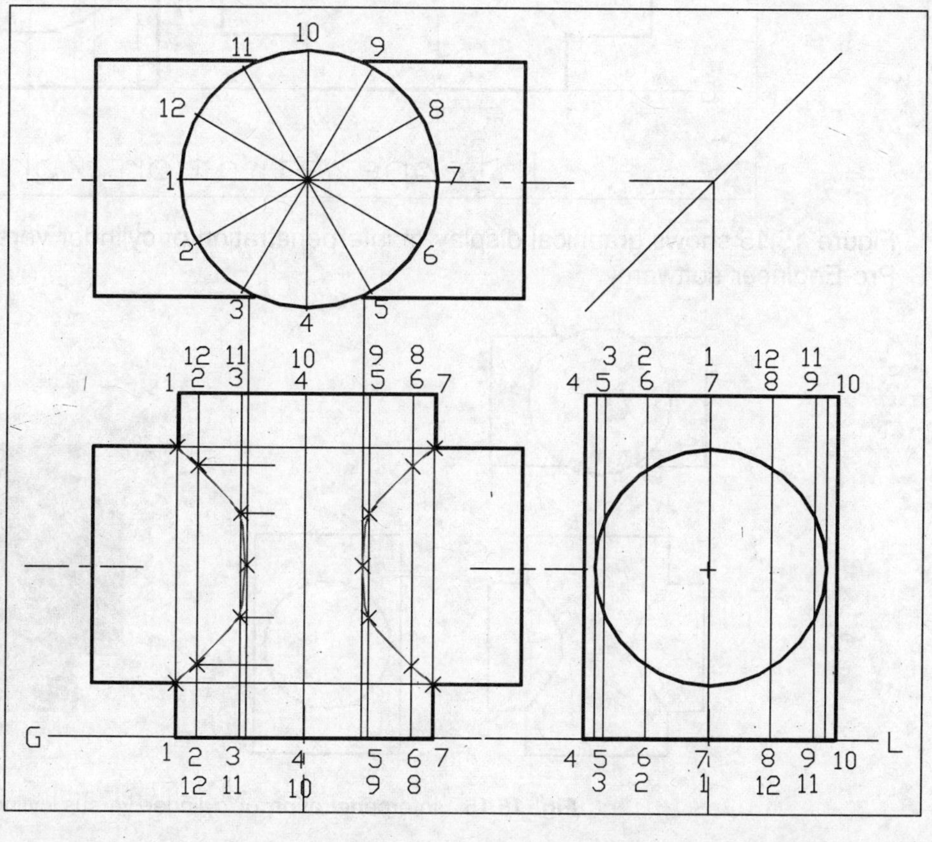

**Entry zone**

10. Join all the key points which are on the right side of axis of the standing cylinder by drawing a PLINE joining key points on {7-7 – 6-6 – 5-5 – guest generator – 5-5 – 6-6 – 7-7}. Mirror the same for the left curve.

(E) **Trimming and visibility**

11. In the interpenetration zone, trim the extreme generators 1-1 and 7-7 between the two key points on them. Similarly, trim the extreme generators of smaller cylinder in the interpenetration zone.

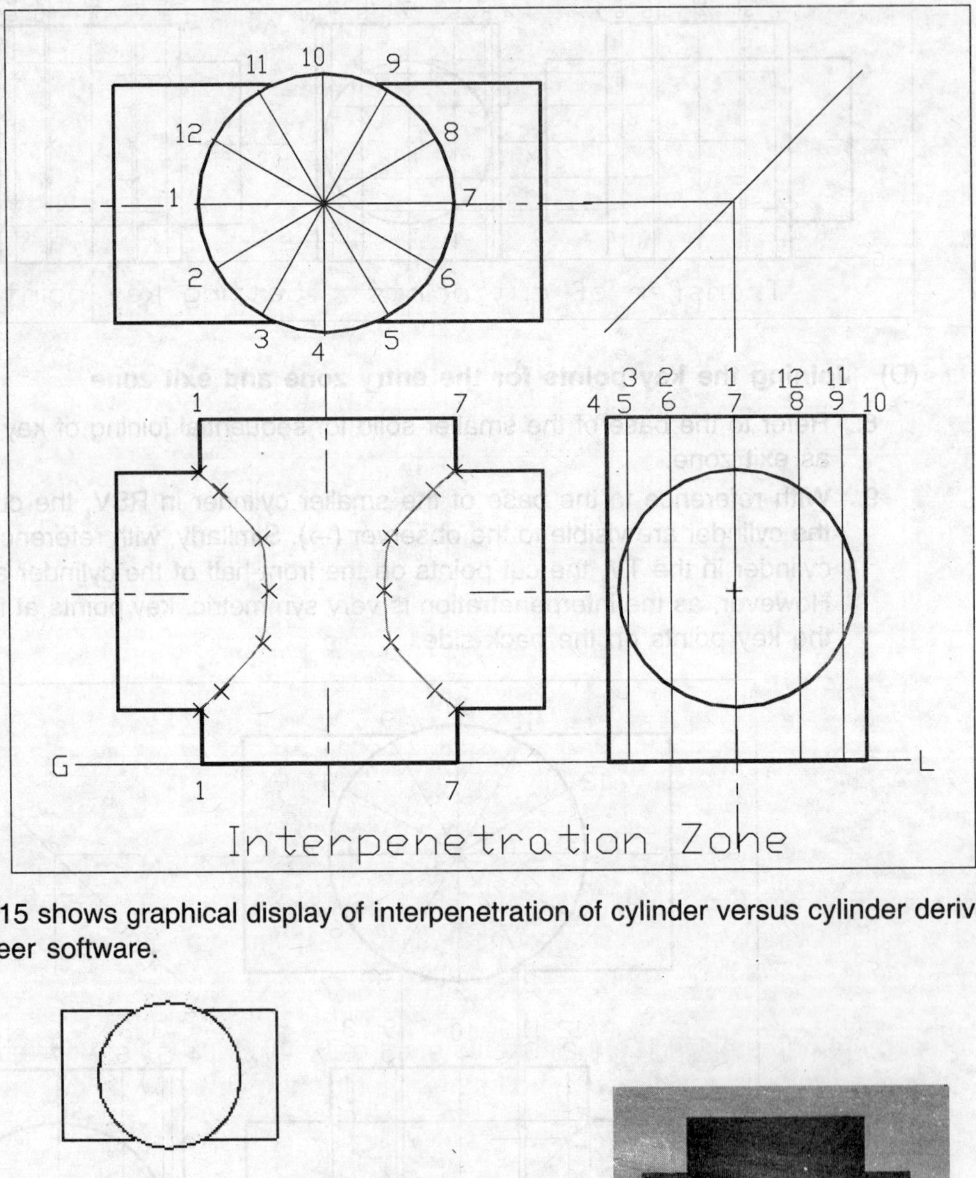


Figure 15.15 shows graphical display of interpenetration of cylinder versus cylinder derived from the Pro-Engineer software.

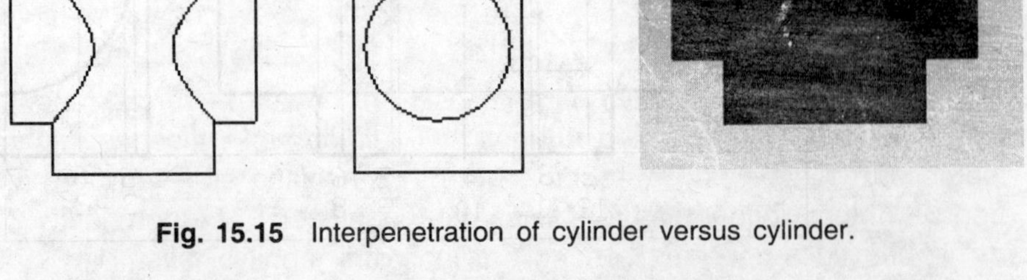

**Fig. 15.15** Interpenetration of cylinder versus cylinder.

Plate 15.5 shows few model solutions.

## Exercise A (Point of Intersection and Line of Intersection)

Solve for the line of intersection in the given problems of Plate 15.1 and 15.2.

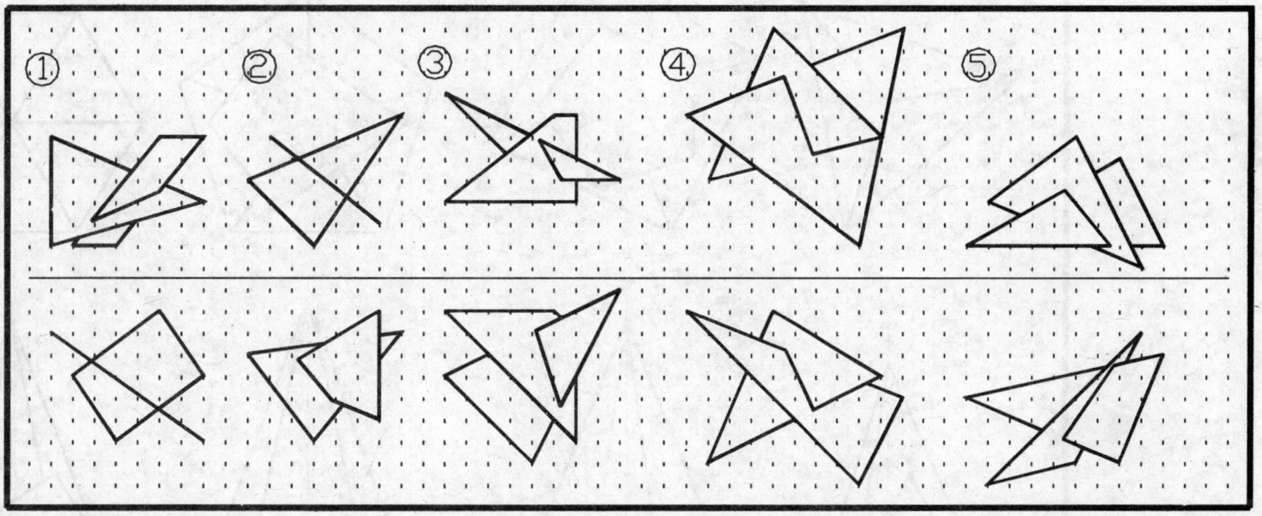

PLATE 15.1

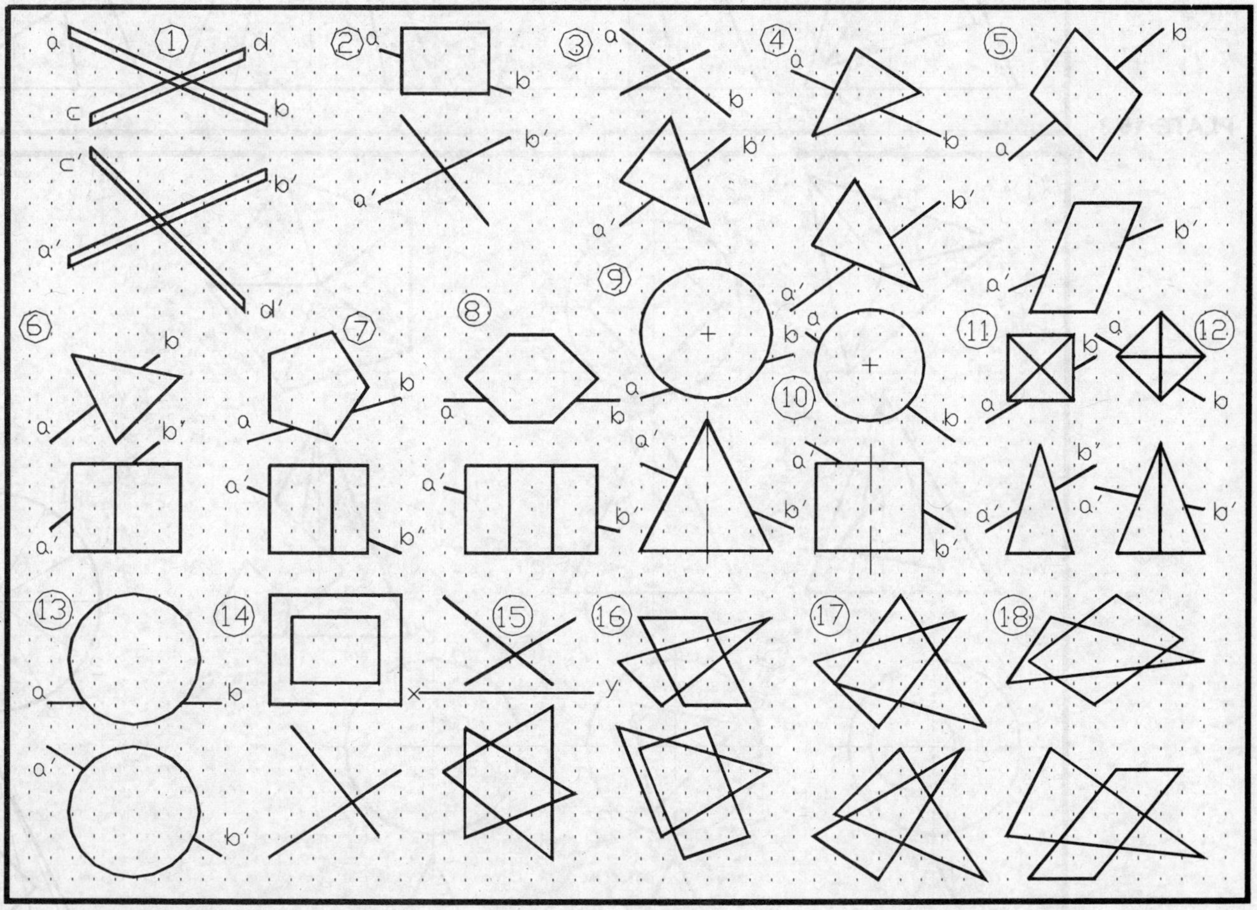

PLATE 15.2

## Exercise B (Intersection of Two Solids)

Solve for the profile of intersection for the engineering solids shown in Plate 15.3. and Plate 15.4.

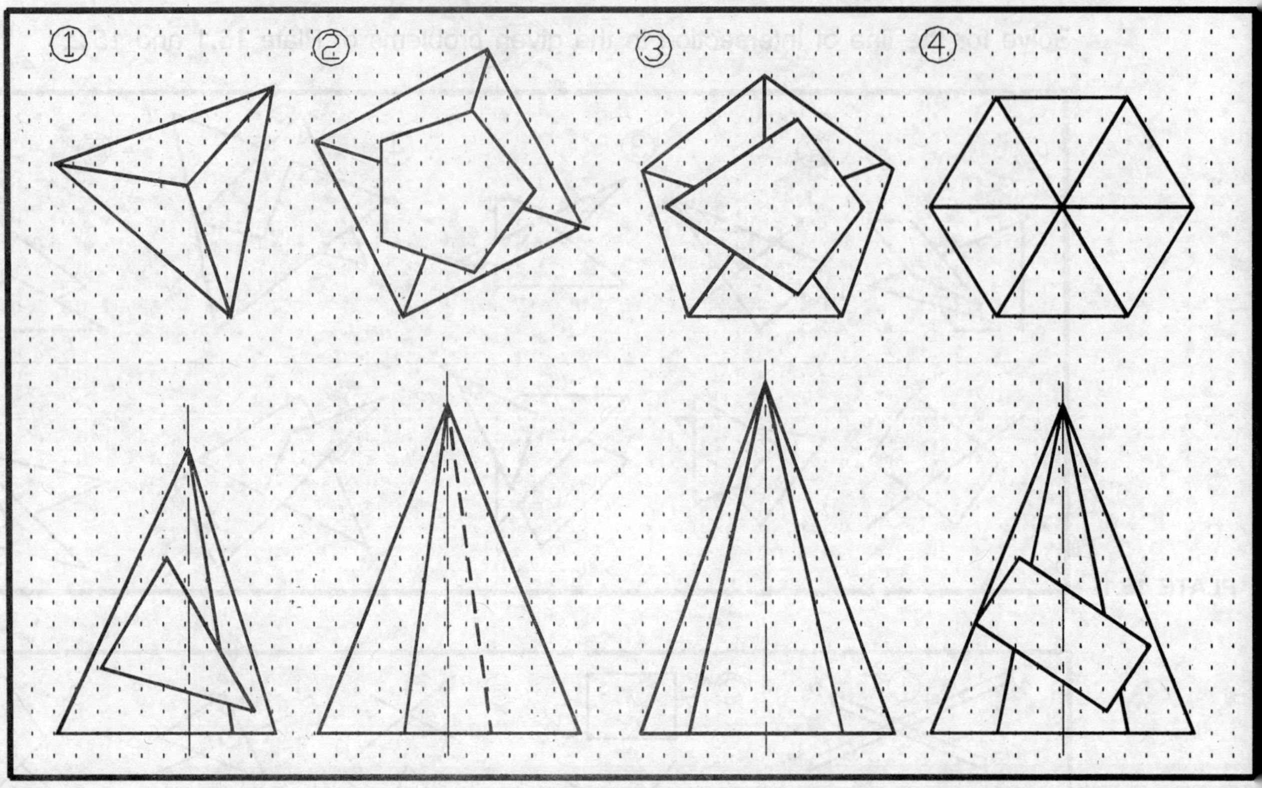


PLATE 15.3

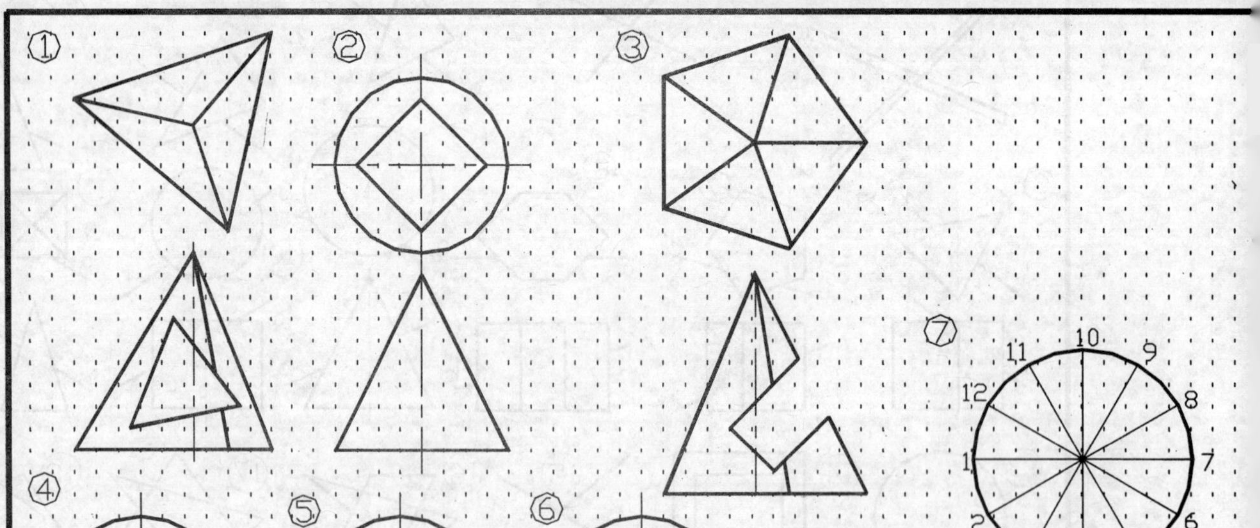


PLATE 15.4

**15.1** A hexagonal pyramid (base edge = 100 mm and height = 200 mm) stands on its base on the ground with one base edge perpendicular to VP. A square prism having base diagonal 75 mm and base parallel to the VP, intersects the pyramid. Axes of both the solids intersect each other at a distance of 75 mm from the ground with the rectangular faces of square prism equally inclined to the ground. Draw a profile of intersection if the prism is removed after the intersection.

**15.2** A pentagonal pyramid (base edge = 125 mm and height = 200 mm) stands on the ground with one base edge perpendicular to the VP. Draw its projections. Draw a square in the FV with diagonal 75 mm, parallel to the ground and centre of the square 125 mm from apex of the pyramid. This represents a through hole in the pyramid. Obtain a profile of this square hole in the TV.

**15.3** A cone (base diameter = 200 mm and height = 200 mm) is standing on its base on the ground. It is intersected by a square prism (base diagonal 150 mm). The axis of the prism coincides with the axis of the cone and rectangular faces of the prism are equally inclined to the VP. Draw identities of the square prism on cone after its removal.

**15.4** A cone (base diameter = 250 mm and height = 200 mm) is standing on its base on the ground. It is intersected by a square prism (base diagonal 150 mm). The axes of both the solids are in the same plane and separated by a distance of 25 mm with the rectangular faces of the prism equally inclined to the VP. Draw identities of the square prism on the cone after its removal.

**15.5** A cone (base diameter = 200 mm and height = 200 mm) is standing on its base on the ground. It is intersected by a cylinder of diameter 150 mm. Axis of the cylinder coincides with axis of the cone. Draw identities of the cylinder on the cone after its removal.

**15.6** A cone (base diameter = 250 mm and height = 200 mm) is standing on its base on the ground. It is intersected by a cylinder of diameter 150 mm. Axes of both the solids are in the same plane and separated by a distance of 25 mm. Draw identities of cylinder on the cone after its removal.

**15.7** A square pyramid (base diagonal = 200 mm and height = 225 mm) is resting on its base on the ground with base edges equally inclined to the VP. A prism having trapezium base (with parallel sides 50 mm and 100 mm separated by 50 mm) intersects the square pyramid completely. While intersecting, axis of the prism is perpendicular to the VP and it is in the plane of axis of the pyramid. The 50 mm side is parallel to the ground and 25 mm from the base of the pyramid. Obtain a profile of this intersection after removal of prism if the 100 mm side is towards the apex of the pyramid.

**15.8** A square pyramid (base diagonal = 200 mm and height = 250 mm) is resting on its base on the ground with base edges equally inclined to the VP. A prism having trapezium base (with parallel sides 50 mm and 100 mm separated by 50 mm) intersects the square pyramid. While intersecting, axis of the prism is perpendicular to the VP and in the plane of axis of the pyramid. The 100 mm side is parallel to the ground and 50 mm from the base of the pyramid. The solid is removed after the intersection. Obtain the profile of this intersection if the 50 mm side is towards the apex of the pyramid.

**15.9** A square prism (base diagonal = 200 mm and height = 250 mm) is resting on its base on the ground with vertical faces equally inclined to the VP. A triangular prism (base edge = 100 mm and height = 300 mm) is intersecting the square prism with two of its rectangular faces equally inclined towards the top base of the square prism and vertical edges perpendicular to the VP. Draw a profile of intersection after removal of the triangular prism. (Assume appropriate location of the intersecting point of both the axes for full coverage of the triangular prism).

**15.10** A hexagonal pyramid (base edge = 100 mm and height = 250 mm) is resting on its base on the ground with two base edges perpendicular to the VP. It is intersected by a square prism (base edge 100 mm and height 250 mm) with two rectangular faces parallel to the VP and the axis coincides with the axis of the pyramid. Draw a profile of intersection after removal of the prism.

**15.11** A hexagonal pyramid (base edge = 100 mm and height = 250 mm) is resting on its base on the ground with two base edges parallel to the VP. It is intersected by a square prism (base edge 100 mm and height 250 mm) with two rectangular faces parallel to VP and axis coincides with the axis of the pyramid. Draw a profile of intersection after removal of the prism.

**15.12** A square pyramid (base diagonal = 200 mm, height = 225 mm) is standing on its base on the ground with the base edges equally inclined to the VP. A square hole (side = 75 mm) with the axis coinciding with the axis of the pyramid is cut through it. Two sides of the hole are perpendicular to the VP. Draw a profile of this hole in the FV.

**15.13** A pentagonal pyramid (base edge = 150 mm and height = 200 mm) is resting on its base on the ground with one of the base edges parallel to the VP and towards the observer. It is completely intersected by a square prism of base diagonal 75 mm with all base edges 45° to the ground. Axis of the square prism is 75 mm above the ground and an imaginary plane containing axes of both the solids is parallel to the PP. Draw a profile of intersection after removal of the prism.

**15.14** A pentagonal pyramid (base edge = 150 mm, height = 250 mm) is resting on its base on the ground, with one of its base edges parallel to the VP and towards the observer. It is intersected by a square prism with base diagonal 100 mm and all base edges equally inclined to the ground. Axis of the prism is 50 mm above the ground and an imaginary plane containing axes of both the solids is parallel to the VP.

Further, the cut pyramid is again completely intersected by a second square prism with base diagonal 100 mm and all base edges equally inclined to the VP. Axis of the second prism passes through the apex and centre of the base of the pyramid. Draw a profile of intersection after removal of the square prism.

## Exercise C (Interpenetration of Two Solids)

Solve for profile of interpenetration of the solids shown in Plate 15.5.

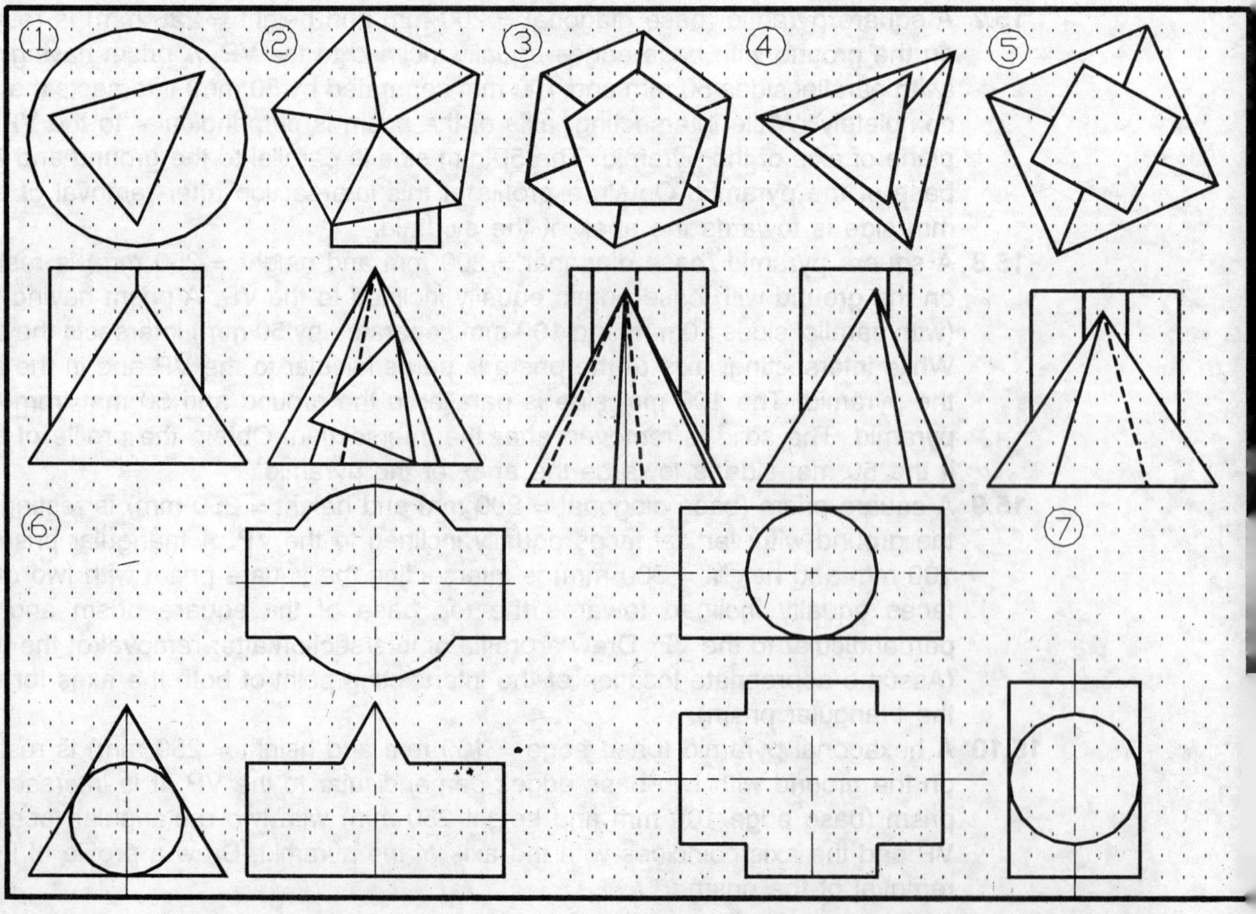


**PLATE 15.5**

**15.1** A vertical square prism base diagonal 200 mm is completely interpenetrated by a horizontal square prism, base diagonal 100 mm so that their axes intersect. The axis of the horizontal prism is parallel to the VP and faces of both the prisms are equally inclined to the VP. Draw the projections of the solids, showing lines of intersection. (Assume suitable lengths of the prism).

**15.2** A square prism (base edge = 125 mm, height = 175 mm) has its base on the ground with a rectangular face inclined at 30° to the VP. It is completely interpenetrated by another square prism having base edge = 75 mm, and height = 150 mm with faces equally inclined to the HP and VP. Axes of the two prisms are offset by 25 mm. The offset distance is away from the observer w.r.t. the axis of the main solid. Draw the projections showing lines of intersection.

**15.3** A square prism (base edge = 125 mm, height = 175 mm), resting on the ground, is interpenetrated by a triangular prism base edge 75 mm. Axis of the triangular prism is parallel to VP and 25 mm from the axis of the square prism towards the observer. Faces of the square prism are equally inclined to the VP, while one face of the triangular prism is inclined 40° (ccw) to the ground. Draw the projections showing lines of intersection. (Take sufficient lengths of both the prisms).

**15.4** A square pyramid of base diagonal = 150 mm and height = 200 mm rests on the ground with the base edges equally inclined to the VP. It is interpenetrated by a horizontal triangular prism of base edge 75 mm. Axis of the prism is 75 mm above the ground intersecting the axis of the pyramid. Draw their projections showing lines of intersections when a rectangular face of the prism is parallel to the VP and away from the observer.

**15.5** A square prism (base diagonal = 200 mm and height = 250 mm) is resting on its base on the ground with vertical faces equally inclined to the VP. A triangular prism (base edge = 100 mm and height = 300 mm) is interpenetrating the square prism with two of its faces equally inclined towards the ground plane and vertical edges parallel to the VP. (Assume appropriate location of the intersecting point of both the axes for full coverage of triangular prism). Obtain a profile of the interpenetration.

**15.6** A hexagonal pyramid (base edge = 100 mm and height = 250 mm) is resting on its base on the ground with two base edges perpendicular to the VP. It is interpenetrated by a square prism (base edge 100 mm and height 250 mm) with two rectangular faces parallel to VP and axis coincides with the axis of the pyramid. Obtain a profile of the interpenetration.

**15.7** A square prism (base diagonal = 250 mm, height = 225 mm) is standing on its base on the ground with faces equally inclined to the VP. It is interpenetrated by a hexagonal prism (base edge = 75 mm and height = 300 mm) with all vertical edges parallel to the VP and two adjacent rectangular faces equally inclined towards the ground. An imaginary plane passing through axes of both the solids is parallel to the VP. Draw a profile of interpenetration.

**15.8** A square prism (base diagonal = 250 mm, height = 225 mm) is standing on its base on the ground with faces equally inclined to the VP. It is interpenetrated by a hexagonal prism (base edge = 75 mm and height = 300 mm) with all vertical edges parallel to the VP and two alternate rectangular faces equally inclined to the ground. An imaginary plane passing through axes of both the solids is parallel to the VP. Draw a profile of interpenetration.

**15.9** A square pyramid (base diagonal = 300 mm, height = 300 mm) is resting on its base on the ground with line joining the opposite base corners parallel to the VP. It is completely interpenetrated by a pentagonal prism (base edge = 75 mm, height = 350 mm). Vertical edges of the prism are perpendicular to the VP with one rectangular face parallel to and 50 mm from the ground plane. An imaginary plane passing through one base corner and the mid-point of the base edge opposite to that base corner of interpenetrating prism also passes through the nearest and farthest slant edges of the pyramid. Draw a profile of interpenetration.

**15.10** A square pyramid (base diagonal = 300 mm, height = 300 mm) is resting on its base on the ground with line joining the opposite base corners parallel to the VP. It is completely interpenetrated by a prism having a trapezium base (with parallel sides 50 and 150 mm separated by 75 mm, height = 400 mm). Vertical edges of the prism are perpendicular to the VP. The parallel side 150 mm is parallel to and 75 mm above the ground near the base of the pyramid. An imaginary plane passing through the mid-points of the parallel base edges of the prism also passes through the nearest and farthest slant edges of the pyramid. Draw a profile of interpenetration.

**15.11** A pentagonal pyramid (base edge = 150 mm, and height = 300 mm) is standing on its base on the ground with line joining any two opposite base corners parallel to the VP and one base corner nearest to the observer. It is completely interpenetrated by a triangular prism (base edge = 125 mm, height = 325 mm) with two rectangular faces equally inclined towards the ground. A third rectangular face, which is parallel to base of the pyramid, is 125 mm from the ground. An imaginary plane passing through the axis of the interpenetrating prism also passes through the nearest base corner and apex of the pentagonal pyramid. Draw a profile of interpenetration.

**15.12** A pentagonal pyramid (base edge = 150 mm, and height = 300 mm) is standing on its base on the ground with the line joining any two opposite base corners parallel to VP and one base corner nearest to the observer. It is completely interpenetrated by a triangular prism (base edge = 125 mm, height = 325 mm) with one rectangular face parallel to the base of the pyramid, is 100 mm from the ground and a vertical edge opposite to this face is towards the apex of the pyramid. An imaginary plane passing through the axis of the intersecting prism also passes through the nearest base corner and apex of the pentagonal pyramid. Draw a profile of interpenetration.

**15.13** A triangular prism (base edge = 200 mm and height = 200 mm) is resting on one of its bases on the ground with one rectangular face parallel to the VP and away from the observer. It is interpenetrated by a square prism (base diagonal = 75 mm and height = 250 mm) with rectangular faces equally inclined to the HP and VP and axes bisecting each other. Draw a profile of interpenetration.

**15.14** A hexagonal pyramid (base edge = 150 mm and height = 250 mm) is standing on its base on the ground with two base edges parallel to the VP. It is interpenetrated by a square prism (base diagonal = 150 mm and height = 350 mm). While intersecting, all rectangular faces of the prism are equally inclined to the ground and perpendicular to the VP with axes intersecting 100 mm above the ground level. Draw a profile of interpenetration.

**15.15** A square prism (base diagonal = 200 mm and height = 250 mm) is resting on one of its base on the ground with line joining any two base corners perpendicular to the VP. It is interpenetrated by a triangular prism (base edge = 150 mm and height = 250 mm) with al vertical edges parallel to the HP and VP and axis bisecting the axis of the square prism. While interpenetrating, one rectangular face of the triangular prism is away from the observer and parallel to the VP. Draw a profile of interpenetration.

**15.16** A pentagonal pyramid (base edge = 150 mm and height = 300 mm) is standing on its base on the ground with one base edge parallel to the VP and farthest from the observer.

It is interpenetrated by a square prism (base diagonal = 150 mm and height = 350 mm) While intersecting, the bases of prism are parallel to the ground. An imaginary plane containing axes of both the solids is perpendicular to the VP and passes through the base diagonals of square prism. Draw an interpenetration profile when the prism axis is offset by a distance of 25 mm towards the observer.

**15.17** A square pyramid (base diagonal = 250 mm and height = 300 mm) is standing on its base on the ground with line joining any two base corners parallel to the VP. It is interpenetrated by a triangular prism (base edge 100 mm and height 300 mm). While interpenetrating, the triangular prism has its two rectangular faces equally inclined towards the ground and bases perpendicular to the VP. Draw an interpenetration profile.

**15.18** A triangular pyramid (base edge = 250 mm and height = 250 mm) is standing on its base on the ground with two base edges equally inclined to the VP and third base edge away from the observer. It is interpenetrated by a square prism (base diagonal 100 mm and height 250 mm). While interpenetrating, all four faces of prism are equally inclined to the VP and bases perpendicular to the ground. Draw an interpenetration profile when their axes bisect each other.

**15.19** A hexagonal pyramid (base edge = 150 mm and height = 300 mm) is resting on its base on the ground with line joining any two opposite base corners parallel to the VP. It is interpenetrated by a square prism with base edge 100 mm and height 300 mm. While interpenetrating, the prism has its two of its adjacent rectangular faces equally inclined to the ground and bases parallel to PP. Draw a profile of interpenetration in the TV and FV.

**15.20** A square pyramid (base edge = 150 mm and height = 300 mm) is standing on its base on the ground with line joining any two opposite base corners parallel to the VP. It is intersected by a square prism with base edge 50 mm and height 300 mm. While interpenetrating, one of the rectangular faces of the square prism is inclined 60° with the VP and perpendicular to the HP. A plane containing axes of both the solids is parallel to the VP. Draw a profile of interpenetration.

**15.21** A hexagonal pyramid (base edge = 150 mm and height = 300 mm) is resting on the ground with line joining any two opposite base corners parallel to the VP. It is interpenetrated by a square prism with base edge 75 mm and height 350 mm. While interpenetrating, the square prism has one of its faces 30° with the ground and perpendicular to the VP. Axis of the prism is 75 mm above the ground intersecting axis of the pyramid. Draw a profile of interpenetration.

# 16

# Freehand Sketching

## Objectives

- To help provide an understanding of the essential elements of freehand sketching.
- To learn and develop the skill in sketching, producing and interpreting drawings.
- To be able to sketch 3-D view of orthographic solids and basic isometric solids.

## Introduction

Freehand sketching is the technique of making drawing without the use of drawing instruments. It is a very convenient and quick method of shape description. Freehand sketching is much emphasised in the industrial technology because this skill is an essential tool in interpreting ideas and visual thinking. Freehand sketching provides the ability to conceptualise and modify design ideas from the first idea through to the final choice of the project. Freehand drawing and sketching is a means of quickly putting your thoughts down on paper. Freehand drawing and sketching differs from instrumental drawing primarily in the amount of time and accuracy required. This chapter presents the fundamentals of freehand sketching. It includes both 2-D and 3-D sketching.

## 16.1 Importance of Freehand Sketches

Freehand sketches are quickly drawn to communicate ideas fast and freely. Try to sketch while you think of ideas. Freehand sketches are sometimes called concept sketches; they are often very basic drawings, which are far from finished but they nevertheless convey a lot of information about a project. These sketches or drawings can formulate record, express design and develop ideas and instructions and reduce the need for long verbal explanations. Good sketching ability enables one to make rapid and clean drawings of ideas that can be given to his/her dealing sections for development into scale working drawings. Sketches are usually drawn freehand, but some ruling and compass work is done, if it facilitates the sketch work.

In appearance, the completed freehand drawing is entirely different from the clean-cut precision of a mechanically ruled or computer generated working drawing, it has a rough and natural appeal, but should be clear, concise and suitable for the intended purpose. A poorly presented freehand drawing is unlikely to be appreciated, understood, or accepted.

## 16.2 Basic Skills

There are a variety of grades of pencil available for sketching. These grades range from a very hard 9H to a very soft EE. A suitable pencil for sketching would be a 2B. Sketch lightly to begin with as you will unavoidably make mistakes and will want to remove some of the errors made. You will need to practice controlling the pressure applied to the pencil. The pressure can control the thickness of the line produced. Shading can also improve the appearance of a simple sketch, by simulating the effect of light and shade on an object. The shade requirement is determined by deciding on the position of a light source.

Freehand sketches and drawings can be quickly prepared on a grid paper. This can help in keeping lines straight and to maintain proportion. Grid paper can come in various sizes and profiles: square grids can support orthogonal projection and allow enlargement of projects and components, 45° and 30° grids allow for pictorial drawing in oblique and isometric. Although full proficiency in freehand drawing is synonymous with complete understanding of the graphic language, and is gained only after acquiring a good knowledge and skill in drawing with instruments, it is advisable that one should be conversant with freehand drawing before starting to make drawing with the instruments.

## 16.3 Proportion in Sketching

Sketches are not made to scale. Although produced quickly, the sketches should be in good proportion, with clear and bold outlines and the dimensioning as complete as possible. While the technical sketches are not scale drawings, their value is greatly enhanced if their general proportions are good. Various components of a sketch should be kept in proportions to those of the actual object, using eye judgement only. This technique takes a lot of practice to master.

Sketches made to supplement the oral description may be rough and incomplete. However, if the sketch is of the medium of conveying important and precise information, it should be executed as carefully as possible under the available time frame. For example, if a dimension of a feature is twice, another dimension of the same or other feature should be also represented in the sketch. Few guidelines are mentioned below:

- If two views are made in orthographic projection they should be placed in their correct relative positions.
- The outlines should be the boldest, being twice or thrice heavy as construction lines, centre lines, dimension lines, leaders, etc. while hidden lines are somewhat thinner than object lines.
- The most common shortcoming in sketching is the tendency to make the sketch too small or too crowded in the drawing. It is, thus, important that all sketches be of a good size and be wisely spaced.

Below are some general rules relating to proportions:

- When sketching, whenever possible, use a graph paper.
- Study the object to be sketched carefully and then mentally break it into its small component parts.
- Estimate the length, width and height of the largest component of the object. Comparir these dimensions to the size of the sketching paper you have, estimate the proportions th will be possible to use in the overall sketch.
- Layout the largest component according to the proportions decided in the previous step, block in irregularly shaped components.

## 16.4 Sketching Straight Lines

Sketches consists of straight and curved lines. With disciplined practice, anyone can obtain the sk and become proficient in creating neat sketches involving straight and curved lines. Before startir sketching of object views, it will be of great help to practice in sketching the lines and curves. Thoug the technique of sketching depends more upon the natural movements of the designer's har holding the pencil, yet below are some suggestions which can help in acquiring the required skill the sketching technique:

1. Place the paper in the rotated position so that the horizontal edges are perpendicular to tl natural position of one's sketching hand, the forearm.
2. When a squared paper or a graph paper is being used for sketching try to locate the ma lines on the ruling of the sketching paper. Sketching straight lines on a graph paper is simple process of guiding the pencil point using the existing graph lines.
3. Hold the pencil with a good grip, but not so tight so as to create a hand fatigue.
4. Hold the pencil approximately 25 mm to 40 mm away from the pencil point.
5. Maintain a comfortable angle between the pencil and the sketching strokes.
6. Sketch all the horizontal lines and lines with small inclination to the horizontal plane from l to right.
7. Sketch all the vertical lines and lines with small inclination to the vertical plane from top bottom.
8. For drawing a long straight line, first fix the endpoints of the line. Add some intermedia points on the line, if the line is sufficiently long. Then sketch the line by a series of overlappi strokes, keeping the eye on the point to which the line goes rather than on the pencil.
9. On a plain drawing paper, pencil dots can be positioned to plot the path of the line. In th way, the sketcher enters a series of pencil dots on the paper which provides a basic outli as to the shape of the object. Then using a series of slightly overlapping strokes, the pen dots are connected.
10. Finger movements, with the wrist stationary, are to be used for short lines, small circles, e whereas stiff forearm movements are more effective for long lines.
11. For any sketch it is advisable to start by sketching light centre lines for circles, hexagons a all other symmetrical figures or objects.

## 16.5 Sketching Circles, Arcs and Fillets

The best method of sketching a circle and an arc is to locate a number of points, through which t curve should pass, by eye judgement. Remembering that all points on the circumference of a circ is equidistant from its centre and that it is symmetrical about any diameter, a circle can easily drawn by using the following steps:

1. Sketch vertical and horizontal centre lines, to locate the centre of the circle.
2. Mark on these centre lines, at distance equal to the circle radius, points on all sides of the centre, to obtain four points lying on the circumference of the required circle.
3. Sketch a square through these four points.
4. Sketch the diagonal of the square.
5. Along these diagonals, plot four more points, one on each side of the centre, at distances of circle radius. These points too lie on the circumference of the circle.
6. Sketch the right half of the circle from top to bottom passing through 4 points. Then similarly, sketch the left half of the circle passing through 4 points to complete the circle as shown in Fig. 16.1.

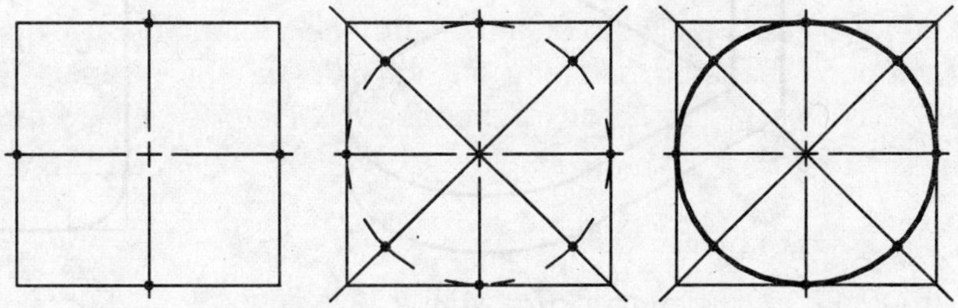

**Fig. 16.1** Using enclosing square.

Another method is to draw radial construction lines lightly, then sketch short arcs across the lines at the estimated radius distance away from the centre, and finally to sketch the full circle as shown in Fig. 16.2.

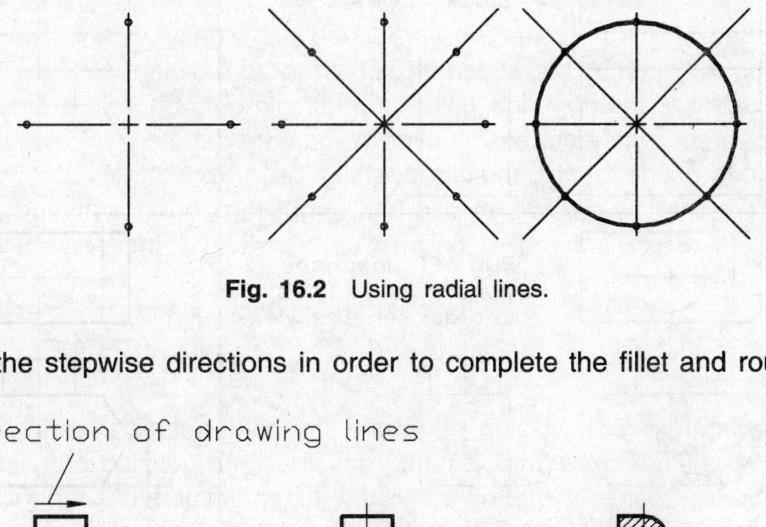

**Fig. 16.2** Using radial lines.

Figure 16.3 gives the stepwise directions in order to complete the fillet and rounds.

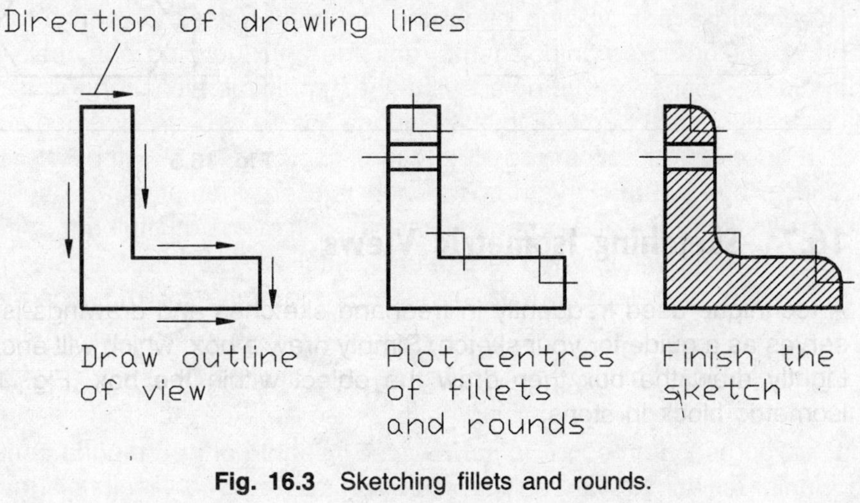


**Fig. 16.3** Sketching fillets and rounds.

## 16.6 Sketching Orthographic Views

Figures 16.4 and 16.5 give the directions of orthographic sketching.

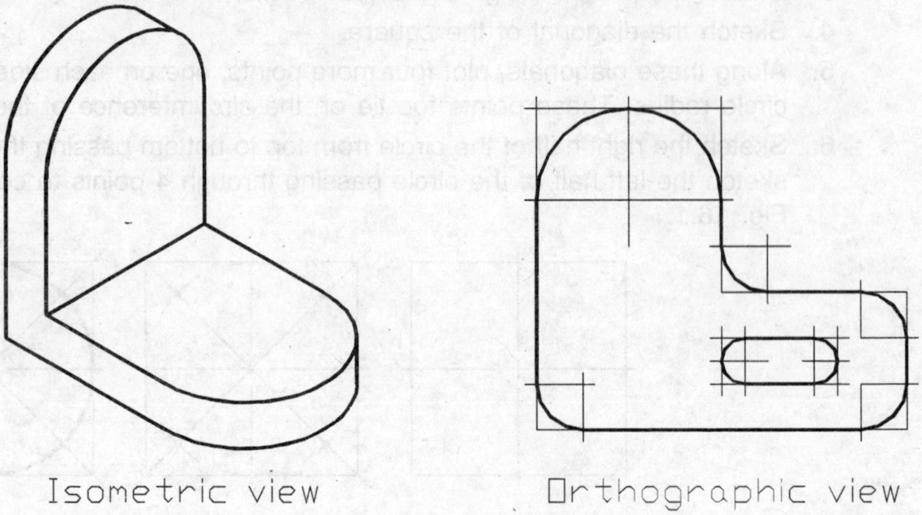


**Fig. 16.4**

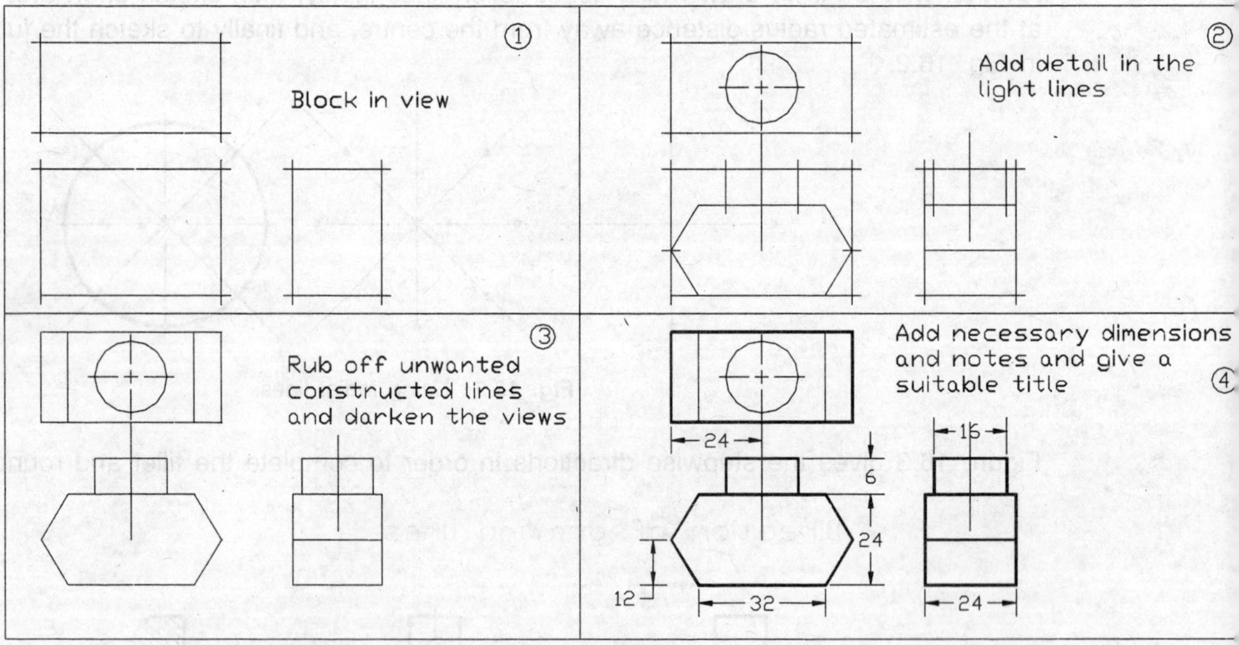


**Fig. 16.5**

## 16.7 Sketching Isometric Views

A technique used frequently in freehand sketches and drawings is to create a "box" or "crate" th serves as a guide for your sketch. Simply draw a box, which will enclose the object you want to drav Lightly draw the box then draw the object within the box. Fig. 16.6 shows the sketching of a isometric block in steps.

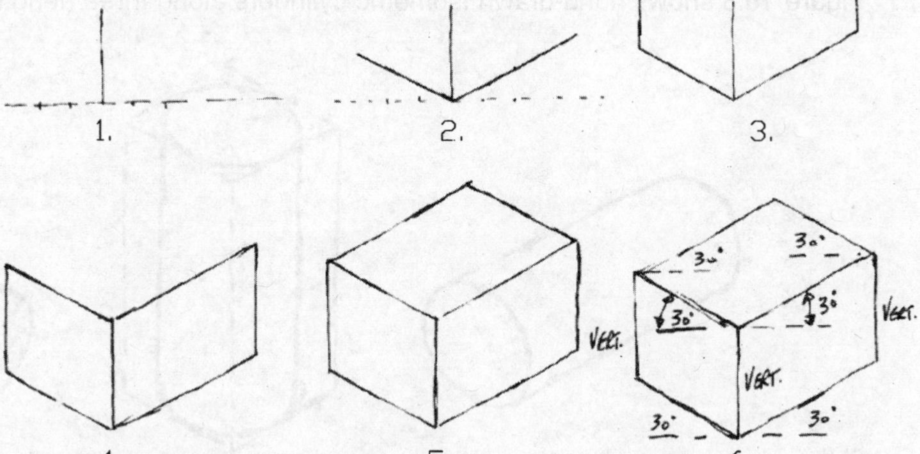


Fig. 16.6

### 16.7.1 Sketching of Isometric Circles in Isometric Block

You can also draw a circle of a given diameter by first sketching the box that it will fit into. Figure 16.7(a) shows the steps to draw an isocircle. Mark the points half way along each side where the circle will touch. Sketch a curve to touch these points and it will form your circle. An ellipse also can be created using this technique, but this time we draw the rectangle box that the ellipse will fit into. This technique is handy for drawing oblique or isometric circles. Figure 16.7(b) and (c) shows hand drawn and AutoCAD drawn isocircles.

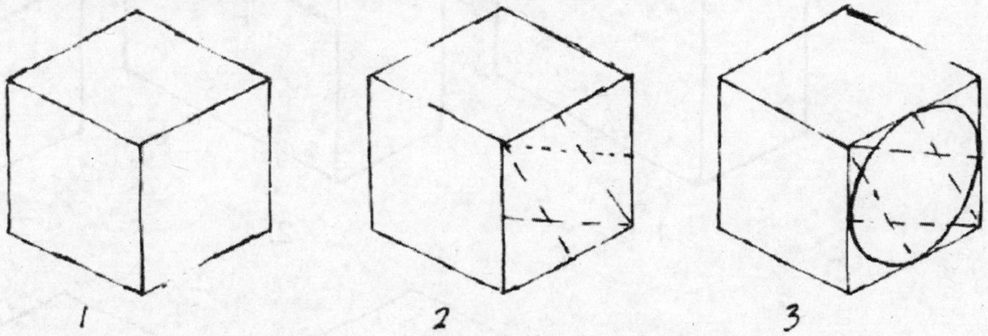


(a)

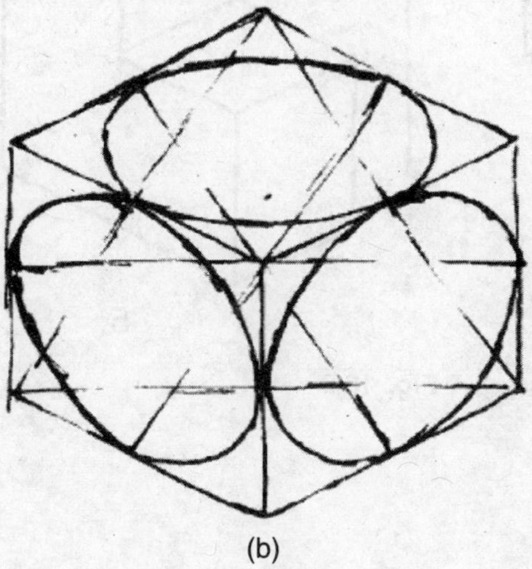

(b)

(c)

Fig. 16.7

### 16.7.2 Sketching of Isometric Cylinders

Figure 16.8 shows hand-drawn isometric cylinders along three (length, depth and height) directions

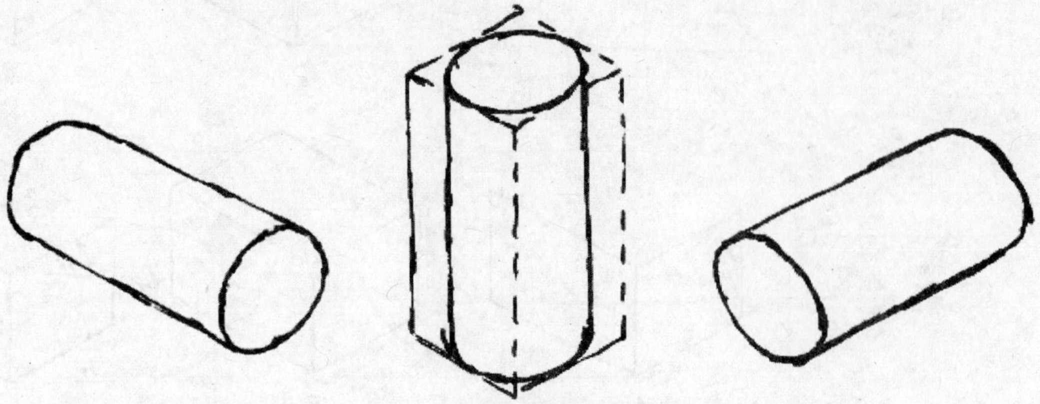

**Fig. 16.8**

**EXAMPLE 16.1** Sketching of isometric view of solid (Fig. 16.9), (Fig. 16.10), (Fig. 16.11) and (Fig. 16.12).

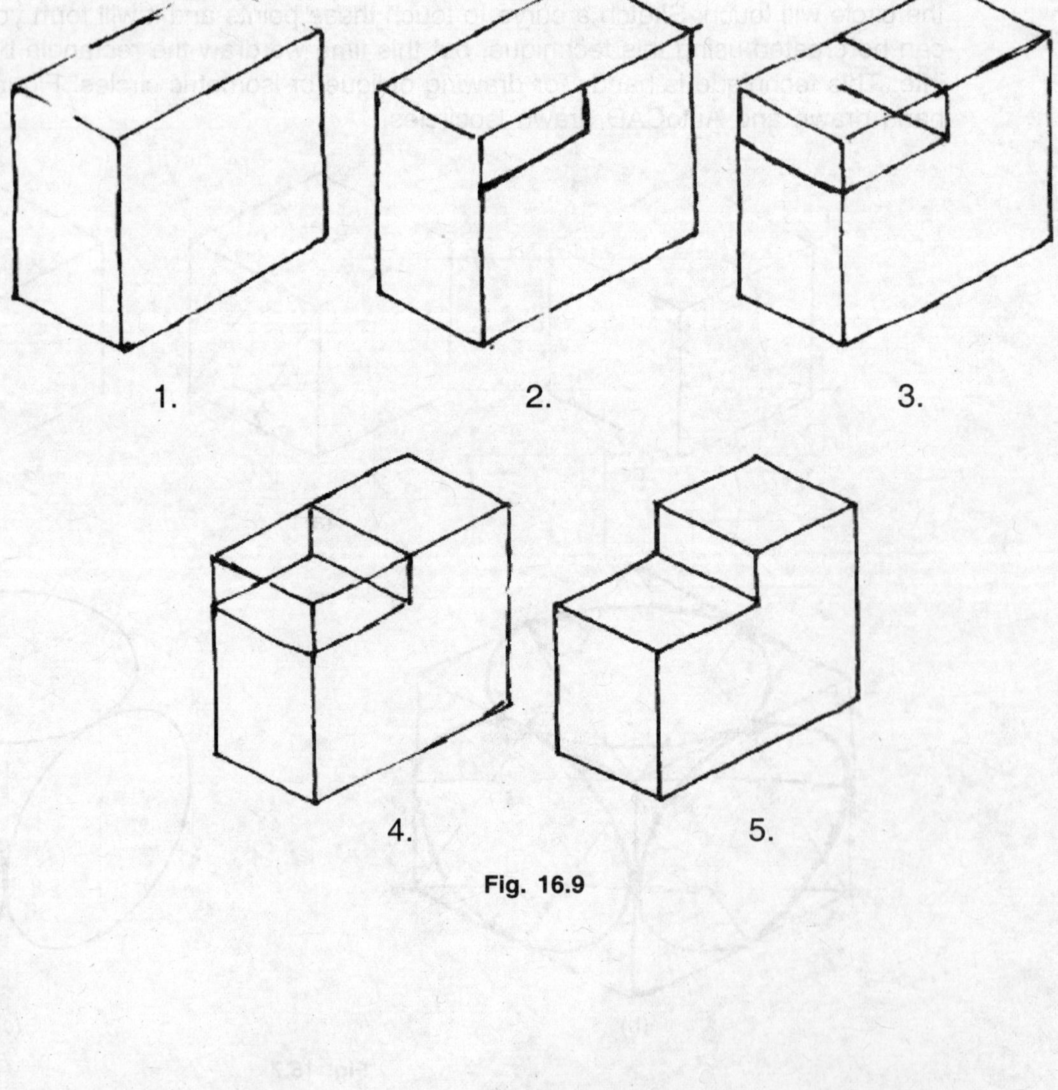


**Fig. 16.9**

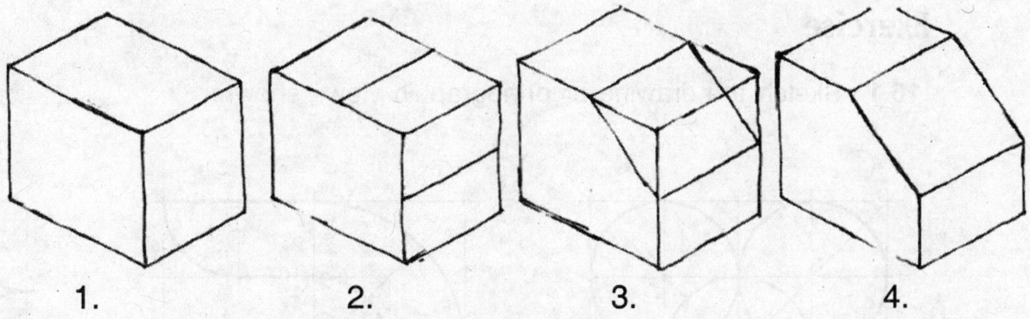


Fig. 16.10

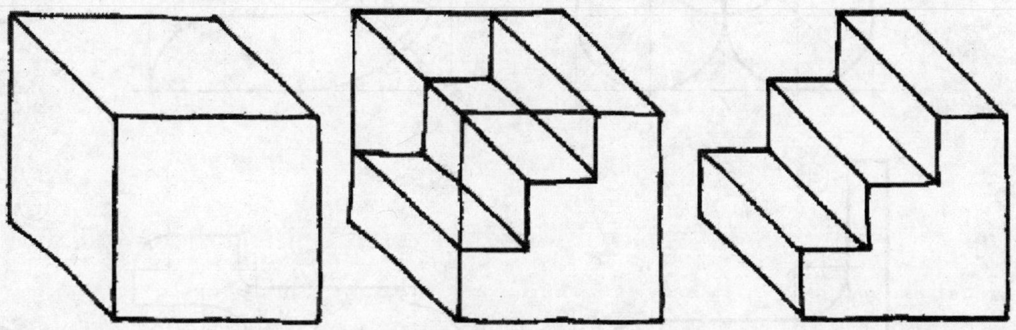

Fig. 16.11

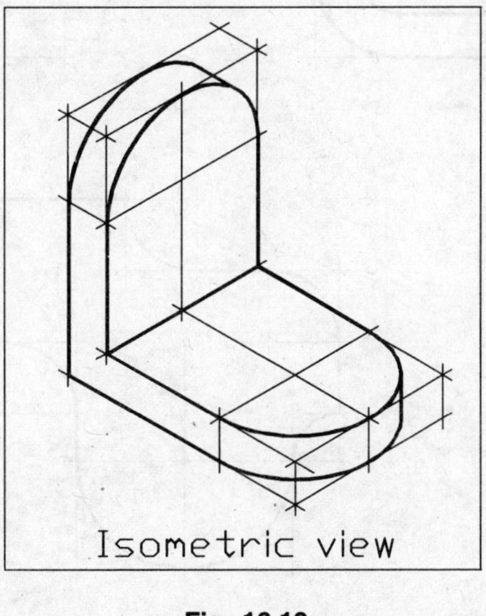

Fig. 16.12

## Exercise

**16.1** Sketch the drawing or orthographic views shown:

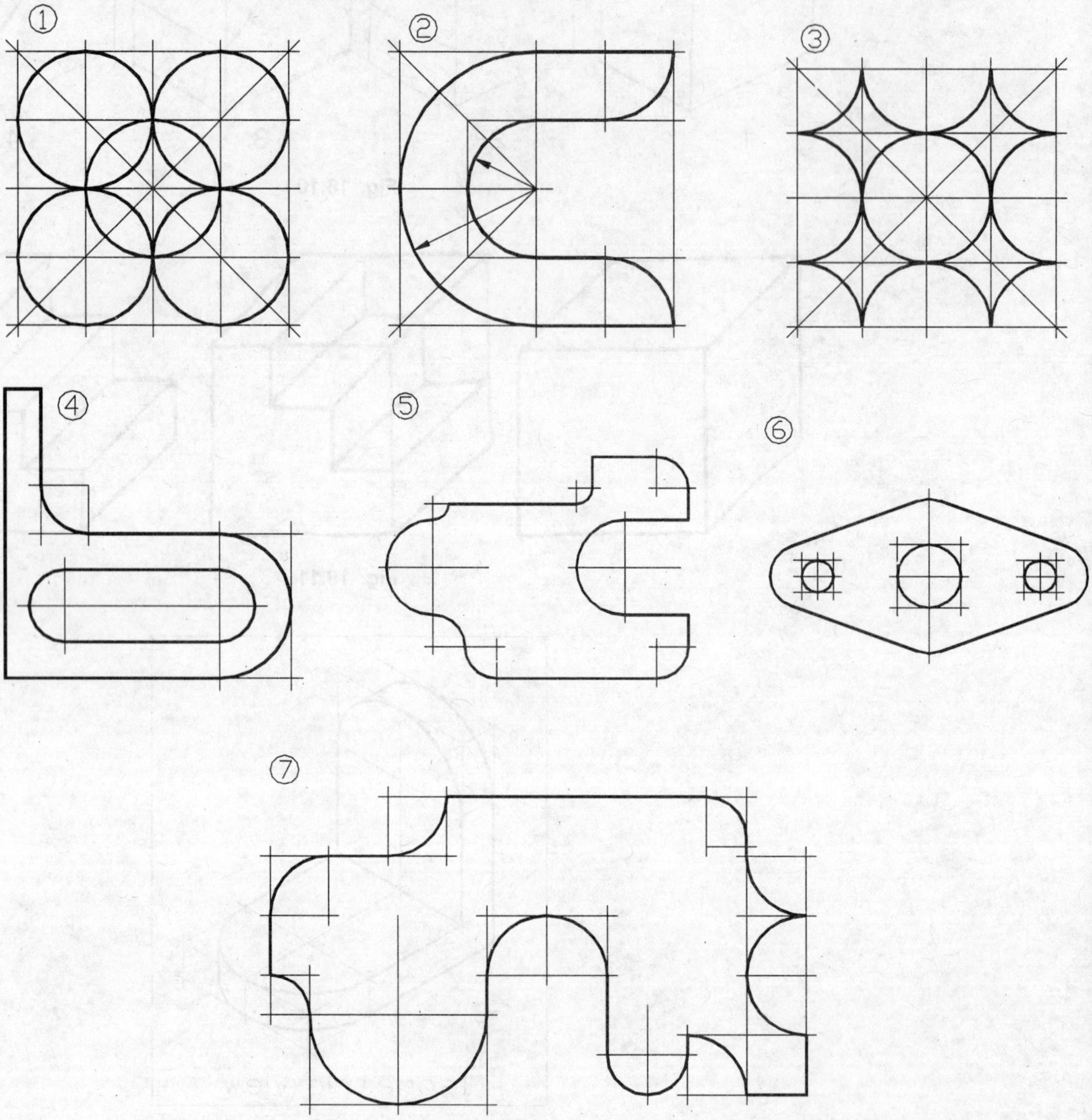

**16.2** Draw sketches for the following machine components:

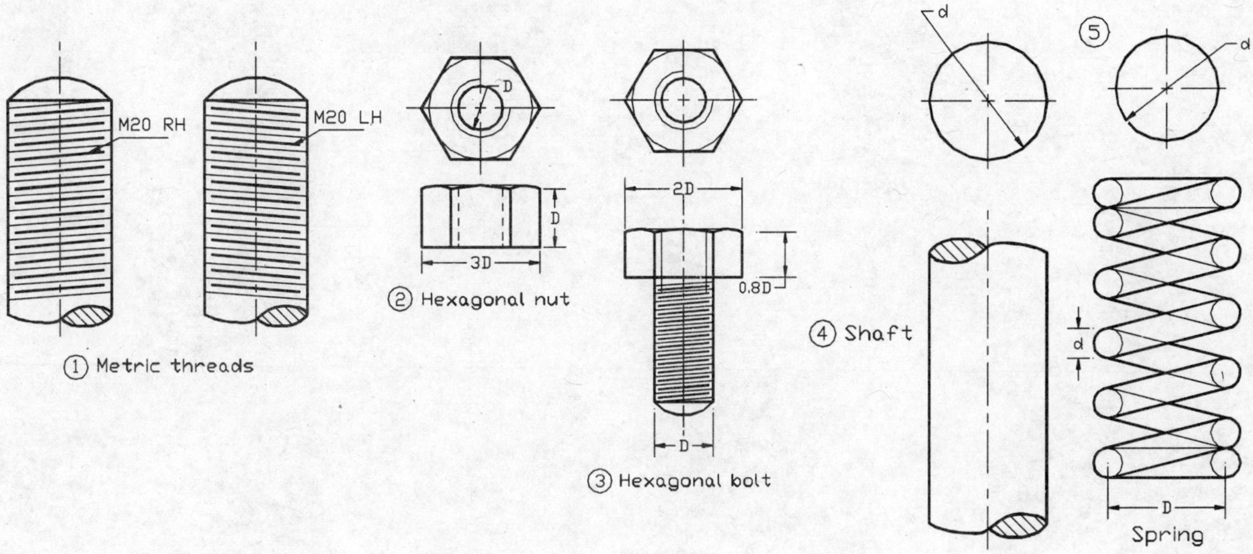


**16.3** Draw sketches for the isometric solids shown:

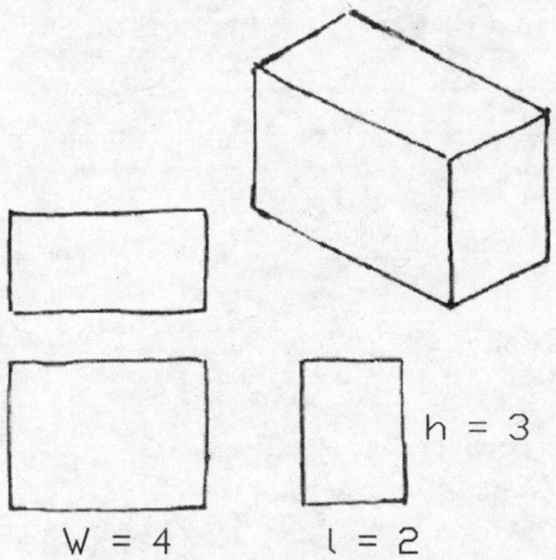


(a)

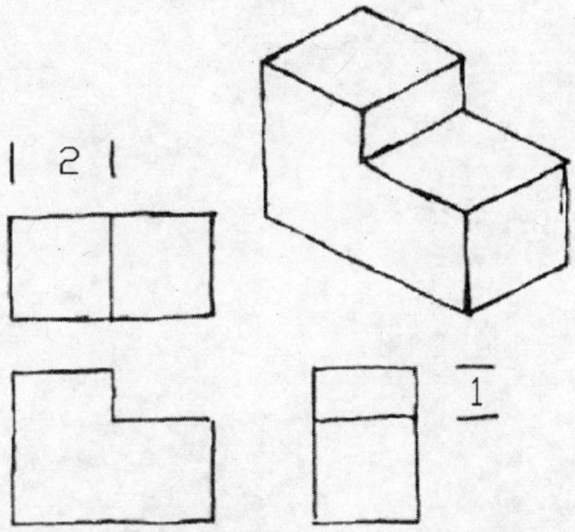


(b)

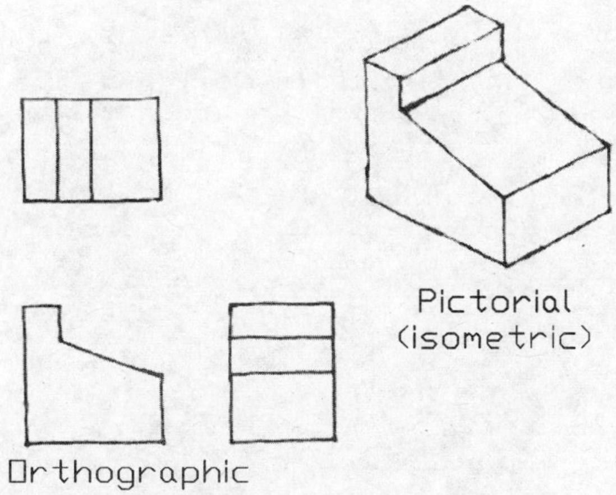


(c)

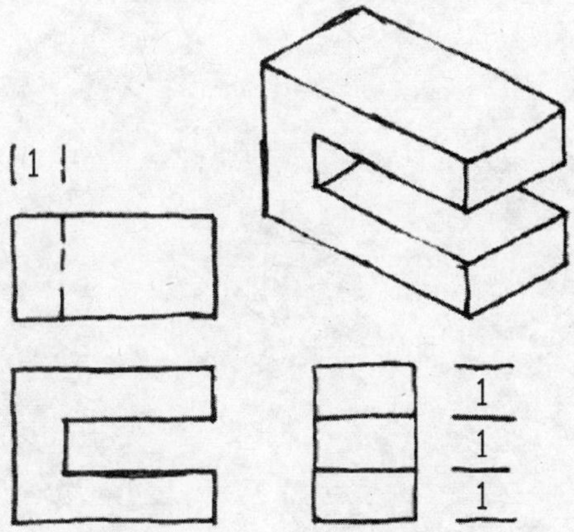


(d)

# Index